Low-Power Analog Techniques, Sensors for Mobile Devices, and Energy Efficient Amplifiers

Kofi A. A. Makinwa • Andrea Baschirotto
Pieter Harpe
Editors

Low-Power Analog Techniques, Sensors for Mobile Devices, and Energy Efficient Amplifiers

Advances in Analog Circuit Design 2018

Springer

Editors
Kofi A. A. Makinwa
Delft University of Technology
Delft, Zuid-Holland
The Netherlands

Andrea Baschirotto
University of Milano-Bicocca
Milan, Italy

Pieter Harpe
Eindhoven University of Technology
Eindhoven, Noord-Brabant
The Netherlands

ISBN 978-3-319-97869-7 ISBN 978-3-319-97870-3 (eBook)
https://doi.org/10.1007/978-3-319-97870-3

Library of Congress Control Number: 2018959116

This Springer imprint is published by the registered company Springer Nature Switzerland AG
The registered company address is: Gewerbestrasse 11, 6330 Cham, Switzerland

Preface

This book is part of the Analog Circuit Design series and contains contributions by all the speakers at the 27th workshop on Advances in Analog Circuit Design (AACD). The aim of the workshop was to bring together a group of expert designers to discuss new developments and future options.

The 27th workshop was held in Edinburgh, Scotland, from May 14 to 16, 2018. The local organizers were Jed Hurwitz (ADI), Paul Lesso (Cirrus Logic), Jim Brown (Dialog Semiconductor), Emma Dixon (Technology Scotland), and Stephen Taylor (Technology Scotland). Analog Devices, Cirrus Logic, and Dialog Semiconductor were the platinum sponsors of the event.

Each AACD workshop is followed by the publication of a book by Springer, which then becomes part of their successful series on Analog Circuit Design. A full list of the previous books and topics covered in this series can be found on subsequent pages. Each book can be seen as a reference work for students and designers interested in advanced analog and mixed-signal circuit design.

This book is the 27th in this series. It consists of three parts, each with six chapters, that cover the following topics that are currently considered of high importance by the analog and mixed-signal circuit design community:

- Analog Techniques for Power Constrained Applications
- Sensors for Mobile Devices
- Energy Efficient Amplifiers and Drivers

We are confident that this book, like its predecessors, will prove to be a valuable contribution to our analog and mixed-signal circuit design community.

Delft, The Netherlands — Kofi A. A. Makinwa
Milan, Italy — Andrea Baschirotto
Eindhoven, The Netherlands — Pieter Harpe

Topics Previously Covered in the Springer Series on Analog Circuit Design

2017	Eindhoven (The Netherlands)	Hybrid ADCs Smart Sensors for the IoT Sub-1V & Advanced-node Analog Circuit Design
2016	Villach (Austria)	Continuous-time $\Sigma\Delta$ Modulators for Transceivers Automotive Electronics Power Management
2015	Neuchâtel (Switzerland)	Efficient Sensor Interfaces Advanced Amplifiers Low Power RF Systems
2014	Lisbon (Portugal)	High-Performance AD and DA Converters IC Design in Scaled Technologies Time-Domain Signal Processing
2013	Grenoble (France)	Frequency References Power Management for SoC Smart Wireless Interfaces
2012	Valkenburg (The Netherlands)	Nyquist A/D Converters Capacitive Sensor Interfaces Beyond Analog Circuit Design
2011	Leuven (Belgium)	Low-Voltage Low-Power Data Converters Short-Range Wireless Front-Ends Power Management and DC-DC
2010	Graz (Austria)	Robust Design Sigma Delta Converters RFID
2009	Lund (Sweden)	Smart Data Converters Filters on Chip Multimode Transmitters

2008	Pavia (Italy)	High-Speed Clock and Data Recovery High-Performance Amplifiers Power Management
2007	Oostende (Belgium)	Sensors, Actuators and Power Drivers for the Automotive and Industrial Environment Integrated PAs from Wireline to RF Very High Frequency Front Ends
2006	Maastricht (The Netherlands)	High-Speed AD Converters Automotive Electronics: EMC issues Ultra Low Power Wireless
2005	Limerick (Ireland)	RF Circuits: Wide Band, Front-Ends, DACs Design Methodology and Verification of RF and Mixed-Signal Systems Low Power and Low Voltage
2004	Montreux (Swiss)	Sensor and Actuator Interface Electronics Integrated High-Voltage Electronics and Power Management Low-Power and High-Resolution ADCs
2003	Graz (Austria)	Fractional-N Synthesizers Design for Robustness Line and Bus Drivers
2002	Spa (Belgium)	Structured Mixed-Mode Design Multi-bit Sigma-Delta Converters Short-Range RF Circuits
2001	Noordwijk (The Netherlands)	Scalable Analog Circuits High-Speed D/A Converters RF Power Amplifiers
2000	Munich (Germany)	High-Speed A/D Converters Mixed-Signal Design PLLs and Synthesizers
1999	Nice (France)	XDSL and Other Communication Systems RF-MOST Models and Behavioural Modelling Integrated Filters and Oscillators
1998	Copenhagen (Denmark)	1-Volt Electronics Mixed-Mode Systems LNAs and RF Power Amps for Telecom
1997	Como (Italy)	RF A/D Converters Sensor and Actuator Interfaces Low-Noise Oscillators, PLLs and Synthesizers
1996	Lausanne (Swiss)	RF CMOS Circuit Design Bandpass Sigma Delta and Other Data Converters Translinear Circuits

1995	Villach (Austria)	Low-Noise/Power/Voltage Mixed-Mode with CAD Tools Voltage, Current and Time References
1994	Eindhoven (The Netherlands)	Low-Power Low-Voltage Integrated Filters Smart Power
1993	Leuven (Belgium)	Mixed-Mode A/D Design Sensor Interfaces Communication Circuits
1992	Scheveningen (The Netherlands)	OpAmps ADCs Analog CAD

Contents

Part I
Analog Techniques for Power Constrained Applications

The first part of this book is dedicated to recent developments in the field of extremely low power circuits and systems. The first papers discuss energy harvesting and power management circuits, followed by chapters on analog and mixed-signal circuits and systems.

The first chapter from Baoxing Chen (Analog Devices International) describes various kinds of energy sources that can be used for energy harvesting and gives an overview of possible energy harvesting transducers, combined with their required power conditioning circuits.

In the second chapter, Wim Kruiskamp (Dialog Semiconductor) presents an example of an energy harvesting system that allows these IoT devices to be powered by alternative energy sources like light, heat, or RF energy. Some circuits and algorithms that are specifically important for energy harvesting are discussed in more detail.

The third chapter by Vadim Ivanov (Texas Instruments) describes how to design analog building blocks and DC/DC converters in a good way with a nanowatt level of power consumption. Starting with a structured design approach, the author further continues with practical design implementations and examples.

Maoqiang Liu (Eindhoven University of Technology) presents various low-power Analog-to-Digital Converters in chapter four. In particular, the combination of low-power references and ADCs is investigated. Various implemented examples are discussed where the reference is co-designed, relaxed, or compensated by means of the ADC.

In chapter five, Ming Ding (Holst Centre/imec) discusses the challenges of low-power clock generation for duty-cycled Internet-of-Things radios. Two design examples are discussed in detail: a nanopower sleep timer and a fast start-up crystal oscillator.

The sixth chapter from David Blaauw (University of Michigan) presents circuits for extremely low power sensor systems. A duty-cycled bridge-to-digital converter for small battery operated pressure sensing systems is presented. Besides presenting the circuit techniques and implementations, the circuit is also demonstrated inside two complete microsystems.

Introduction to Energy Harvesting Transducers and Their Power Conditioning Circuits

Baoxing Chen

1 Introduction

Wireless sensor nodes are usually powered by batteries; however, maintenance for them can become a significant hassle. Batteries have limited life span and they need to be replaced. In some applications, the wireless sensor nodes can be installed in a remote and tough-to-reach area. Maintenance-free wireless sensor nodes are essential to drive the continuous adoption of wireless sensor network for wide range of IoT applications, from machine health monitoring, building automation, to smart wearables. Energy harvesting, i.e., harvesting energy from the ambient environment, is an ideal solution to enable maintenance-free wireless sensor network. An energy harvesting system for wireless sensors consists of the following as shown in Fig. 1.: (1) one or multiple energy harvesters; (2) some kind of energy storage device; (3) power management device to ensure efficient energy extraction from the harvesters, to produce the output voltages suitable for use by other devices, and to manage the storage device; (4) various sensors to sense a variety of environmental data such as temperature, pressure, or gas; (5) the signal conditioning circuits and microcontrollers to interface with the sensors,; and (6) the wireless transceivers. In this chapter, we will mainly focus on 1, 2, and 3.

B. Chen (✉)
Analog Devices, Inc., Wilmington, MA, USA
e-mail: Baoxing.Chen@analog.com

K. A. A. Makinwa et al. (eds.), *Low-Power Analog Techniques, Sensors for Mobile Devices, and Energy Efficient Amplifiers*,
https://doi.org/10.1007/978-3-319-97870-3_1

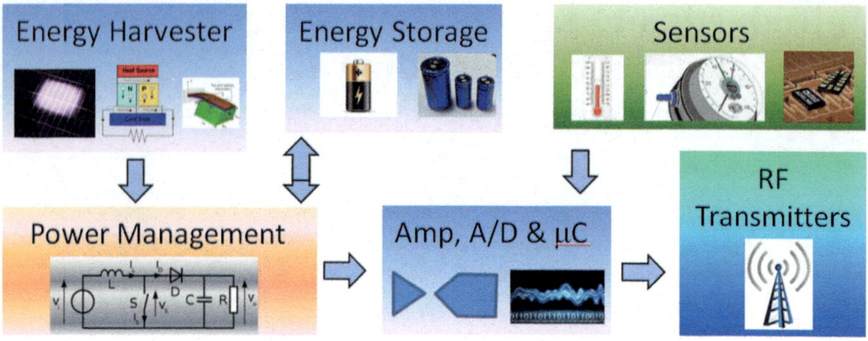

Fig. 1 Typical energy harvesting system

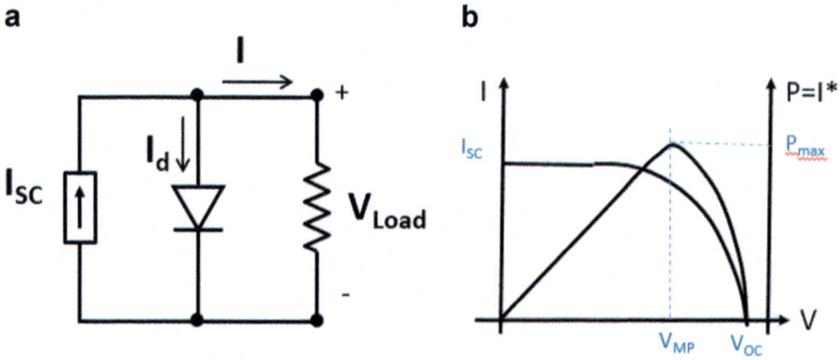

Fig. 2 (**a**) Circuit model and (**b**) I-V curve for photovoltaic cell

2 Photovoltaic Harvesting

PV panels can usually generate around 100 W/m^2 or 10 mW/cm^2 outdoors with 10% solar cell efficiency, but this number can drop 2–3 orders of magnitude indoors dependent on lighting condition. For area-constraint application, the solar cell efficiency is key. Low solar cell efficiency is due to low quantum efficiency where only the photons with energy bigger than the bandgap can be absorbed and excessive photon energy above the bandgap is also lost to heat. The solar cell efficiency can be improved with multi-junction device, where junctions with different bandgaps are stacked so that photons with different energy are absorbed more efficiently by different junctions.

Figure 2a shows the circuit model for a PV cell where the current through the load is the short circuit current minus the diode current. Figure 2b shows typical I-V curve where V_{OC} is the open circuit voltage and I_{SC} is the short circuit current. The power it generates has a peak at certainly voltage V_{MP}, fraction of the V_{OC}, and this is the maximum power point with P_{max}. For the PV power conditioning circuit, it is important that solar cell operates at this point. With low power application, we need to balance the power gain from the additional efficiency gain from accurate

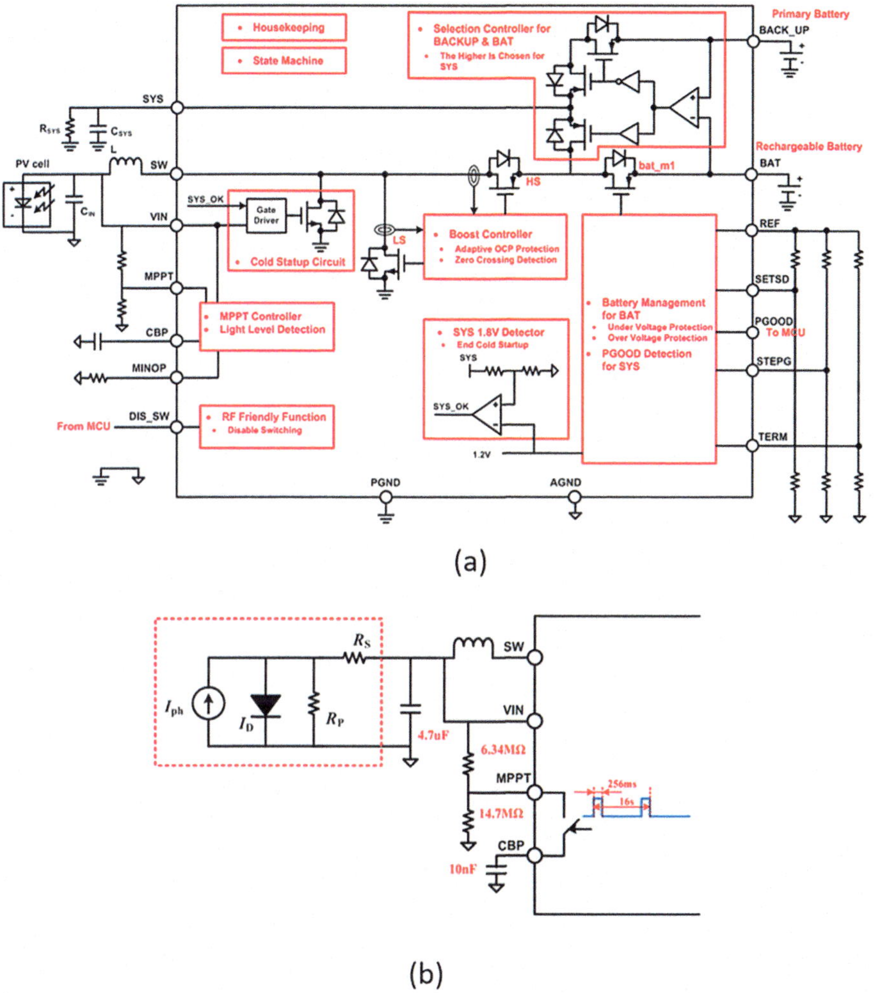

Fig. 3 (**a**) Boost converter architecture (ADP5090) and (**b**) VOC sense implementation

maximum power point tracking and the power loss for implementing the MPPT algorithms. One common approach is to approximate the point using some fraction of the open circuit voltage (FOV). While the V_{OC} changes with different lighting conditions, FOV stays relatively constant. This can be easily implemented with a comparator and voltage divider. The load can be either disrupted to get V_{OC} or a dummy reference cell is used to get V_{OC}.

An example implementation for how V_{OC} is sampled is shown in Fig. 3 [1]. Figure 3a is the overall circuit architecture for an ultralow power boost converter, and Fig. 3b shows the V_{OC} sampling method. In every 16 s, the converter is interrupted for 256 mS to allow FOV to be sampled and stored in C_{BP} through a

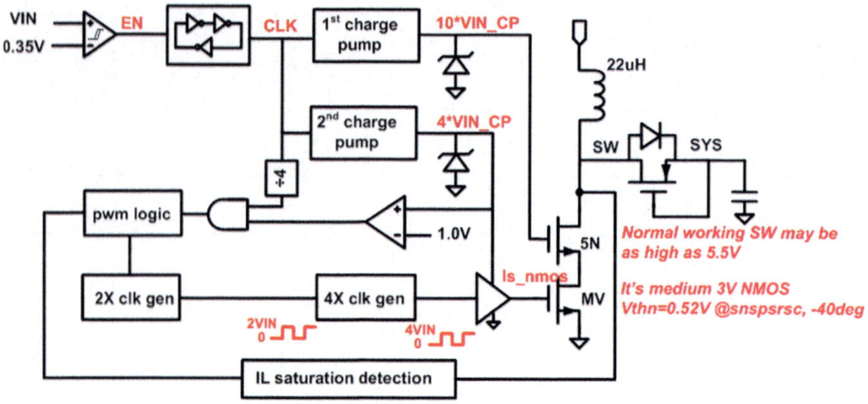

Fig. 4 Cold startup circuit implementation

Table 1 Comparison for startup voltage and low load efficiency

Supplier features	ADI ADP5090	TI bq25504	Linear Tech LTC3105	Maxim Max17710
Start-up input voltage	380 mV	330 mV	ZSOmV	750 mV
Efficiency (Vin@0.5 V; Vout@3.0 V)	58% (10 uA) 79% (100 uA)	35% (10 uA) 75% (100 uA)	30% (10 uA) 50% (100 uA)	Not supported
lq	≈350 nA	570 nA	10 uA	625 nA

voltage divider of V_{OC}. Since a single cell output voltage can be low, an important piece of the PV harvesting is the cold startup circuit. An example implementation is shown in Fig. 4 [2].

The first charge pump 10*VIN_CP is used to control a cascade device to protect the 3 V medium VT startup switch MV from the switching node voltages, and a comparator is used to make sure the second charge pump output voltage 4*VIN_CP is good. Inductor current saturation is detected to minimize startup current as long startup period from a regular oscillator can easily make inductor current saturated. Table 1 is the summary of startup voltage and efficiencies for a few ultralow power boost converters on the market. As much as 58% efficiency can be achieved with only 10uA load, and the quiescent current is only about 350 nA.

3 Thermoelectric Harvesting

Thermoelectric harvesters rely on a temperature gradient to generate power based on one of the thermoelectric effects, the Seeback effect, which is the direct conversion of temperature differences to electric voltages. The efficiency for a thermoelectric generator depends not only on the temperatures at the hot side and cold side but also on the figure of merit ZT for the thermoelectric material used, where $ZT = \frac{S^2 \sigma T}{\kappa}$. S is the Seeback coefficient or the thermopower, σ is the electrical conductivity, and κ is the thermal conductivity. The best bulk thermoelectric material at room temperature is Bi_2Te_3, which has a ZT $\sim$1. Besides materials, there has been progress in enhancing ZT using low-dimensional structures such as quantum wells or nano-wires where larger S and/or lower thermal conductivity can be achieved.

Thermoelectric harvesters are scalable and are suitable for integration. The efficiency of a thermoelectric element does not depend on the size, and the heat that can conduct through a certain cross-section area for a given thermoelectric element increases with the reduction in its length. However, the reduction in length will lead to reduction in thermal resistance which can pose challenges in applications with high external thermal resistances where the available temperature gradient is divided between external and internal thermal resistances and only the gradient across the thermoelectric element contributes to the power generation. A thermoelectric harvester usually consists of multiple thermoelectric legs with positive thermoelectric power (p-type) and negative thermoelectric power (n-type) connected in series electrically, but in parallel thermally, to build enough voltage that can be used. The voltage across a single element can be quite small, and in many cases, a booster converter is used to further raise the voltage and ensures impedance matching to extract the maximum power from the harvester. To reduce the thermal shunt by the ambient air surrounding the P and N legs, wafer capping can be used to seal the thermoelectric legs in a vacuum as shown in Fig. 5a [3]. To improve device thermal resistance without the need for thick films, the thermoelectric films can be deposited along a thick polyimide island as shown in Fig. 5b.

A thermoelectric harvester delivers the maximum power to the load when the load resistance matches the internal device resistance. Similar circuit can be adapted from FOV for the PV cell harvesting but setting FOV equal to 0.5. Besides load matching for maximum power, another common need is the startup circuit due to the low voltage output from the thermoelectric harvester with limited DT. Depleted NMOS together with a step-up transformer can form self-oscillation to build up

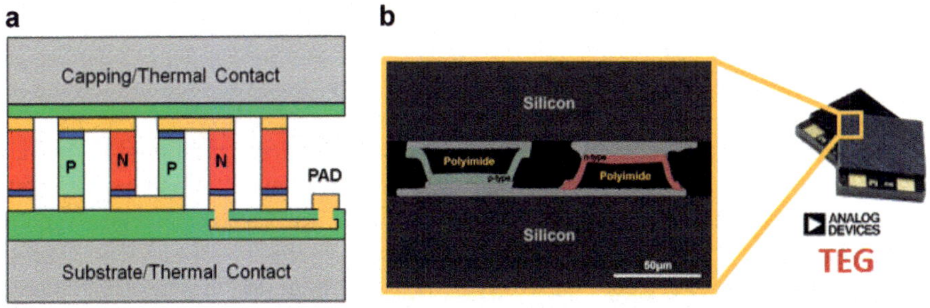

Fig. 5 (**a**) Structure for a vertical thermoelectric harvester with vacuum capping (**b**) pyramid-shaped thermoelectric harvester

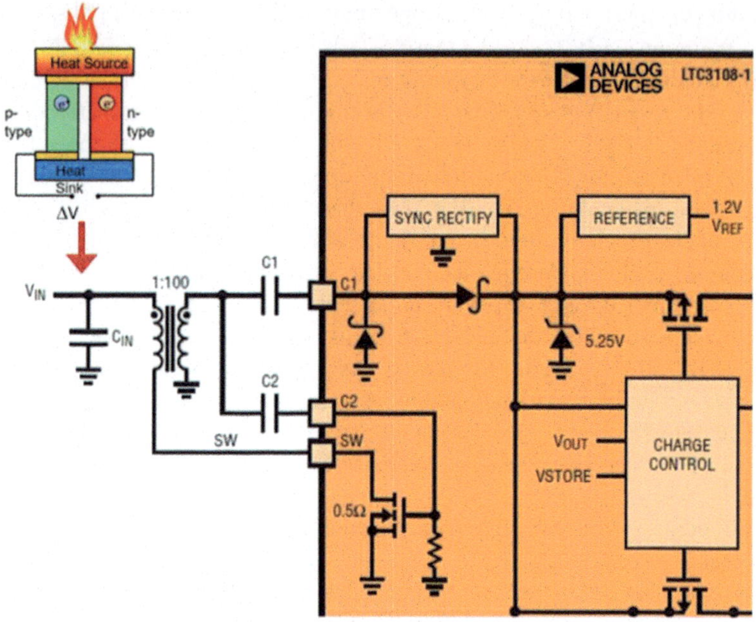

Fig. 6 Low voltage startup using step-up transformers

startup voltage as shown in Fig. 6 [4]. With 1:100 step-up transformers, the converter can startup from input as low as 20 mV. In some applications, the harvester also needs to harvest energy from both positive input voltages and negative input voltages. For example, a ground spike can be used to harvest energy from the temperature differences between the surface and the soil under, and the polarity can change depending on the day or night. A full bridge circuit with parallel diodes and switches can be used, and the switches are turned on or off by looking at the voltage polarities at the input. For bipolar startup, two startup transformers can be used [5].

4 Vibrational Harvesting

Vibrational harvesters usually rely on certain mechanical structures to convert external vibration to the kinetic energy, and the mechanical structures are coupled to energy transducers, such as piezoelectric transducers or electromagnetic transducers, to convert kinetic energy to electricity. Figure 7a shows a typical cantilever-based piezoelectric harvester. The beam operates in a bending mode, strains the piezoelectric films, and generates charges from the piezoelectric effect. While a cantilever provides low resonant frequency, a proof mass at the end further reduces the resonant frequency, more suitable for low-frequency vibrational

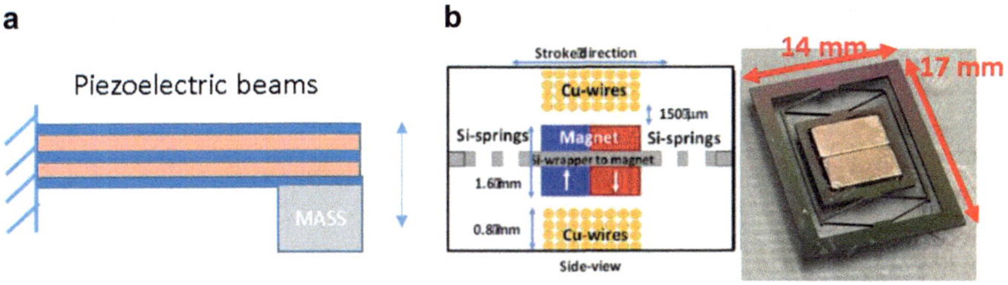

Fig. 7 (**a**) Piezoelectric harvester (**b**) electromagnetic harvester

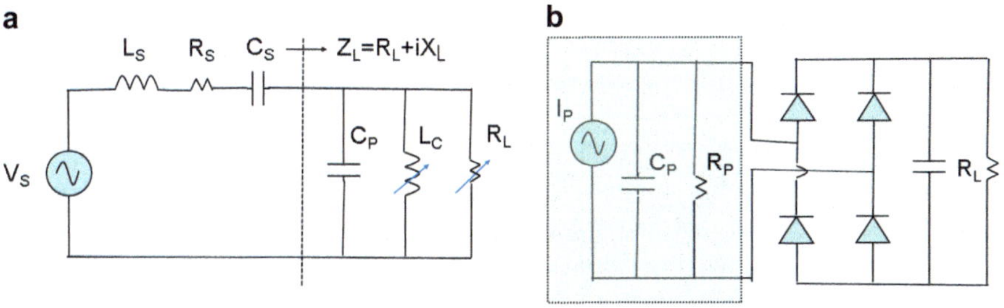

Fig. 8 (**a**) Piezoelectric harvester model (**b**) full bridge rectifier for the harvester

harvesting. Figure 7b shows an example MEMS electromagnetic harvester [6]. The magnets mount on a silicon spring, and the magnetic field will cut across the coils mount above and below the magnet once the spring oscillates. Vibrational harvesters can be analyzed using damped mass-spring systems [7]. It is desired to maximize the mass displacement with the power increasing with the square of the amplitude but the mass displacement is limited by the size of the system.

Equivalent circuit model for a piezoelectric harvester is shown in Fig. 8a. Its source impedance is equivalently a serial RLC network [8], where RS $= \eta/\Theta^2$, LS $=$ M/Θ^2, CS $= \Theta^2/K$. η is the mechanical damping coefficient, Θ is the piezoelectric coefficient, M is the mass, and K is the effective stiffness. This can be derived from the vibration and transducing equations. The dotted line represents electromechanical interface. To extract maximum power into the electrical domain, it is desired that the loading impedance $Z_L = R_L + jX_L$ be the complex conjugate of the source impedance, $Z_S{}^* = R_S - jX_S$, where $X_S = \omega L_S - 1/(\omega C_S)$. With conjugate matching, the source basically sees source impedance R_S and R_L, and current and voltage waveforms are synchronized. In principle, Z_L can be adjusted with variable L_C and R_L but L_C can be large, tens to hundreds of Henry. In many practical systems, simple full bridge rectifier is used without conjugate matching as shown in Fig. 8b; however, the efficiency can be low as its ideal efficiency is only $4/\pi Q_P$, where $Q_P = \omega_P C_P R_P$ and Q_P is usually bigger than 10. Its efficiency can be improved to $8/\pi Q_P$ with a bias flip switch without inductors, and the harvesting efficiency can be dramatically boosted with synchronized switching harvesting on

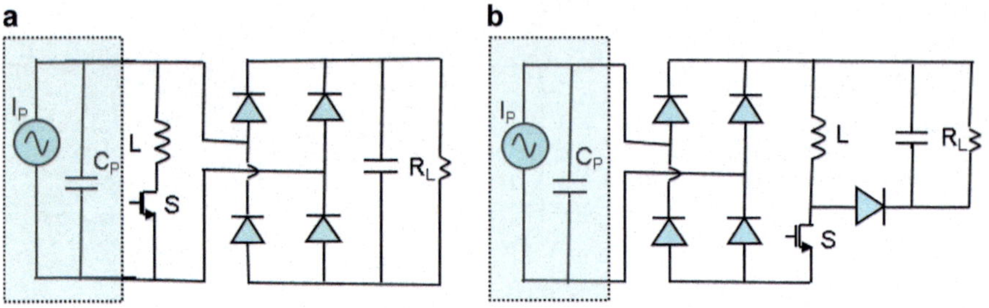

Fig. 9 (**a**) SSHI diagram (**b**) SECE diagram

inductors (SSHI) [9, 10] or synchronous electrical charge extraction (SECE) [11] as shown in Fig. 9.

SSHI adds an inductive switch path, *L* and *S*, to avoid the energy wasted for charging and discharging internal capacitance C_P. SECE has a switch *S* that will turn on each time the rectified voltage reaches maximum and turn off each time it reaches 0. This allows the stored charge to be removed completely and for the transducer to deliver the power to the load through L. Full wave rectified diodes, as shown in Fig. 9, can lead to significant rectification loss with the finite diode drop. Active NMOS with cross-coupled PMOS or active PMOS with cross-coupled NMOS can be used to reduce rectification loss considerably. With limited harvested energy, the comparators used to control the active switches need to be designed with low quiescent power.

While synchronized switching can boost harvester efficiency, it relies on the harvesters operating at resonant frequencies. However, the manufactured harvesters can have certain percentage of variations in their resonant frequencies because of manufacturing tolerances, and extracted power can be significantly lower if these resonant frequencies do not match those of the vibrational sources. Off-resonance efficiency can be improved by introducing switching delays for synchronized switching techniques based on the conjugate impedance matching principle [12].

As shown in Fig. 10a, synchronized switching can be analyzed with a simple current source with parallel capacitance in parallel with a serial-connected inductor and switches. At zero crossing of the source current, the switch turns on for half of the period of the LC resonance to allow V_S to be flipped. V_S is not an ideal square waveform with the loss in switches and inductors. Similarly, the synchronized switching waveforms with delays can be shown in Fig. 11. If the delay is positive, the load appears capacitive as shown in Fig. 11a, and if the delay is negative, the load appears inductive as shown in Fig. 11b. A delay to the voltage waveform basically introduces a quadrature term besides the fundamental term. The equivalent impedance seen at the electromechanical interface becomes complex. By adjusting the delay, the equivalent complex impedance can be tuned to match the conjugate source impedance for maximum power transfer.

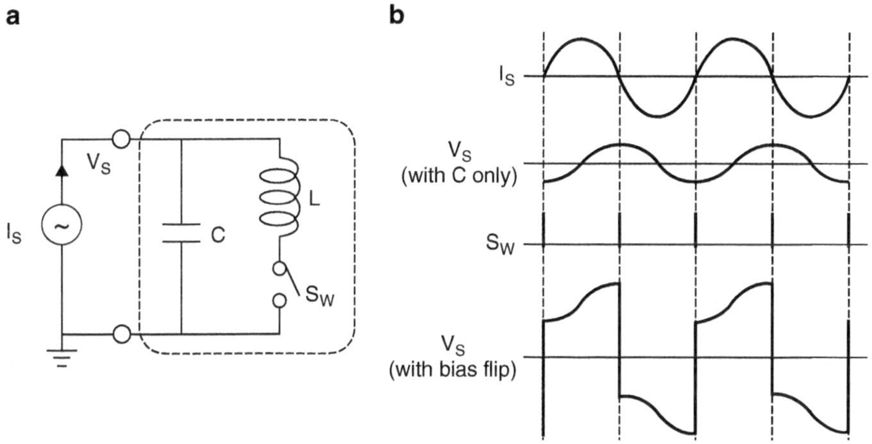

Fig. 10 (**a**) Synchronized switching schematic (**b**) waveforms

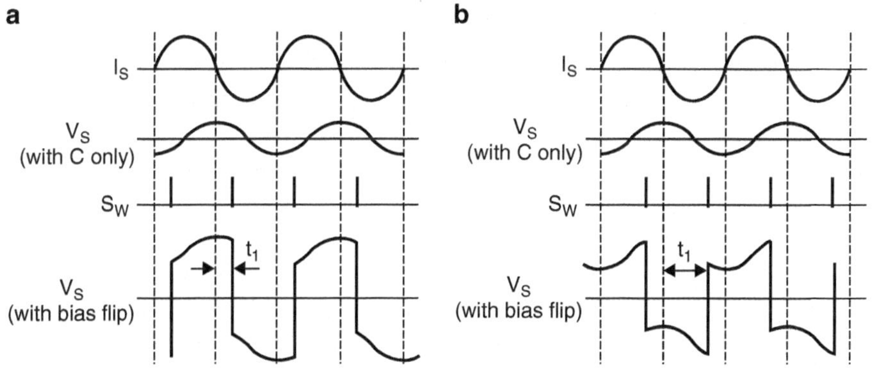

Fig. 11 Synchronized switching with (**a**) positive delay (**b**) negative delay

5 Conclusions

Various energy harvesting transducers and their power conditioning circuits have been reviewed and discussed. Both the transducers and power conditioning circuits need to be optimized to maximize the power delivered to the load. While nonlinear harvesting circuits such as SSHI and SECE can significantly boost the output available to the load, it is important to minimize the circuit overhead for implementing these techniques. To accommodate manufacturing tolerances of the resonant frequencies for the vibrational harvester, off resonance output power can be improved through conjugate impedance matching or through introducing delays in synchronized switching.

Acknowledgment The author would like to acknowledge contributions from members of the energy harvesting team at Analog Devices, Inc. and our university collaborators.

References

1. ADP5090 datasheet. http://www.analog.com/media/en/technical-documentation/data-sheets/ADP5090.pdf.
2. Lu Y, Yao S, Shao B, Brokaw P. A 200nA single inductor dual-input-triple-output (DITO) converter with two-stage charging and process-limit cold-start voltage for photovoltaic and thermoelectric energy harvesting. ISSCC Dig. Tech. Papers, Feb. 2016, pp 368–70.
3. Cornett J, Lane B, Dunham M, Asheghi M, Goodson K, Gao Y, Sun N, Chen B. Chip-scale thermal energy harvester using Bi_2Te_3. IECON 2015-Yokahama, 41st Annual Conference of the IEEE Industrial Electronics Society, 2015, pp. 3326–9.
4. LTC3108 datasheet. http://www.linear.com/product/LTC3108.
5. LTC3109 datasheet. http://www.linear.com/product/LTC3109.
6. Shin A, Radhakrishna U, Yang Y, Zhang Q, Gu L, Riehl P, Chandrakasan AP, Lang JH. A MEMS magnetic-based vibration energy harvester. Power MEMS Proceedings, 2017, pp. 363–6.
7. Beeby S, Tudor M, White N. Energy harvesting vibration sources for microsystems applications. Meas Sci Technol. 2006;17:175–95.
8. Lien IC, Shu YC, Wu WJ, Shiu SM, Lin HC. Revisit of series-SSHI with comparison to other interface circuits in piezoelectric energy harvesting. Smart Mater Struct. 2010;19:125009–20.
9. Guyomar D, Badel A, Lefeuvre E, Richard C. Toward energy harvesting using active materials and conversion improvement by nonlinear processing. IEEE Trans Ultrason Ferroelectr Freq Control. 2005;52(4):584–95.
10. Ramadass Y, Chandraksan A. An efficient piezoelectric energy harvesting interface circuit using a bias-flip rectifier and shared inductor. IEEE J Solid State Circuits. 2010;45(1):189–204.
11. Lefeuvre E, Badel A, Richard C, Guyomar D. Piezoelectric energy harvesting device optimization by synchronous electric charge extraction. J Intell Mater Syst Struct. 2005;16(10):865–76.
12. Hsieh P-H, Chen C-H, Chen H-C. Improving the scavenged power of nonlinear piezoelectric energy harvesting interface at off-resonance by introducing switching delay. IEEE Trans Power Electron. 2015;30(6):3142–55.

From Bluetooth Low-Energy to Bluetooth No-Energy: System and Circuit Aspects of Energy Harvesting for IoT Applications

Wim Kruiskamp

1 Introduction

The Internet of Things (IoT) is the network of physical devices, vehicles, home appliances, and other items embedded with electronics, software, sensors, actuators, and network connectivity which enables these objects to connect and exchange data [1]. The "Things" in IoT can be anything and the variety in implementations is enormous, but many of them can be simplified to the system as depicted in Fig. 1. On the left side, there is the "Thing," which often is a small battery-powered device that includes a transceiver to connect wirelessly to the internet. It also often includes a sensor and some processing power.

The radio connection can be Bluetooth low-energy (BLE), Wi-Fi, cellular, or any other wireless standard. The main topic of this chapter is the energy supply of the IoT devices. The majority of today's devices are powered by a battery, either a 3 V coin-cell or a rechargeable battery. These batteries usually do not contain enough energy to supply the IoT devices for their entire lifetime, so they must be replaced or recharged on a regular basis. That might be acceptable today with only a few devices per person, but with the expected fast growth of the IoT, this will not be rational anymore in the near future. Either the power consumption of IoT devices needs to be reduced in orders of magnitude or they need to be powered by alternative sources. This is where energy harvesting will come into play. The only realistic scenario for everyone having tens of IoT devices will be that these devices are self-sustainable. This can be done by making use of available ambient energy

W. Kruiskamp (✉)
Dialog Semiconductor, 's-Hertogenbosch, The Netherlands
e-mail: wim.kruiskamp@diasemi.com

K. A. A. Makinwa et al. (eds.), *Low-Power Analog Techniques, Sensors for Mobile Devices, and Energy Efficient Amplifiers*,
https://doi.org/10.1007/978-3-319-97870-3_2

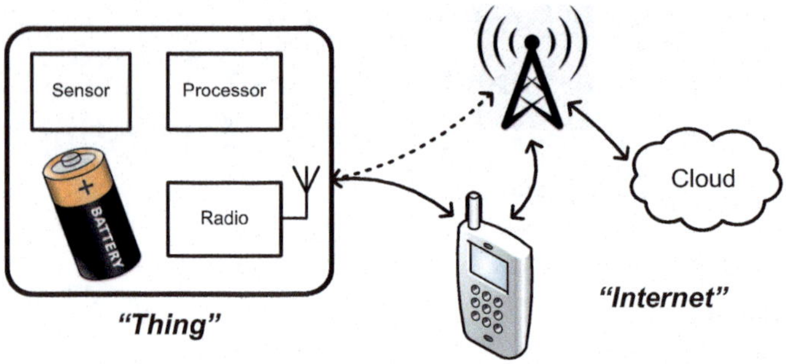

Fig. 1 Internet of Things device

like light, heat, movement, or RF energy. Thus the main driver for energy harvesting will not be cost or an environmental reason, it will be ease-of-use and install-and-forget.

The remainder of this chapter is structured as follows: In Sect. 2, we will look into the average power consumption of a typical BLE IoT device. In Sect. 3, we will briefly discuss alternative energy sources that can deliver this power. In Sect. 4, we will propose a power management unit (PMU) to connect these sources to an IoT device.

2 Average Power Consumption

A popular battery for IoT devices is a 3 V CR2032 coin-cell battery containing about 225mAh of charge. For applications that consume in the order of 5µW, this results in a battery lifetime of well over 10 years:

$$\text{lifetime} = \frac{225mAh \times 3V}{5\mu W \times 24\,\text{hours} \times 365\,\text{days}\,\mu Wd} = 15\text{ years}$$

This kind of battery lifetime is long enough for an acceptable product lifetime without the need to replace the battery. For such applications, there is hardly any reason not to use a coin-cell primary battery; they are reliable, cheap, and relatively small.

However, the majority of today's IoT devices consume at least an order of magnitude more power. This power consumption of an IoT device is due to the following actions:

- Sensor data acquisition and processing.
- Wireless communication to a host.
- Timekeeping during sleep.

Each of these components will be discussed in the next sections, followed by two application examples.

2.1 *Sensor Data Acquisition and Processing*

The power consumed to acquire and process information is very application dependent. For functions like temperature measurement, sensors are available which consume less than 1 μW [2–4], and more power-hungry temperature sensors can be operated at a low duty-cycle to achieve similar power consumptions. For applications that only perform these kinds of measurements, the power consumption for data acquisition and data processing will not be the limiting factor for a 10-year battery lifetime. For BLE beacons and tags, there is no sensor available hence no power consumption associated with it.

There are, however, also measurements that consume significantly more power. A well-known example is location via GPS which typically consumes in the mW range. Also optical heart-rate monitoring typically consumes a significant amount of power; today ranging from more than 100 μW [5, 6] to several mWs. These kinds of measurement functions dominate the total power consumption of fitness trackers for example.

2.2 *Wireless Communication (BLE)*

Wireless communication from an IoT device to a host (i.e., a smartphone) via BLE consumes a significant amount of energy. A typical power profile, measured on a DA14580 chip [9], known to be best in class for power consumption, is depicted in Fig. 2. A connection event consists of a receive action and transmit action plus some overhead for processing and startup of the crystal oscillator. The total energy associated with a connection event is about 14 μJ measured at the battery. With a connection interval of 1 second, that would correspond to an average power consumption of 14 μW, plus 1–3 μW associated with the power consumption during sleep between the connection events.

2.3 *Timekeeping During Sleep*

BLE devices are usually asleep most of the time, counting to the next communication event. In order to maintain a synchronous BLE link, the timing accuracy needs to be better than 500 ppm. Traditionally, this timekeeping is done with a 32.768 kHz Quartz oscillator. Such an oscillator plus counter can easily be achieved with a power consumption below 1 μW and can have an accuracy in the single-digit

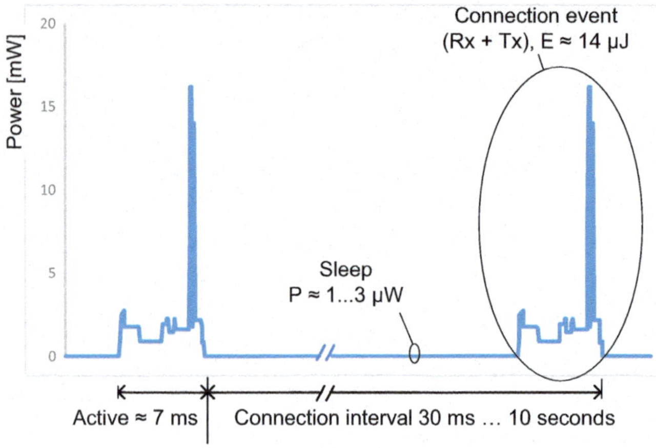

Fig. 2 Typical BLE power consumption

ppm range. Academic research has even reported a power consumption of only 2 nW for a Quartz oscillator [7]. Apart from this low-frequency reference, a BLE chip also needs a reference for the 2.4 GHz RF signal. This is typically done with another Quartz oscillator, operating at a higher frequency, usually 16 MHz or 32 MHz. The power consumption of this higher frequency oscillator is in the order of 100 μW and is therefore only enabled when the radio is active and can therefore unfortunately not be used for timekeeping. For that reason, BLE devices traditionally used to have two Quartz oscillators.

Despite their very good accuracy, very good stability and low power consumption, there is a clear trend in BLE to replace the low-frequency Quartz oscillator by an on-chip relaxation oscillator. These oscillators can also be designed with a power consumption of less than 1 μW [8] and reasonably good accuracy. The accuracy over temperature of such oscillators is usually not sufficient for the 500 ppm BLE requirements, but by calibrating against the high-frequency Quartz oscillator each time the radio is active, a timing accuracy of better than 500 ppm can be achieved. The benefit of this configuration over the traditional two-crystal configuration is lower cost and smaller size.

Although the power consumption of a 32 kHz relaxation oscillator in itself is comparable to a 32 kHz Quartz oscillator, the limited accuracy of relaxation oscillators comes with a power consumption penalty at a system level. If the BLE device is synchronized with a smartphone, the receiver must be enabled at the moment the smartphone is transmitting its data. With a very accurate Quartz oscillator, this is not a problem and can be done just in time. In case of a relaxation oscillator, this might vary by 500 ppm, the receiver must be enabled early enough not to miss the smartphone signal. This is depicted in Fig. 3. The power penalty will be equal to:

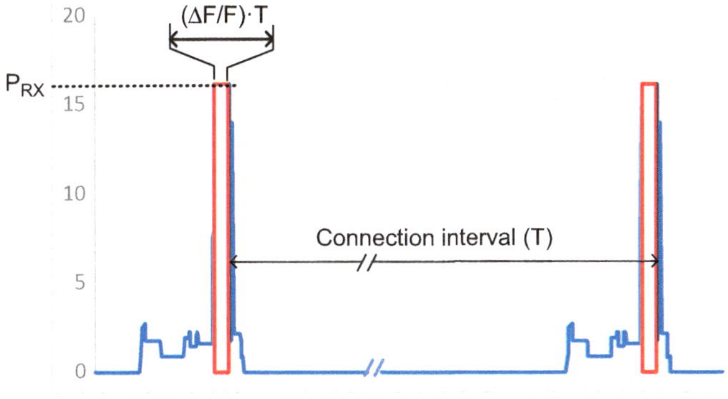

Fig. 3 Effect of timer inaccuracy on average power consumption

$$P_{\text{extra}} = \frac{(\Delta F)_{\max}}{F} \cdot P_{RX}$$

This extra power consumption is independent of the connection interval: If the connection interval is larger, the absolute time the receiver has to be enabled is larger, but it happens less often.

When we consider a typical receiver power consumption of 10 mW and a relaxation oscillator with a maximum inaccuracy of 500 ppm, the power consumption penalty is:

$$P_{\text{extra}} = 500ppm \cdot 10mW = 5\mu W$$

This 5 μW is about equal to the maximum allowed power consumption if we want to operate for more than 10 years on a coin-cell battery and is therefore a very significant contribution.

2.4 Application Example: BLE Beacon

A BLE beacon is like a lighthouse. It repeatedly transmits the same signal to surrounding BLE-enabled devices such as smartphones and tablets. The transmitted data includes a unique identifier of the beacon and might also include a small amount of other data like a link to a website. The signal is transmitted multiple times a second on each of the three advertisement channels of BLE. For a beacon, there is no power required for sensor or data-processing, and inaccuracy of the sleep-timer is irrelevant since a beacon is only one-way communication and therefore not synchronized to a smartphone. The power consumption is therefore dominated by the advertisement event (today, typically 20 μJ for advertising in

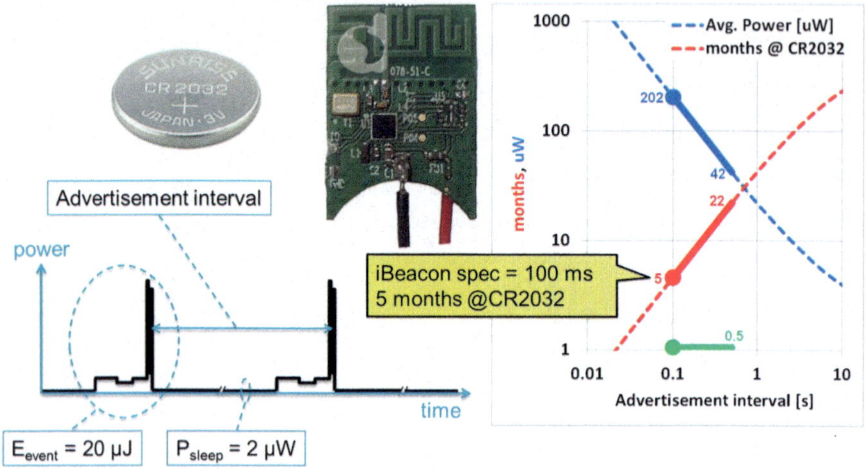

Fig. 4 BLE beacon: Typical power-consumption and battery-lifetime

all three advertisement channels [9, 10]) with a power consumption between the advertisement events in the order of 2 μW. Due to this, the battery lifetime is proportional to the advertisement interval as depicted in Fig. 3. In its iBeacon specification, Apple recommends an advertisement interval of 100 ms in order to have high chances that a passing smartphone will catch the beacon signal. This performance might not be needed in all applications and often beacons have longer advertisement intervals, in the order of 300 ms, to save power.

As can be seen in the Fig. 4, the battery lifetime of beacons with the recommended 100 ms advertisement interval is currently well below 1 year. Even if the advertisement interval is stretched to a questionably long interval of 500 ms, the batteries will have to be replaced every 2 years. This power consumption will decrease in next generation BLE devices but is still far away from the 5 μW power consumption target to run for more than 10 years on a coin-cell battery.

2.5 *Application Example: Fitness Band*

Fitness bands and activity trackers are often powered by a rechargeable Li-ion battery with a battery lifetime ranging from less than a week to more than a month. If we assume a battery lifetime of 1 month from a 60 mAh Li-ion battery, we can calculate that the average power consumption has to be in the order of 300 μW.

These activity trackers are synchronized with the smartphone with one up to a few connections per second. With an energy consumption of 15 μJ per connection event [9], the radio connection only explains about 10% of the total energy consumption. Power consumption due to inaccuracy of the timer is in the order of 5 μW and is therefore negligible compared to the 300 μW total power consumption. We can therefore conclude that the power consumption is dominated by the sensors, the data-processing, and notifications to the user via a display and/or buzzer. This power consumption per function is expected to decrease in next-generation devices, but most likely new features will be added to bring the power consumption back up again. Today, we are still far away from the 5 μW target to run many years on a small primary battery, and this is not expected to change in the near future.

3 Alternative Energy Sources

In the previous paragraphs, we learned that there is a need for alternative energy sources with an average power in the order of several 10s to several 100 s of μW. The possible options are light, movement, heat, RF-energy, and bio-fuel.

One of the easiest sources to use is light. Photovoltaic (PV) cells are cheap and thin and the available power ranges from 15 μW/cm^2 indoors to 15 mW/cm^2 in full sunlight. Even indoors, a few cm^2 could already be sufficient to power an IoT device.

Another option is human body heat. An average human consumes in the order of 2000 kcal per day, which is equal to an average power consumption of 100 W. This power is eventually transformed into heat, and with a typical skin area of about 2 m^2, this results in an average thermal power of 5 mW/cm^2. With a Thermo Electric Generator (TEG), this thermal power can be converted into electrical power. The efficiency of this power transfer has a theoretical upper limit equal to the Carnot efficiency:

$$\eta_{\text{Carnot}} = 1 - \frac{T_{\text{cold}}\ [K]}{T_{\text{hot}}\ [K]}$$

This means we can only achieve 0.3% efficiency per degree Celsius temperature difference over the TEG. Practical TEGs are often made from the material Bismuth Telluride and can achieve an efficiency up to 18% of the theoretical limit. This results in an average electrical power from human heat in the order of 3μW/cm^2 per degrees Celsius temperature difference. Another option might be ambient RF energy. However, this is only useful in the near vicinity of an RF source (Table 1).

3.1 *Maximum Power Point*

In order to harvest as much power as possible from an alternative energy source, the voltage across the source must be kept at the value corresponding to the maximum power point (MPP). For a TEG that is half the open-clamp voltage (V_{OC}), for a PV-cell it is somewhere between 70% and 80% of its open-clamp voltage.

Table 1 Typical characteristics of alternative energy sources

Source	Conditions	Harvested power	Open-clamp voltage
Light	Outdoors, sunny Outdoors, shade Indoors	15 mW/cm^2 500 μW/cm^2 15 μW/cm^2	0.6 V 0.5 V 0.4 V
Human heat	ΔT = 5 °C ΔT = 1 °C	15 μW/cm^2 3 μW/cm^2	0.4 V 80 mV
RF (20 dBm) (=100 mW)	Distance = 0.3 m Distance = 1 m	9 μW/cm^2 0.8 μW/cm^2	

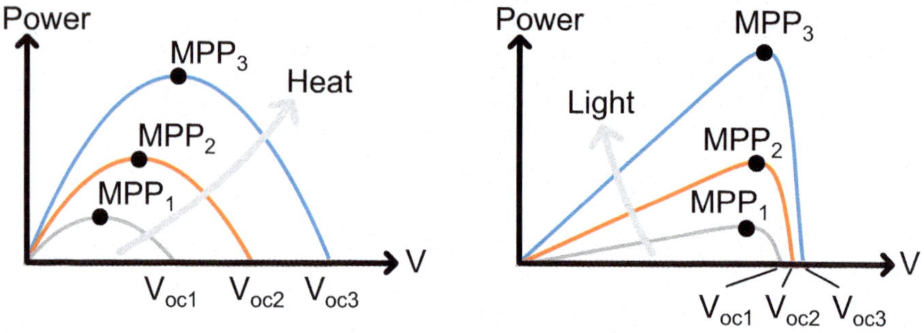

Fig. 5 Maximum power point of TEG (left) and PV-cell (right)

A practical challenge for an energy-harvesting system is the fact that the power density of alternative energy sources depends on environmental conditions such as illumination density and temperature. Therefore, a maximum power point tracking (MPPT) system is needed to constantly harvest the maximum amount of energy (Fig. 5).

A popular way to do MPPT is to periodically sample the open-clamp voltage and regulate to a fixed fraction of that voltage [13, 14]. The main drawback of this approach is that the energy harvesting must be interrupted periodically to measure the open-clamp voltage.

Another option is to measure the harvested power and apply a search algorithm (perturb and observe) to stay close to the MPP [15, 16]. This approach does not require the harvesting to be interrupted but does require additional circuitry to measure the harvested power.

4 Proposed Architecture

To supply an IoT device from alternative energy sources, the architecture as depicted in Fig. 6 is proposed. The main task of this circuit is to transfer energy from the sources to a storage capacitor or rechargeable battery connected to pin STORAGE. Like many other harvesting circuits [12, 13, 15, 16], this is done by an inductive

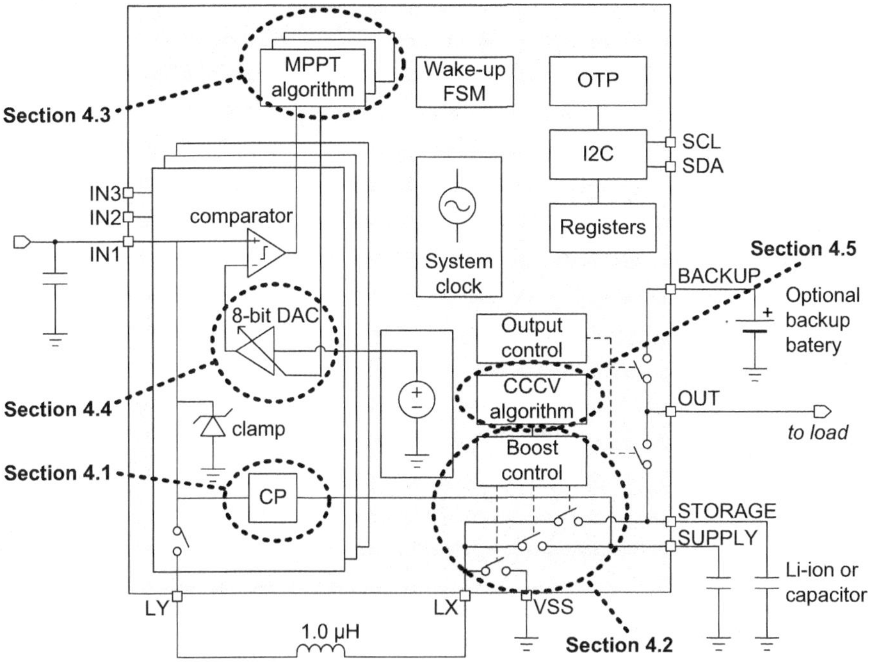

Fig. 6 Proposed architecture

boost converter. The control circuits for this boost converter are supplied by the other output of the boost converter: pin SUPPLY. This voltage at the output of the boost converter is not yet available during startup. Therefore, a charge pump is added to allow cold-start at voltages as low as 230 mV.

The system includes a digitally implemented MPPT at each of the three inputs as well as a digitally implemented Constant-Current Constant-Voltage (CCCV) charging algorithm at the output.

In the next sub-sections, the circuits and aspects which are typical for energy harvesting systems are discussed in more detail.

4.1 Startup Circuit

The startup circuit consists of a 7-stage ring-oscillator, clock-buffers, and a 14-stage charge-pump. The charge-pump stage is a modified version of the two-phase voltage doubler, presented in [11] and shown in the right-hand side of Fig. 7.

The chip is processed in standard TSMC 55 nm technology, without using low threshold voltage transistors. This process choice was made for easy integration in a BLE chip but is not the ideal choice for extremely low-voltage operation. In order

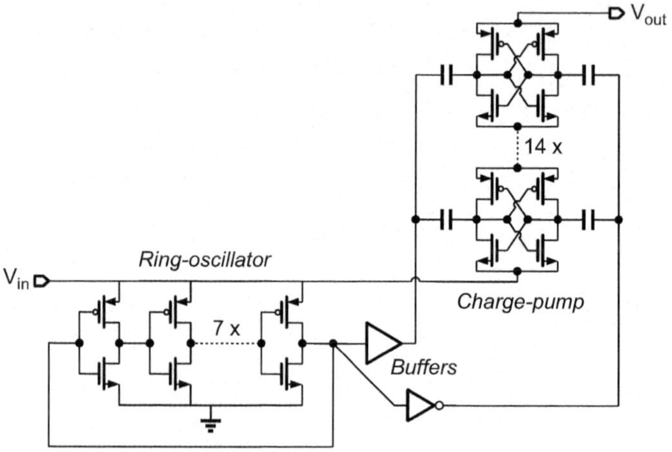

Fig. 7 Startup circuit principle

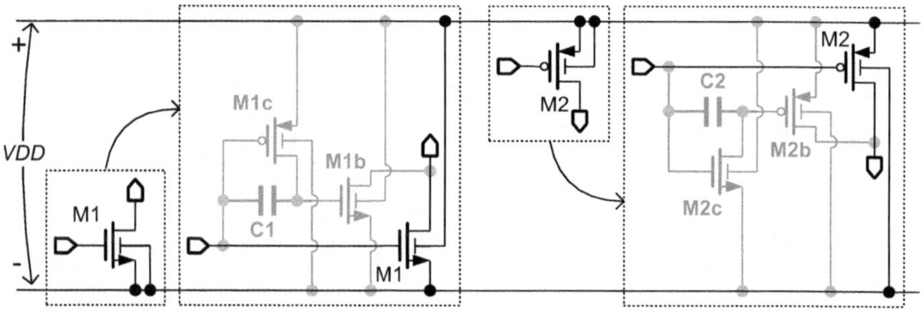

Fig. 8 Bootstrapped transistors (gray) added to original transistors

to use the circuit of Fig. 5 at input voltages as low as 230 mV, reverse body biasing was applied to lower the threshold voltage. Furthermore, each transistor is assisted by a transistor with a bootstrapped gate voltage. This is depicted in Fig. 8.

Each nMOS transistor (M1 in Fig. 8) gets a parallel connected transistor M1b, of which the gate is connected via a capacitor C1. Capacitor C1 is charged via transistor M1c and acts like a floating voltage source, increasing the value of Vgs. While the gate of M1 is switching between 0 V and VDD, the gate of M1b is switching between VDD and 2xVDD. The same is done for each pMOS transistor (M2 in Fig. 8). Furthermore, the bulks of all nMOS transistors are connected to the higher rail and the bulks of all pMOS transistors are connected to the lower rail. These techniques are applied in the ring-oscillator, the buffers, and in the charge-pump stages.

Initially the bootstrap capacitors are uncharged and the circuit will operate at a very low frequency with very low drive capability. This slow oscillation will however be sufficient to charge the capacitors in the bootstrapped transistors, which will give these transistors an increased conductivity. This will cause the frequency of the oscillator to rise significantly and will also lower the on-resistance in the switches of the charge-pump stages significantly.

The drawback of this modification, apart from increased complexity, is increased leakage in the off-state and the need to limit the input voltage of the charge pump. This is overcome by disabling the circuit when cold-start is completed and by adding a voltage limiter in front of the startup circuit.

4.2 Boost Converter

The boost converter is operating in discontinuous conduction mode (DCM) as depicted in Fig. 9. In order to reduce pcb space, a small inductor of 1 μH is used, which is an order of magnitude smaller than most other energy-harvesting chips [13, 15, 16].

The control of the boost converter is done by comparators: The input and output voltages are monitored by dynamic (clocked) comparators, which does not consume static power. If both the input voltage is available and the output voltage is below its maximum, a DCM-pulse is started. The inductor is connected to ground and the inductor current will rise linearly, storing energy in the inductor. When a continuous comparator detects that the current has reached a certain value I_{max}, the inductor is connected to the output and the inductor current drops linearly, releasing its energy to the output. When another continuous comparator detects that the current is zero, the output switch is opened and the procedure starts all over again. The DCM-mode is very suitable for multiple-input operation, since after each DCM-pulse, the boost converter returns to its idle state with zero current in the inductor, which is an ideal situation to change to another input.

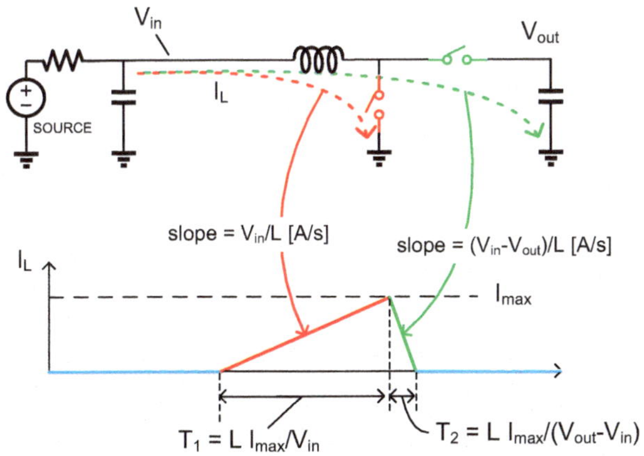

Fig. 9 DCM-mode operation of boost converter

For cost and size reasons, the used inductor is a small 1μH inductor. The consequence of this small inductor value is short ON-times of the switches: With the used I_{max} of 280 mA, the ON-time (T_2) of the output switch can be as low as 70 ns. The two comparators (peak-current detection and zero-cross detection) need to react significantly faster than the ON-time and yet be accurate. To keep the quiescent power of these comparators within acceptable levels, automatic calibration is applied as described in [9], and the comparators are only enabled when needed.

4.3 MPPT Algorithm

The applied MPPT algorithm is "perturb-and-observe": The input DAC is set to a certain value and the harvested energy is measured for a fixed amount of time. Then the DAC is changed by a small amount and the harvested energy is again measured for a same amount of time. If the harvested energy has increased, the DAC is again changed in the same direction, otherwise, the DAC is changed in the opposite direction.

The measuring of the harvested energy is done by counting the DCM pulses to the output and digitally scaling that number by a factor E_{DCM}, which is an estimate of the energy per DCM pulse. The factor E_{DCM} can be estimated from Fig. 8 as given below:

$$E_{DCM} = V \cdot I \cdot T \approx \frac{L \cdot (I_{max})^2}{2} \cdot \frac{V_{STORAGE}}{(V_{STORAGE} - V_{in})}$$

In which L and I_{max} are known constants, V_{in} is set by the DAC and therefore available in a digital form, and $V_{STORAGE}$ is a slowly changing voltage that is measured by a SAR ADC each time the DAC is updated. Therefore, this algorithm can be implemented with little analog overhead; mainly a slow and low-resolution ADC.

This MPPT algorithm has several advantages over the more commonly used *fractional open-clamp-voltage* approach, mentioned in Sect. 3.1:

1. The harvesting process does not have to be interrupted to measure the open-clamp voltage. This interruption would typically be in the order of 2% of the time, reducing the harvested energy by the same amount.
2. This MPPT algorithm searches for the maximum energy transferred to the storage and therefore automatically takes into account voltage-dependent efficiency of the boost-converter. The boost-converter is less efficient at very low input voltages, so the overall optimal voltage from an output power perspective might be higher than the optimal voltage for maximum power delivered by the source.

3. The MPP voltage is by definition lower than the open-clamp voltage. The fact that this algorithm does not need to measure the open-clamp voltage therefore allows the use of better transistors (with lower voltage rating) or to accept sources with higher open-clamp voltages.

4.4 Low-Power DAC

The boost-converter is only enabled when there is energy to be harvested and therefore does not have to be designed for very low quiescent power while active. The DACs which set the MPP threshold for the input voltages are always on and therefore has to be very low power, especially since there are three of them, one for each input.

The DACs are realized as switched capacitor circuits, with a single transistor (M_1) as gain-stage, biased with 10 nA as depicted in Fig. 10. The capacitors C_1 and C_2 are made of 0.6 fF units, C_1 being 100 units and C_2 scalable from 0 to 255 units. During the beginning of phase 2, voltage over C_1 makes a step equal to the reference voltage (1 V). The current to charge C_1 flows through C_2, therefore increasing the voltage over C_2 with a voltage equal to the reference, multiplied by the capacitor ratio. This voltage is sampled on a capacitor and used as reference for the input comparators.

The bias current can be as low as 10 nA since the speed of the DAC can be low. The MPPT algorithm updates the DAC value at a rate of several 10s of Hz, so a clock speed for the DAC of 1 kHz is more than sufficient.

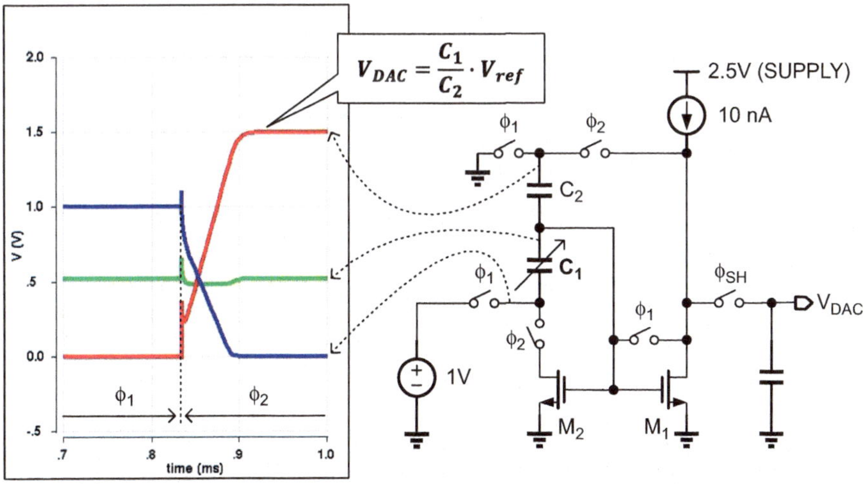

Fig. 10 Low-Iq DAC

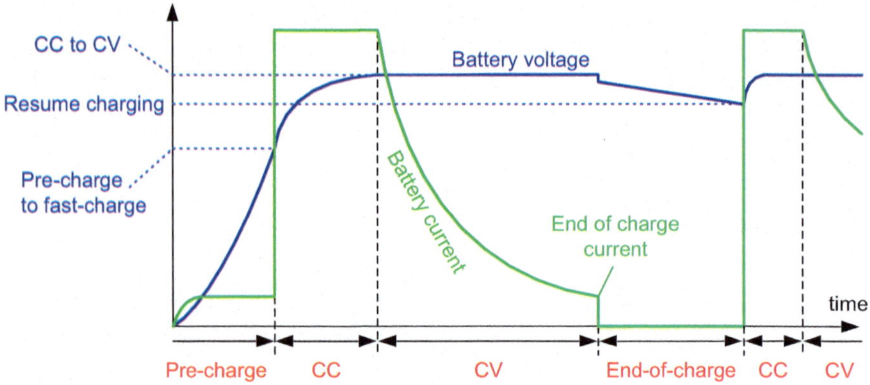

Fig. 11 CCCV charging profile

4.5 CCCV Charging

In case a Li-ion battery is used for energy storage, charging should be done according to the well-known Constant-Current, Constant-Voltage (CCCV) profile as depicted in Fig. 11. Four different phases of the charging profile can be recognized: “pre-charge,” “constant-current (CC),” “constant-voltage (CV),” and “end-of-charge.”

When the battery is almost empty, it is pre-charged by a small current until its voltage reaches a certain level. As soon as the battery voltage has reached that level, the charge current limit is increased to a maximum value and charging continues in CC-mode. The battery voltage will rise further and eventually reach its maximum allowed level. The charging will then be done with a current at which the battery voltage does not exceed the maximum voltage (CV mode). The charging current will drop and eventually falls below the end-of-charge limit. Charging will be stopped completely until the battery voltage has dropped below a certain threshold.

In order to charge a battery according to a CCCV charging profile, the charge current must be controlled and measured. In this proposal, it is done in a digital way. The charge current can be limited by setting a minimum time between each boost-converter DCM pulse relative to the time the inductor is connected to the output (P-switch closed). This is depicted in Fig. 12.

By measuring the on-time T_{out} of the output switch with an oscillator and a counter, the minimum period T_{CC} to meet the maximum constant-current limit I_{CC} can be calculated as follows:

$$T_{CC} = T_{out} \cdot \frac{I_{max}}{2 \cdot I_{CC}}$$

In the same way, the period T_{end} corresponding to an end-of-charge current I_{end} can be calculated:

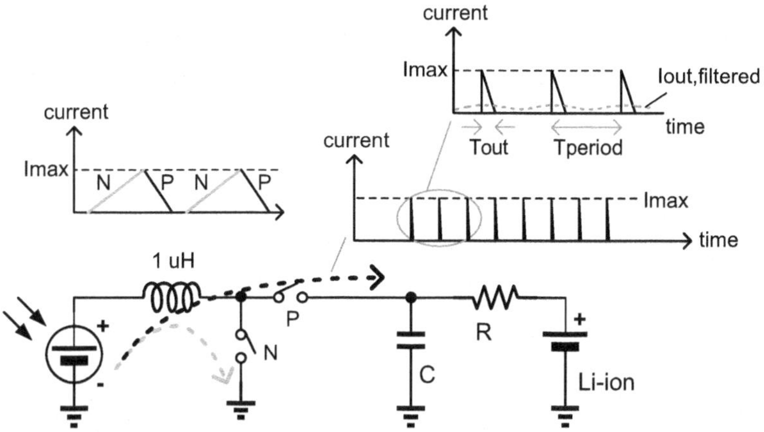

Fig. 12 Boost converter current limit

$$T_{\text{end}} = T_{\text{out}} \cdot \frac{I_{\text{max}}}{2 \cdot I_{\text{end}}}$$

The procedure to meet the CCCV charging curve is now as follows: If the output voltage is below its maximum allowed level and the time from the previous DCM pulse has exceeded the time T_{CC}, a next DCM pulse is allowed. When a DCM pulse is blocked by the condition that the output voltage is above the maximum allowed level for more than T_{end}, the end-of-charge condition has been reached and charging is stopped completely. In order to reduce large ripple currents into the Li-ion battery, an external RC filter with typical values of 1 Ω and 10 μF is added as depicted in Fig. 12.

5 Measurement Results

The chip was fabricated in TSMC 55 nm. The output multiplexer and the boost-converter switches consume a large part of the area due to their relatively high voltage tolerance to allow the use of Li-ion batteries. The charge pumps also consume a significant amount of area due to the many stages and on-chip capacitors. The input multiplexer and the digital control are relatively small (Fig. 13).

The chip was tested with a voltage source with fixed source resistance. Even though the input circuits cannot withstand more than 2.5 V, the chip can be used with sources with a much higher open-clamp voltage as depicted in Fig. 14. The MPPT algorithm will set the input threshold at a value between 0 V and 2.5 V, which makes the boost converter under normal conditions to prevent the inputs to rise too high. Internal clamp circuits are added to prevent over-voltage at the input when the boost converter pulses are blocked by the CCCV algorithm.

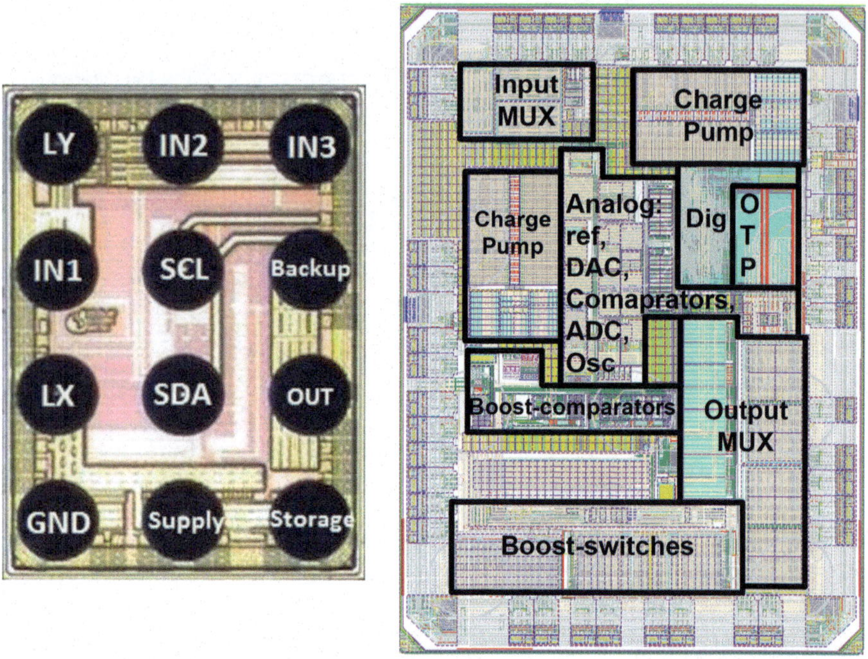

Fig. 13 Die photo and floorplan (1.5 × 2 mm)

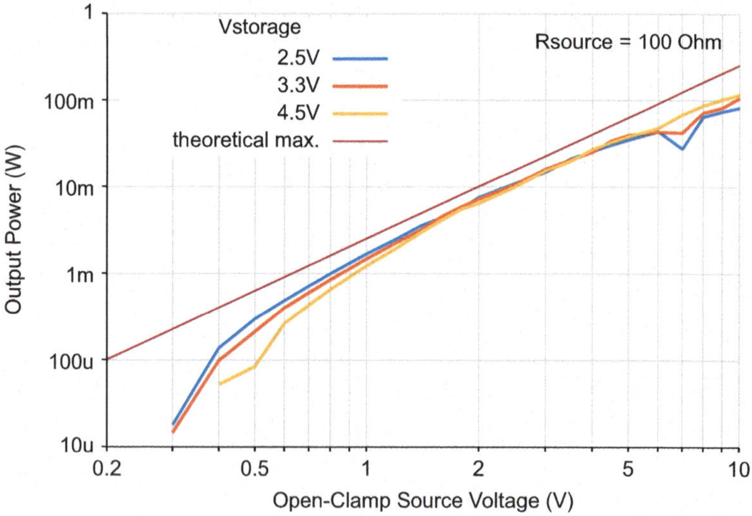

Fig. 14 Measured output power versus open-clamp voltage

6 Conclusions

Many of today's IoT devices consume too much power to run on the same nonrechargeable battery for the entire product lifetime. It is not to be expected that this will change in the near future. With the anticipated growth in number of IoT devices per person, changing batteries or charging batteries regularly is not a practical situation. Alternative energy sources are available that can deliver the required amount of power. Since these sources are not present continuously, it might be needed to use more than one of these sources and to add a rechargeable battery to the system to bridge longer periods without available energy from the sources.

This chapter shows an example of a power management system that is required for such an energy-harvesting IoT device. It includes circuits that can operate at low voltages, circuits that operate with very low quiescent power, and control algorithms to maximize the harvested power and to guarantee safe charging of batteries.

References

1. Wikipedia. https://en.wikipedia.org/wiki/Internet_of_things.
2. Makinwa KAA. Temperature Sensor Performance Survey. [Online]. Available: http://ei.ewi.tudelft.nl/docs/TSensor_survey.xls.
3. Lin Y-S, Sylvester D, Blaauw D.An ultra low power 1V, 220nW temperature sensor for passive wireless applications. 2008 IEEE Custom Integrated Circuits Conference, San Jose, CA, 2008, pp. 507–10.
4. Souri K, Chae Y, Thus F, Makinwa K. 12.7 A 0.85V 600nW all-CMOS temperature sensor with an inaccuracy of $\pm 0.4°C$ (3σ) from -40 to 125°C. In 2014 IEEE International Solid-State Circuits Conference Digest of Technical Papers (ISSCC), San Francisco, CA, 2014, pp. 222–3.
5. Konijnenburg M, et al. A multi(BIO)sensor acquisition system with integrated processor, power management, 8 times 8 LED drivers, and simultaneously synchronized ECG, BIO-Z, GSR, and two PPG readouts. IEEE J Solid-State Circuits. 2016;51(11):2584–95.
6. Rajesh PV, et al.. 22.4 A 172μW compressive sampling photoplethysmographic readout with embedded direct heart-rate and variability extraction from compressively sampled data. In 2016 IEEE International Solid-State Circuits Conference (ISSCC), San Francisco, CA, 2016, pp. 386–7.
7. Hsiao KJ. 17.7 A 1.89nW/0.15V self-charged XO for real-time clock generation. In 2014 IEEE International Solid-State Circuits Conference Digest of Technical Papers (ISSCC), San Francisco, CA, 2014, pp. 298–9.
8. Griffith D, Røine PT, Murdock J, Smith R. 17.8 A 190nW 33kHz RC oscillator with $\pm 0.21\%$ temperature stability and 4ppm long-term stability. In 2014 IEEE International Solid-State Circuits Conference Digest of Technical Papers (ISSCC), San Francisco, CA, 2014, pp. 300–1.
9. Prummel J, et al. A 10 mW Bluetooth low-energy transceiver with on-chip matching. IEEE J Solid-State Circuits. 2015;50(12):3077–88.
10. Intaschi L, Bruschi P, Iannaccone G, Dalena F. A 220-mV input, 8.6 step-up voltage conversion ratio, 10.45-μW output power, fully integrated switched-capacitor converter for energy harvesting. In 2017 IEEE Custom Integrated Circuits Conference (CICC), Austin, TX, 2017, pp. 1–4.

11. Nakagome Y, Tanaka H, et al. An experimental 1.5V 64Mb DRAM. IEEE J Solid-State Circuits. 1991;26(4):465–72.
12. Wu HH, Chen LY, Wei CL. Wide-input-voltage-range and high-efficiency energy harvester with a 155-mV startup voltage for solar power. In ESSCIRC 2017 - 43rd IEEE European Solid State Circuits Conference, Leuven, 2017, pp. 295–8.
13. Lu Y, Yao S, Shao B, Brokaw P. 21.3 A 200nA single-inductor dual-input-triple-output (DITO) converter with two-stage charging and process-limit cold-start voltage for photovoltaic and thermoelectric energy harvesting. In 2016 IEEE International Solid-State Circuits Conference (ISSCC), San Francisco, CA, 2016, pp. 368–9.
14. Simjee FI, Chou PH. Efficient charging of supercapacitors for extended lifetime of wireless sensor nodes. IEEE Trans Power Electron. 2008;23(3):1526–36.
15. Bandyopadhyay S, Chandrakasan AP. Platform architecture for solar, thermal, and vibration energy combining with MPPT and single inductor. IEEE J Solid-State Circuits. 2012;47(9):2199–215.
16. Yu G, Chew KWR, Sun ZC, Tang H, Siek L. A 400 nW single-inductor dual-input–tri-output DC–DC buck–boost converter with maximum power point tracking for indoor photovoltaic energy harvesting. IEEE J Solid-State Circuits. 2015;50(11):2758–72.

Design of Powerful DCDC Converters with Nanopower Consumption

Vadim Ivanov

1 Introduction

Design of the integrated systems with nanopower consumption is quite different from the standard practice and by no means is business as usual. It starts with process selection that often require unusual options; use of altered system structure and operation manner; meticulous attention to the secondary procedures like startup, power sequencing, etc.; transistor sizing and new nano-specific circuit cells. Power management of such systems should be equally efficient when it is sleeping and when it is operating at full throttle and with instant switching from one mode to another at unpredictable timing. Common concept of operation mode switching is very inefficient: every additional operation mode triples design labor, as we have to create two systems instead of one along with transition procedure; production testing becomes a nightmare; operation and behavior of such systems is almost impossible to explain to customer or somebody without years of deep dive in the subject. Hence, choice of operating mode is narrow. We have to move away from digital options with high-frequency clocks from fixed-frequency DCDC converters and instead concentrate on variable-frequency operation modes with new techniques of the adaptive error- and load-dependent biasing. Another limitation specific to industrial design is selection of external components—such DCDC converter should be operational with cheapest monolithic inductors and ceramic capacitors, which vary 70–90% in current and voltage range [1], as well as operation temperature range from −40 to 85, 125, or even 150C, and robustness to process variation of component parameters.

V. Ivanov (✉)
Texas Instruments Inc., Tucson, AZ, USA
e-mail: ivanov_vadim@ti.com

K. A. A. Makinwa et al. (eds.), *Low-Power Analog Techniques, Sensors for Mobile Devices, and Energy Efficient Amplifiers*,
https://doi.org/10.1007/978-3-319-97870-3_3

New applications and requirements demand new, nano-specific circuit techniques and cells. Cells shown below are created with structural design methodology [2]—powerful tool in any design. Viability of methodology, cells, and structures has been proven by multiple ICs in mass production.

2 Structural Design Methodology of Analog Circuits

There are 18,000 different amplifiers that can be created from just two transistors. This number is derived from a multiple of options: NMOS/PMOS, common gate/source/drain, four kinds of feedback for each transistor, and for the amplifier as a whole. With the typical analog circuit containing more than a 100 transistors, the number of variants is greater than the number of atoms in the galaxy—and only a few can solve a designer's problem. As a result, most of analog designers are using a cookbook approach, creating a new circuit from the existing one with the fewest changes possible. Radically new solutions are rare, and they are considered to be the major intellectual property by designers and their employers. A method to invent circuit solutions is needed. The structural methodology is one such technique that can be deployed to find a set of acceptable application solutions to weed out bad or inferior circuits instantly. It has a long success record in the design of operational amplifiers, references, power amplifiers, and DCDC converters. By following the steps described below, a designer can find a set of satisfactory solutions, some of which are known and some are new. Then designer can finally choose the one based on personal preference and secondary parameters of importance.

2.1 Graphic Presentation of the System

The first step in the circuit design should be a presentation of the problem to be solved in a graphic form. The graphical representation is much more informative and easier for comprehension then text description or set of equations. The most common language for such presentation is a structural diagram. Another option is the signal flow graph, which has the advantages of existing formal rules for equivalent transformations [3] and drawing simplicity. Almost forgotten, but preferred by founders of the control theory, like Mason or Bode, the signal flow graphs have recently started to gain popularity [4].

An example of signal flow graphs of the differential pair is given in Fig. 1. The differential stage can be presented in the simple form of a single g_m link or in more details as illustrated in Fig. 1. Graph in Fig. 1 includes the transconductance of each transistor and a common-mode feedback. The graph in Fig. 1 is called the "general structure with common-mode feedback." Properties of this graph can be extrapolated to any multiloop, multidimensional structure (or multidimensional

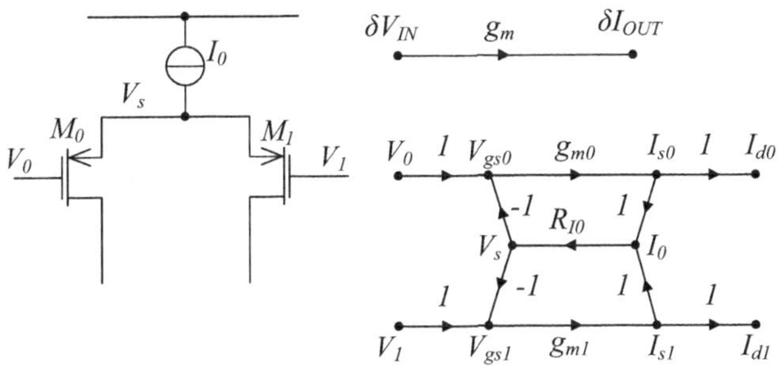

Fig. 1 Signal flow graph of differential stage

structure can be equivalently transformed to this graph), just as complex numbers represent properties of the n-dimensional space.

An analysis of the differential structure with common-mode feedback [2, supplement A] is instrumental in the design of circuits with multiple input/output variables, such as class AB stages or multiple output DCDC converters. It also helps in the single-glance estimation and selection of the circuit within the set of possible options.

2.2 *Dedicated Feedback Control for Each Important Parameter*

The next step in the circuit design is a transformation of system structure to the form where every important variable is controlled by a dedicated feedback loop. Circuits without such feedbacks should be weeded out without any further consideration. The advantage of the system where all significant parameters are controlled is obvious; however, the main obstacle to the universal application of this rule is the problem of stability in the resulting multiloop structure.

Although not necessary, but sufficient, the condition for the whole system's stability is the stability in each and every loop within this system [5]. A feedback loop can be unconditionally stable (with any load and signal source impedance) if its open-loop transfer function has only one pole. Consequently, the easy way to ensure system stability is designing each loop with the single-stage (single-pole) amplifiers only.

This approach to stability immensely simplifies the design process. Although in some cases the exclusive use of the single-stage amplifiers is not possible; here, conventional compensation techniques need to be applied and stability has to be carefully verified.

Standard verification of the stability using the merit of phase margin requires a break in the feedback loop and is not suitable for the multiloop system (which one of the loops to break?). Method of the multiloop system small-signal stability verification by using AC simulations has been described in [6]. Due to the

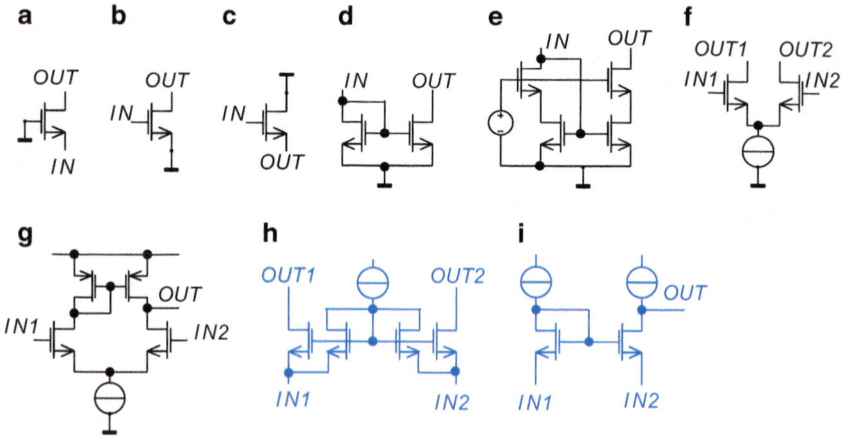

Fig. 2 Elementary cells library

unavoidable presence of the nonlinear effects in the circuit, the small-signal only stability verification is not sufficient. The small- and large-step response transient simulations followed by extraction of the overshoot and damping factor could be used instead.

2.3 *Library of Elementary Cells*

The next design step is implementation of the system structure with elementary cells. The library of these cells includes circuits described in every textbook on analog design, as shown in Fig. 2.

It also includes lesser known cells of the current-input amplifiers shown in Fig. 2h, i, which should be a part of every designer's arsenal. In the circuit of Fig. 3a, *M0* and *M1* currents are matched, as well as the currents of *M2/M3*. Consequently, input currents do not depend on the common-mode input voltage, so the common-mode input impedance is high. The differential input impedance is small and equals to $1/g_m$. Dependence of the output current vs. input voltage (Fig. 3c) is similar to the standard differential stage of Fig. 2f.

The single-output version of this amplifier is shown in Fig. 3b. In this cell, the current sinking from output is unlimited and output current vs. input voltage curve (Fig. 3d) is nonsymmetrical.

Use of the current-input amplifier cell inside the local feedback loops improves the speed of these loops at least five times for any given current budget. It simplifies the frequency compensation, allowing replacement of the common-source gain stages in the signal path with the common-gate ones, which have much smaller delay.

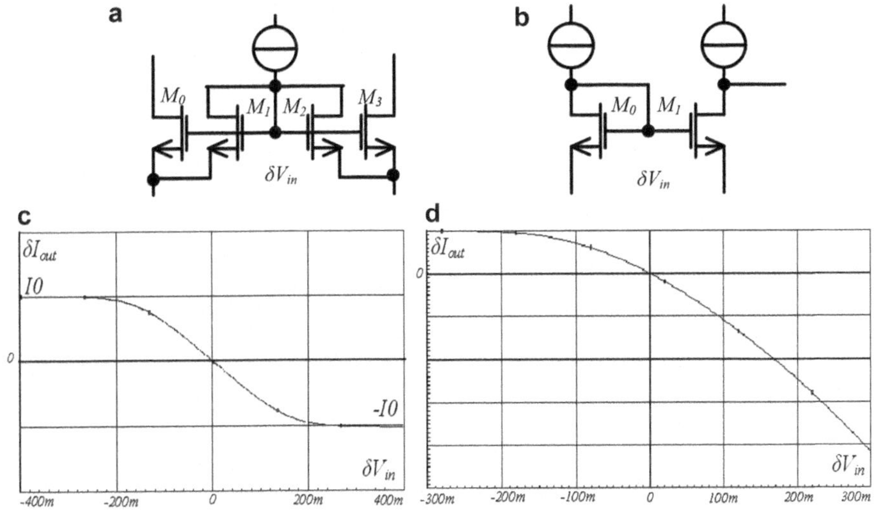

Fig. 3 The current-input amplifier cells

The elimination of all or most compensation capacitors becomes possible. For example, an operational amplifier described in [7] comprises more than 25 feedback loops, but the only compensation capacitors on its chip are the two Miller capacitors in the main signal path.

2.4 Features of a Good Circuit

With structural methodology, we restrict a set of circuits to be considered "good" circuits only.

Features of the Good Circuit

1. Good circuit has a dedicated feedback loop controlling each parameter which is important for the reaching of system goals.
2. Dynamically each local loop and system as a whole are stable and their step response looks like the response of the system with first- or second-order transfer function.
3. Good circuit is robust to the variation of the component parameters, process, and temperature.
4. Nonlinear effects (startup, power glitch, input/output overload, etc.) have been considered and necessary clamps/limiters added.
5. For being embedded in SoC designs, a good circuit should not be sensitive to substrate noise.

Acceptable application solutions can and sometimes do exist outside of the "good" circuit domain. However, after 30 years of experience, these "no good" circuits could never outperform circuits from the chosen set.

Nesting of the feedback loops inside the system has been discussed above, as well as the requirement of stability in the each loop. Requirement of the circuit robustness makes parametric optimization efforts practically useless. If optimum of the goal function is smooth, then based on a common-sense choice, the parameters should be good enough; if this optimum is sharp, then this circuit is not robust and consequently is inadequate.

Designing a circuit for a nominal mode of operation normally occupies no more than 20% of total design time. The rest is taken up in consideration of nonlinear effects and in creation of various protective measures. There is no general way to predict such effects. All we can do is study the application and play multiple "what if?" scenarios.

In the SoC design, interaction of different units through substrate and supply should be taken into account from the very beginning. The taken measures can be in the layout and process (unit placement, isolation rings, double-well process, separate supply wiring, and wirebonds), in the choice of components not sensitive to substrate noise, in the circuit techniques (differential signal processing), and in the choice of the system architecture.

The problem-solving approach in structural design is close to the one described in [8] and to the modern philosophy called "systems thinking" [9].

3 Process Choice and Transistor Sizing

Nanopower IC design poses few requirements to the process:

1. The most important and obvious requisite is availability of the low-leakage transistors. In some cases, it is feasible to decrease leakage by transistor body bias outside of supply rails (NMOS below VSS, PMOS above VDD) with dedicated charge pumps, but this significantly complicates design and may create startup and dynamic problems.
2. In nanopower design, we do not need large resistors—we need huge ones. 100–200MOhm per die is the necessity; 5–10 GOhm is desirable. Hence, process should provide high sheet R and narrow resistors.
3. During process development, transistor's parameters are usually measured very thoroughly and at wide range of bias currents starting in fA values. But typical models are fitted to measure data at the commonly used range of bias starting from tens of nA; therefore, the model's parameter adjustment is necessary for better nA-current fit. The only parameter which is assessed poorly at low currents is noise—for ease of measurement, it is normally taken when transistors are biased in very strong inversion. Flicker noise in weak inversion can be 100 times lower than strong inversion estimation, so these measurements—and models—have to be retaken for correct design of noise-sensitive circuits.

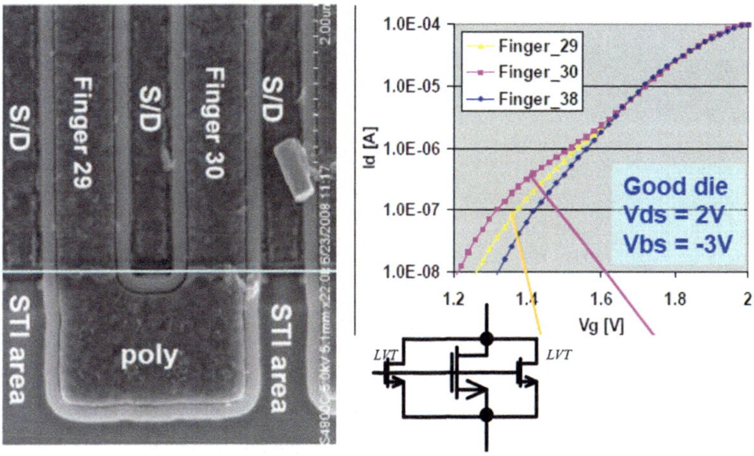

Matching is defined by size of low-Vth microfractions

Fig. 4 Low-Vth NMOS at the edge of regular NMOS

4. Often forgotten during process development, we need component for the bias startup. All existing bias cores require some small trickle current to guarantee operation at low temperatures, after supply glitch, etc. Capacitive startup, very popular in university papers, is no good for industrial part. This trickle current should be larger than leakage but much smaller than IC bias budget. Depletion-mode device, FET, leaky low-Vth transistor can be sufficient—but should be available on the process.
5. The matching of PMOS transistors at very weak inversion and small currents is in line with anticipations. But NMOS devices very commonly can have unexpected—and very large—mismatch due to process flaw: presence of natural low-Vth NMOS micro fractions at the edge, as shown in Fig. 4.

Problem can be solved by slight change in design rules so the shallow trench isolation (STI) edge would not be coinciding with the NMOS gate edge.

3.1 Transistor Sizing

It is long known that V_{gs} matching and transconductance at the given bias current are best when transistor is operating in weak inversion [10], Fig. 5.

Less known is the fact that flicker noise is also significantly smaller in this region. As a general guideline, every MOS device should operate in weak inversion in almost any circuit; fortunately, in nanopower design, there is no choice and transistor will be in weak inversion with any realistic W and L.

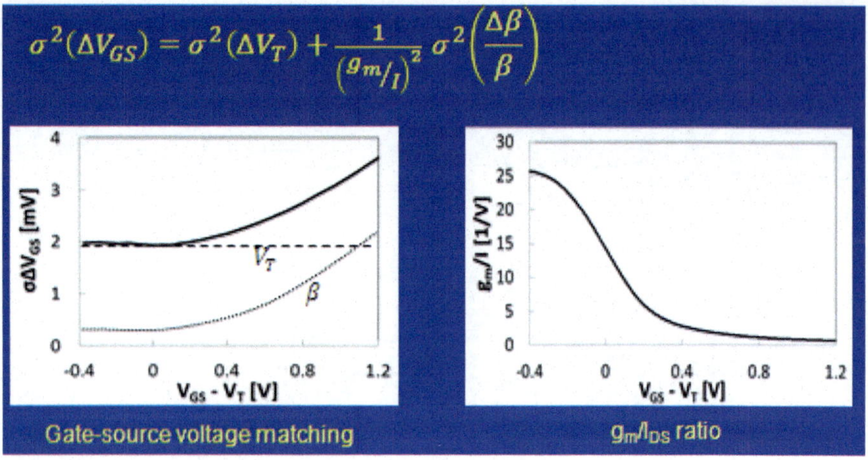

Fig. 5 Gain and V_{gs} matching vs. $(V_{gs} - V_{th})$

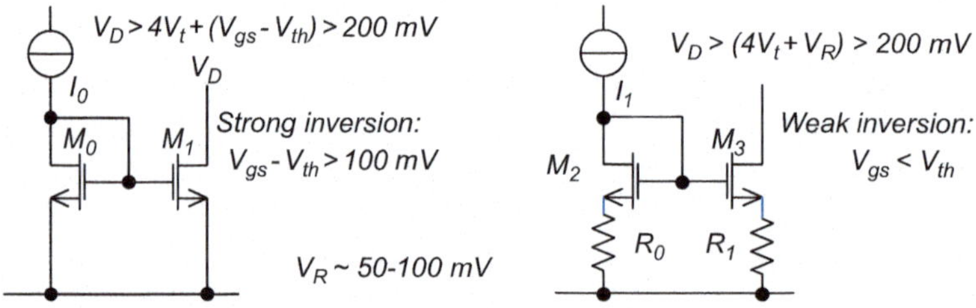

Fig. 6 Current mirror in weak inversion

Main objection to weak inversion is in current mirror design. V_{gs} mismatch is low, but transconductance g_m is high; hence, the current mismatch in mirror operating in weak inversion is higher than when inversion is strong. Solution is to add degeneration resistors as shown in Fig. 6.

As a result, weak inversion mirror with resistors has much better accuracy (resistor matching is ~100 times better than transistor matching for the same area) and much less noise as smaller transistor noise is further attenuated by smaller transconductance (*1/R* instead of g_{m}). Resistor voltage dropout should be at least 50–70 mV; transistor mismatch becomes insignificant for mirror current mismatch when dropout exceeds 100 mV. Transistor in weak inversion starts trioding when $V_{ds} \leq 4V_t = 100$ mV at room temperature. In strong inversion, trioding $V_{ds} \sim (4V_t + (V_{gs} - V_{th}))$, thus the minimum output voltage is approximately the same for both mirrors.

This technique should be used in every mirror where accuracy is important; its only drawback is requirement of very large resistors: 100 mV dropout with 1 nA bias means 100MOhm resistor.

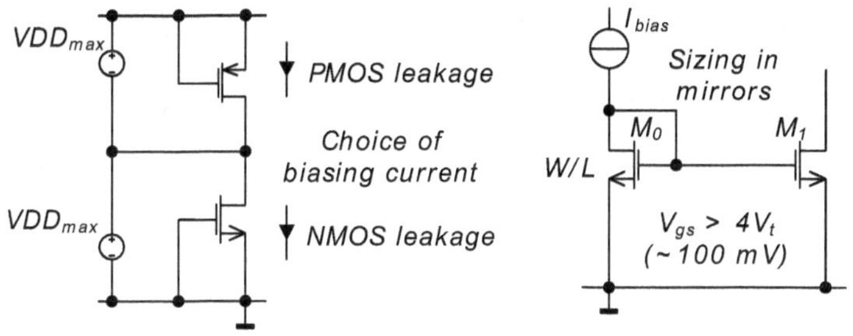

Fig. 7 Mirror transistor sizing and minimum bias current choice

When sizing transistors in mirror, another consideration is $V_{gs} = V_{ds}$ of the diode-connected device (Fig. 7).

When V_{gs} is $\leq$ saturation V_{ds}, current ratio of the mirror becomes unpredictable, consequently *W/L* of transistors in the mirror should ensure $V_{gs} > 150$ mV vs. all temperatures and process corners.

Another consideration in transistor sizing and bias current selection is leakage: when completely OFF, transistor of the picked size should leak at least 10 times less than its minimum bias current at highest operating temperature and strong process corner. Actual bias current value varies from process to process; the lowest unit current used by author in industrial 5 V IC on 0.35 μm process was 0.5 nA.

4 Nanoampere-Specific Cells

Some circuit blocks cannot be adjusted from traditional designs by usual way of parametric optimization and resistor's value increase. These include:

- Biasing core and mirror tree.
- Very large ratio current mirrors (up to 10,000).
- Power-on reset (POR) circuits.
- Current-efficient oscillators.
- Accurate and low noise voltage references with nA consumption.

4.1 Biasing and POR

The largest problem of the nanoampere biasing is reliable startup. Each and every existing bias core has at least two stable operating points, where the first one is what this core was designed for and second when all currents are very close to zero. To prevent operation at the second point, some small current should be added to the core, usually disconnected after core starts operating normally. This small

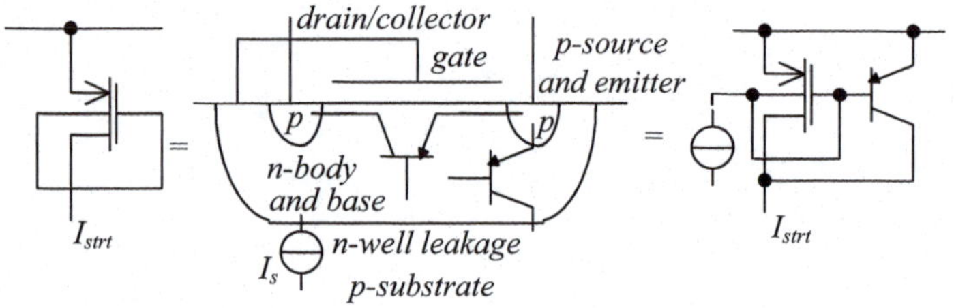

Fig. 8 Accelerated leakage component

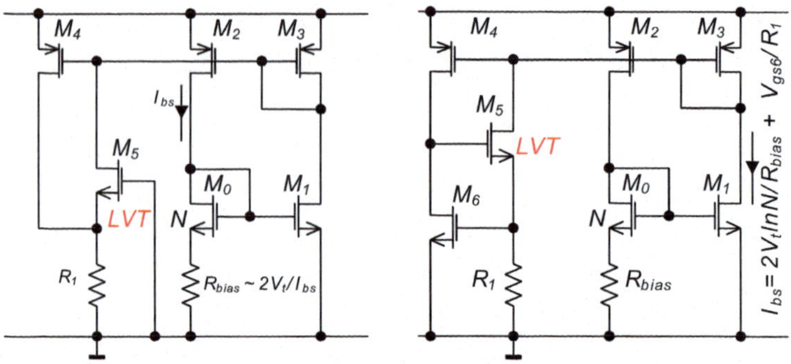

Fig. 9 Biasing cores

current should be larger than core transistor leakages at all operating points. In usual circuit, we can use large resistor for this startup current, but for nanopower ICs value of this resistor becomes unacceptably large. Here we can use depletion MOS transistor, FET, and leaky low-V_{th} device (i.e., natural NMOS). In absence of these components, the MOST structure with accelerated leakage (Fig. 8) can be implemented.

In this structure, the leakage current I_s from n-well to substrate is multiplied by the lateral PNP β and amended by the increased, due to the body bias by V_{be} of PNP, drain to source leakage current of PMOS. Resulting I_{strt} is guaranteed to be larger than similar-sized MOS leakage, but it is very much dependent on temperature and needs to be characterized.

Temptations to use capacitive startup should be avoided as supply ramp rate and its glitch behavior are unknown.

Simplest nanoampere biasing cores with PTAT and zero-TC output are shown in Fig. 9.

M_5 provides startup; feedback loop controlling M_5 current has only one inversion and is unconditionally stable. Stability in startup loop is often overlooked and is critically important as its oscillations are very hard to reveal in simulations but very often show up in silicon. In the left circuit, voltage drop across R_1 should be large enough to completely shut down M_5 during normal operation.

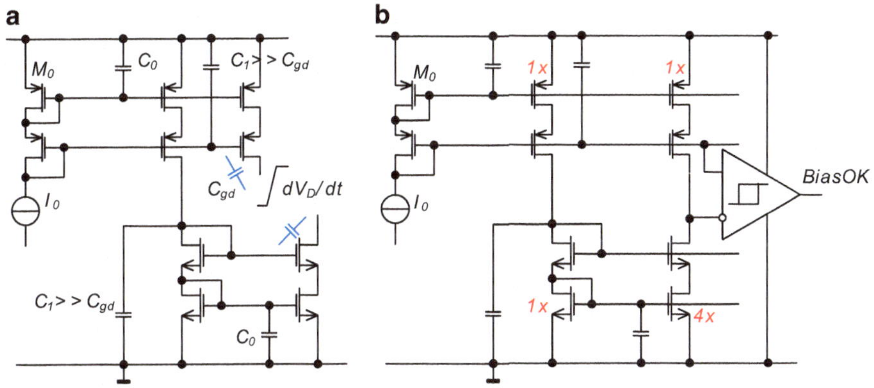

Fig. 10 Bias mirror tree

The difference of the current distribution mirror tree in nanopower circuits is addition of the capacitors at the tree gates (Fig. 10a).

Every fast voltage transitions at one of the current source drains propagates to the gate through parasitic gate-drain capacitance; with the initial tree current I_0 being large, this charge would be attenuated by low impedance at the M_0 gate ($1/g_m$) and would not affect tree operation. With 1–2 nA I_0, impedance of M_0 is large (10–20 MOhm), thus every fast voltage transition can propagate to the rest of current sources. Capacitors C_1/C_0 of Fig. 10a ($\sim 0.5-1$ pF) filter these charges and prevent parasitic signal propagation between legs of the bias mirror tree.

These capacitors are initially charged by tiny I_0, thus delaying the bias startup by 100–200 μs, and plenty of catastrophic things can happen in power management system during this time. System should be disabled as long as it takes to start biasing, which can be detected with N-side/P-side current comparator as shown in Fig. 10b. Its *BiasOK* signal is an additional condition for power-on reset (POR) and enabling of the system operation. Another POR condition is the value of supply, which should be large enough for logic operation (most critical is operation of flip-flops containing T-gates).

In addition to the nA range constantly running currents, bias cell should provide periodically enabled much larger currents in tens of μA range, for example, reference for current limit, inductor current hysteresis, etc. Scaling by 10,000 with traditional current mirror having $1\times$ M_0 and $10{,}000\times$ M_1 (Fig. 11a) would be very large in size and inaccurate due to the M_0 small size.

Accuracy and area could be dramatically improved when implementing M_0 as series and M_1 as parallel connection of identical cells (Fig. 11b). In the example shown, the area decreases 50 times and matching improves by $\sqrt{200} \sim 14$ times. This technique works as long as M_1 drain voltage is low and short-channel effects are insignificant.

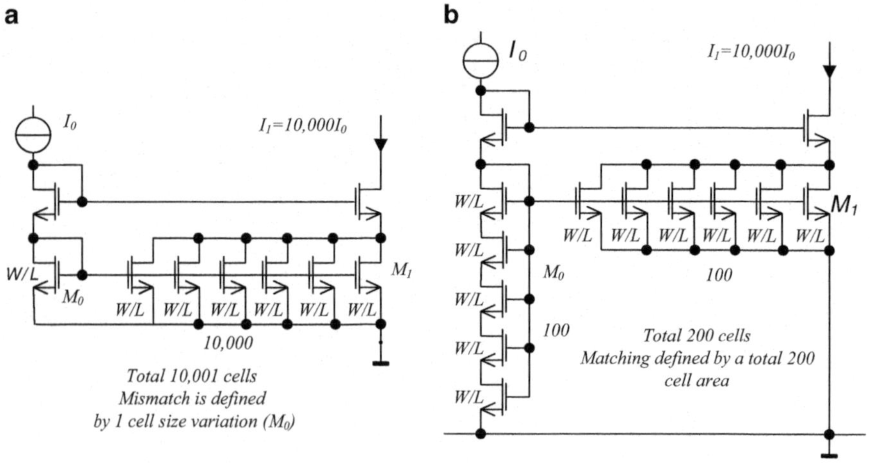

Fig. 11 Large ratio current mirror

4.2 Current-Efficient Oscillators and Logic Supply

An oscillator is an essential part of any nanopower system required, among other functions, for timing of sleep and activity periods. Its achievable accuracy is limited by temperature and time drift of available components; when using gate oxide capacitors and poly-Si resistors, frequency variation can be as large as 30%. Anything better would require trim and/or use of the off-chip components like quartz, MEMS, BAW, etc.

Frequency of this oscillator should be as low as possible, as every switching means energy loss for recharging of gate capacitors, logic shoot-through currents, etc. In *RC* oscillators, we have the choice between ring and relaxation oscillators. Ring oscillator takes less design effort but its efficiency is impaired by shoot-through currents in invertors during slow signal edges. The relaxation oscillator consists of current source, capacitor, and comparator. To save energy, charge-in capacitor should be recycled and comparator should be powered only when needed, that is, when its input voltages are getting close to each other. Such oscillator is shown in Fig. 12a. Recycling of the capacitor charge is provided by its flipping with S_0–S_3, comparator bias current increases exponentially with voltage on its input (M_0–M_1 mirror).

Oscillator frequency is $f_{\text{osc}} = I_0/C_0V_0$. V_0 can be implemented as comparator input transistor mismatch (PTAT) which in combination with PTAT I_0 from bias core ensures zero-TC f_{osc}. This oscillator should be aided by a startup circuit preventing V_{CMP} being high for more than few ns. Efficient comparator design is discussed in the next chapter. When implemented on 0.35 μm process, such comparator consumes ~ 4 nA/kHz including 1 nA I_0, comparator, and digital consumption.

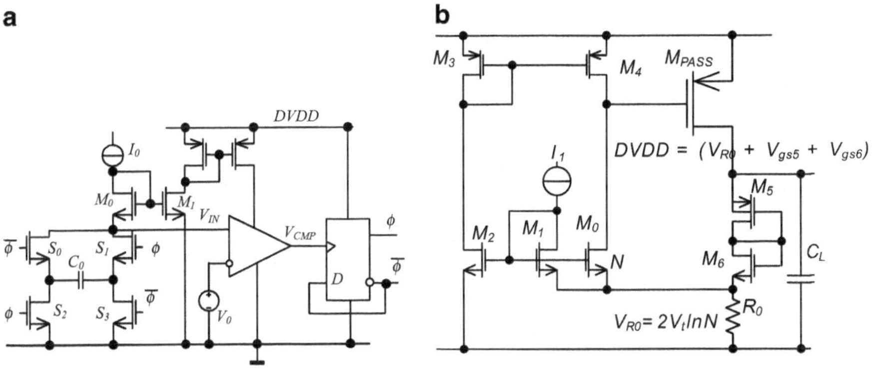

Fig. 12 Oscillator and digital supply subregulator

Consumption of the digital cells is proportional to $f_{\text{clk}}(\text{DVDD} - V_{\text{thNMOS}} - V_{\text{thPMOS}})^2$, hence digital supply LDO can significantly decrease overall system consumption. At low side, this LDO should keep DVDD above ($V_{\text{thNMOS}} - V_{\text{thPMOS}}$) to support operation of T-gates inside flip-flops. One of the simplest circuits for such LDO is shown in Fig. 12b, which keeps supply just right combining minimal consumption of the digital cells while supporting flip-flop operation, as defined by its reference consisting of M_5–M_6 diodes. Single inversion feedback loop (M_{pass}–M_6–M_0–M_{pass}) improves stability, and this LDO may not need any compensation, which ensures the fast load current step response while consuming very small current of 5–6 nA.

4.3 Voltage Reference

The PM system includes multiple voltage references, that is, reference for error amplifier, reference for under-voltage lockout, etc. In high-current systems, reference of choice is bandgap having untrimmed accuracy of 2–3%, which can be improved below 1% if trimmed. Current consumption of such bandgap should be 0.5 μA and above. Attempts to get accurate bandgap operational in industrial temperature range with consumption below 100 nA were not successful due to large noise, slow startup, and poor power supply dynamic rejection.

CMOS voltage references use poorly controlled and modeled process parameter (V_{th}), thus have 5–10 times worse accuracy than bandgaps, while having similar noise and dynamic problems if designed with low I_q.

Another option are floating gate voltage references [11], which can be accurate and low consuming but are not feasible on the most existing processes.

High accuracy, low noise, and very low average consumption can be achieved with structure of Fig. 13.

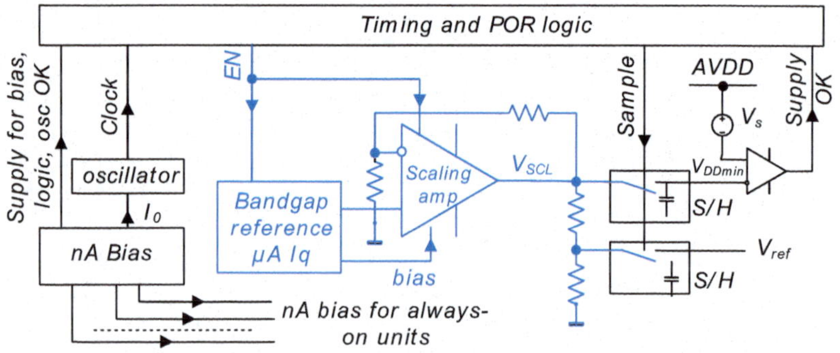

Fig. 13 Sampled voltage reference

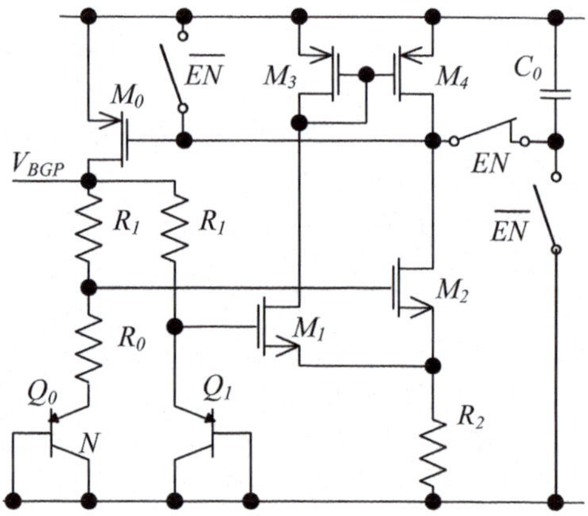

Fig. 14 Bandgap with fast startup

During operation, bandgap reference with 1−2 μA I_q is periodically enabled. Its output is scaled up and divided to create replicas of all required voltages: output voltage of the PM DCDC converter or LDO, reference for under-voltage lockout, etc., which are then sampled on capacitors. Capacitors hold charges till the next bandgap enable period. The average consumption of the sampled reference with 100 μs/10 μA enable period and 1 s hold time is only 1 nA—good enough for any nanopower system while supporting the industry-standard accuracy and temperature range.

Key cells for such reference are bandgap with fast (2–4 μs) startup and sample/very long hold unit. Example of bandgap with fast startup is shown in Fig. 14, where compensation capacitor C_0 is overcharged during OFF period. When enabled, charge in C_0 jumpstarts bandgap core with large currents and large output voltage. Settling of the bandgap to nominal value is accelerated due to large gain and bandwidth of the amplifier M_1-M_4.

Figure 15 presents evolution from simple sample/hold cell of Fig. 15a, having few ms before sampled across C_0 voltage changes by 5–10 mV due to the drain-source and drain-body leakage of M_0, to the circuit with control of the voltage across

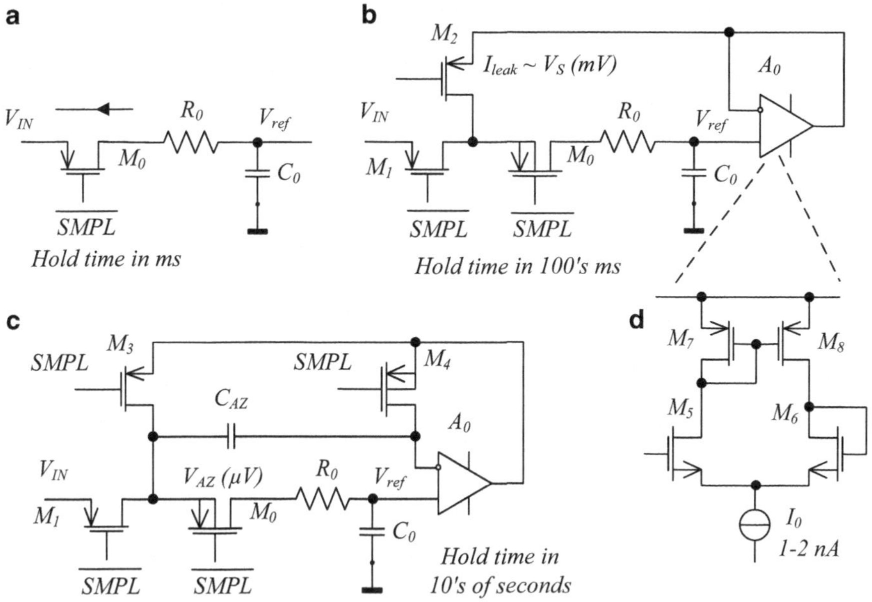

Fig. 15 Sample/long hold cells

M_0 of Fig. 15b. Now, voltage across M_0 is only few mV = offset of A_0. It improves hold time to 10's or even 100's of ms, depending on sampling capacitor size and process.

During sampling time, amplifier A_0 is idle and this time can be used for its autozeroing, integrating offset on C_{AZ} through M_4 (Fig. 15c). This circuit can hold C_0 charge for 10s of seconds. During such a long time, temperature of the power management IC providing high load current can significantly change, causing C_0 voltage to shift due to its temperature coefficient. Hold time of 0.1–0.2 s is large enough to ensure few nA average reference consumption while avoiding large errors from die temperature change.

5 Amplifiers with Adaptive Consumption and Unlimited Output

Clearly, current in the nanopower system's cell should be consumed only where necessary and when it is necessary. The main continuous current user in power management system is the error amplifier or comparator. Speed of this amplifier or propagation delay of the comparator defines system stability and dynamics of reaction on the load change (undershoot/overshoot after load step, etc.). At small LDO or DCDC converter load current, output is supported by the load bypass capacitor C_L, change rate is low, and error amplifier or comparator can be slow

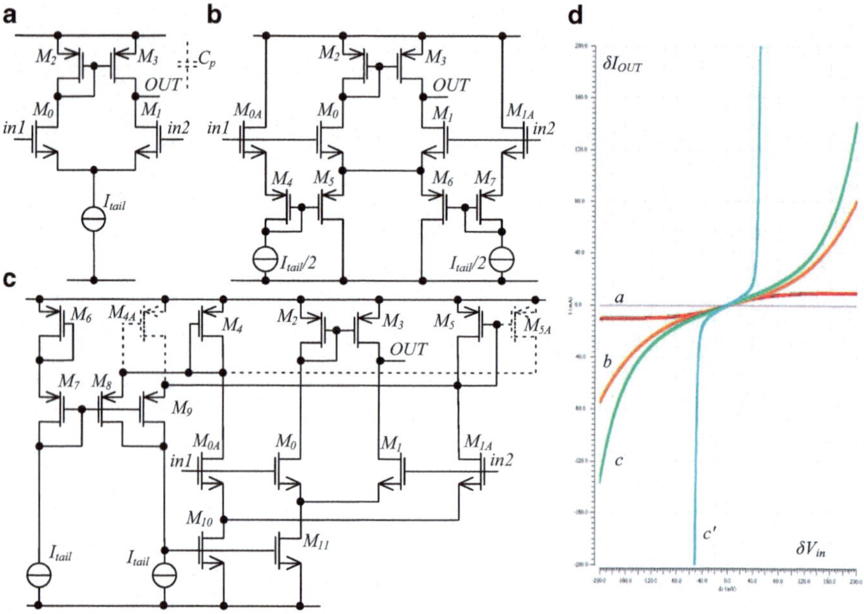

Fig. 16 Amplifiers with error-dependent gain and consumption

to decrease consumption and small-load efficiency. At high I_L, the speed and I_q of error amplifier can be increased as its consumption is still a small fraction of the total load current. Adaptive load-dependent biasing [12] has been used in multiple industry designs and can be instrumental to keep dynamic error small—but only when load current is large enough. Common problem of these circuits is large error when load steps from zero to high value.

Adaptive biasing is more beneficial when it is dependent on the error (input voltage of the error amplifier). This error increases during rapid changes and at large load currents—exactly when high speed is needed and when high I_q is not an issue.

Bandwidth of the simplest error amplifier in Fig. 16a, consisting of differential stage M_0, M_1, and current mirror M_2, M_3, is equal to $g_m/2\pi C_p$, where g_m is proportional to I_{tail} (assuming transistor operation in weak inversion) and C_p is the total capacitance at its output. To increase speed, we have to increase I_{tail}. Output current is limited by I_{tail} value (curve *a* at Fig. 16d).

Amplifier with error-dependent biasing in Fig. 16b [13] has negative feedback loops controlling minimum current through each of input devices: $M_{0\text{A}}-M_4-M_5$ for M_0 and $M_{1\text{A}}-M_7-M_6$ for M_1. It has the same transconductance g_{m} when $\delta V_{\text{in}} = 0$, but at larger δV_{in} g_{m} (and bandwidth) increases and output current is unlimited (curve *b* at Fig. 16d).

Adaptive bias amplifier of Fig. 16c also has feedback loops controlling minimum of M_0, M_1 currents, employing the minimum current selector $M_7-M_8-M_9$:

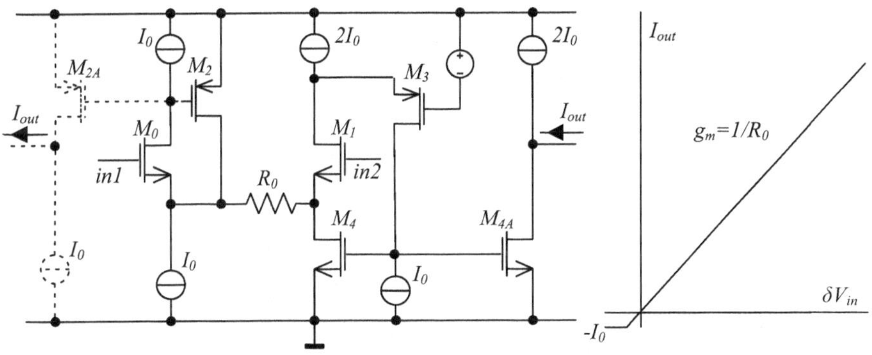

Fig. 17 Amplifier with large g_m, nA-range consumption and unlimited output

$M_{0A}-M_8-M_{10}/M_{11}$ for M_0 and $M_{1A}-M_9-M_{10}/M_{11}$ for M_1. It has the same g_m at $\delta V_{in} = 0$, but nonlinearity and g_m rise at larger δV_{in} are stronger due to larger gain in feedback loops (curve *c* at Fig. 16d). This nonlinearity can be sharpened by a weak positive feedback (M_{4A} and M_{5A}) which increases the minimum current of M_0/M_1 with δV_{in} rise as well. Curve *c'* of Fig. 16d shows the output current nonlinearity with width of M_{4A}/M_{5A} sized at ¼ of M_4/M_5. Another advantage of this amplifier is the wider common-mode input range, starting at V_{gs} from negative supply.

Amplifiers in Fig. 16 have push-pull outputs and are perfect to drive large gate of LDO pass device. In the current-mode DCDC converter, output signal of the error amplifier is subtracted from the output of inductor current sensor. This current flows in one direction, so there is no need in symmetric push-pull error amplifier. Also, it may require much larger values of g_m (see Chap. 7) not achievable in differential stage when running nA-range bias.

In the amplifier of Fig. 17 both input devices M_0/M_1 are in feedback loops which keep their current stable: M_0-M_2 and $M_1-M_3-M_4$. As long as these loops keep control, $V_{gsM0} = V_{gsM1}$ and voltage $V_{R0} = (V_{in1} - V_{in2})$; therefore, currents of M_2 and M_4 are equal to $(V_{in1} - V_{in2})/R_0$. Output current can be mirrored from M_4 or M_2 and equivalent $g_m = 1/R_0$.

Bandwidth of this amplifier is defined by lowest bandwidth of the feedback loops $g_{M0}/2\pi C_{gM2}$ and $g_{M1}/2\pi C_{gM4}$, where C_{gM2} and C_{gM4} are total capacitances at M_2 and M_4 gates. With 50 nA I_q and when implemented on 0.35 μm process, such amplifier has 10–20 kHz bandwidth.

6 Efficient Comparators

Every DCDC converter comprises multiple comparators. Comparator in the error amplifier signal path is always on and its consumption is a large part of no-load system I_q, which makes traditional multistage comparators [14] improper due to low speed vs. consumption ratio.

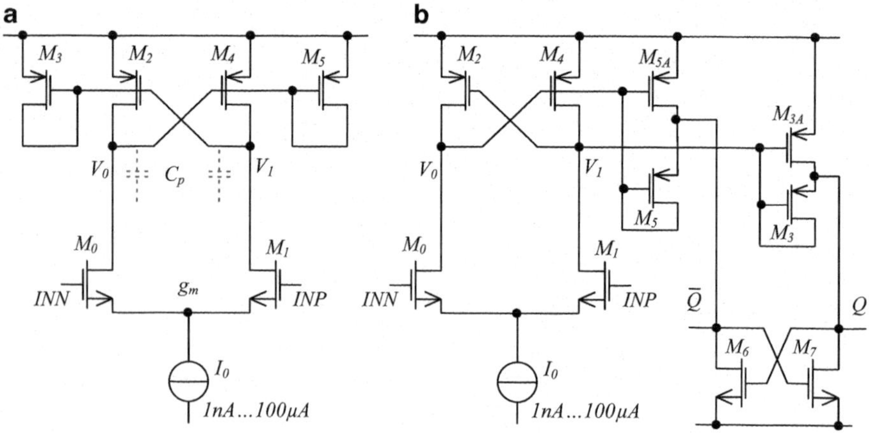

Fig. 18 Fast comparator with variable biasing and delay

Propagation delay of the comparator is dominated by $(V_{\text{swing}}C_p)/(\delta V_{\text{IN}} g_{\text{m}})$, where g_{m} is the transconductance of the input stage, V_{swing} is the voltage swing at the output of input stage required to flip the comparator output, δV_{IN} is input overdrive, and C_p is the total parasitic capacitance at the output of input stage. Input stage g_{m} is proportional to its tail current (I_q), minimum possible C_p is limited by a process, and the only circuit technique to improve I_q/propagation delay ratio is to decrease V_{swing} to as small as possible value. The minimal denomination of V_{swing} to turn on or off comparator output devices is V_{th}. Note that correctly designed comparator should have hysteresis.

The comparator input stage in Fig. 18a is close to ideal: it is simple, has V_{th} swing of the voltage at its output nodes V_0/V_1 and has hysteresis which is defined by the area ratios of M_2/M_3 and M_4/M_5.

To convert V_0/V_1 to rail-to-rail swing, currents of M_5 and M_3 can be mirrored around [15], but this would double consumption and multiply propagation delay. To avoid these problems, circuit in Fig. 18b [16] is using the series connected $M_5/M_{5\text{a}}$ and $M_3/M_{3\text{A}}$ with latch M_6/M_7, which is controlled by $M_{3\text{A}}$ and $M_{5\text{A}}$. In order to flip latch M_6/M_7, current of $M_{3\text{A}}$ or $M_{5\text{A}}$ should be stronger than fully on M_6 or M_7. When comparator output Q is high, M_3 and $M_{3\text{A}}$ form a diode whose size in relation to M_2 defines input hysteresis in the same way as in Fig. 18a. When Q is low, transistor M_3 is off, consequently V_1 is not limited by V_{th} and V_{gsM3A} can become as large as necessary to flip over the M_6/M_7 latch, regardless of how small is I_0. V_0/V_1 voltage swing becomes larger by 200–300 mV, slightly increasing propagation delay but to much less extent than a mirror-based level converter [14].

Static consumption of this comparator is equal to the tail current I_0. Its propagation delay on 0.35 μm process with 1 nA I_0 and 50 mV input overdrive is 10–20 μs

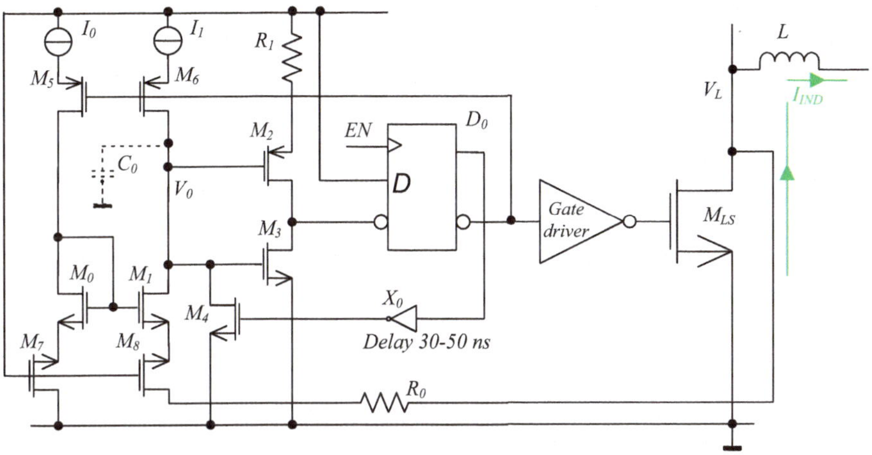

Fig. 19 Current-input comparator for synchronous rectifier

and decreases to 10–20 ns when tail current is raised to 1 μA. This is 10–100 times better than traditional multistage [14] or mirror-based [15] comparators.

When used in the error amplification path of DCDC converter, this comparator can be adaptively biased by the replica current of M_4 from Fig. 17. Thus, its propagation delay can be longer (μs's) when load is small and converter V_{OUT} changes are slow, decreasing to ns's when load (and error) rise and faster dynamics is necessary.

Another essential comparator for DCDC converters is the one in synchronous rectifier. It is enabled only as long as inductor current flows through rectifier switch and its consumption does not affect no-load I_q. It may have some built-in offset to compensate for inductor current change during its propagation delay. This propagation delay should be at least 10 times less than minimum ON time of the rectifier, that is, less than 10 ns for 100 ns minimum pulse width of the DCDC. To decrease it, this comparator should have current inputs—on the same process and with the same consumption, current-input comparator is 5 times faster than voltage-input one due to larger power consumption from the inputs. For 10–20 μA Iq and on 0.35 μm process, propagation delay is 7–10 ns and can be 3–5 times less for 130 nm implementation.

Comparator in Fig. 19 has current input stage on low voltage fast transistors M_0/M_1, protected from large V_L by extended-drain M_7/M_8. Swing of V_0 at the M_0/M_1 stage output is limited to $\sim V_{\mathrm{th}}$ by limiting current through $M2$ with R_1. Comparator is enabled with some delay after power switch M_{LS} is turned ON (by *EN* signal) to allow inductor current to switch and V_L to swing below GND. I_1R_0 creates the pre-determined comparator offset to compensate for propagation delay. After comparator flips, D_0/M_4/M_5/M_6 turns it off till the next DCDC switching cycle.

7 Inductor Current Measurement

The current-mode operation dramatically improves dynamics of DCDC converter [17]. Required for this mode, measurement of the inductor current can be done with resistor connected in series with inductor (Fig. 20a) or with power switches M_H and M_L, by integration of the voltage across inductor (Fig. 20b) or by mirroring current though power switches (Fig. 20c).

Days when the external low-value resistor for current measurement was acceptable in DCDC design are over, thus circuit of Fig. 20a can be ruled out.

Circuit of Fig. 20b became quite popular recently (g_{ms} is often replaced with resistors [19]), but it has unsolvable flaws:

- It does not work when supply voltage is very close to V_{OUT} due to error from voltage drop on inductor resistance, g_{ms} offset, etc.
- It severely underestimates I_{ind} when inductance value varies significantly with I_{ind} (low-cost monolithic inductors drop by up to 80% with current increase in datasheet range, as shown in Fig. 21 [1]), it cannot be used for DC I_{ind} measurements required for current limit, which is the part of every DCDC converter.

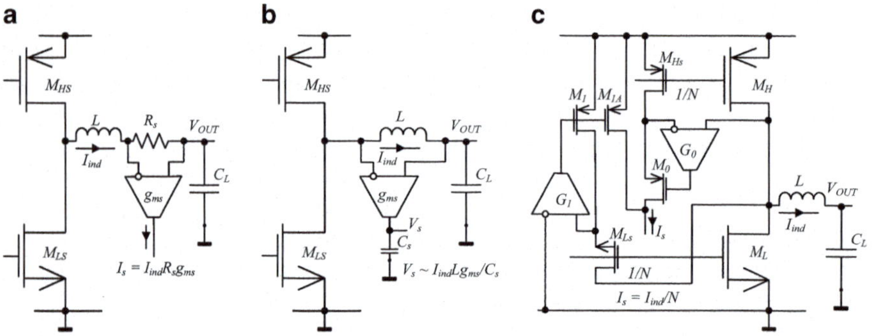

Fig. 20 Inductor current measurement

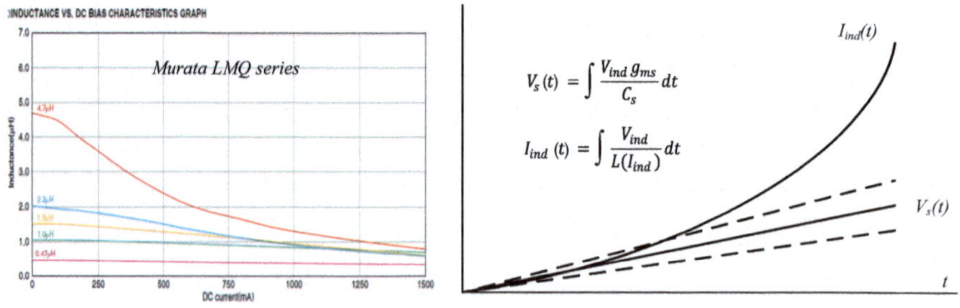

Fig. 21 Monolithic inductor value versus current

It leaves us with circuit on Fig. 20c. It comprises power switches with built-in sensors M_H/M_{Hs} and M_N/M_{Ns} with large area ratio $1/N$ (10,000 or more). To improve accuracy while having large ratio, multiple sense cells distributed inside power switch can be connected in series similar to mirror in Fig. 11b. Operation of the $M_{Hs}-G_0-M_0$ and $M_{Ls}-G_1-M_1$ feedback loops guarantees equality of the V_{ds} between power switch and its sensor, hence currents of M_0 or M_1 are equal to I_{ind}/N. By summing M_0 current with replica current of M_1, continuous-time measurement of I_{ind} is achieved. In some designs, in order to decrease die area and cost, M_L is undersized, its Vds can exceed Vbe and body diode can get active. For such cases, continuous current measurement with the Fig. 20c circuit is not possible.

Settling time of $M_{Hs}-G_0-M_0$ and $M_{Ls}-G_1-M_1$ feedback loops should be 5–10 times faster than the minimum DCDC switching pulse duration. That is, in converters operating from Li battery the minimal pulse width is 100 ns, thus, settling time should be 10–20 ns.

Simple and fast loops can be created with current-input amplifiers as shown at Fig. 22. G_0 consists of M_1/M_2 with current sources I_0/I_1 and G_1 consists of M_3/M_4 with current sources I_2/I_3. Gates of the sense transistors M_{Hs}/M_{Ls} can be constantly connected to the subsequent gate drive supplies, decreasing switching losses. Switches M_8-M_{11} disable current measurement when subsequent power switch is off, therefore, current is consumed only as long as there is an inductor current.

Feedback loops using single-pole amplifiers G_0 and G_1 are unconditionally stable with any biasing and do not require compensation capacitors and are fastest

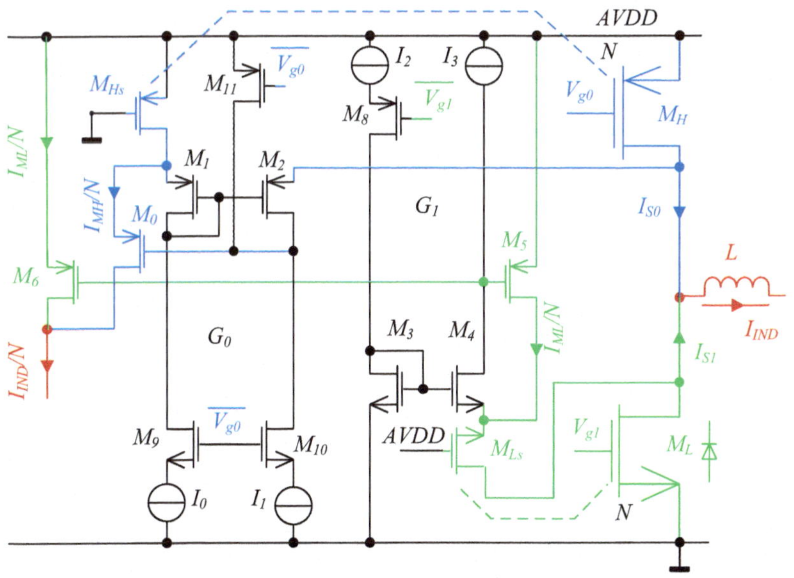

Fig. 22 Inductor current measurement implementation

possible for any chosen I_0–I_3 values. Bandwidth of these loops is equal to $g_{m2}/2\pi C_{p2}$ and $g_{m4}/2\pi C_{p4}$, where C_{p2}/C_{p4} are parasitic capacitances at M_2/M_4 drains. On 0.35 μm process, I_0–I_3 values for 10 ns settling are ~ 5 μA. At nm-scale processes, switching frequencies are higher, pulse width smaller, and settling time should be shorter. But C_ps are also decreasing with process scaling and loop bandwidth increases, hence consumption of the current measurement circuit stays approximately the same independent on the process. At no load state $I_{ind} = 0$, both switches are off and current measurement loops are off as well. Current consumed by this circuit affects DCDC converter efficiency but does not affect its I_q.

Current measurement error of this circuit is ~ 20%. At low currents, error is dominated by G_0/G_1 offset, at large currents by M_L–M_{Ls} and M_N–M_{Ns} mismatch. This accuracy is sufficient both for use in current-mode compensation and in inductor current limit.

8 Selection of the DCDC Converter Operation Mode

Despite of large number of existing DCDC operating modes [17], choice for converters with nanopower consumption is very limited. First, we have to rule out all voltage-mode variants as these are not only slow but also require continuously on, power-hungry active filters for compensation. All fixed-frequency DCDC converters constantly waste energy on switching, which is prohibitive for low-load efficiency. The idea of the mode switching between high- and low-power triples the design effort—and circuit complexity—as both modes and transition procedure have to be well thought-out. Operation in the discontinuous-only mode limits output current range, decreases efficiency, and increases die size/cost due to larger I_{ind} ripples (see Fig. 23).

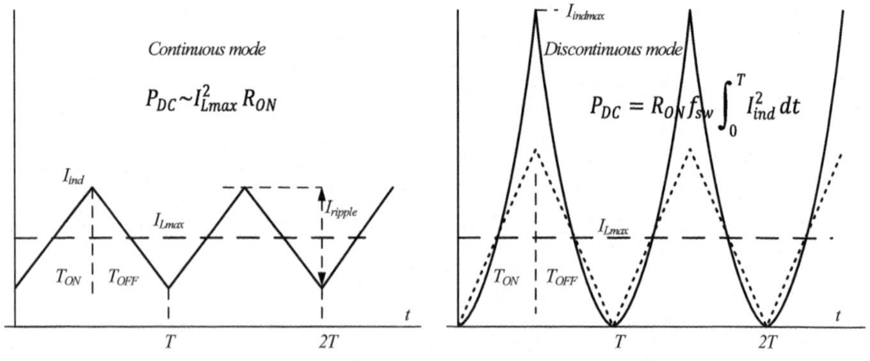

Fig. 23 Inductor current and power loss in continuous/discontinuous modes

In continuous-mode, DCDC converter and at maximum load, the DC power loss is $P_{DC} \sim I_{L\,\max}^2 R_{ON}$, assuming $I_{L\max} > I_{rpl}$, and R_{ON} is the power switch ON resistance. In discontinuous mode, $I_{rpl} \gg I_{L\max}$ due to strong $L(I_{ind})$ dependence (Fig. 20), which increases DC power loss $P_{DC} = R_{ON} f_{sw} \int_0^T I_{ind}^2 dt$ and demands larger switches, occupying larger die area and consuming more energy for switching. Inductor sizes for both modes are close, as for continuous mode inductance value is larger but $I_{L\max}$ is smaller, vice versa for discontinuous mode. Ideally, converter should transfer from continuous to discontinuous operation without any structural or compensation changes.

With these considerations, we can choose between:

- Voltage hysteretic mode.
- Fixed timing (fixed T_{ON}, fixed T_{OFF} or minimum T_{ON} and T_{OFF}).
- Current hysteretic mode.

The voltage hysteretic mode is applicable for control of buck converters only; it is also relatively high consuming as it requires continuously on comparator of the V_{out} ripples with small (3–5 mV) amplitude.

Fixed-timing current modes can be very efficient with both low- and high-current load but require high-cost wire-wound inductors. Such DCDC converter, perfectly operational in the lab, loses efficiency or even fails when customer uses the low-cost chip inductor having enormous L variation vs. I_{load} (Fig. 21) due to the I_{indmax} violation (similar to Fig. 23).

The only operation mode which can be used in any kind of DCDC converter (buck, boost, buck-boost, inverting, single inductor-multiple output, etc.), very efficient with low I_{load} and tolerant to the L variation, is the current hysteretic mode.

It does not require any mode switching when moving from continuous to discontinuous operation. High I_{load}, when I_{ind} is approaching the maximum datasheet value, can cause only some increase in switching frequency (Fig. 24). It also can be very simply implemented (see below). Switching frequency in continuous mode is defined by L, I_{hyst} and input/output voltages. If the constant switching frequency is required, I_{hyst} value can be controlled by phase-locked loop. Frequency in discontinuous mode is defined by I_{load}.

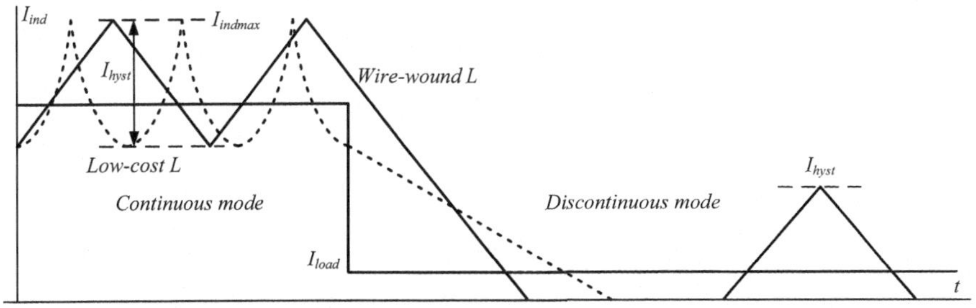

Fig. 24 Current hysteretic mode operation

9 Design Example

The circuits and structures described above have been used in many industrial designs: boost battery charger from energy harvesters [18], stand-alone buck converters, and embedded power management units. The depicted below buck converter with 100 nA no-load consumption and > 80% efficiency in 10 μA to 30 mA load range was done on 130 nm process for IoT radio PM.

Structure of this converter is in Fig. 25. It comprises:

- Always-on error amplifier G_{m0} from Fig. 17, $R_0 = 300$ ohm, amended by 3 μA current limit for I_e (total 60 nA I_q budget).
- Error comparator A_0 of Fig. 18b biased by the I_c replica of I_e plus another 10 μA through S_c when S_0 is ON (2 nA no-load I_q).
- Power switches S_0, S_1 (~ 3 ohm when ON) with inductor current sensor of Fig. 22 with $N = 10{,}000$.
- Synchronous rectifier S_1/A_1 with comparator A_1 from Fig. 19.
- Time-dependent current source I_h with 1 μA value, defining $N{*}I_h = 10$ mA I_{hyst}, connected to error summing point by S_h when S_0 is ON (Fig. 26).
- Sampled voltage reference V_{REF} as described in chap. 4 (30 nA I_q budget including biasing core and POR).

Time limit of I_{hyst} is required to eliminate the only problem of the current-hysteretic mode: when input and output voltages are close to each other, ON time is undefined. This may become a problem in scenario of V_{IN} glitch to V_{out} or below as

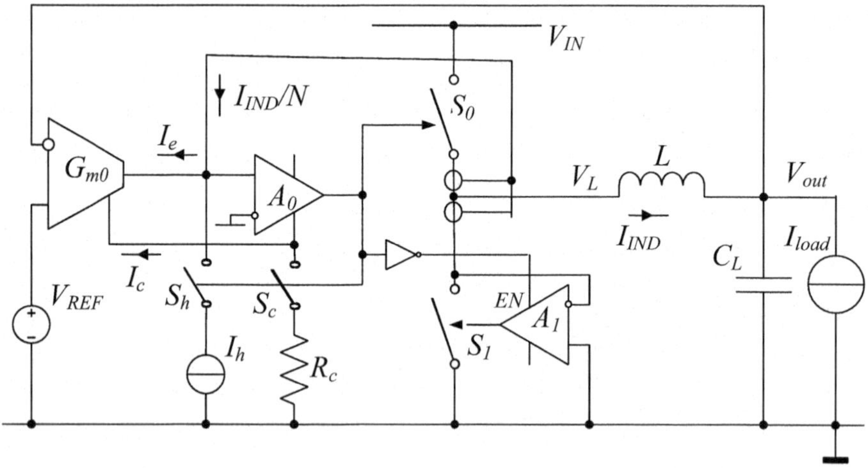

Fig. 25 Structure of the buck converter in current hysteretic mode

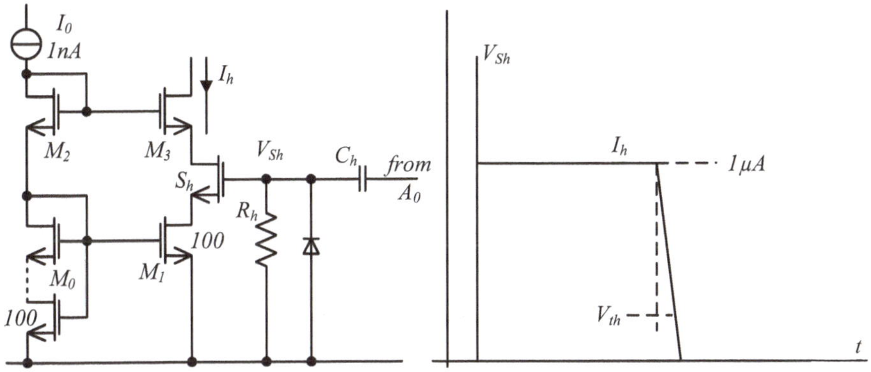

Fig. 26 Time-dependent current which defines I_{hyst}

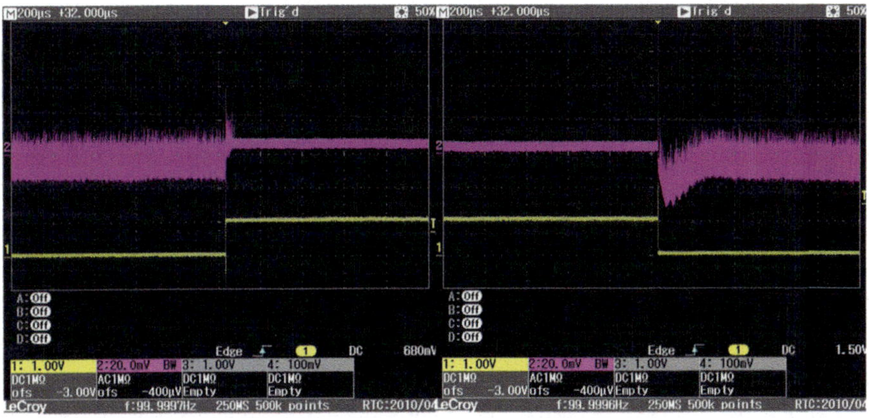

Fig. 27 Buck converter response on 0-10 mA load step

S_0 would turn ON during a glitch and never turn OFF if $I_{load} < I_{hyst}$ since inductor current will be equal to I_{load}.

In the 10,000 ratio current mirror of Fig. 26, I_h is turned ON as soon as A_0 flips high and stays ON as long as S_h gate is above V_{th}, which is defined by R_hC_h time constant and supply voltage.

Figure 27 depicts this buck converter step response on 0–10 mA load step with 10 μH inductor and 10 μF C_L, illustrating fast first-order dynamics, equivalent to the best large-I_q counterparts.

Power measurements show 80–85% efficiency of the converter from 10 μA to the maximum current of 30 mA (Fig. 28). Even at 5 μA load efficiency is above 70%.

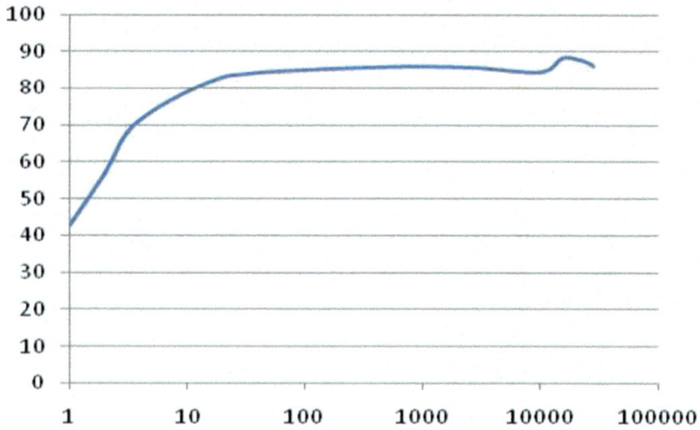

Fig. 28 Buck converter efficiency versus load

10 Conclusions

Design of systems with nanopower consumption does require special approach and different circuit techniques. There is an intensive university research in this area as well as numerous industrial parts. When thinking in nanopower space, one can find out that almost every existing IC or electronic system consumes more power than necessary by orders of magnitude—this approach can save enormous amount of energy when extrapolated outside of IoT-only area.

Structural design methodology is a very efficient tool to create solution to almost every task in hand or significantly improve existing circuits. All circuits above have been originated from this methodology. Apply feedback loop to control every important parameter, use current-input amplifiers wherever possible, make local feedback loops unconditionally stable which ensures the overall system stability.

References

1. Murata catalog. https://www.murata.com/en-us/products.
2. Ivanov V, Filanovsky I. Operational amplifier speed and accuracy improvement. Kluwer; 2004. https://www.springer.com/gp/book/9781402077722.
3. Mason S. Feedback theory – further properties of the signal flow graphs. Proc IRE. 1956;44(7):920–6.
4. Shmid H-P. Circuit transposition using signal-flow graphs. Proc ISCAS. 2002;2:25–8.
5. Попов Е. Теория линейных систем автоматического управления. (E. Popov, “Linear system control theory”), Moscow, Nauka, 1988, in Russian.
6. Milev M, Burt R. Tool and methodology for AC-stability analysis of continuous-time closed-loop systems. Proceedings of DATE-2005.
7. Ivanov V, Filanovsky I. A 110 dB PSRR/CMRR/gain CMOS micropower operational amplifier. ISCAS-2005.

8. Polya G. How to solve it. Princeton University Press; 1971, Princeton, New Jersey, USA.
9. O'Connor J, McDermott I. The art of systems thinking. Thorsons; 1997, London, UK.
10. Williams J, editor. Analog circuit design. Oxford, UK: Butterworth-Heinemann; 1991.
11. Ahuja B, Vu H, Laber C, Oven W. A very high precision 500-nA CMOS floating-gate analog voltage reference. JSSC. 2005;40(12):2364–72.
12. Rincon-Mora G, Allen P. A low-voltage, low quiescent current. Low drop-out regulator. JSSC. 1998;33(1):36–44.
13. Ivanov V, Baum D. Slew rate boost circuitry and method, US patent 6,437,645, 2002.
14. Allen P, Holberg D. CMOS analog circuit design. 3rd ed. Oxford; 2012.
15. Nanda S, Panda A, Moganti G. A Novel design of a high speed hysteresis-based comparator in 90-nm CMOS technology. ICIP 2015, IEEE.
16. Ivanov V, Venkataraman H, King D. Adjustable speed comparator, US patent 8482317, 2013.
17. Kislovsky A, Redl R, Sokal N. Dynamic analysis of switching mode DC/DC converters. Van Nostrand Reinhold; 1991, https://www.springer.com/gp/book/9789401178518.
18. Kadirvel K, et al. A 330nA energy-harvesting charger with battery management for solar and thermoelectric energy harvesting. ISSCC 2012, pp. 106–8.
19. Solis C, Rincon-Mora G. 0.6-μm CMOS-switched-inductor dual-supply hysteretic current-mode buck converter. IEEE TPE. 2017;32(3):2387–94.

Nanopower SAR ADCs with Reference Voltage Generation

Maoqiang Liu, Kevin Pelzers, Rainier van Dommele, Arthur van Roermund, and Pieter Harpe

1 Introduction

SAR ADC has been extensively acknowledged as a very power-efficient architecture. However, most SAR ADC publications do not take the power consumption of the reference voltage generation into consideration, which can be much higher than the SAR ADC itself [1, 2]. Power supply is frequently used as the reference in publications while this solution suffers from an instable power supply and interference from other blocks through the shared power supply. One example of a SAR ADC with integrated reference generation is shown in Fig. 1. A voltage reference generates a well-defined stable voltage to be multiplied by a LDO or buffer to drive the DAC in the SAR ADC.

In terms of voltage reference, bandgap reference (BGR) circuits based on BJTs are favorable thanks to their high accuracy over process corners as the output reference voltage is determined by the bandgap voltage of silicon. Conventional BGR generates a reference voltage around 1.2 V and requires a power supply above 1.4 V [3], making it incompatible with low-power sub-1 V systems. Alternatively, CMOS voltage references can more easily operate at sub-1 V VDD and consume low power. However, the output reference voltage of many CMOS voltage references [4, 5] highly depends on the absolute value of the MOSFET threshold voltage, quite sensitive to process variations. It is claimed that a DTMOST-based voltage reference could reduce this dependency by a factor of two [6], this advantage may become smaller with technology scaling as the back gate tends to have less impact on the MOSFETs depletion region.

M. Liu (✉) · K. Pelzers · R. van Dommele · A. van Roermund · P. Harpe
Eindhoven University of Technology, Eindhoven, The Netherlands
e-mail: m.liu@tue.nl

K. A. A. Makinwa et al. (eds.), *Low-Power Analog Techniques, Sensors for Mobile Devices, and Energy Efficient Amplifiers*,
https://doi.org/10.1007/978-3-319-97870-3_4

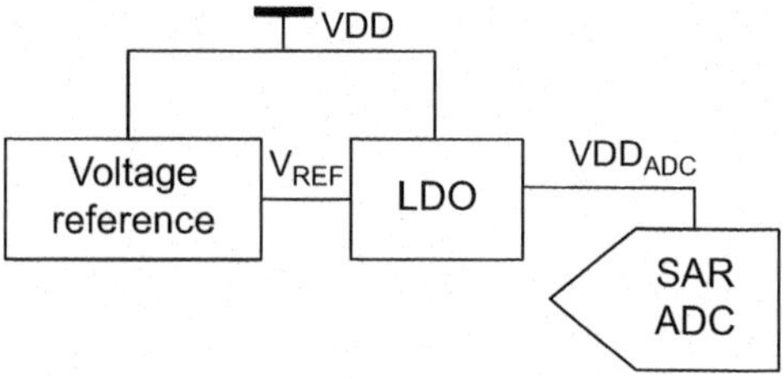

Fig. 1 Architecture of a SAR ADC with reference generation

Since the generated voltage by a voltage reference usually does not have driving capability, a LDO is needed. Due to the frequent DAC switching steps in one SAR conversion, the bandwidth of the LDO has to be high enough to recover the reference voltage in a short time, consuming significant energy consumption (>9× the power of SAR ADC in [1, 2]). Alternatively, a large decoupling capacitor can be used to stabilize the reference voltage at the expense of a large chip area (200× the area of SAR ADC in [7]).

In order to achieve an overall low-power operation of a SAR ADC with reference generation, the power of the SAR ADC should also be optimized. In a SAR ADC, the power consumption is usually dominated by the comparator and DAC [8, 9]. Due to the noise requirements, the power of above two sub-blocks has to be quadrupled when increasing the resolution by 1 more bit. In the popular dynamic comparator [8], the parasitic capacitances are pre-charged to VDD and discharged to ground for each comparison while only one of these two processes is used for comparison. In many DAC schemes for SAR ADCs, the reset operation consumes similar or even two times the energy during conversion [10–12] while most works only focus on the reduction of conversion energy.

In this chapter, corresponding to Fig. 1 and above discussion, different low-power techniques are introduced to realize low-power operation from three aspects: the voltage reference, the SAR ADC, and the LDO. In Chap. 2, a 106nW 10b 80 kS/s SAR ADC with duty-cycled reference generation is presented [13]. A 0.62 V-VDD 25nW CMOS voltage reference is introduced for the SAR ADC. In order to reduce the static power, the voltage reference is duty-cycled with a three-stage sample-and-hold. A low-power bidirectional comparator is presented to reduce the SAR ADC power consumption. The overall SAR ADC with reference voltage generation achieves a FoM of 2.4fJ/conv.-step. In Chap. 3, the energy-free "swap-to-reset" scheme is illustrated to eliminate the large reset energy existing in many DAC switching Schemes [14]. Instead of the conventional "switch-to-reset," which releases the charge in the DAC and draw new charge from the reference voltage, consuming significant energy, the DAC can be reset by swapping the positions of capacitors between P and N sides while the charge can be preserved. In Chap. 4, a discrete-time passive DAC driving scheme is presented where a decoupling capacitor is pre-charged to the reference voltage during tracking phase and drives the DAC passively during conversion [15]. The signal-dependent energy consumption of each DAC switching causes nonbinary steps due to the charge sharing between the DAC and the decoupling capacitor, which is compensated by a small auxiliary DAC. The discrete-time passive driving avoids an always-on power-hungry buffer or a large decoupling capacitor, achieving low power, and small chip area simultaneously.

2 A 106nW 80 kS/s SAR ADC with Duty-Cycled Reference Generation

This chapter presents a low-power SAR ADC with duty-cycled reference generation. The architecture of the overall circuit blocks is shown in Fig. 2 [13].

The whole system operates at 0.8 V VDD. The CMOS voltage reference generates 0.4 V reference voltage. The reference voltage is sampled and held by the three-stage duty-cycling block and then multiplied to 0.6 V to drive the DAC in the SAR ADC by the LDO. An off-chip capacitor is used in the LDO to stabilize the reference voltage. A low-power bidirectional comparator is adopted in the SAR ADC.

2.1 Low-Power Low-VDD CMOS Voltage Reference

A voltage reference is required to output stable voltage over process, power supply, and temperature, of which the process-independence is the most challenging as process variations are usually inevitable and there is not much to do about it from the aspect of circuit design. For CMOS voltage references, an obvious voltage unit is the threshold voltage of MOSFETs, which has a poor process-dependency. As shown in Fig. 3, the process variation of different voltage units is simulated at three corners (typical, slow, fast) and is divided by the average value to evaluate the effect of process variations for various transistor sizes. The absolute threshold voltage reveals the largest variation (15%) over corners in the adopted 65 nm CMOS technology.

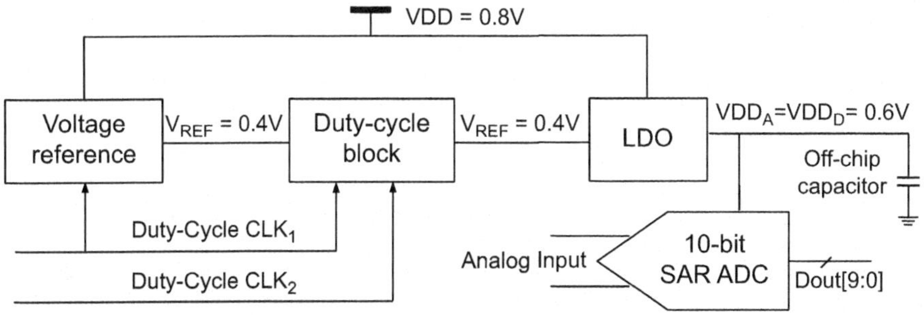

Fig. 2 The presented SAR ADC with duty-cycled reference generation

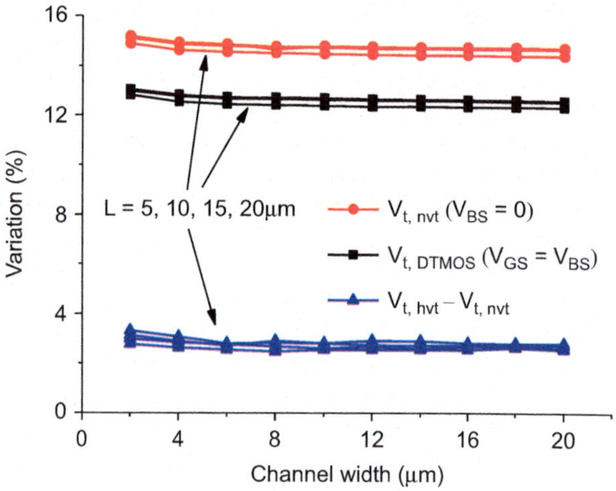

Fig. 3 The process variations of different voltage units in CMOS technology

It is claimed that DTMOS (a MOS with the back gate connected to the gate) could improve the threshold voltage stability over process corners by a factor of 2 in [6], but in the simulations based on the utilized 65 nm technology, the improvement is only about 20% (Fig. 3, $V_{t,\,DTMOS}$). In many technologies, the foundry provides transistors with higher or lower threshold voltage by an extra ion-implantation step while maintaining the same oxide thickness. The threshold voltage increment by the ion implantation is easier to control so that it has much smaller process variation [16]. As shown in Fig. 3, ΔV_t ($=V_{t,\,hvt} - V_{t,\,nvt}$) reveals five times smaller variations than the absolute threshold voltage. It can be concluded that the threshold voltage difference is a more suitable voltage unit for CMOS voltage references thanks to the excellent process stability.

The temperature dependency of ΔV_t is also investigated in [17]. For a shallower implanted depth than the depletion edge, ΔV_t is CTAT (Complementary To Absolute Temperature). A temperature-insensitive voltage can be generated by summing a PTAT (Proportional To Absolute Temperature) voltage with ΔV_t. As shown in Fig. 4, M_2 and M_1 are high-V_t and nominal-V_t transistors, respectively. They are diode-connected and biased in sub-threshold region. The drain-source current can be expressed as follows when $V_{GS} >> V_T$ (thermal voltage):

$$I_d = \mu_n C_{ox} \frac{W}{L} (\eta - 1) V_T^2 \exp\left(\frac{V_{GS} - V_t}{\eta V_T}\right) \tag{1}$$

where μ_n is the carrier mobility, C_{OX} is the gate-oxide capacitance, W and L are the width and length of the transistor, respectively, η is the sub-threshold slope factor, V_T is the thermal voltage, V_{GS} is the gate-source voltage, and V_t is the threshold voltage. Hence, the gate voltage difference of M_1 and M_2 could be expressed as:

$$\Delta V_G = \left(V_{t,hvt} - V_{t,nvt}\right) + \ln\left(\frac{I_2}{I_1} \cdot \frac{\mu_{n1}}{\mu_{n2}} \cdot \frac{(W/L)_1}{(W/L)_2}\right) \cdot \eta V_T \tag{2}$$

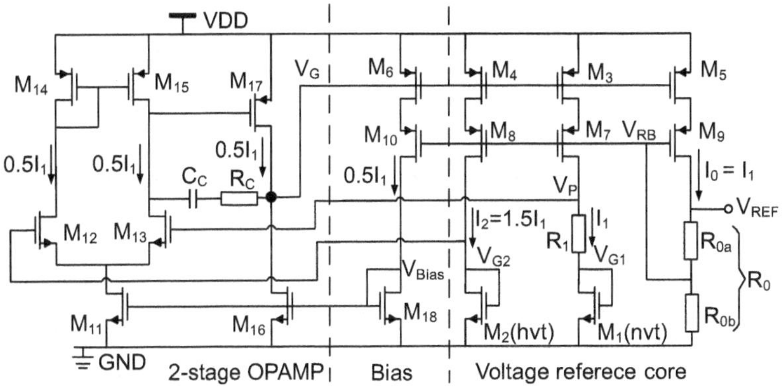

Fig. 4 CMOS voltage reference without duty-cycling

where the first term is CTAT and the second term is PTAT. By tuning the current ratio I_2/I_1, the TCs (Temperature coefficient) of these two terms can cancel out each other, leading to a temperature-insensitive ΔV_G.

Now that a process- and temperature-insensitive ΔV_G can be achieved, the next step is to generate a reference voltage relative to ground. As shown in Fig. 4, the negative feedback loop using the OPAMP forces V_P to equal V_{G2}, hence applying ΔV_G to the terminals of R_1. By copying I_1, a reference voltage relative to ground can be realized:

$$V_{REF} = I_0 \cdot R_0 = \frac{I_0}{I_1} \cdot \frac{R_0}{R_1} \cdot \Delta V_G \tag{3}$$

where R_0 equals the sum of R_{0a} and R_{0b}.

The temperature performance of the voltage reference at five process corners is simulated and shown in Fig. 5. The V_{REF} only has a variation range of $\pm 1.5\%$ over process corners, well corresponding to Fig. 3.

2.2 *Three-Stage Duty-Cycling Technique*

In order to reduce the power consumption, the voltage reference can be duty-cycled. In this work, the LDO is always on, hence requiring a continuously stable reference voltage. One-stage sample-and-hold can be used to sample and hold the reference voltage and feed it to the LDO. As shown in Fig. 6, the voltage reference and the switched capacitor network (S_A, C_A) are duty-cycled by the same clock, CLK_1. This scheme suffers from two aspects. When CLK_1 switches everything on, V_A will experience a large startup ripple of the voltage reference. When CLK_1 switches

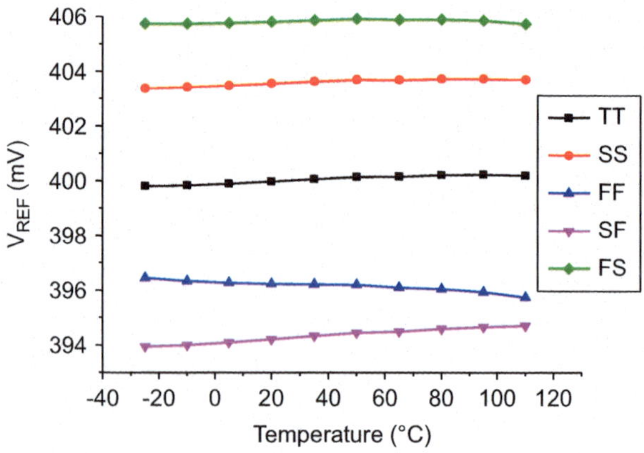

Fig. 5 Simulated V_{REF} temperature dependency at five process corners

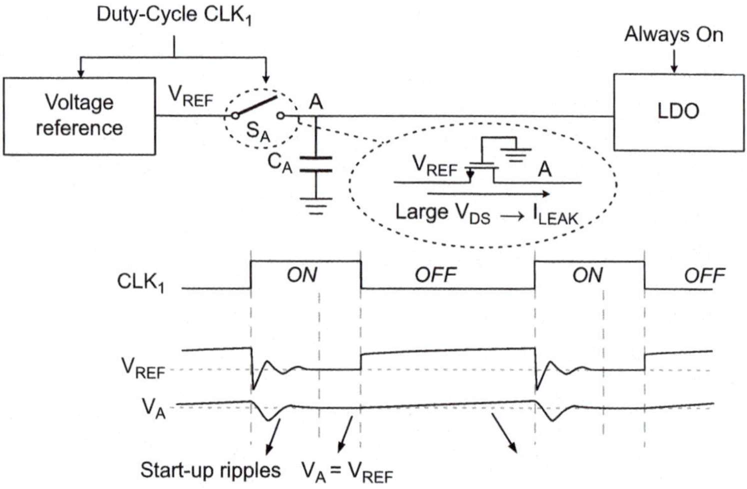

Fig. 6 Architecture and waveforms of a one-stage sample-and-hold

everything off and C_A begins to hold the sampled voltage, the large voltage drop over the NMOS switch, S_A, will cause a leakage current through S_A, leading to the drop of V_A. The instable reference voltage V_A will degrade the performance of the SAR ADC.

To solve above two problems, a three-stage dual-clock sample-and-hold is presented as shown in Fig. 7. Compared with Fig. 6, two more switched capacitors are inserted between node A and the LDO, controlled by CLK_2. CLK_1 and CLK_2 have the same frequency but different pulse width. CLK_1 switches the voltage reference and C_A on as before and V_A will be stable after some startup ripples. After that, CLK_2 switches C_B and C_C on to refresh the voltage on them. In this way, V_C

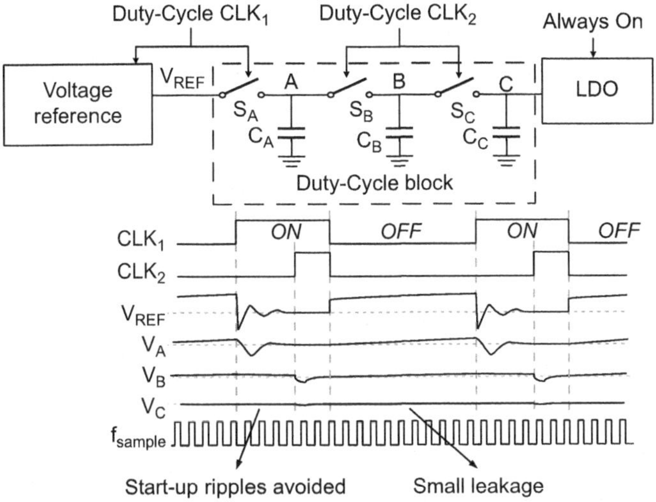

Fig. 7 Architecture and waveforms of the three-stage sample-and-hold

is isolated from the startup behavior of the voltage reference. Subsequently, CLK_1 and CLK_2 switch everything off and the holding phase begins. Leakage through S_A still exists and causes V_A drop. However, the voltage drop over S_B is much smaller than S_A, leading to a more stable V_B than V_A. Furthermore, the voltage drop over S_C is smaller than S_B, leading to a more stable voltage at node C. As a result, V_C is stable throughout the entire clock period. The simulated maximum ripple of V_C is merely 0.1 mV. For the targeted 10b SAR ADC with an LSB of 1 mV, this small voltage drop is totally acceptable.

In this work, the duty-cycling clocks, CLK_1 and CLK_2, are generated from an FPGA externally. The frequency is 20 Hz and the pulse widths are 5 ms and 2.5 ms, respectively. Since the frequency is low and the accuracy requirement for the pulse width of the clock signals is relaxed, the clock signals are also be generated on-chip with little power consumption. The simulated power for the clock generator is only 2nW, occupying a chip area of only 50 μm $\times$ 12 μm.

2.3 *Low-Power Bidirectional Dynamic Comparator*

The typical dynamic comparator [18] is shown in Fig. 8. It comprises an integration-based preamplifier and a latch. During the reset phase, the parasitic capacitors $C_{PP,\ PN}$ are charged to VDD. The comparison starts when CLK goes high. $C_{PP,\ PN}$ are discharged and $V_{AP,\ AN}$ will drop according to the input voltages $V_{INP,\ INN}$. In time, the common mode of $V_{AP,\ AN}$ gradually decreases while the differential input signal is gradually amplified. When the common mode of $V_{AP,\ AN}$ reaches the threshold

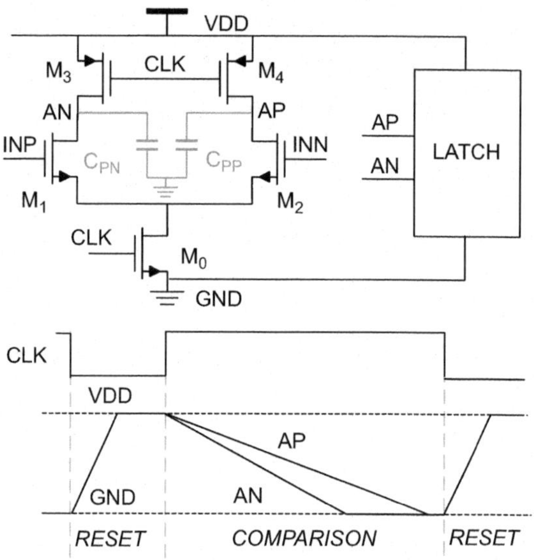

Fig. 8 Typical dynamic comparator [18] and the waveforms of key nodes

of the latch, the latch will take over and output the comparison result. Next, CLK will go low, and the preamplifier and latch are reset to their initial conditions. The preamplifier dominates the noise and power efficiency of the comparator due to the amplification. The equivalent input noise of the preamplifier can be expressed as [18]:

$$\sigma_{\mathrm{V}} \approx \mathrm{k} \cdot \mathrm{T} \cdot \sqrt{\frac{8}{q}} \cdot \frac{1}{\sqrt{|C_{PP} \cdot (\mathrm{V_{thlatch}} - \mathrm{V_{DD}})|}} \sim \frac{1}{\sqrt{Q_C}} \tag{4}$$

where V_{thlatch} represents the threshold voltage of the latch and Q_{C} is the absolute value of average transferred charge on parasitic capacitor $C_{\mathrm{PP,\,PN}}$ before the latch takes over. The preamplifier energy dissipation of one comparison can be expressed:

$$\mathrm{E_C} = 2 \cdot C_{PP} \cdot V_{DD}^2 \tag{5}$$

It can be observed from (4) and (5) that the noise performance and the energy consumption have to be traded off. In order to reduce the noise by a factor of 2, the energy consumption has to be quadrupled.

Figure 8 illustrates that only the discharging phase is used for the comparison while the charging is just to reset the parasitic capacitance. The presented bidirectional comparator makes use of both the charging and discharging slopes to perform the amplification and thus can improve the energy efficiency.

As shown in Fig. 9, a pair of PMOS ($\mathrm{M}_{3,4}$) is inserted. Before the comparison starts, $V_{\mathrm{AN,AP}}$ is reset to ground. When the comparison starts, $\mathrm{M}_{3,4}$ are enabled first

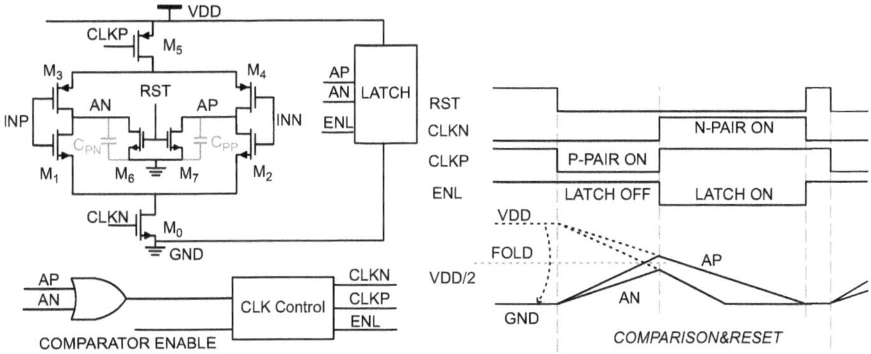

Fig. 9 Bidirectional dynamic comparator and key nodes waveforms

by switching on the current tail (M_5) and $V_{AN,AP}$ starts to increase according to the inputs. An OR gate is detecting when V_{AP} or V_{AN} reaches half VDD. At that moment, $M_{3,4}$ are switched off and $M_{1,2}$ are enabled together with the latch. From this point onward, the comparator operates as the typical structure and achieves the same gain. Another advantage of this architecture is that $V_{AN,AP}$ returns to ground inherently after each comparison while a reset signal (RST) is still present to avoid floating. For this bidirectional comparator, the sum of the absolute value of average transferred charge on $C_{PP,\,PN}$ during charging and discharging equals:

$$\mathrm{Q_{CB}} = \left|C_{PP}\cdot\left(\frac{V_{DD}}{2}-0\right)\right| + \left|C_{PP}\cdot\left(V_{\text{thlatch}}-\frac{V_{DD}}{2}\right)\right| = C_{PP}\cdot(V_{DD}-V_{\text{thlatch}}) = Q_C \tag{6}$$

which means that the bidirectional comparator achieves the same noise performance with the typical architecture, while $C_{PP,\,PN}$ are only charged to about half VDD. The energy consumption of the presented comparator can be expressed:

$$\mathrm{E_B} = 2\cdot C_{PP}\cdot\frac{V_{DD}}{2}\cdot V_{DD} = C_{PP}\cdot V_{DD}^2 \tag{7}$$

which is only half of (5).

For the comparator in the work, the power consumption ratio between the preamplifier and the latch is 3:1, if typical dynamic comparator architecture is used. By replacing the preamplifier with the presented bidirectional preamplifier, 37.5% of the comparator power is saved. In practice, the additional logic introduces 4.5% overhead, which is mostly the short circuit current in the OR gate, thus saving 33% overall.

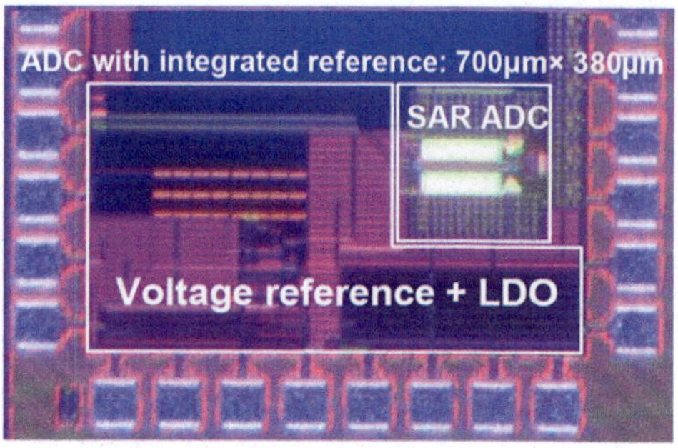

Fig. 10 Die photo in 65 nm CMOS technology

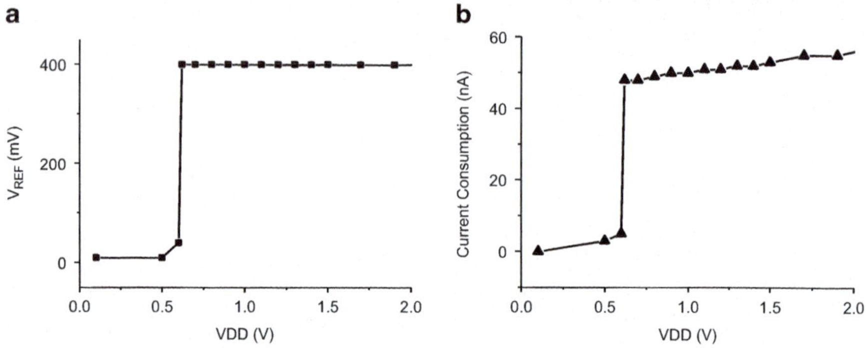

Fig. 11 (**a**) Measured V_{REF} versus VDD. (**b**) Measured current consumption versus VDD

2.4 Measurement Results

The SAR ADC with duty-cycled reference generation is implemented in 65 nm CMOS technology and occupies a chip area of 0.266mm^2 due to the large resistors used in the CMOS voltage reference and the LDO as shown in Fig. 10. Both the stand-alone voltage reference without duty-cycling and the SAR ADC with duty-cycled reference are measured.

A total of 15 voltage reference samples are measured. The minimum operational VDD is within the range of 0.60–0.62 V at room temperature. The V_{REF} and current consumption of one sample are displayed in Fig. 11 as a function of VDD. The line sensitivity is 0.07%/V from 0.62 V to 2.0 V. At 0.8 V VDD, the measured power consumption is only 38 nW. At the minimum 0.62 V VDD, the power consumption is 25 nW.

The PSRR of the voltage reference is also measured at 0.8 V VDD and maintains less than -49 dB up to the near Nyquist frequency as shown in Fig. 12.

Figure 13 shows the temperature dependency of the 15 samples. From −25 °C to 110 °C, the temperature coefficients vary from 44 to 248 ppm/°C with an average TC of 108 ppm/°C. At room temperature, the average V_{REF} is 389.9 mV with a standard deviation of 4.0 mV. According to Monte-Carlo simulations, the dominant

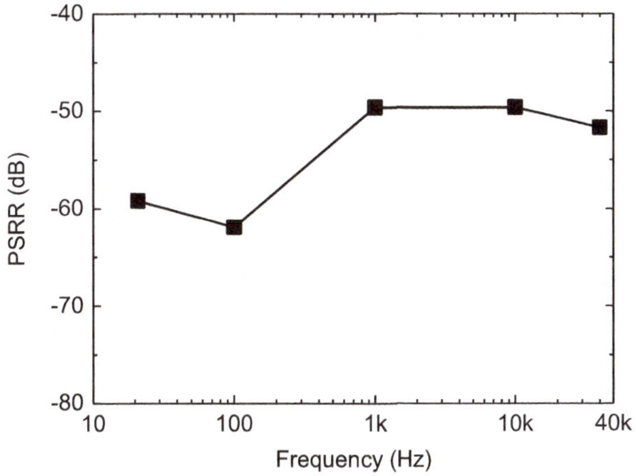

Fig. 12 Measured PSRR of the voltage reference

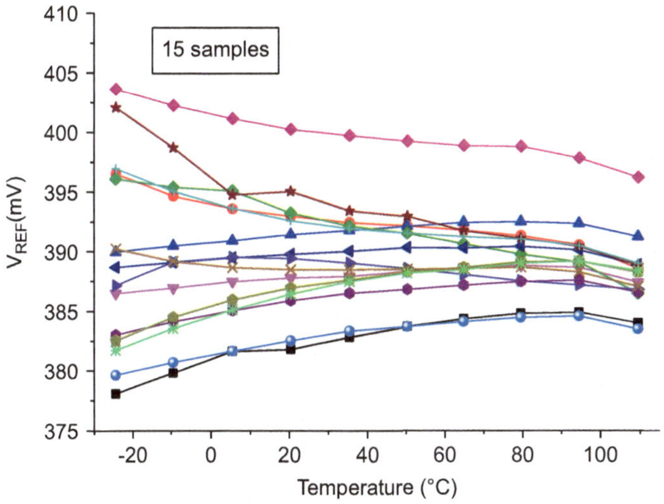

Fig. 13 Temperature dependency of 15 voltage reference samples

reason for part-to-part variation is mismatch in the OPAMP and the current mirror $M_{3-5,7-9}$. Table 1 summarizes the presented voltage reference and compares it with other low power works. The power consumption of the reference voltage can be further reduced with duty-cycling technique. Besides, the sample-to-sample variation of this work is comparable to BJT voltage references and substantially better than V_t-based CMOS voltage references.

The power breakdown and the ENOB of the reference-included ADC with different duty-cycling rates are shown in Fig. 14. At 10% duty-cycling of the voltage reference, the power consumption is only 3.7 nW, which is far less than that of the

Table 1 Voltage reference performance summary and comparison

		[19]	[20]	[21]	[4]	[5]	This work
Type		BJT	BJT	BJT	V_t	V_t	ΔV_t
VDD_{min}		0.75	0.7	0.5	0.9	0.45	0.62
Area(mm^2)		0.070	0.025	0.026	0.045	0.043	0.077
TC(ppm/°C) [T range (°C)]		40 [−20:85]	114 [−40:120]	75 [0:100]	10 [0:80]	165 [0:125]	108 [−25:110]
Power (nW)	W/o DC	170	52.5	32	36	2.6	25
	With DC	–	–	–	–	–	2.5
Sample-to-sample V_{REF} variation: σ/μ (%)		1.0	1.05	0.67	3.1	3.9	1.0
Line sensitivity (%/V)		0.005	–	–	0.27	0.44	0.07
PSRR @100 Hz (dB)		–	−62	–	−47	−45	−62

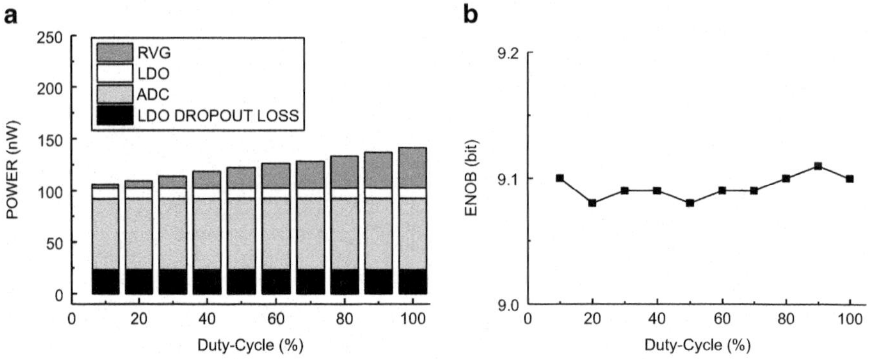

Fig. 14 (**a**) Power breakdown and (**b**) ENOB versus duty-cycling rates

SAR ADC core. Meanwhile, the ENOB with near Nyquist input remains 9.1bit regardless of the duty-cycling, resulting in a FoM of 2.4 fJ/conv.-step. The always-on LDO and the drop-out loss take about one-third of the total power consumption. The spectrum and the INL/DNL at 10% duty-cycling are shown in Figs. 15 and 16. At near Nyquist input, the SNDR is 56.6 dB and the SFDR is 65.0 dB.

Compared with other low-power SAR ADCs (Table 2), this work is the only one to integrate the reference generator and LDO with an ADC. Meanwhile, it has a comparable ENOB and FoM.

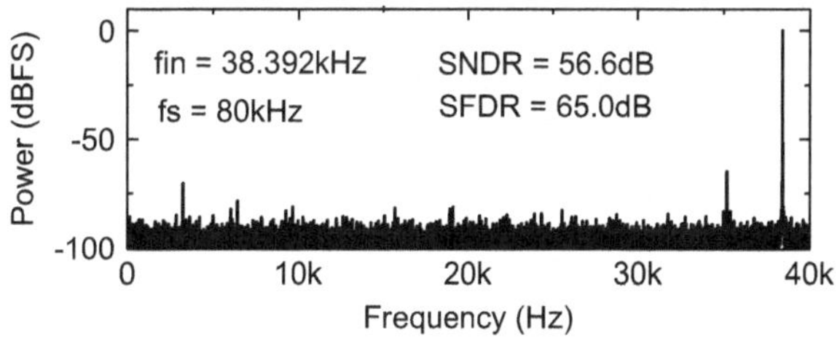

Fig. 15 Spectrum of the SAR ADC with 10% duty-cycled reference generation

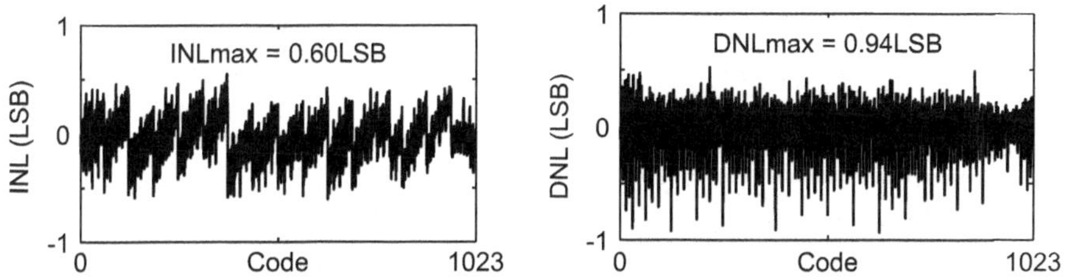

Fig. 16 INL/DNL of the SAR ADC with 10% duty-cycled reference generation

Table 2 SAR ADC performance summary and comparison

	[22]	[12]	[23]	[24]	[25]	This work
Technology (nm)	90	40	90	65	65	65
Resolution (bit)	10	10	10	10	10	10
VDD (V)	0.4	0.45	0.4	0.6	0.6	0.8
Sampling rate (S/s)	250 k	200 k	500 k	40 k	100 k	80 k
Area (mm^2)	0.04	0.065	0.042	0.076	0.053	0.26
INL (LSB)	0.67	0.45	0.62	0.48	0.87	0.60
DNL(LSB)	0.43	0.44	0.34	0.32	0.96	0.94
Power (μW)	0.2	0.084	0.5	0.072	0.088	0.106
Including reference	No	No	No	No	No	Yes
ENOB (bit)	8.6	8.95	8.72	9.4	9.2	9.1
FoM (FJ/conv.-step)	2.02	0.85	2.47	2.7	1.5	2.4

2.5 Conclusions

In this chapter, a 10b SAR ADC with duty-cycled reference generation is presented. The low-VDD low-power CMOS voltage reference is based on the threshold voltage difference between high-V_t and nominal-V_t transistors, which has 4x better process stability than the absolute threshold voltage of a single transistor or a DTMOS. The subthreshold operation enables the low-power (25 nW) low-VDD (0.62 V) operation at room temperature. Besides, the power supply and temperature dependency of the voltage reference are also small. The power consumption of the voltage reference is further reduced by a three-stage dual-clock duty-cycling block.

The multistage architecture reduces the leakage and enables 10% duty-cycling of the voltage reference. In the SAR ADC, a low-power bidirectional dynamic comparator is utilized which makes use of both the charging and discharging phase to perform the comparison, achieving better power efficiency. Compared to a typical architecture [18], this bidirectional comparator reduces the power by 33% while maintaining the same noise performance. With the above techniques, the SAR ADC with reference generation achieves a FoM of 2.4fJ/conv.-step.

3 Energy-Free "Swap-to-Reset" for the DAC in a SAR ADC

3.1 DAC Energy Consumption in a SAR ADC

Charge-redistribution DAC in a SAR ADC usually consumes significant energy, and many switching schemes are presented to reduce the energy consumption [10–12, 26–28]. As shown in Fig. 17, assuming the same total capacitance, [26–28] can save conversion energy compared to conventional switching scheme while bringing new problems like DAC output common mode shift or a third reference required. The common mode shift introduces signal-dependent offset errors, and a third reference voltage increases the energy consumption of the reference generation and circuit complexity. [10–12] can also reduce conversion energy without above problems. However, they require a large reset energy that is 1∼2× the conversion energy, which diminishes the total energy saving.

Take the split monotonic as an example [11], where one pair of differential capacitor is split into two pairs. The two pairs are reset differentially and at each side, the two capacitors are connected to V_{REF} and GND, respectively, as shown in Fig. 18a. For the i-th switching during conversion phase (Fig. 18b), one of the two pairs will be switched differentially, causing DAC output voltage shift and

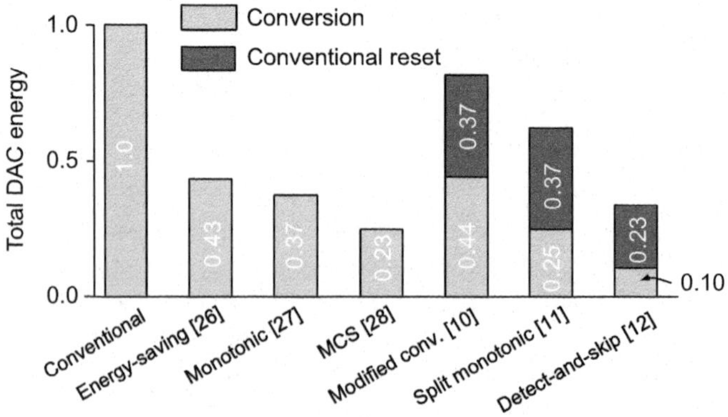

Fig. 17 Normalized energy consumption of different switching schemes

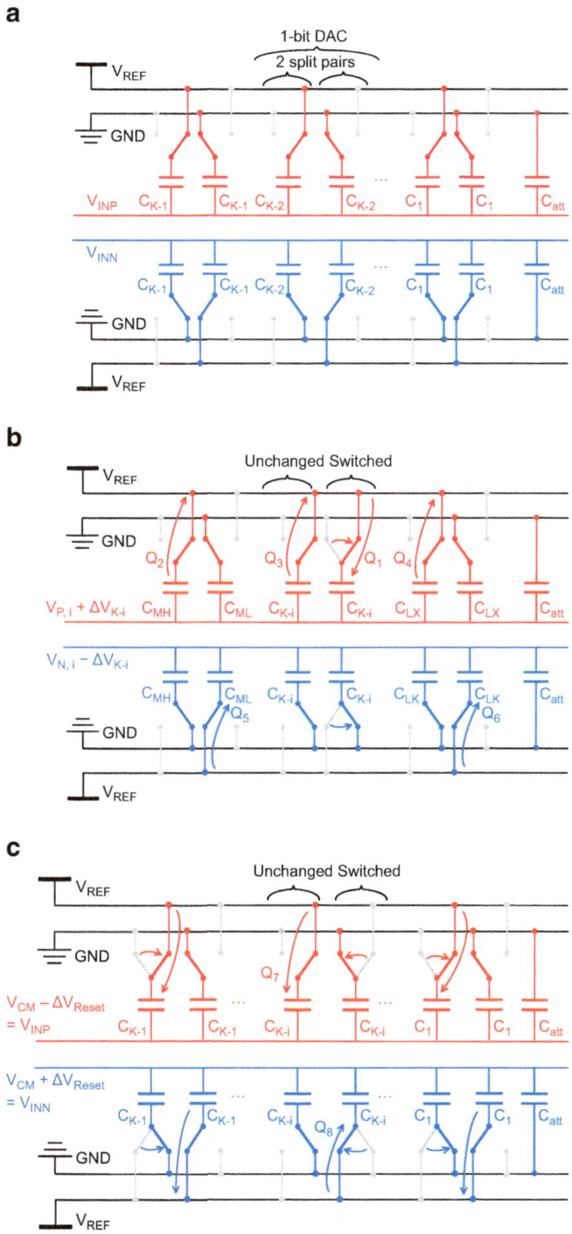

Fig. 18 (**a**) Tracking phase, (**b**) conversion phase, and (**c**) reset phase of a split monotonic DAC

charge redistribution. Assuming all the codes have the same probability, the average conversion energy consumption can be derived:

$$E_{Conversion} \approx 0.33 C_T \cdot V_{REF}^2 \tag{8}$$

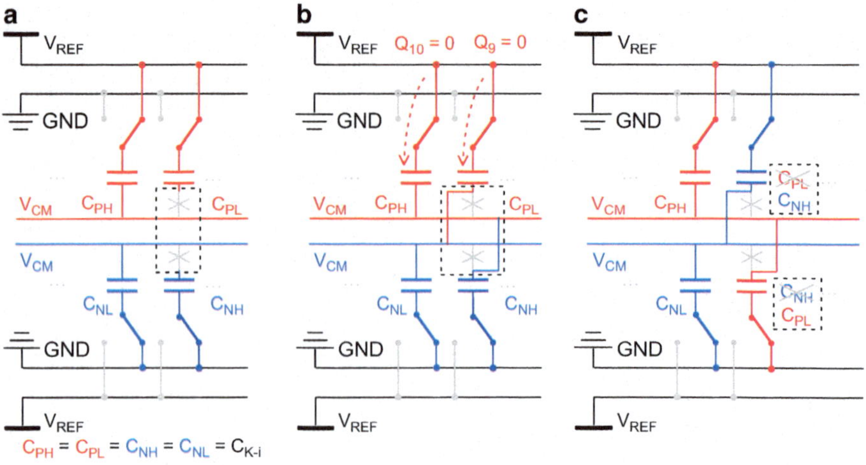

Fig. 19 Operation and charge flow of the "swap-to-reset" scheme

After each conversion, the DAC needs to be recover to the original configuration, and the conventional way to realize this is simply to switch the DAC conversely with the conversion phase (Fig. 18c), which consumes significant energy. It can be derived that the reset energy equals:

$$E_{Reset} \approx 0.5C_T \cdot V_{REF}^2 = 1.5 \cdot E_{Conversion} \tag{9}$$

3.2 *Energy-Free "Swap-to-Reset"*

From the previous discussion, it is noted that the number of capacitors and the total capacitance value connected to V_{REF} always remains the same. Each time a capacitor is disconnected from V_{REF}, its opposite element will be connected to V_{REF}. Due to the fully differential architecture, the DAC outputs converge to V_{CM} at the end of each conversion. Based on the above observations, the energy-free "swap-to-reset" can be introduced as shown in Fig. 19. The first step is to disconnect C_{PL} and C_{NH} from the DAC outputs. Subsequently, the P/N connections of C_{PL} and C_{NH} are swapped by connecting them to the opposite sides. Since the DAC outputs are both V_{CM}, the swap operation causes no charge redistribution and the DAC is reset successfully without any energy consumption. In the very next tracking and conversion phases, the previous C_{PL}/C_{NH} will act as the new C_{NH}/C_{PL}.

The energy-free "swap-to-reset" can be adopted in such switching schemes that are reset and switched differentially and the DAC outputs converge to V_{CM} at the end of each conversion. For instance, [10–12] can be swapped to reset.

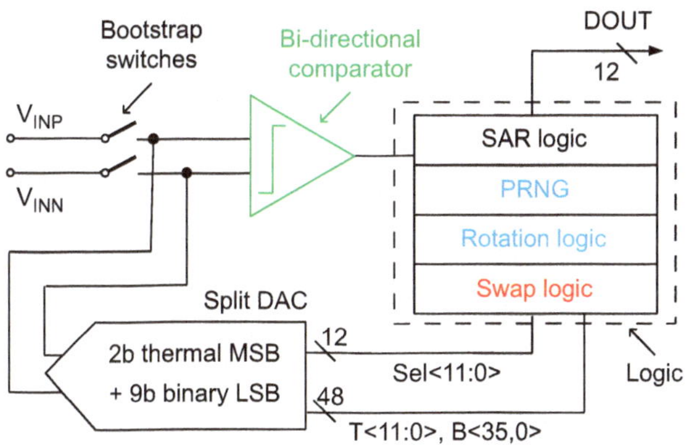

Fig. 20 Architecture of the 12b SAR ADC with "swap-to-reset" and rotation

3.3 A 12b SAR ADC with 2b "Swap-to-Reset"

The "swap-to-reset" is used for the 2 MSBs in the DAC of a 12b SAR ADC where the rotation [29] is also adopted for the 2 MSBs to enhance the linearity as shown in Fig. 20. The swap logic controls the swap operation based on the connections of the 2 MSBs after the very previous conversion. The swap operation is realized with bootstrapped switches for the sake of a good linearity. A pseudo random number generator (PRNG) and the rotation logic control the rotation operation. Furthermore, the bidirectional comparator is adopted to reduce the power consumption.

3.4 Measurement Results

The prototype is implemented in 65 nm CMOS technology and occupies a chip area of 0.105 mm^2. The SAR ADC operates at 40 kS/s and the power consumption is shown in Fig. 21. By enabling "swap-to-reset," 33% of the DAC power is saved while the remainder of the SAR ADC consumes 7% more power due to the auxiliary circuits for the "swap-to-reset," resulting in a total power saving of 18% for the SAR ADC. Since rotation has no impact on power saving, similar power saving is achieved when repeating above measurements with enabled rotation despite the little additional power of the rotation logic. The near Nyquist spectrums with different "swap-to-reset" and rotation setup are shown in Fig. 22. Regardless of "swap-to-reset," rotation maintains the same SNDR ($\sim$64.2 dB) and improves the SFDR by around 15.5 dB. Regardless of rotation, enabling "swap-to-reset" will maintain the same SNDR and SFDR.

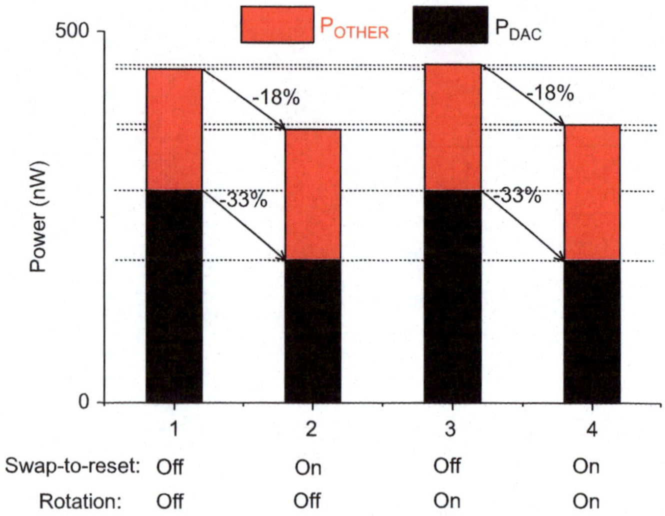

Fig. 21 Measured ADC power with conventional reset and "swap-to-reset"

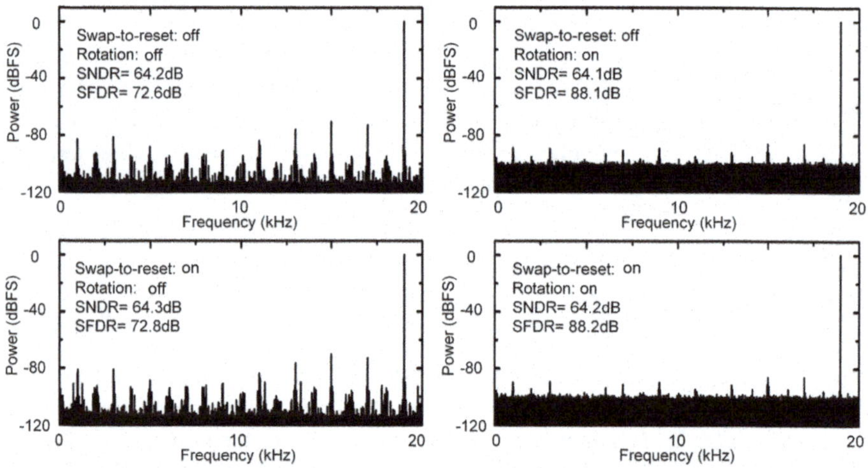

Fig. 22 Spectrums with different setup

3.5 Conclusions

Some DAC switching Schemes [10–12] save conversion energy while bringing significant reset energy. In this chapter, energy-free DAC switching schemes are presented that can eliminate the large reset energy for [10–12]. The swap operation has no impact on the dynamic performance of the SAR ADC. Along with rotation, a DAC is realized where each capacitor of the 2 MSBs can change position between

P and N sides (swap-to-reset) and along P or N side (rotation). This improves the SFDR (+15.4 dB) and the energy efficiency (18%) of the SAR ADC at the same time.

4 A Low-Power and Area-Efficient Discrete-Time Reference Driver

4.1 DAC-Compensated Reference Driver

For a charge-redistribution SAR ADC, the DAC is usually binary scaled and the reference voltage needs to be constant to perform an ideal binary search as shown in Table 3 (leftmost three columns). Considering a decoupling capacitor (C_{DEC}) pre-charged to V_{REF} driving the DAC of a SAR ADC passively, the charge sharing between the DAC and the C_{DEC} will cause signal-dependent V_{REF} drop and nonbinary searching steps (ΔV_{DAC}) as shown in Table 3 (middle three columns). For a given switching scheme, the energy consumption for each code is predictable. As a result, the DAC switching steps can be made binary again by switching extra compensation capacitance ($C_{a2,\,a3}$) as shown in Table 3 (rightmost three columns) [15]. The first step consumes constant energy and fixed V_{REF} drop (Table 3), regardless of the code for a symmetric DAC. Hence, ΔV_{DAC} of the first switching can be used as the reference for the later compensated steps. The capacitance of the compensation capacitors needs to be calculated before implementation and chosen to switch according to a certain code. During tracking phase, the C_{DEC} should be pre-charged to V_{REF}.

The SNDR and SFDR of a 10b SAR ADC driven by different C_{DEC} with different number of compensation bits are simulated in Matlab and shown in Fig. 23. Without compensation, the C_{DEC} has to be larger than 2^9C_T to achieve 62 dB SNDR, occupying significant chip area. With more and more compensated bits, the required C_{DEC} for 62 dB SNDR can be reduced greatly. Furthermore, the passive driving does not require an always-on power-hungry buffer, achieving low power at the same time.

Table 3 First three DAC switching steps with different drivers

Ideal reference driver			Passive driver w/o compensation			Passive driver with compensation		
V_{REF}	C_i	ΔV_{DAC}	V_{REF}	C_i	ΔV_{DAC}	V_{REF}	C_i	ΔV_{DAC}
1	8C	0.5	1→ 0.96	8C	0.48	1→ 0.96	8C	0.48
1	4C	0.25	0.96→ 0.89	4C	0.21	0.96→ 0.90	4C + C_{a2}	0.24
1	2C	0.125	0.89→ 0.84	2C	0.10	0.90→ 0.85	2C + C_{a3}	0.12

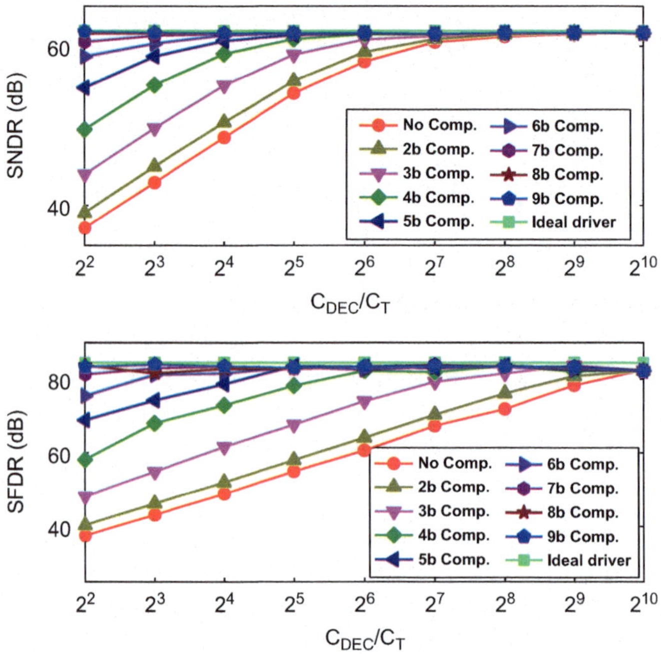

Fig. 23 SNDR/SFDR of a 10b SAR ADC driven by different C_{DEC} with different number of compensation bits

4.2 10b SAR ADC with DAC-Compensated Reference Driver

The above reference driver is adopted for a 10b SAR ADC, where the first three switching steps are compensated. As shown in Fig. 24, an on-chip C_{DEC} equal to $20C_T$ ($C_T \approx 1$ pF) is used to drive the DAC passively. It is pre-charged by a pre-charger during tracking phase, controlled by the sampling clock. The compensation DAC is connected in parallel with the binary DAC, controlled by a logic block to select the correct compensation capacitor. The total capacitance of the compensation DAC is only 1% of C_T, introducing little hardware overhead.

4.3 Experiment Results

The prototype is implemented in 65 nm CMOS technology and occupies a chip area of 0.076mm^2 (Fig. 25), where only 10.1% is occupied by the reference driver (pre-charger, C_{DEC}, compensation DAC, and logic). The SAR ADC operates at 20 MHz sampling frequency. The spectrums before and after enabling the compensation is shown in Fig. 26. After enabling 3b compensation, the SNDR/SFDR are improved

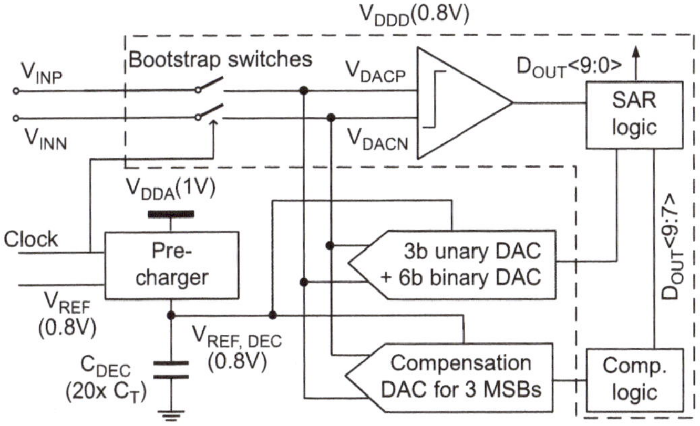

Fig. 24 SAR ADC driven by a DAC-compensated reference driver

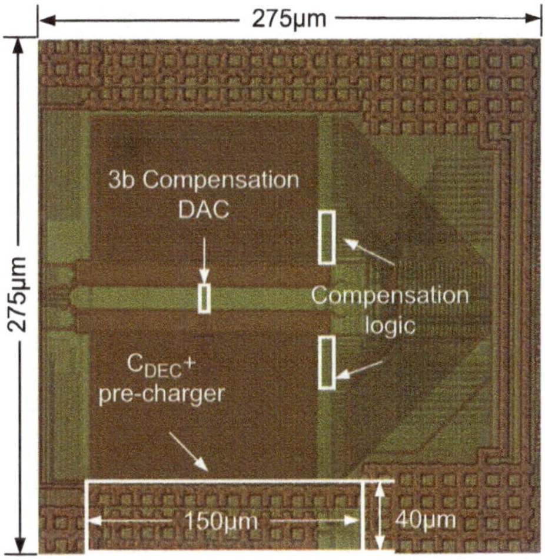

Fig. 25 Die photo in 65 nm CMOS

by 2.7 dB/11.6 dB, respectively. The power breakdown is shown in Table 4. After enabling the compensation, the power consumption of the DAC is only increased by 3% due to the small compensation capacitance. The extra digital logic adds only 1.6% more power. The pre-charger consumes 7.8 μW, and 4 μW is lost due to the dropout. In total, the DAC-compensated driver consumes 10.8% more power compared with external VDD driving.

4.4 Conclusions

Charge-redistribution SAR ADCs are energy-efficient while the reference driver can consume significant power or occupy a large chip area. In this chapter, a low-

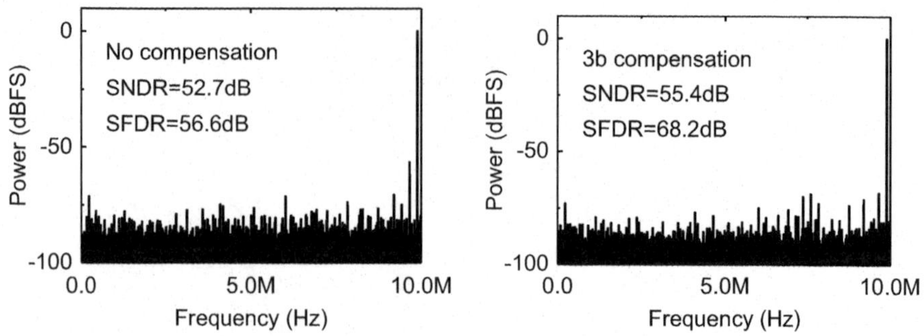

Fig. 26 Spectrums of the SAR ADC without and with compensation

Table 4 Power breakdown of the SAR ADC with/without compensation

Power (μW)	External VDD	C_{DEC} w/o comp.	C_{DEC} with comp.
$P_{pre\text{-}charger}$	0	7.8	7.8
P_{DAC}	36.7	36.7	37.9
$P_{dropout}$	0	4.1	4.2
P_{DIG}	99.8	99.8	101.4
P_{total}	136.5	148.4	151.3

power and area-efficient passive DAC-compensated reference driver is presented. The continuous-time buffer is replaced to save power. Moreover, the DAC switching can be compensated with small capacitance to reduce the chip area of C_{DEC} substantially. The presented driver combines the advantage of a charge-redistribution DAC (simple and insensitive to parasitic capacitance) and a charge-sharing DAC (only using reference driver during tracking).

5 Conclusions

SAR ADCs are well known for their power-efficiency while the power and chip area of the reference driver must be taken into account as well in practice. In this chapter, several low-power techniques for nanopower SAR ADC with integrated reference driver are presented and validated by experiments from three aspects: the voltage reference, the SAR ADC, and the driver. First, A 0.62 V-VDD 25 nW CMOS voltage reference is presented in 65 nm CMOS based on the threshold voltage difference between different types of transistors, achieving much better (>4×) process stability that other CMOS voltage reference topologies. In combination with a three-stage sample-and-hold, the power of the voltage reference can be further reduced (by 10×) while providing an accurate continuous-time reference voltage. Second, as the DAC and comparator usually dominate the total power of SAR ADCs, an energy-free "swap-to-reset" scheme is presented to eliminate the large reset energy existing in many DAC switching Schemes (33% DAC energy saving in the prototype), and a bidirectional dynamic comparator is presented to reduce the comparator power (by 33%). Finally, a discrete-time passive reference driver is introduced where a small capacitor driving the DAC is tolerated by compensating the DAC switching steps, consuming low power (10.8% of the ADC core) and occupying small area (0.007mm^2).

References

1. Borghetti F, Nielsen JH, Ferragina V, Malcovati P, Andreani P, Baschirotto A. A programmable 10b up-to-6MS/s SAR-ADC featuring constant-FoM with on-chip reference voltage buffers. In Proc. ESSCIRC, Sep. 2006, pp. 500–3.
2. Harikumar P, Wikner JJ. Design of a reference voltage buffer for a 10-bit 50 MS/s SAR ADC in 65 nm CMOS. In Proc. IEEE ISCAS, May 2015, pp. 249–52.
3. Razavi B. Design of analog CMOS integrated circuits. Columbus: McGraw-Hill Education; 2000.
4. Vita GD, Iannaccone G. A sub-1-V, 10 ppm/°C, nanopower voltage reference generator. IEEE J Solid State Circuits. 2007;42(7):1536–42.
5. Magnelli L, Crupi F, Corsonello P, Pace C, Iannaccone G. A 2.6 nW, 0.45 V temperature-compensated subthreshold CMOS voltage reference. IEEE J Solid State Circuits. 2011;46(2):465–74.
6. Souri K, Chae Y, Ponomarev Y, Makinwa KAA. A precision DTMOST-based temperature sensor. ESSCIRC. 2011;12–16:279–82.
7. Vence A, Chittori C, Bosi A, Nani C. A 0.076 mm2 12 b 26.5 mW 600 MS/s 4-way interleaved subranging SAR-$\Delta\sum$ ADC with on-chip buffer in 28 nm CMOS. IEEE J Solid State Circuits. 2016;51(12):2951–62.
8. Harpe P, Cantatore E, van Roermund A. An oversampled 12/14b SAR ADC with noise reduction and linearity enhancements achieving up to 79.1dB SNDR. ISSCC Dig Tech Papers, 2014, pp. 194–195.
9. Liu S, Shen Y, Zhu Z. A 12-Bit 10 MS/s SAR ADC With High Linearity and Energy-Efficient Switching. IEEE Trans Circuits Syst I, Reg Papers. 2016;63(10):1616–27.
10. Harpe P, Zhou C, van der Meijs NP, Wang X, Philips K, Dolmans G, de Groot H. A 26 μW 8-bit 10 MS/s asynchronous SAR ADC for low energy radios. IEEE J Solid State Circuits. 2011;46(7):1585–95.
11. Liu C-C, Chang S-J, Huang G-Y, Lin Y-Z, Huang C-M. A 1V 11fJ/conversion-step 10bit 10MS/s asynchronous SAR ADC in 0.18μm CMOS. Symp On VLSI Circuits, June 2010, pp. 241–2.
12. Tai H-Y, Hu Y-S, Chen H-W, Chen H-S. A 0.85fJ/conversion-step 10b 200kS/s Subranging SAR ADC in 40nm CMOS. ISSCC Dig. Tech. Papers, Feb. 2014, pp. 196–7.
13. Liu M, Pelzers K, van Dommele R, van Roermund A, Harpe P. A 106nW 10 b 80 kS/s SAR ADC with duty-cycled reference generation in 65 nm CMOS. IEEE J Solid-State Circuits. 2016;51(10):2435–45.
14. Liu M, van Roermund A, Harpe P. A 7.1-fJ/conversion-step 88-dB SFDR SAR ADC with energy-free "swap to reset". IEEE J Solid State Circuits. 2017;52(11):2979–90.
15. Liu M, van Roermund A, Harpe P. A 10b 20MS/s SAR ADC with a low-power and area-efficient DAC-compensated reference. ESSCIRC, Sep. 2017, pp. 231–4.
16. Schemmert W, Zimmer G. Threshold-voltage sensitivity of ion-implanted m.o.s. transistors due to process variations. Electron Lett. 1974;10:151–2.
17. Song B-S, Gray PR. Threshold-voltage temperature drift in ion-implanted MOS transistors. IEEE J Solid State Circuits. 1982;17(2):291–8.

18. van Elzakker M, van Tuijl E, Geraedts P, Schinkel D, Klumperink EAM, Nauta B. A 10-bit charge-redistribution ADC consuming 1.9μW at 1MS/s. IEEE J Solid State Circuits. 2010;45(5):1007–15.
19. Ivanov V, Brederlow R, Gerber J. An ultra-low power bandgap operational at supply from 0.75 V. IEEE J Solid State Circuits. 2012;47(7):1515–23.
20. Osaki Y, Hirose T, Kuroki N, Numa M. 1.2-V supply, 100-nW, 1.09-V bandgap and 0.7-V supply, 52.5-nW, 0.55-V subbandgap reference circuits for nanowatt CMOS LSIs. IEEE J Solid State Circuits. 2013;48(6):1530–8.
21. Shrivastava A, Craig K, Roberts NE, Wentzloff DD, Calhoun BH. A 32nW bandgap reference voltage operational from 0.5V supply for ultra-low power systems. ISSCC Dig Tech Papers, Feb. 2015, pp. 94–5.
22. Chen Y-J, Hsieh C-C. A 0.4V 2.02fJ/conversion-step 10-bit hybrid SAR ADC with time-domain Quantizer in 90nm CMOS. IEEE Symp VLSI Circuits, Jun. 2014, pp. 35–6.
23. Liou C-Y, Hsieh C-C. A 2.4-to-5.2fJ/conversion-step 10b 0.5-to-4MS/s SAR ADC with charge average switching DAC in 90nm CMOS. ISSCC Dig. Tech. Papers, Feb. 2013, pp. 280–1.
24. Harpe P, Cantatore E, van Roermund A. A 2.2/2.7fJ/conversion-step 10/12b 40kS/s SAR ADC with data-driven noise reduction. ISSCC Dig. Tech. Papers, Feb. 2013, pp. 270–1.
25. Harpe P, Gao H, van Dommele R, Cantatore E, van Roermund A. A 3nW signal-acquisition IC integrating an amplifier with 2.1 NEF and a 1.5fJ/conv-step ADC. ISSCC Dig. Tech. Papers, Feb. 2015, pp. 382–3.
26. Chang Y-K, Wang C-S, Wang C-K. A 8-bit 500 kS/s low power SAR ADC for bio-medical application. ASSCC Dig Tech Papers, Nov. 2007, pp. 228–31.
27. Liu C-C, Chang S-J, Huang G-Y, Lin Y-Z. A 10-bit 50-MS/s SAR ADC with a monotonic capacitor switching procedure. IEEE J Solid State Circuits. 2010;45(4):731–40.
28. Hariprasath V, Guerber J, Lee S-H, Moon U-K. Merged capacitor switching based SAR ADC with highest switching energy-efficiency. Electron Lett. 2010;46(9):620–1.
29. Lin Y-Z, Chang S-J, Shyu Y-T, Huang G-Y, Liu C-C. A 0.9-V 11-bit 25-MS/s binary-search SAR ADC in 90-nm CMOS. ASSCC Dig. Tech. Papers, Nov. 2011, pp. 69–72.

Ultra-Low-Power Clock Generation for IoT Radios

Ming Ding, Pieter Harpe, Zhihao Zhou, Yao-Hong Liu, Christian Bachmann, Kathleen Philips, Fabio Sebastiano, and Arthur van Roermund

1 Introduction

Many remote wireless-sensor-nodes (WSNs) in Internet-of-Things (IoT) applications are battery supplied. Since replacing the batteries frequently increases the cost and is not convenient or not possible for many applications, it is crucial to reduce the power consumption of the WSNs to extend battery lifetime. Wireless radios often consume in the order of a few mW or beyond and dominate the power consumption of the WSNs. In the past decade, significant efforts have been made to successfully reduce the power consumption of the wireless radios. For example, the power consumption of Bluetooth/Bluetooth Low Energy (BT/BLE) radios has been reduced by more than 10× in the past 10 years, down to only a few mW [1, 2]. However, this is not sufficient because the battery for IoT applications is often very small for better integration and, consequently, has a small capacity. As shown in Fig. 1, for a coin cell with 100 mAh capacity, the overall power consumption of a radio has to be as low as a few μW to enable a 10-year lifetime. This requires that power consumption of wireless radios should be reduced by more than 1000×

M. Ding (✉)
Holst Centre/imec, Eindhoven, Netherlands

Eindhoven University of Technology, Eindhoven, Netherlands
e-mail: Ming.Ding@imec-nl.nl

P. Harpe · A. van Roermund
Eindhoven University of Technology, Eindhoven, The Netherlands

Z. Zhou · F. Sebastiano
Delft University of Technology, Delft, The Netherlands

Y.-H. Liu · C. Bachmann · K. Philips
Holst Centre/imec, Eindhoven, Netherlands

K. A. A. Makinwa et al. (eds.), *Low-Power Analog Techniques, Sensors for Mobile Devices, and Energy Efficient Amplifiers*,
https://doi.org/10.1007/978-3-319-97870-3_5

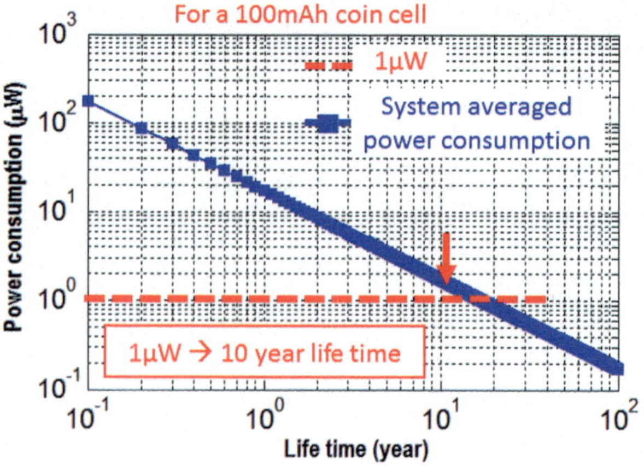

Fig. 1 Estimated relationship between system average power consumption and lifetime for a 100 mAh coin cell

without sacrificing their performances, which is very difficult. As a result, burst-mode operation has been introduced to reduce the average power consumption of wireless radios while maintaining the communication quality.

In a burst-mode operated radio, the radio is only activated when needed and disabled most of the time to save power. In this way, the total system power consumption is averaged down to the μW level, thus meeting the required battery lifetime. In some event-driven applications, the wake-up function is performed by a wake-up receiver [3, 4], while in many other duty-cycled IoT applications, the wake-up function is performed by a sleep timer [5–9]. In a duty-cycled IoT radio, there are three clocks in total as shown in Fig. 2: a system clock, a sleep timer, and a high-frequency oscillator, such as a DCO (digitally controlled oscillator) or a VCO (voltage-controlled oscillator). The system clock operates as the frequency reference for the phase-locked loop (PLL), as sampling clock for the data converters, and as clock for the digital circuits. The sleep timer is continuously running and performs the wake-up/sleep control. The DCO/VCO outputs an RF frequency and is less relevant for the duty-cycling operation. In this work, the performance requirements for system clock and sleep timer are discussed. In addition, the possible candidates for the clock generation circuits are also provided and discussed. Furthermore, two low-power circuit examples are provided, respectively: one system clock and one sleep timer.

2 Requirements for IoT Clock Generation Circuits

Overall, the system clock must be an accurate, low-noise, and stable frequency reference (∼MHz), which is used as reference clock for the phase-locked loop (PLL) to synthesize the RF carrier (Fig. 2) and to derive clocks for all other parts of the

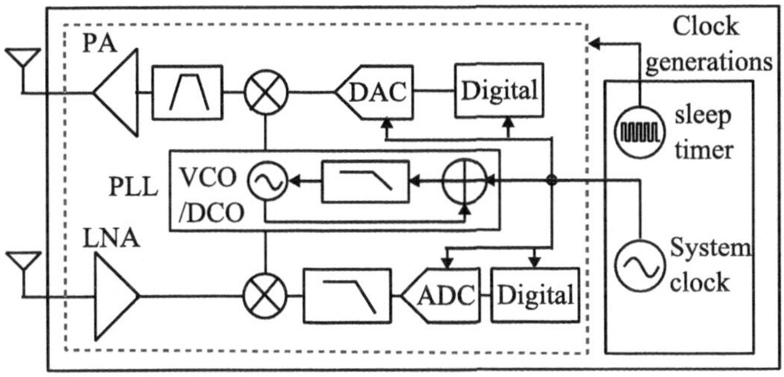

Fig. 2 Simplified block diagram for a wireless radio in IoT applications

transceiver SoC, e.g., the ADC and the digital baseband section. The sleep time is in the order of hundreds of ms or even longer, and therefore, the sleep timer requires low-power, long-term stability, and small frequency variations against temperature and supply variations. The detailed design considerations for the two circuits are discussed in this section.

2.1 Design Considerations for the Sleep Timer

As discussed, in many duty-cycled IoT radios, a sleep timer is always on to synchronize the transmitting and receiving burst. The frequency of the sleep timer is typically in the kHz range due to the long sleep time, and the power consumption has to be in sub-μW range to enable the long battery lifetime (Fig. 1). The main radio is disabled to save power when no data is transmitted and only active when receiving the wake-up signal generated by the sleep timer. In this case, it is important that the receiver and the transmitter are well synchronized; otherwise the transmitted information can be lost. In practice, any clock generation circuit, including the sleep timer, has frequency errors. Constant frequency error is relatively easy to correct. However, frequency errors due to jitter or frequency variations due to environmental changes are more difficult to tackle. As a result of the frequency error in the sleep timer, the wake-up signal might be generated too late. Consequently, the receiver will be switched on too late and fail to receive the transmitted packet. To avoid this situation, the receiver has to be activated for a longer time (the so-called guard time), ensuring that the transmitted packet can be received (Fig. 3). This, however, degrades the overall power efficiency. The average power consumption of a duty-cycled radio can be estimated as:

$$P_{\text{avg}} = \frac{P_{\text{slp}}T_{\text{slp}} + P_{\text{on}}(T_{\text{on}} + T_{\text{gt}})}{T_{\text{slp}} + T_{\text{on}} + T_{\text{gt}}}, \quad (1)$$

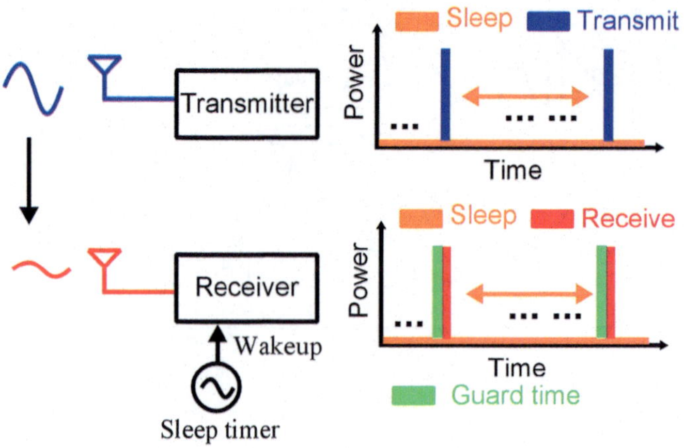

Fig. 3 Illustration of a duty-cycled transceiver with guard time

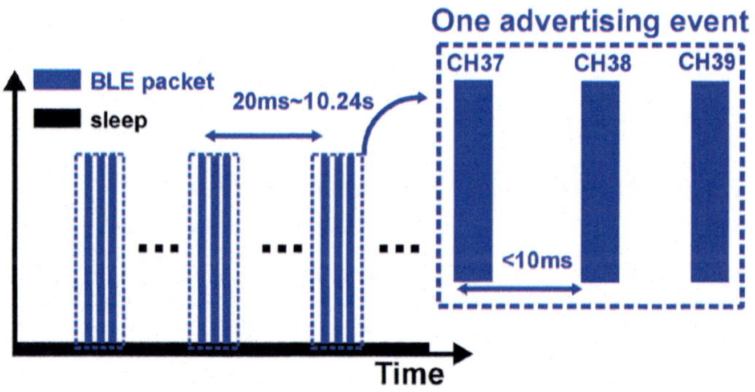

Fig. 4 One advertising event of a BLE radio

where P_{slp} and P_{on} are the power consumption in sleep state and active state and T_{slp}, T_{on}, and T_{gt} are the sleep time duration, the transmitting/receiving duration, and the guard time duration, respectively. To minimize the overall average power consumption, the power consumption and the frequency stability of the sleep timer should be properly optimized.

To have a better overview for the requirement of the sleep timer, a power consumption analysis of a popular BLE radio for IoT applications is provided. In a typical BLE advertising scenario, one sensor node is broadcasting as shown in Fig. 4, and the other nodes are listening without acknowledgment. The time interval between two consecutive advertising events ranges from 20 ms to 10.24 s, during which the main radio is switched off to save power. The main radio is only switched on during each advertising event, where the advertising packet is transmitted three

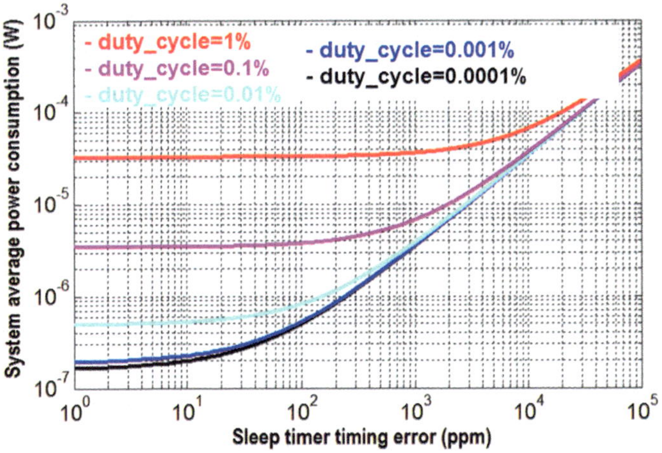

Fig. 5 The overall power consumption of a duty-cycled BLE radio [10] as function of the sleep timer timing accuracy for various duty-cycling ratios

times over three advertising channels, respectively. In sleep mode, the total power consumption of the RF transceiver comes from the sleep timer's power and the leakage of the whole chip. Noting that the leakage of the chip can be reduced down to a negligible level using techniques such as power gating [6], the sleep timer's power can be the biggest contributor to the total power budget in the sleep state.

As shown in Fig. 5, the system average power consumption can be reduced proportionally to the duty-cycling ratio up to a certain extent (0.01%). When the duty-cycling ratio further reduces below 0.01%, there is limited additional reduction of average system power. If the sleep timer is sufficiently accurate, with a very low duty-cycling ratio (0.0001%), the system average power consumption will be limited by the total power consumption of the chip blocks in the always-on domain, which consists of the sleep timer and the leakage of the chip. In practice, a guard time is required due to the sleep timer's timing inaccuracy (Fig. 3). As mentioned, the resulted guard time degrades the power efficiency of a duty-cycled radio. Therefore, a small timing error is crucial to reduce the overall power consumption as shown in Fig. 5. The more accurate the sleep timer is, the shorter the guard time will be. Since the sleep time duration is in the order of hundreds of mini-seconds or even longer, the long-term stability of the sleep timer (typically limited by the flicker noise) dominates the timing error. In addition, due to the long sleep time, the frequency of the sleep timer could fluctuate due to environmental changes (e.g., temperature, supply voltage). This also results in timing error for the sleep timer.

The key requirements for the sleep timer can be summarized as ultra-low-power (sub-μW), and small timing error ensured by good long-term stability and frequency stability over temperature and supply changes.

2.2 *Design Considerations for the System Clock*

To meet the requirements of IoT radios, high performance is desired for the PLL and thus also for the system clock. To meet the requirement of adjacent channel interference and also transmission modulation quality, low-phase noise is required. In a PLL, the phase noise of the system clock is up-converted and contributes to the total phase noise of the PLL (Fig. 6). The out-of-band phase noise of the PLL is usually dominated by the VCO, because the phase noise of the system clock at these frequencies is greatly attenuated. However, the in-band phase noise of the PLL can be dominated by the system clock because it cannot be attenuated. The contribution of the in-band PLL phase noise due to the system clock can be described by Eq. 2, in which $\mathrm{PN}_{\mathrm{pll,in-band}}$ and $\mathrm{PN}_{\mathrm{ref,in-band}}$ are the PLL and reference in-band phase noise, and N is the ratio between the PLL frequency and reference frequency. To ensure that the phase noise of the system clock will not dominate the PLL in-band phase noise, the frequency of the system clock is usually in the order of a few MHz to tens of MHz, so as to keep N in (2) below 100.

$$PN_{\mathrm{pll,in-band}} = PN_{\mathrm{ref,in-band}} + 20\log_{10}(N) \tag{2}$$

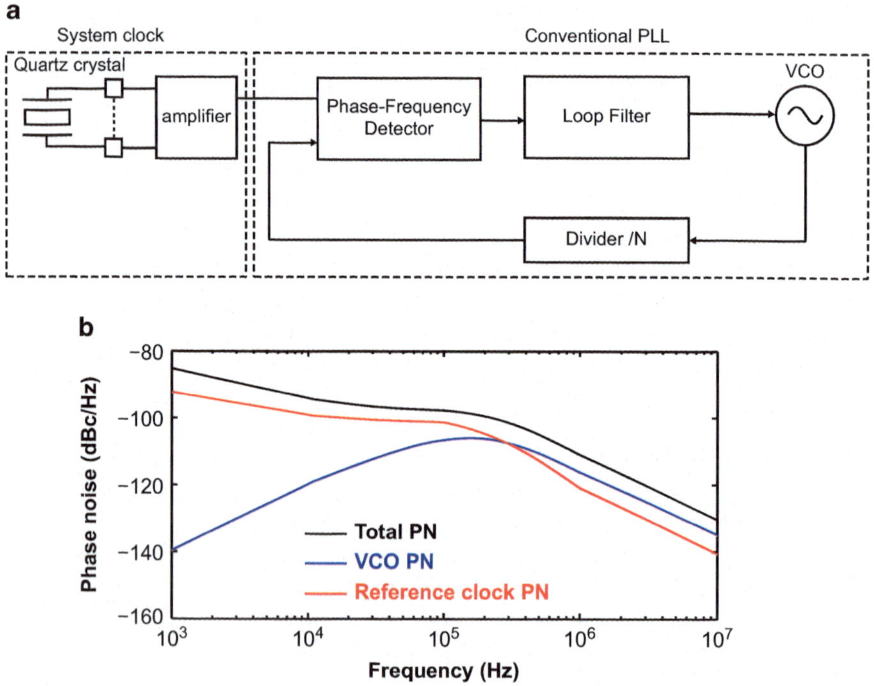

Fig. 6 Traditional PLL with system clock (**a**) and the phase noise distribution of the PLL (**b**)

In addition, the jitter of the system clock has to be sufficiently low to meet the requirement for the ADC sampling. The jitter of the system clock is calculated as the root-mean-square (rms) of the integrated phase noise. The jitter of the ADC sampling clock results in a sampled voltage error, thus degrading ADC performance. The impact of the system clock jitter on the ADC SNR can be described by Eq. 3, where $\mathrm{SNR}_{\mathrm{adc}}$ is the ADC signal-to-noise ratio, f_{in} is the frequency of the ADC input signal, and σ_{jitter} is the jitter of the system clock. For an ADC with 80 dB SNR and 1 MHz bandwidth, the required jitter for the sampling clock should be smaller than 10 ps. In many applications, this requirement can be easily achieved and is not a limiting factor.

$$SNR_{\mathrm{adc}} = -20\log_{10}(2\pi f_{\mathrm{in}}\sigma_{\mathrm{jitter}}) \tag{3}$$

Besides the phase noise and jitter requirement of the system clock, the frequency offset error of the system clock is also important. A frequency offset in the system clock results in a frequency error of the PLL. This is harmful because the transmitter and the receiver are expected to communicate in the same frequency channel. BLE standard requires that the transmission frequency error should be within ±41 ppm of the desired frequency [11]. Thus, this frequency offset error should be small and stable over environmental variations to ensure a robust communication. Furthermore, the power consumption of the system clock should be relatively small compared to the total power of the RF transceiver, which is in the order of a few mW [1, 2, 10].

In addition, to support the duty-cycling operation to reduce the average power consumption, additional challenges exist for the system clock. As mentioned, in a duty-cycled radio, the transceiver is only activated when transmitting or receiving and disabled to save power when not needed. This requires swift start-up behavior for the whole transceiver. This is not a problem for the transmitter/receiver, the phase-locked loop (PLL), and the power management unit (PMU), whose start-up times are typically below a few μs [10]. But this can be a problem for the system clock, since it often uses a crystal oscillator (XO) as shown later because of the high performance required. Due to the high-quality factor of the quartz crystal, the start-up time of the crystal oscillator is relatively long (namely, ~ms). In a duty-cycled application where the sleep time interval is similar to the XO start-up time, the XO can therefore not be switched on and off in time. Thus, it has to be always active, thereby significantly increasing the off-state power of the whole system significantly. In some duty-cycled applications, the XO can be switched on and off, but the extra power due to the start-up process cannot be neglected, e.g., in a duty-cycled radio with a relatively short packet length, the packet length can be as short as 128 μs in BLE and the power overhead due to the XO start-up process can go beyond 25% [12]. Therefore, a reduction of the start-up time of the XO is necessary, and at the same time, the energy overhead to enable a fast start-up should be minimized in order to reduce the overall energy consumption.

The key requirements for the system clock in a IoT radio are low-power, low-noise, stability, small frequency error, and fast and efficient start-up.

2.3 *Candidates for Clock Generation*

The typical performance of the system clock and the sleep timer is shown in Table 1. The frequency of the sleep timer is in the kHz range, with sub-μW power consumption and with relaxed noise and stability performance. It is continuously running. The frequency of the system clock is in the MHz range, with tight requirement for noise and stability, and also relatively higher power consumption. It is duty-cycled, and the start-up time of the system clock should be minimized to μs level to enable the burst-mode operation.

Once the requirements for the system clock and the sleep timer are clear, their circuit implementation should be chosen. The popular candidates for clock generation circuits are summarized in Table 2. Overall, they can be divided into six categories: LC-based, TD-based (thermal-diffusivity), RC-based, MOS-based, MEMS-based (microelectromechanical systems), and XO-based oscillators, which will be discussed below.

LC-based oscillators are widely used in RF circuits with GHz frequency, e.g., PLLs [10]. They consist of an on-chip LC resonator tank, which defines the resonance frequency, and an amplifying circuit, which compensates the loss in the LC tank and assures the oscillation. The typical frequency of LC-based oscillators is between few hundreds of MHz and few GHz. When implemented for lower-

Table 1 The typical performance for the system clock and the sleep timer

	System clock	Sleep timer
Frequency	~MHz	~kHz
Power consumption	Sub-mW	Sub-μW
Temperature stability	(1~tens of) ppm	(1~hundreds of) ppm
Jitter	~ps	ps~ns
Long-term stability	~ppb	~ppm
Start-up time	Tens~hundreds μs	–
Power state	Duty-cycled	Always on

Table 2 The popular architectures for clock generation circuits

	LC	TD*	RC	MOS	MEMS	XO
Frequency	GHz	MHz	kHz~MHz	Hz~GHz	kHz~GHz	kHz~MHz
CMOS process	☺	☺	☺	☺	☹	☹
Temp. stability	☹	😐	☹☹	☹☹☹	☺☺	☺☺
Power	1mW	1–10mW	<1mW	pW~mW	nW~mW	nW~mW
Noise	☹	😐	☹☹	☹☹	☺☺	☺☺☺
Off-chip components	No	No	No	No	Yes	Yes
Suitable for system clock	☹	☹	☹	☹	☺	☺
Suitable for sleep timer	☹	☹	☺	☺	☺	☺

frequency oscillators, the inductor size will be too large to be acceptable. Moreover, the quality factor of an on-chip inductor is usually small (~10) compared to that of a quartz crystal (tens of thousands). Consequently, the resulting phase noise is relatively poor. This is not a problem if it is used as the RF oscillator in a PLL, because its noise is greatly reduced by the closed-loop operation of the PLL and is usually not dominant, as shown in Fig. 6b. However, if it is used as the system clock, it will degrade the jitter performance of a PLL, because the in-band noise of the reference clock cannot be reduced at all. Therefore, it is not suitable for the system clock, and neither it is for the sleep timer.

TD-based oscillators utilize the effect of the thermal diffusivity of IC grade silicon to generate a well-defined frequency [13], where an FLL locks the frequency of the oscillator to a process-insensitive phase shift of an electrothermal filter. The TD oscillator frequency is in the MHz range with an acceptable chip area. However, it usually consumes more than 1 mW of power and has a poor jitter performance (~hundreds of ps), which is too high for a wireless radio. Therefore, it is also not preferred for the system clock nor the sleep timer.

RC-based oscillators can be fully integrated on-chip. The frequency of the oscillator is determined by the product of the resistor R and capacitor C value. The frequency range can vary from the kHz to the MHz range, with a corresponding power consumption from nW to mW. The disadvantage of this oscillator is that the noise performance is poor. As mentioned, the system clock requires low noise and high stability. This leads to a great challenge for on-chip RC oscillators, since the noise performance is several orders worse [14, 15] compared to that of a crystal-based oscillator [16]. To improve the performance, significant more power consumption (~mW) has to be put [17], which is too much for an IoT radio [10]. Therefore, an RC-based oscillator is not a good candidate for the system clock. However, it can be a candidate for the sleep timer, which has a relatively relaxed noise performance requirement.

A MOS-based oscillator can be integrated on-chip. Similar to the RC-based oscillator, the noise performance is not sufficient for system clock. Its frequency can range from Hz to GHz. Therefore, it can be a candidate for the sleep timer [5, 18, 19], but not for the system clock.

MEMS-based oscillators [20–22] show better performance than the aforementioned oscillators because they adopt an off-chip MEMS resonator with a high-quality factor (>10,000). As a sleep timer, the MEMS-based oscillators manage to achieve comparable frequency stability performance to crystal-based oscillator [20, 22] while consuming few μW power. As a system clock, the phase noise of the MEMS-based oscillators is much better than the aforementioned oscillators [21] but still lower than that of a crystal-based oscillator [16]. Therefore, a MEMS-based oscillator is suitable to be used as a sleep timer clock if the system allows to integrate an off-chip MEMS device. It can be used as a system clock in applications where the phase noise requirement is less critical.

XO-based oscillators [16, 23] have better phase noise performance compared to MEMS-based oscillators and are easily accessible. This makes them currently the best candidate for the system clock, which requires low-noise performance. The frequency range can vary from kHz to MHz, with a scaled power consumption from nW to hundreds of μW. This makes them possible for both the system clock and the sleep timer. At the moment, XO-based oscillators are the most popular choice for the system clock in an IoT radio as it is most mature and easily accessible. It can also be used as a sleep timer because of its good performance, if the system allows to integrate an additional off-chip quartz crystal.

For many IoT sensor nodes, it is preferred not to have two quartz crystals as it increases the cost. Therefore, in our case study, the system clock will use a crystal-based oscillator to assure sufficient noise performance for communication [16, 23], while the sleep timer uses a fully integrated RC-based oscillator with a relaxed performance.

3 Low-Power Sleep Timer

This section reviews the design of a low-power sleep timer, which was published in [24]. Aiming at designing a fully integrated ultra-low-power sleep timer operated at a low-supply voltage, a 0.7 V 0.43-pJ/cycle bang-bang digital-intensive frequency-locked loop (DFLL) for IoT applications is presented. First, the architecture of this design will be introduced. After that, the implementation details of the circuit will be provided, to show how the challenges are addressed. Measurement results will also be shown.

3.1 Introduction

As mentioned, for IoT applications with size, cost, and power limitations, RC oscillators are a preferred choice. However, convectional RC relaxation oscillators require continuous-time comparators, of which the delay is sensitive to PVT variations [6, 25, 26]. This limits their frequency stability. It has been shown in [13, 27] that frequency-locked loops (FLL)-based RC oscillator can overcome this limitation by avoiding the continuous comparator. However, it still requires analog-intensive circuits (e.g., operational amplifier), and thus, they are not friendly to technology scaling in terms of area and required supply voltage.

Alternatively, in this work, a digital-intensive FLL (DFLL) architecture is introduced for sleep timer, allowing low area, low power, and low-supply voltage.

3.2 Architecture of the Sleep Timer

Figure 7 shows the architecture of the proposed DFLL, and Fig. 8 illustrates its timing diagram. The DFLL consists of a frequency detector (FD), a dynamic

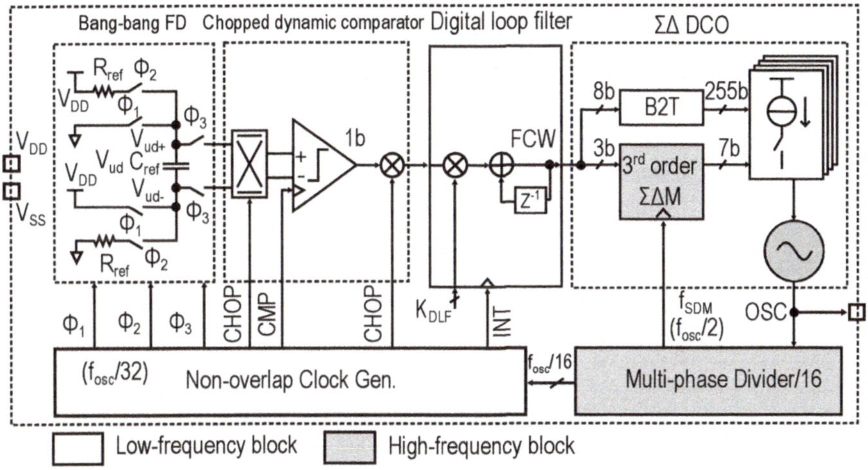

Fig. 7 Architecture of the proposed DFLL wakeup timer

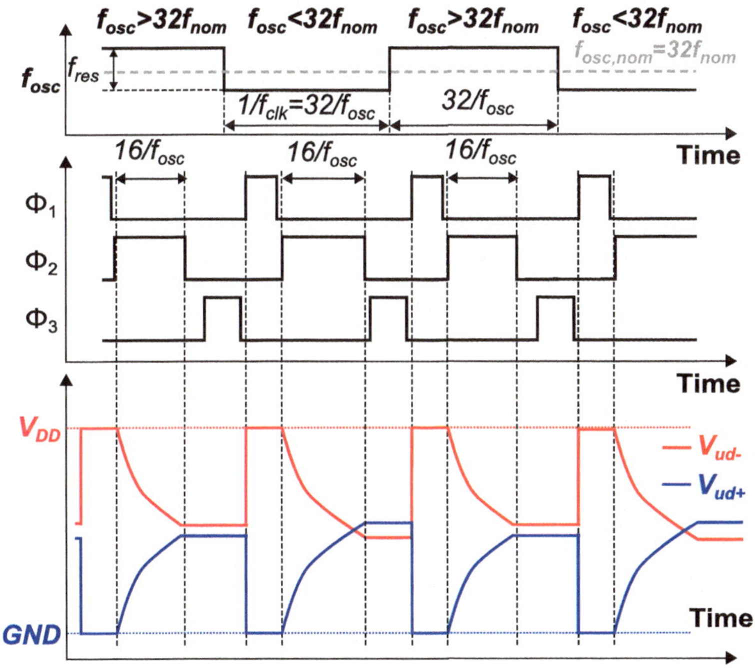

Fig. 8 Timing diagram of the DFLL and its frequency-locking behavior

comparator, a digital loop filter (DLF), a DCO, and two clock generation circuits. The FD, the comparator and the DLF operate at a clock frequency $f_{\text{clk}}=f_{\text{osc}}/32$ derived from the output frequency of the DCO f_{osc}. In the FD, during phase Φ_1, the capacitor C_{ud} is charged to $V_{ud} = V_{ud+} - V_{ud-} = V_{DD}$ and during phase

Φ_2 [27] discharged via resistors R_{ref}. At the end of Φ_2, the FD output voltage can be calculated as $V_{ud} = V_{DD}[1 - 2e^{-1/(4R_{ref}C_{ref}f_{clk})}]$. This equation indicates the relation between the clock frequency f_{clk} and the FD nominal frequency $f_{nom} = 1/(4R_{ref}C_{ref} \ln 2)$. For example, as shown in Fig. 8, when $f_{clk} < f_{nom}$, V_{ud} will be positive and otherwise negative. The sign of the FD output V_{ud} is detected by a dynamic comparator. The DLF accumulates the comparator output and generates the 11b output driving the DCO. When the DFLL reaches its stable state through the negative feedback loop, $V_{ud} = 0$ and $f_{osc}/32 = f_{nom}$.

3.3 Circuit Implementation Details

Thanks to the digital-intensive architecture, the only analog components in the DFLL are a switching passive RC network for the FD, a comparator and a DCO. Such analog circuits can be implemented using switches and inverter-based structures, so that they can be easily integrated in a nanometer CMOS process with a low-power consumption, a low-supply voltage, and a small area.

A multiphase clock divider generates all the clocks required in this self-clocked FLL from the DCO output (Fig. 8). In addition, a frequency division factor (32×) is adopted in the divider. Thanks to the multiple phases, this assures that Φ_2 and, consequently, the output frequency (f_{osc}) can be accurately set. Also, most of the circuit in Fig. 7 runs at a 32× lower frequency, which greatly reduces the power consumption. Moreover, by operating at a lower frequency, the design requirements for the analog circuits are relaxed. For example, a fixed and relatively long comparator delay ($\approx$4.8 μs) can be allowed compared to the ~ns delay of continuous-time comparators [6]. This enables the comparator to be optimized for power instead of speed.

According to $f_{osc,nom} = 32 f_{nom} = 8/(R_{ref}C_{ref} \ln 2) \approx 417$ kHz, the DFLL output frequency is set. $R_{ref} = 5.5$ MΩ and $C_{ref} = 5$ pF are chosen to minimize the required area. For the resistor R_{ref}, a series combination of non-silicided p-poly and n-poly resistors is used with opposite temperature coefficients (TC). This provides a first-order compensation of the TC of f_{osc} as shown later.

The comparator is implemented as a dynamic StrongARM latch to save power. However, the offset and flicker noise of the comparator can still result in a PVT-dependent frequency offset error and degraded long-term stability, respectively. To cope with these two effects, the dynamic comparator is chopped at a frequency of $f_{osc}/256$ and then processed by a DLF. The chopper consists of an analog and a digital modulator at the comparator input and output, respectively. Thanks to the single-bit comparator output, the DLF is implemented in a compact and low-power form by a bit-shifter and an up/down counter. In addition, the gain of the DLF (K_{DLF} in Fig. 7) can be programmed. This allows the possibility to easily configure and predict the overall bandwidth of the DFLL to flexibly trade-off bandwidth and noise for different IoT scenarios.

Due to the bang-bang operation of the DFLL, the DCO output frequency is continuously toggling in the steady state. Ideally, when the random noise in the

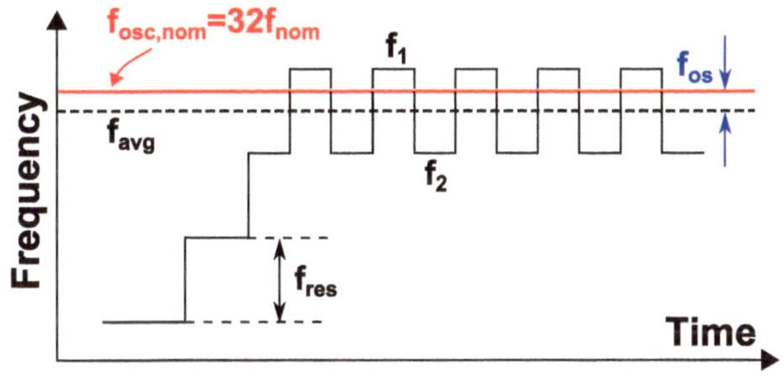

Fig. 9 Illustration of the frequency error due to the DCO finite resolution

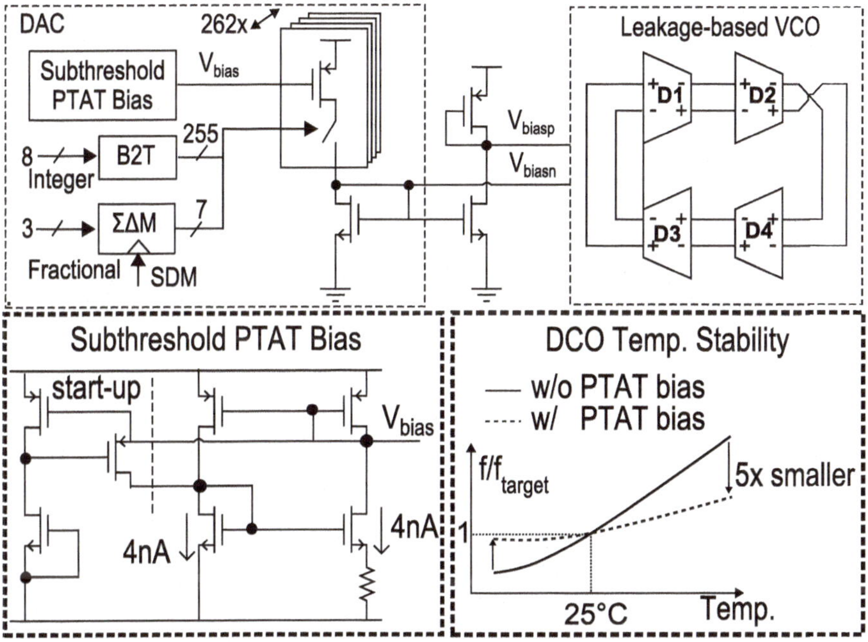

Fig. 10 Implementation of the self-biased Σ Δ DCO

loop is not considered, the DCO control word will toggle between two consecutive values as shown in Fig. 9. Since such a locking condition is satisfied for any $f_{\rm nom}$ between f_1 and f_2, this results in a frequency offset error f_{os} (Fig. 9). In the worst case, this frequency offset error $|f_{os}|$ is close to $\frac{f_1-f_2}{2} = \frac{f_{\rm res}}{2}$, where $f_{\rm res}$ is the DCO resolution. Although this inaccuracy can be partially mitigated by the dithering effect of random noise, it is important to improve the DCO resolution not to degrade the timer accuracy. Moreover, lower DCO resolution gives better long-term stability as shown in Fig. 11 [24]. Consequently, $f_{\rm res} = 250\,{\rm Hz}$ was chosen for the DCO.

Figure 10 shows the circuit of the DCO. An ultra-low-power leakage-based delay cell is used to construct a four-stage differential ring oscillator [28]. Thanks to

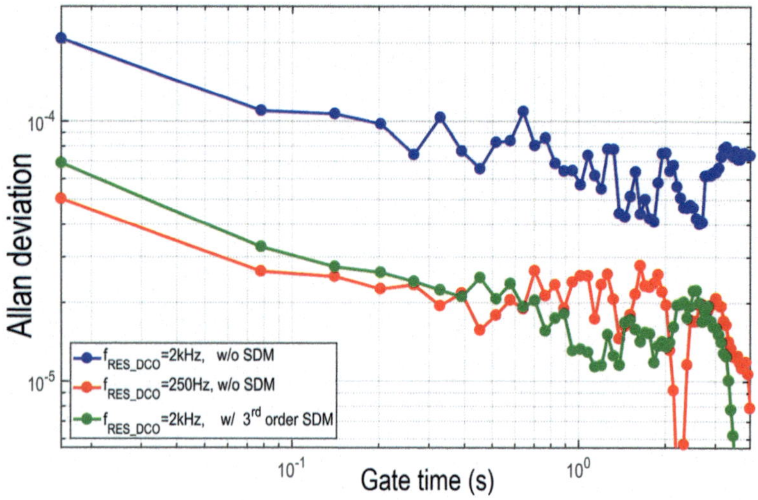

Fig. 11 Impact of DCO resolution on the Allan deviation

the low-power delay cell, the oscillator power consumption is kept below 60nW. However, the frequency drift of the nW oscillator over PVT is also quite large. For this reason, the DCO also requires a sufficient tuning range to tackle its frequency drift over PVT. As a result, a large dynamic range for the DCO is required. This is very challenging with the very limited power budget in the wake-up timer ($\ll$1 μW). To cope with this difficulty, a temperature compensation technique facilitated by a local proportional-to-absolute-temperature (PTAT) current bias is introduced (Fig. 10), which successfully reduces the temperature drift by 5× and so that smaller DCO range is required. In addition, a ΣΔ DAC is employed to improve the DCO resolution. The ΣΔ DAC consists of 255 + 7 = 262 unary elements: an integer 8-b thermometric DAC and a 3-b fractional thermometric DAC. The integer DAC is clocked at $f_{osc}/32$, while the fractional DAC is processed by a 3rd-order digital ΣΔ modulator clocked at $f_{osc}/2$. In this way, an oversampling ratio of 16× is achieved to further improve the DCO resolution from 2 kHz to below 250 Hz. The enhancement in resolution given by the ΣΔ operation improves the Allan deviation floor in the same way as a standard DCO with the same equivalent resolution, as illustrated in Fig. 11. Thanks to the feedback loop, no strict linearity requirements are required for the DAC other than the monotonicity necessary for loop stability, which is ensured by the unary-coded DAC.

3.4 Measurement Results

The prototype was fabricated in 40-nm CMOS technology, and it occupies 0.07-mm^2 (Fig. 12). Operated at 0.7 V, the DFLL consumes 259 nA at 417 kHz, whereas

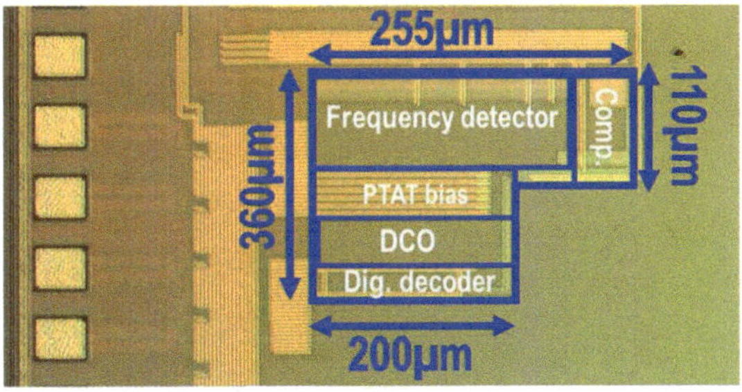

Fig. 12 Chip micrograph

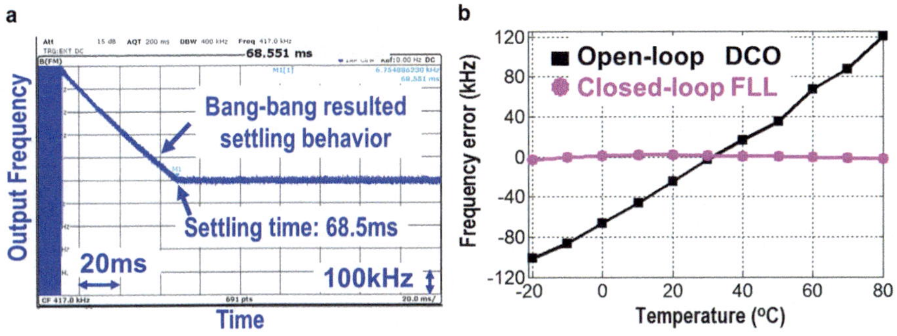

Fig. 13 Measured DFLL settling (KDLF = 1/8) (**a**) and open/closed loop performance (**b**)

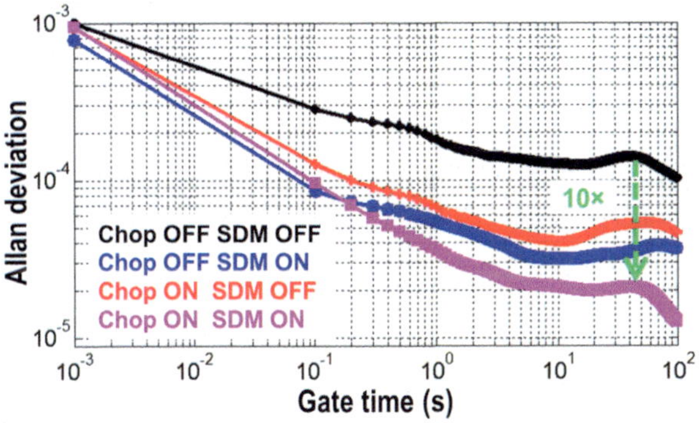

Fig. 14 Measured long-term stability

32% is consumed by FD/comparator, 38% by digital, and 30% by DCO. This corresponds to a state-of-the-art energy efficiency of 0.43 pJ/cycle.

The settling behavior of the timer is shown in Fig. 13a, where it can be clearly seen that the frequency increments or decrements toward the steady-state frequency due to the bang-bang operation. The measured DCO output frequency over temperature drift in open-loop and closed-loop configuration are compared as shown in Fig. 13a, which confirms the locking of the DFLL. As shown in Fig. 14, the measured long-term stability (Allan deviation floor) is improved by 10×, resulting

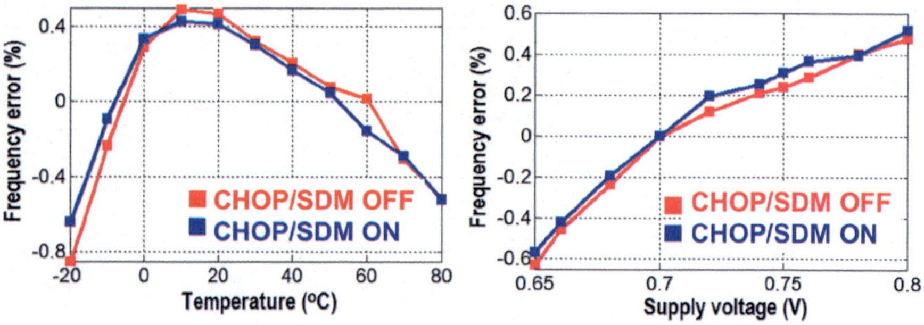

Fig. 15 Measured DFLL frequency stability against temperature and supply voltage variation

Table 3 Performance summary and comparison

	This work	Savanth ISSCC'17	Paidimarri JSSC'16	Jang ISSCC'16	Choi JSSC'16	Wang JSSC'16	Griffith ISSCC'14	Tokairin VLSI'12
Architecture	**DFLL**	Relaxation oscillator	Relaxation oscillator	Analog FLL	Analog FLL	Capacitive discharging	RC oscillator	Relaxation oscillator
Process (nm)	**40**	65	65	180	180	250	65	90
Frequency (Hz)	**417k**	1350k	18.5k	3K	70.4k	6.4k	33k	100k
VDD (V)	0.7	1.4	1	0.85–1.4	1.3	0.8	1.15–1.45	0.8
Power (nW)	**181**	920	130	4.7	110	75.6	190	280
Freq. Var. to VDD (%)	**±0.6 @ 0.65–0.8V**	±0.54@ 0.9–2.0V	<±0.25@ 0.95–1.05V	±0.14@ 0.85–1.4V	±0.23@ 1.2–1.8V	±0.27@ 0.6–0.9V	<±0.14@ 1.15–1.45V	±0.3@ 0.5–1.0V
TC (ppm/°C)	**106 @ -20–80°C**	96@ 0–145 °C	85@ -40–90 °C	13.8@ -25–85 °C	34.3@ -40–80 °C	144@ -20–80°C	38@ -20–90 °C	105@ -40–90°C
Allan deviation floor (ppm)	12 (>100s)	–	20(>100s)	63(>100s)	7(>12s)	60 (>100s)	4 (>2s)	–
Energy/Cycle (pJ/Cycle)	0.43	0.68	6.5	1.6	1.56	11.8	5.8	2.8
Area (mm^2)	**0.07**	0.005	0.032	0.5	0.26	1.08	0.015	0.12

in a 12 ppm beyond 100 s integration time. When enabling the chopping and the $\Sigma\Delta$ modulation, DCO resolution and the comparator offset-induced error are mitigated. As a result, the temperature sensitivity of the output frequency is improved from 134 to 106 ppm/°C. When operated from 0.65 to 0.8 V supply voltage, the measured frequency deviation is ±0.6% (Fig. 15). Although such temperature and supply sensitivities are sufficient for typical IoT applications and are on par with state-of-the-art designs (see Table 3), simulations show that they are limited by the on-resistance of the FD switches at such low supply, and this can be improved by proper redesign.

The performance of the timer is summarized and compared with other sub-μW state-of-the-art designs in Table 3 [6–8, 25–29]. This work achieves the best power efficiency (0.43 pJ/cycle) at the lowest operating supply voltage (0.7 V) among state-of-the-art sub-μW timing references.

4 Fast Start-Up Crystal Oscillator

This section reviews the design of a low-power fast start-up technique for crystal oscillators, which was published in [30]. Aiming to reduce both the start-up time and start-up energy, a dynamically adjusted load (DAL) is proposed with negligible overhead in power and area. First, the concept of the proposed DAL will be described and analyzed together with a few other prior works. After that, an implementation of the method will also be provided. Thanks to the proposed DAL, the negative resistance is boosted, achieving a >13 start-up time reduction and 6.9× start-up energy reduction and an overall power of 95 μW at 1 V.

4.1 *Background for Fast Start-Up Crystal Oscillator*

The start-up time of the crystal oscillator (T_s) is determined by many factors, among which the quality factor of the crystal oscillator is an important one. As shown in Fig. 16, the crystal quartz can be modeled with lumped L_m, C_m, R_m, and C_p as the motional inductor, capacitor, resistor, and parallel parasitic capacitor. The quality factor of a quartz crystal can be estimated as $\omega_{osc}L_m/R_m$, where ω_{osc} is the angular resonance frequency of the quartz crystal. Besides the quality factor of the crystal oscillator, there are a few more factors that can influence the start-up time of a crystal oscillator, as shown in Table 4. Some of them are fixed by the system (ω_{osc} and C_L) or the technology (V_{DD}),where C_L is the load capacitance, which equals to $C_1C_2/(C_1 + C_2)$, while others are not fixed and can be optimized for a shorter start-up time. In [16], the start-up time of a crystal oscillator was calculated as:

$$T_s = -\frac{2L_m}{R_m - R_N} ln\left(\frac{0.9\omega_{osc}(C_L + C_p)V_{DD}}{|i_{M(0)}|}\right), \tag{4}$$

where R_N is the negative resistance and $i_M(0)$ is the initial state current that flows into the quartz crystal at the beginning of the start-up. Over the past decades, efforts

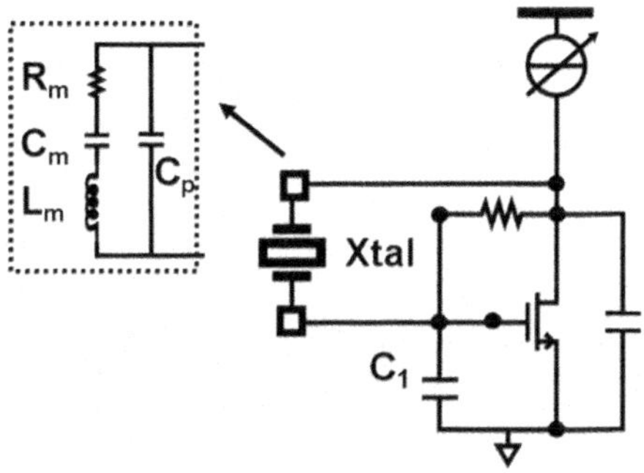

Fig. 16 A simplified crystal oscillator with a lumped model for the quartz crystal

Table 4 The factors that can influence the start-up time of a crystal oscillator

<table>
<tr><td></td><td colspan="2">Start-up time factors</td></tr>
<tr><td rowspan="4">Fixed by system requirements</td><td colspan="2">Quality factor ($\omega_{osc}L_M/R_M$)</td></tr>
<tr><td colspan="2">Frequency (ω_{osc})</td></tr>
<tr><td colspan="2">Supply voltage (V_{DD})</td></tr>
<tr><td colspan="2">Load capacitance (C_L)</td></tr>
<tr><td rowspan="3">Room for design</td><td rowspan="2">Negative resistance (R_N)</td><td>g_m</td></tr>
<tr><td>Initial load capacitance (C_L)</td></tr>
<tr><td colspan="2">Initial state current ($i_M(0)$)</td></tr>
</table>

to reduce the start-up time of a crystal oscillator have been put into two aspects, increasing g_m for higher $|R_N|$ and increasing $i_M(0)$, which we are going to be discussed respectively in the following.

4.1.1 Higher g_m Based Fast Start-Up Method

The negative resistance $|R_N|$ of a popular Pierce crystal oscillator can be calculated as [31]:

$$|R_N| = \frac{g_m}{(g_m C_p)^2 + \omega^2 (C_1 C_2 + C_2 C_p + C_1 C_p)^2}. \tag{5}$$

where g_m is the transconductance of the NMOS. To increase $|R_N|$, increasing g_m has been widely used [16, 32]. Higher g_m can be achieved in two ways: putting more bias current in the NMOS or reducing the gate length of the NMOS device. Since reducing the gate length of NMOS will increase the flicker noise, putting more bias current has been more popularly used. However, a higher bias current increases the current consumption at the same time. As a result, although the start-up time is reduced, the start-up energy, which is the product of start-up time and power consumption, is not reduced (Fig. 17). This is not desired for many IoT applications in which low power consumption is crucial.

4.1.2 Frequency Injection-Based Fast Start-Up Method

It has been shown that frequency injection method can effectively reduce the start-up time of a crystal oscillator by increasing $i_M(0)$ [12, 16, 33, 34]. In this method, certain energy is injected into the quartz crystal at the beginning of the start-up by adopting a separate oscillator (Fig. 18). However, it requires accurate calibration for

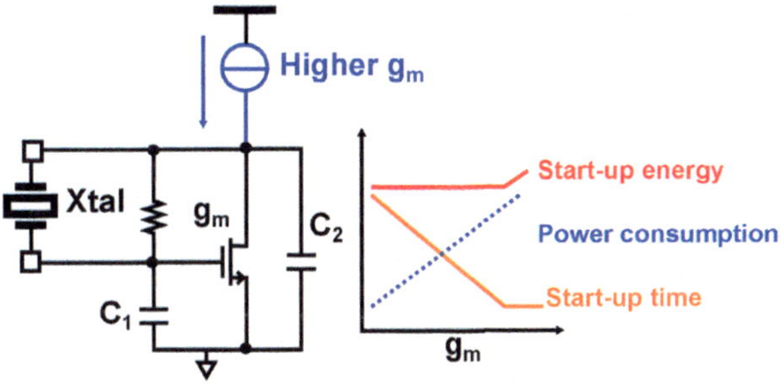

Fig. 17 Illustration of start-up time reduction techniques: increasing g_m

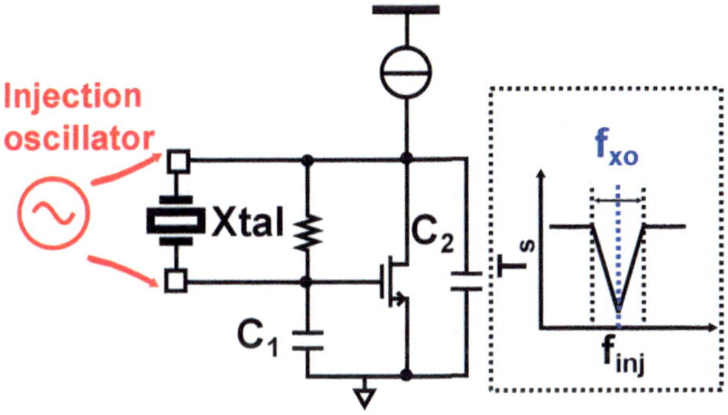

Fig. 18 Illustration of start-up time reduction techniques: increasing $i_M(0)$

the injection oscillator frequency and the injection time duration. Otherwise, the start-up time reduction will not be optimal.

Constant frequency injection (CFI) has been proven to effectively reduce the start-up time in [33]. A separate oscillator injects certain energy into the quartz crystal at the beginning of the start-up. The frequency of the injection oscillator is calibrated within the required accuracy using the crystal oscillator clock. This is not convenient for many applications since frequent calibration is needed to cope with the time-varying temperature-induced frequency drift.

To relax the frequency calibration requirement for the injection oscillator, a dithered frequency injection (DFI) method has been introduced [12]. Instead of injecting a single tone, the injection frequency is dithered around the crystal oscillator frequency with 1LSB of the injection oscillator tuning step. In this way, as long as the crystal oscillator frequency is within the dithered frequency window, the start-up time can be reduced. This is more robust against PVT variations compared to CFI. However, once the frequency variation exceeds the dithered frequency range, it cannot recover, and calibration for the injection oscillator is needed again.

To make the frequency injection method effective, not only the frequency of the injection oscillator has to be accurate, but the injection time duration also has to be accurately controlled [34]. If the injection time duration is too short or too long,

Table 5 Comparison of fast start-up techniques for crystal oscillators

Method	Start-up time reduction	Start-up energy	Robustness to PVT variations	Calibration overhead
Higher g_m	☺☺	😐	😐	😐
CFI/DFI/PTI	☺☺☺	☹	☹	☹☹/☹/☹
CI	☺	☹	☺	☺
Desired method	☺☺☺	☺	☺	☺

the start-up time cannot be optimally reduced. The work in [34] uses a constant frequency injection method with a precisely timed injection (PTI) control. Thus, it also suffers from the frequency variation due to PVT variations. In addition, the power consumption of the start-up circuits is too high (>10 mW), degrading the start-up energy.

To further relax the frequency calibration requirement of the injection oscillator to make it more robust against PVT variations, a chirp injection method was reported in [16]. It adopts a voltage-controlled oscillator (VCO) as the injection oscillator, of which the frequency is swept at the beginning of the start-up with an even larger range by tuning the control voltage of the VCO. In this way, it is much more robust against temperature and supply voltage variations compared to the abovementioned three-frequency injection techniques. However, by using the CI only, the reduction of the start-up time is very limited (2.3×). Moreover, due to the power hungry injection circuits, although the start-up time is reduced, the start-up energy is increased.

As shown in Table 5, the higher g_m method compromises between power consumption and start-up time. As a result, although the start-up time is reduced, the start-up energy is not reduced. The injection method is not very power efficient due to the extra injection oscillator. Besides, the calibration overhead cannot be neglected since the frequency of the injection oscillator is sensitive to PVT variations. Therefore, the requirements for the desired start-up method for low-power low duty-cycled IoT applications are effective reduction of start-up time and low-power, robust, and low calibration overhead.

4.2 *Proposed Method: Dynamically Adjusted Load (DAL)*

There are roughly two phases for the start-up process of a crystal oscillator: start-up phase and stable phase. In the start-up phase, the amplitude of the oscillator increases gradually. In the stable state, the performance of the crystal oscillator (e.g., frequency error, phase noise) should be sufficient for the RF blocks. In this work, to reduce both the start-up time and the start-up energy without affecting the performance in the stable phase, a dynamically adjusted load (DAL) method is proposed as follows.

The negative resistance R_N can be computed according to Eq. 5. Since for most quartz crystals, $C_{1(2)} \gg C_p$, R_N can be approximated by:

$$|R_N| \approx \frac{g_m}{(2\omega C_L)^2}, \tag{6}$$

where ω is the oscillation angular frequency and C_L equals $0.5C_{1(2)}$. It is obvious that R_N is approximately quadratic to $1/C_L$. Thus, reducing C_L seems very attractive to obtain a high $|R_N|$ at start-up. In addition, according to [31], the required minimum g_m for a crystal oscillator to start-up is proportional to $R_m(2\omega C_L)^2$. This indicates that for crystal oscillators with smaller C_L, less power is sufficient to maintain oscillation. Therefore, in the start-up phase, a smaller C_L is desired for both start-up time reduction and low power consumption.

In the contrary to the start-up phase, in the stable phase, a large load capacitance C_L is desired, as it gives more stable operation. According to [31], the frequency pulling factor ($\sim \frac{C_m}{2C_p+2C_L}$) indicates the difference between the real oscillation frequency and the intrinsic resonant frequency of a quartz crystal. A large C_L results in a small frequency pulling factor and thus a small frequency error, which makes the crystal oscillator less sensitive to environmental variations.

In this work, to optimize both the performance at start-up phase and stable phase, an XO with a dynamically adjusted load (DAL) technique is proposed, minimizing the C_L in the start-up phase at first for fast start-up and thereafter incrementing C_L for stable operation in the stable phase. Thanks to the proposed method, both the start-up time and energy are greatly reduced, while the overhead is minimized as shown later.

4.3 Circuit Implementation

Figure 19 shows the block diagram of the fast start-up XO with the proposed DAL as well as an illustration of its start-up behavior in the time domain. The core of the oscillator is a Pierce oscillator, and the bias current can be reconfigured to achieve various g_m values. The load capacitors are implemented through two switched capacitor banks ($C_{1(2)}$).

The proposed dynamically adjusted load method works as follows (Fig. 19). $C_{1(2)}$ is minimized to boost the negative resistance R_N, facilitating a fast start-up. A feedback loop, which includes an envelope detector, a comparator, and a finite-state machine (FSM), is continuously monitoring the amplitude of the oscillator. Once the crystal oscillator has a sufficient output swing, a finite-state machine (FSM) will automatically increment $C_{1(2)}$ to the nominal value. During this start-up process, the frequency of the crystal oscillator first deviates from the nominal frequency (maximum hundreds of ppm as shown later) because of the frequency pulling effect due to the varying capacitor value ($C_{1(2)}$) and settles to the desired value eventually. All the digital circuits are clocked at a relatively lower frequency (divide by four),

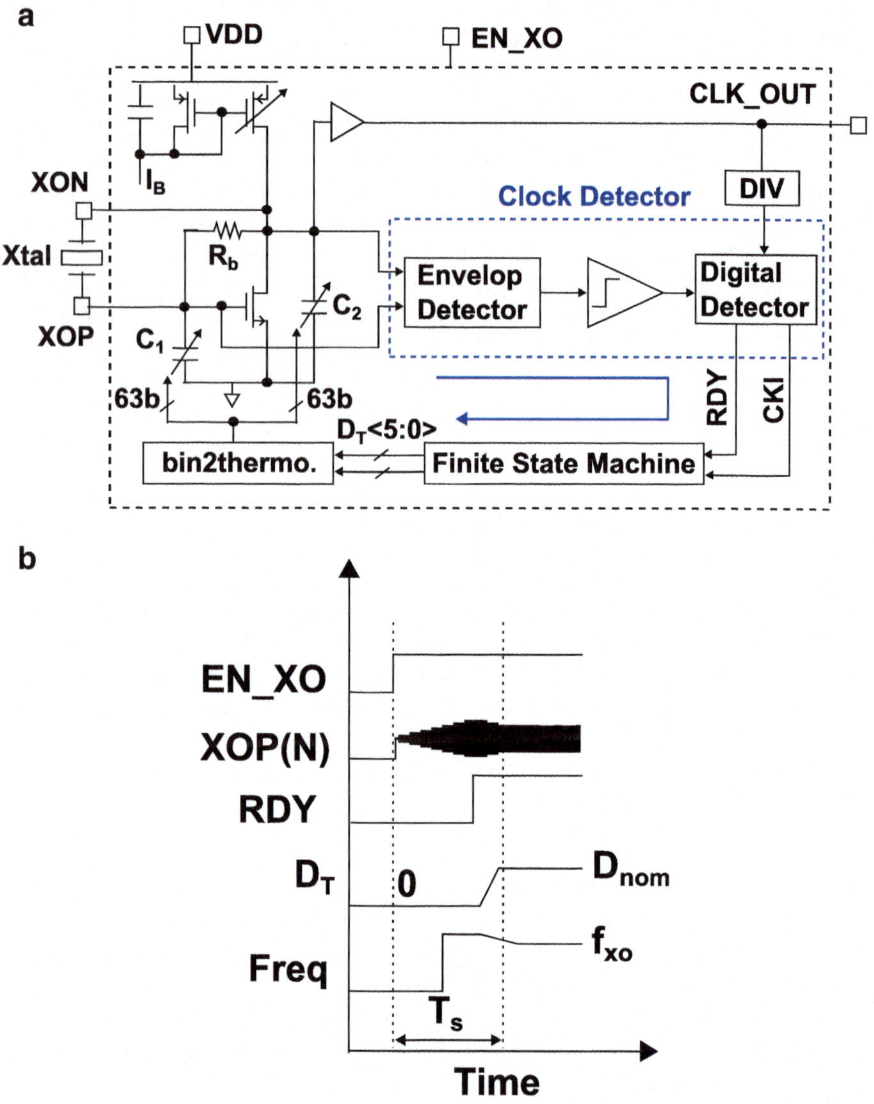

Fig. 19 Block diagram of the crystal oscillator with autonomous dynamically adjusted load (DAL) (**a**) and an illustration waveform (**b**)

and the 6b capacitor banks are thermometer-encoded, to assure a smooth transition when the load capacitors are varying.

The whole start-up process is fully autonomous without requiring an extra oscillator nor sequence, making it possible to implement the DAL method with only an enable signal. In addition, the proposed DAL is compatible with the implementation of a digitally controlled XO (DCXO) thanks to the switched capacitor banks. These features are desired for many IoT applications.

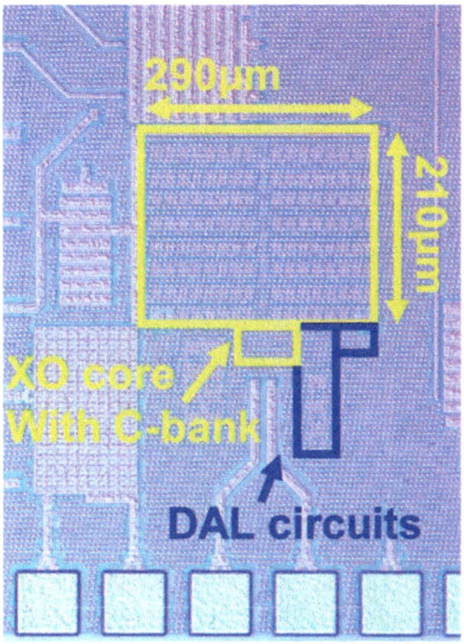

Fig. 20 Chip photo

4.4 Measurement Results

The prototype of the test chip is implemented in 90 nm LP CMOS as shown in Fig. 20. The core area is 0.072 mm^2, including the load capacitor (84%) and the feedback loop to implement the DAL method (10%). A 24 MHz quartz crystal is used to verify the functionality of the chip. At 1 V, the chip consumes 95 µW in the steady state, and the circuits to implement the DAL method consume 9 µW extra at the start-up.

Figure 21 shows the measured amplitude (left part) and frequency (right part) of the XO output CLK_OUT during the start-up. The start-up time is defined as the duration for the frequency to settle within ±20 ppm from the target frequency, which is well within the requirement of many IoT standards, e.g., BLE (±41 ppm) and IEEE802.15.4 (±40 ppm) [11]. Without any start-up technique, the oscillator takes 2.66 ms to start up. With the DAL technique only, 7× T_s reduction start-up time is achieved. Thanks to the low-power start-up circuits, the start-up energy is reduced with a similar factor (6.5×). By slightly increasing the power to 146 µW for higher g_m, the start-up time reduction can be further improved to 13.3× T_s.

In addition, the start-up time measurements are repeated with temperature variations, supply voltage variations, and load variations (Fig. 22). Relatively stable performance is achieved for temperature variation (13%) and supply voltage variation (66%). It also achieves very stable start-up performance (3%) against load capacitance variation.

This work achieves a state-of-the-art start-up time reduction ratio (13.3×) while showing the best start-up energy reduction ratio (6.9×) among prior art (Table 6). In addition, the overhead of the DAL circuitry in area and power consumption is negligible since no extra injection oscillator is needed. More importantly, the

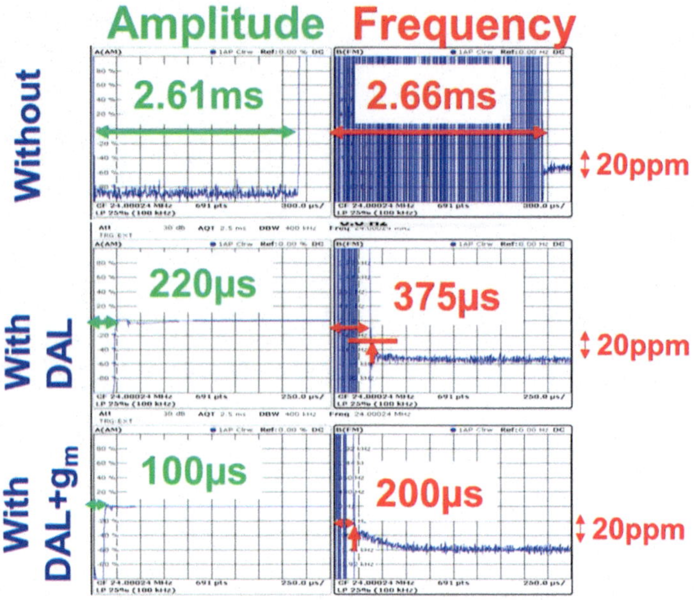

Fig. 21 Measured start-up behavior in three scenarios: without, with DAL, and with DAL and gm

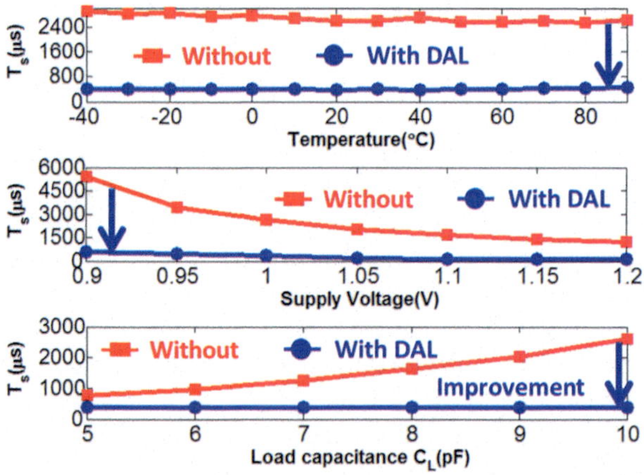

Fig. 22 Measured start-up time with respect to temperature, supply voltage, and load capacitance variations

whole circuit is operating fully autonomous with only an enable signal, avoiding an additional calibrated higher frequency oscillator or a foreground control sequence, reducing the system complexity significantly.

Table 6 Performance summary and comparison with state-of-the-art

	JSSC'12	VLSI'14		ISSCC'16		This work	
Technology (nm)	65	180		65		90	
Core area (mm^2)	0.15	0.12		0.08		0.072	
Supply (V)	1	1.5		1.68		1.0	
Frequency (MHz)	26	39		24		24	
Load capacitance, C_L (pF)	8	6		6	9	10	
Steady state power (μW)	2180	181		393	693	95	
Start-up time (μs)	3200	880	158	64	435	375	200
Total start-up energy (nJ)	6976	453	434	–[b]	–[b]	39	36.7
Start-up time reduction ratio	1×	2.3×	13.3×	6.7×	6.2×	7×	13.3×
Start-up energy reduction ratio	1×	0.84×	0.88×	–[b]	–[b]	6.5×	6.9×
Area of start-up circuits (mm^2)	–	0.1		0.017		0.0073	
Energy of start-up circuits (nJ)	–	294[a]		–[b]		3.3[c]	
T_s variation with temp.	–	<31%	<7%	±35%	±20%	26.6%	27.5%
Temperature range (°C)	–	−30 to 125	−30 to 125	−40 to 90	−40 to 90	−40 to 90	
Fast start-up technique	None	CI only	CI + g_m	Dithered injection		DAL	DAL + g_m
With an separate oscillator	–	Yes		Yes		No	

[a]Chirp injection circuits energy
[b]Injection circuits energy not provided
[c]DAL circuits energy

5 Conclusions

In this work, the requirements and challenges of the clock generation circuits for IoT applications have been discussed. To enable μW system average power consumption by aggressive duty-cycling, it requires a ultra-low-power sleep timer with good frequency stability and a low-power system clock with very short start-up time. In addition, two low-power clock generation circuit examples are provided: a ultra-low-power DFLL-based sleep timer and a low-power fast start-up technique for crystal oscillators.

References

1. Liu H et al. An ADPLL-centric bluetooth low-energy transceiver with 2.3mW interference-tolerant hybrid-loop receiver and 2.9mw single-point polar transmitter in 65nm CMOS. In: IEEE ISSCC Digest of Technical Papers; San Francisco 2018. p. 444–5.
2. Ding M et al. A 0.8V 0.8mm^2 bluetooth 5/BLE digital-intensive transceiver with a 2.3mW phase-tracking RX utilizing a hybrid loop filter for interference resilience in 40nm CMOS. In: IEEE ISSCC Digest of Technical Papers; San Francisco 2018. p. 446–7.
3. Jiang H et al. A 4.5nW wake-up radio with −69dBm sensitivity. In: IEEE ISSCC Digest of Technical Papers; San Francisco 2017. p. 416–7.
4. Ding M, et al. A 2.4GHz BLE-compliant fully-integrated wakeup receiver for latency-critical IoT applications using a 2-dimensional wakeup pattern in 90nm CMOS. In: IEEE RFIC; Hawaii 2017. p. 168–71.
5. Griffith D et al. A 190nW 33kHz RC oscillator with 0.21% temperature stability and 4ppm long-term stability. In: IEEE ISSCC Digest of Technical Papers; San Francisco 2014. p. 300–1.
6. Paidimarri A, et al. A +10dBm 2.4GHz transmitter with sub-400pW leakage and 43.7% system efficiency. In: IEEE ISSCC Digest of Technical Papers; San Francisco 2015. p. 246–7.

7. Savanth T et al. A 280nW, 100kHz, 1-cycle start-up time, on-chip CMOS relaxation oscillator employing a feedforward period control scheme. In: IEEE VLSI Symposium; Hawaii 2016. p. 16–7.
8. Wang H et al. A reference-free capacitive-discharging oscillator architecture consuming 44.4pW/75.6nW at 2.8Hz/6.4kHz. IEEE J Solid-State Circuits. 2016;51(6):1423–35.
9. Drago S et al. Impulse-based scheme for crystal-less ULP radios. IEEE Trans Circuits Syst I. 2009;56(5):1041–52.
10. Liu YH et al. A 3.7mW-RX 4.4mW-TX fully integrated bluetooth low-energy/ IEEE802.15.4/proprietary SoC with an ADPLL-based fast frequency offset compensation in 40nm CMOS. In: IEEE ISSCC Digest of Technical Papers; San Francisco 2015. p. 236–7.
11. B. S. I. G. (SIG). 2016 Specification of the bluetooth system, core package version 5.0, bluetooth specifications, bluetooth special interest group (sig). [Online]. Available: https://www.bluetooth.org/en-us/specification.
12. Griffith D et al. A 24MHz crystal oscillator with robust fast start-up using dithered injection. In: IEEE ISSCC Digest of Technical Papers; San Francisco 2016. p. 104–5.
13. Kashmiri SM, Pertijs MAP, Makinwa KAA. A thermal-diffusivity-based frequency reference in standard CMOS with an absolute inaccuracy of ±0.1% from −55°C to 125°C. IEEE J Solid-State Circuits. 2010;45(12):2510–20.
14. Cao Y, Leroux P, Cock WD, Steyaert M. A 63,000 Q-factor relaxation oscillator with switched-capacitor integrated error feedback. In: IEEE ISSCC Digest of Technical Papers; San Francisco 2013. p. 186–7.
15. Gurleyuk C et al. A CMOS Dual-RC frequency reference with ±250ppm inaccuracy from −45°C to 85°C. In: IEEE ISSCC Digest of Technical Papers; San Francisco 2018. p. 54–5.
16. Iguchi S et al. Variation-tolerant quick-start-up CMOS crystal oscillator with chirp injection and negative resistance booster. IEEE J Solid-State Circuits. 2016;51(2):496–507.
17. Satoh Y, Kobayashi H, Miyaba T, Kousai S. A 2.9mW, +/−85ppm accuracy reference clock generator based on RC oscillator with on-chip temperature calibration. In: VLSI Symposium; Hawaii 2014. p. 1–2.
18. Sebastiano F et al. A 65-nm CMOS temperature-compensated mobility-based frequency reference for wireless sensor networks. IEEE J Solid-State Circuits. 2011;46(7):1544–52.
19. Sebastiano F et al. Mobility-based time references for wireless sensor networks. In: Ismail M, Sawan M, editors. Analog circuits and signal processing. New York: Springer; 2013.
20. Ruffieux D, Pengg F, Scolari N, Giroud F, Severac D, Le T, Piazza SD, Aubry O. A $3.2\times1.5\times0.8mm^3$ 240nA 1.25-to-5.5V 32kHz-DTCXO RTC module with an overall accuracy of ±1ppm and an all-digital 0.1ppm compensation-resolution scheme at 1Hz. In: IEEE ISSCC Digest of Technical Papers; San Francisco 2016. p. 208–9.
21. Perrott M et al. A temperature-to-digital converter for a MEMS-based programmable oscillator with < ±0.5-ppm frequency stability and < 1-ps integrated jitter. IEEE J Solid-State Circuits. 2013;48(1):276–91.
22. Zaliasl S et al. A 3 ppm $1.5\times0.8mm^2$ 1.0μA 32.768kHz MEMS-based oscillator. IEEE J Solid-State Circuits. 2015;50(1):291–302.
23. Griffith D et al. A 37μW dual-mode crystal oscillator for single-crystal radios. In IEEE ISSCC Digest of Technical Papers; San Francisco 2015. p. 104–5.
24. Ding M et al. A 0.7-V 0.43-pJ/cycle wakeup timer based on a bang-bang digital-intensive frequency-locked-loop for IoT applications. IEEE Solid-State Circuits Lett (SSCL). 2018;1(2):30–3.
25. Koo J et al. A quadrature relaxation oscillator with a process-induced frequency-error compensation loop. In: IEEE ISSCC Digest of Technical Papers; San Francisco 2017. p. 94–5.
26. Savanth A et al. A 0.68nW/kHz supply-independent relaxation oscillator with 0.49% /V and 96ppm/°C stability. In: IEEE ISSCC Digest of Technical Papers; San Francisco 2017. p. 96–7.

27. Lee J et al. A 4.7MHz 53μW fully differential CMOS reference clock oscillator with 22dB worst-case PSNR for miniaturized SoCs. In: IEEE ISSCC Digest of Technical Papers; San Francisco 2015. p. 106–7.
28. Jang T et al. A 4.7nW 13.8ppm/ °C self-biased wakeup timer using a switched-resistor scheme. In: IEEE ISSCC Digest of Technical Papers; San Francisco 2016. p. 102–3.
29. Choi M et al. A 110nW resistive frequency locked on-chip oscillator with 34.3ppm/ °C temperature stability for system-on-chip designs. IEEE J Solid-State Circuits. 2016;51(9):2106–18.
30. Ding M, et al. A 95μW 24MHz digitally controlled crystal oscillator for IoT applications with 36nJ start-up energy and >13× start-up time reduction using a fully-autonomous dynamically-adjusted load. In: IEEE ISSCC Digest of Technical Papers; San Francisco 2017. p. 90–1.
31. Vittoz EA, Degrauwe MGR, Bitz S. High-performance crystal oscillator circuits: theory and application. IEEE J Solid-State Circuits. 1988;23(3):774–83.
32. Karthaus U. A differential two-pin crystal oscillator-concept, analysis, and implementation. IEEE Trans Circuits Syst II. 2006;53(10):1073–77.
33. Kwon Y-I, Park S-G, Park T-J, Cho K-S, Lee H-Y. An ultra low-power CMOS transceiver using various low-power techniques for LR-WPAN applications. IEEE Trans Circuits Syst I. 2012;59(2):324–36.
34. Esmaeelzadeh H, Pamarti S. A precisely-timed energy injection technique achieving 58/10/2μs start-up in 1.84/10/50MHz crystal oscillators. In: IEEE Custom Integrated Circuits Conference (CICC); Austin, USA 2017, p. 1–4.

Low-Power Resistive Bridge Readout Circuit Integrated in Two Millimeter-Scale Pressure-Sensing Systems

Sechang Oh, Yao Shi, Gyouho Kim, Yejoong Kim, Taewook Kang, Seokhyeon Jeong, Dennis Sylvester, and David Blaauw

1 Introduction

Pressure-sensing systems have applications in a broad range of fields, such as automotive, industrial [1, 2], and medical [3, 4]. Each application is associated with a different pressure range. For medical applications, the pressure is typically low to medium (1–100 kPa), whereas the hydraulic pressure in automotive and industrial applications can be as high as 100 MPa. Piezoresistive strain gauges, which are often configured as a Wheatstone bridge (Fig. 1), are most commonly used to sense pressure because of their simplicity, high sensitivity, and low cost [5]. Various piezoresistive bridge sensors that have diverse pressure ranges with similar resistance are available, and hence the same sensor readout circuit can be used for various applications. When the external pressure changes, the sensor transduces the mechanical strain into a resistance change of the bridge sensor. Their resistance is typically 1–10 kΩ and their size is approximately 0.5–3 mm.

Recently, small Internet of Things (IoT) systems [6–9] have become popular, and incorporating pressure sensing into these devices would open up many new applications. One unique element of millimeter-scale systems is that they have a very small battery. Typically, these small batteries have a capacity in the range of 10 μAh, and their internal resistance can be 10 kΩ [10]. A common method to realize a digital output from the piezoresistive MEMS bridge involves biasing the bridge with a DC voltage source and using a low-noise amplifier followed by an ADC. This bridge measurement is very power hungry because of the low bridge resistance. Moreover, critical battery IR drop will prevent the circuit from being functional, because the small battery resistance is comparable with the bridge resistance.

S. Oh · Y. Shi · G. Kim · Y. Kim · T. Kang · S. Jeong · D. Sylvester · D. Blaauw (✉)
University of Michigan, Ann Arbor, MI, USA
e-mail: blaauw@umich.edu

K. A. A. Makinwa et al. (eds.), *Low-Power Analog Techniques, Sensors for Mobile Devices, and Energy Efficient Amplifiers*,
https://doi.org/10.1007/978-3-319-97870-3_6

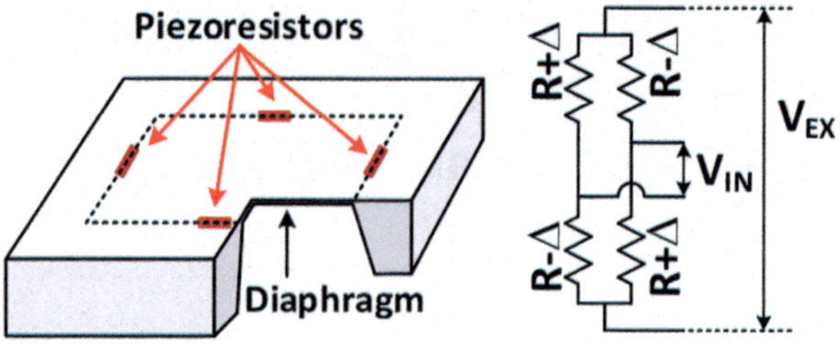

Fig. 1 Piezoresistive MEMS bridge sensor

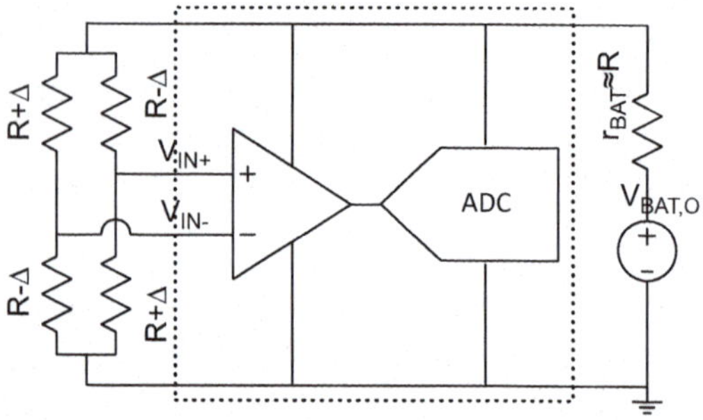

Fig. 2 Conventional bridge readout interface

Therefore, a pressure readout circuit needs to have low power consumption to enable long battery lifetime, and it must maintain a low instantaneous current to avoid dangerous battery IR drop.

In the following sections, a low-power resistive bridge readout circuit is introduced and demonstrated in millimeter-scale systems. Section 2 introduces a conventional bridge readout interface circuit and discusses problems associated with employing it in small systems. Section 3 introduces the proposed duty-cycled bridge-to-digital converter (BDC) and describes the detailed design implementation. Section 4 discusses the BDC measured results. Millimeter-scale system integration for barometric pressure range and hydraulic pressure range are described in Sects. 5 and 6, respectively. Finally, Sect. 7 concludes the chapter.

2 Conventional Bridge Readout Interface Circuits

Figure 2 shows a conventional BDC scheme. An external DC bias excitation voltage is provided to the bridge, and a small differential voltage proportional to the pressure is built up. The differential input voltages are amplified in the preamp and digitized in the ADC [4, 11–13]. A conversion time in milliseconds is desirable to achieve the proper circuit power level when excluding the bridge. However, the excitation current is high because of the low bridge resistance, and combined with the millisecond conversion time, the BDC will consume micro joules per conversion, which is quite high for mm-scale IoT systems. In addition, the battery resistance of these small systems is comparable to the resistance of the bridge. Therefore, there will be a very large voltage drop, approximately half of the battery open circuit voltage, during bridge biasing. Both the high energy and large voltage drop make it unsuitable as a sensing interface in miniaturized microsystems.

3 Duty-Cycled Bridge-to-Digital Converter

Duty-cycled excitation was proposed in [2] to reduce power in moderate dynamic range (DR) applications, lowering bridge excitation by up to 125× compared to static biasing. However, the excitation energy consumption (~250 nJ) is still much larger than the interface circuit conversion energy and therefore limits overall sensor energy efficiency. To address this challenge, we propose a BDC that uses an extremely duty-cycled excitation [14] (Fig. 3). The BDC energy consumption is lowered by having a very short excitation, and the sampling switches are ON only during the 170 ns. Instantaneous battery current is reduced by using an on-chip decoupling capacitor. In the sampling phase, V_{IN} and V_{EX} are placed on the capacitors, and the large current of the bridge sensor is drawn from the decoupling capacitor instead of from the battery. In the A-to-D phase, the ADC operates based on the sampled voltages, and the decoupling capacitors are recharged from the battery. As a result, the proposed BDC consumes 6000× less excitation energy than conventional DC biasing and maintains low instantaneous battery current.

3.1 *Sampling Pulse Generation*

Figure 4 shows a more detailed BDC circuit implementation. A sampling pulse generator drives the sensor excitation and sampling switches. The design samples not only the inputs ($V_{IN+/-}$) but also V_{EX} since this voltage will differ from the battery voltage. The values of C_S and SPL pulse width are determined by the input resolution requirement. We target 200 μV $V_{IN+/-}$ resolution at 3.6 V V_{EX}. C_S is

Fig. 3 Duty-cycled bridge-to-digital converter

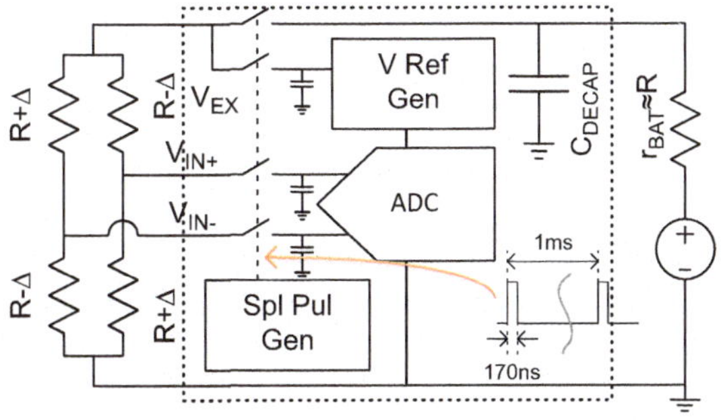

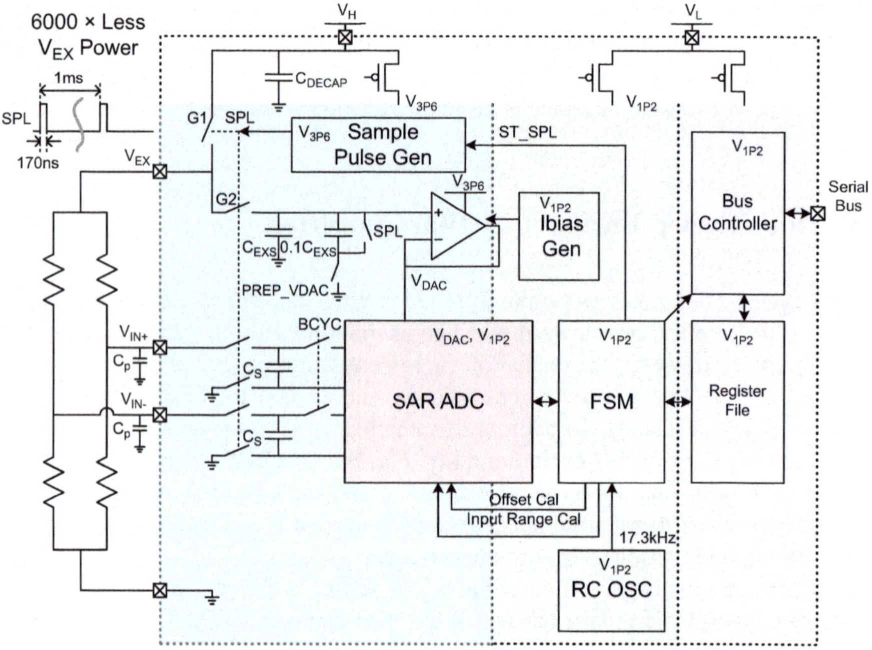

Fig. 4 BDC circuit implementation

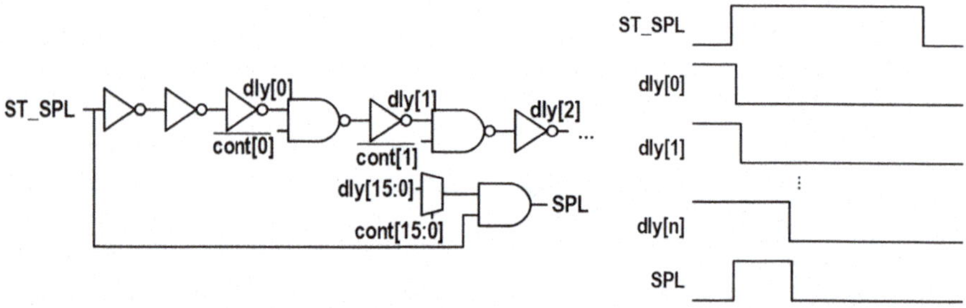

Fig. 5 Sampling pulse generator implementation

set to 4 pF so that kT/C noise is <50 μV, and hence, the RC time constant of the bridge and the sampling capacitor remains small at 12 ns. The SPL width is set to 170 ns to satisfy >16 bit sampling resolution with the RC settling of $V_{IN+/-}$. The bridge is exposed to the supply voltage for only 170 ns within the 1 ms total conversion time, enabling bridge power consumption to be 6000× less than that obtained with conventional DC biasing. The sampling generator is shown in Fig. 5 and is composed of an inverter delay chain that is 4-bit programmable from 60 to 240 ns. Once propagation reaches a selected stage, the remainder of the delay chain is gated to reduce energy consumption.

Fig. 6 V_H drop during excitation

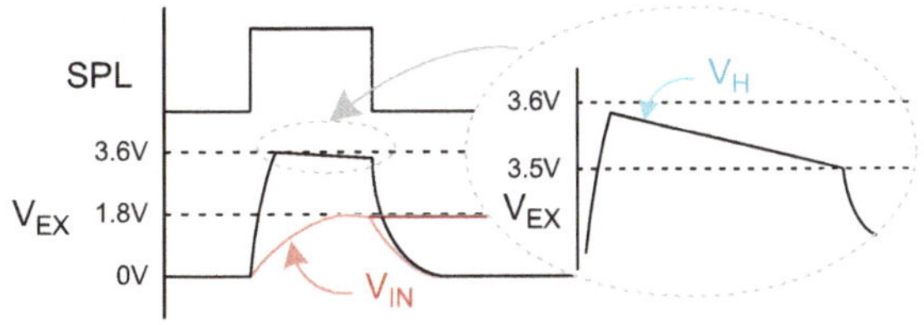

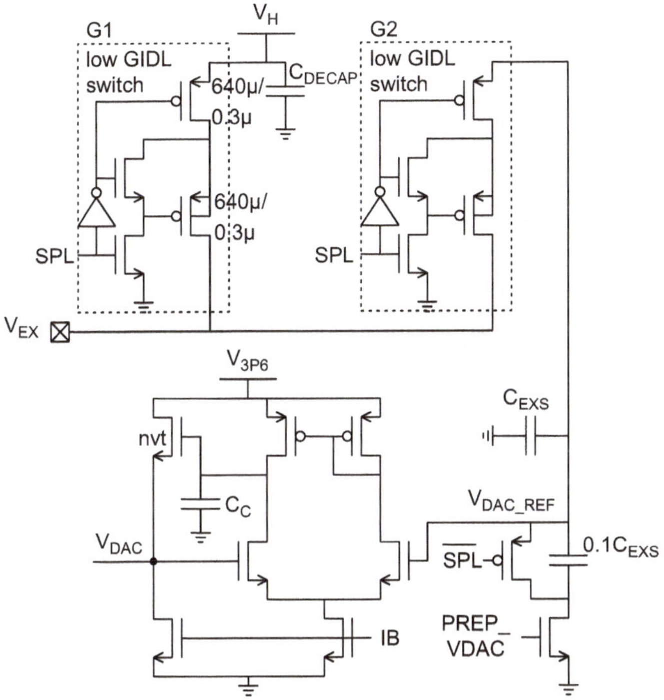

Fig. 7 Detailed implementation of V_{DAC} generation

3.2 DAC Reference Voltage Generation

Due to the high battery resistance, the current through the bridge sensor is pulled almost exclusively from the 1.2 nF on-chip decoupling capacitor, whose voltage drops ~100 mV during excitation and is then slowly recharged by the battery between sensor readings (Fig. 6). To avoid accuracy degradation due to this supply voltage fluctuation during ADC operation, we need to dynamically adjust the DAC reference voltage (V_{DAC}) to the reduced V_{EX} at the end of excitation. To achieve this, we use a V_{EX} sampling circuit and a reference buffer (Fig. 7). In the V_{EX} sampling circuit, V_{DAC_REF} is multiplied by 10/11 through charge sharing to provide more than 200 mV V_{DS} to ensure all transistors are in the saturation region within the amplifier that generates the final regulated output V_{DAC}. The simulated amplifier

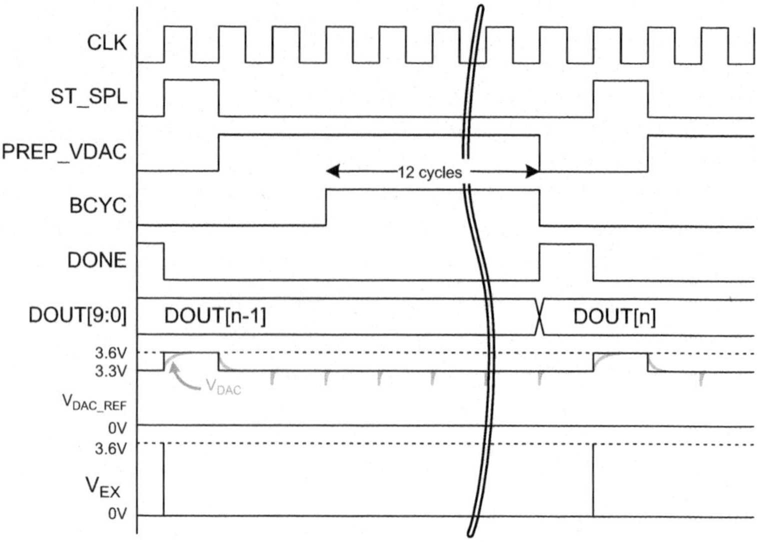

Fig. 8 BDC timing diagram

PSRR is −56 dB. The amplifier is designed for 25 kHz bandwidth and 50 μV integrated noise and draws 170 nA to achieve 12 b accuracy with the SAR ADC load at 1 kS/s. By sampling the excitation voltage in this way, the BDC is also insensitive to supply variation, which is important for sensor nodes operating on small batteries and hence often unstable supplies. Since V_{EX} is at ground for most of the conversion time, and its large V_{SD} and V_{GD} incur significant GIDL current, these circuits use GIDL reduction devices G1 and G2 [15] (Fig. 7). The BDC timing diagram is shown in Fig. 8. After ST_SPL pulses, PREP_VDAC is on and acts to multiply V_{DAC_REF} by 10/11. V_{DAC} settles during the on period of PREP_VDAC, after which the bit cycle phase is entered.

3.3 ADC Implementation

Figure 9 shows the proposed SAR ADC with input range matching and offset calibration features. In conventional SAR ADCs, the input voltage is sampled to a binary DAC array. However, in this implementation, such an approach would require $V_{IN+/-}$ to drive >12 pF, increasing the sampling time constant and V_{EX} energy by 3×. Targeting a 4 pF sampling capacitor instead (as determined by kT/C constraints), we separate C_S from the DAC [16] as shown. By doing so, $V_{IN+/-}$ do not have to drive the full DAC load in the left, and the excitation time and energy improve. It also reduces V_{DAC} amplifier current, which is proportional to the total DAC capacitance load for the same bandwidth constraint as the DAC

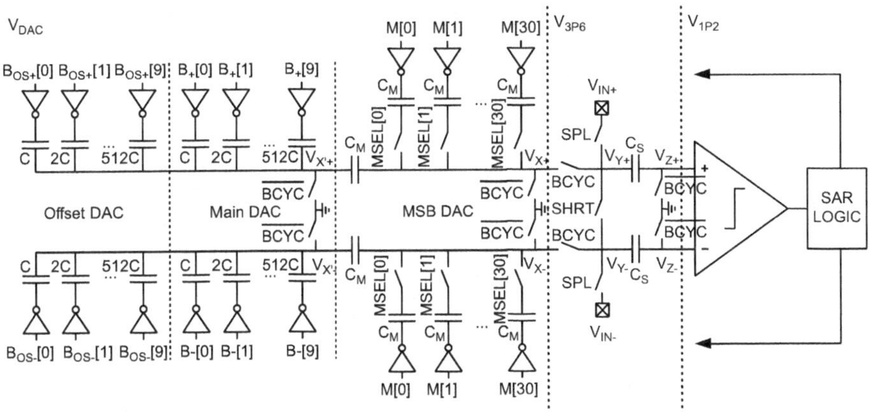

Fig. 9 Implementation of 10b ADC with range matching and offset calibration

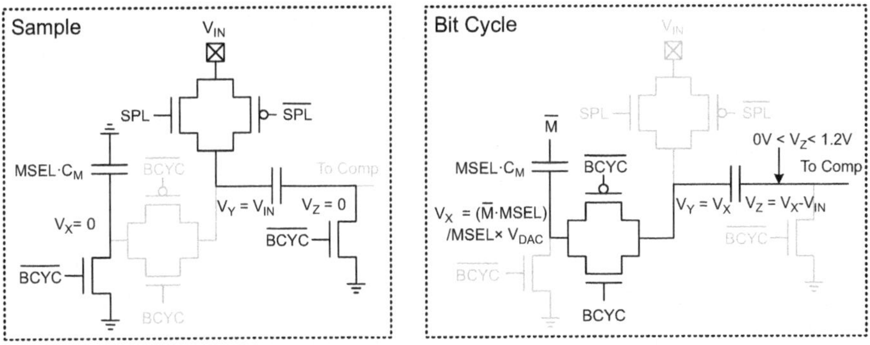

Fig. 10 Sampling and bit-cycle switch connections of the ADC

settling constraint. Bridge sensor resistance changes at most a few % at full-scale input. To match the input range of the bridge, we use an additional programmable MSB DAC. To accommodate an input range from ±50 to 100 mV, the MSB DAC is implemented with 31-bit unary capacitors with selection switches. The MSB DAC uses a split-DAC structure [17] to reduce total DAC capacitance and further improves the excitation time and energy.

During the sampling phase, the DAC purges all its charges while the input is sampled on C_S. The DAC uses only V_{DAC} and ground to avoid the power consumption of a common mode reference voltage generation, which requires very high bandwidth because of the short sampling time. At the beginning of the bit-cycling phase, the MSB of the main DAC flips, so $V_{X'}$ goes to half of VDAC. In the MSB DAC, C_M is selected by the MSEL amount and flipped by M amount (Figs. 10 and 11). Also, the DAC top plate (V_X) connects to the left plate of C_S (V_Y). As a result, C_S charge is conserved during the bit-cycling phases, and V_Y change is directly coupled to the comparator input. The remainder of the conversion process

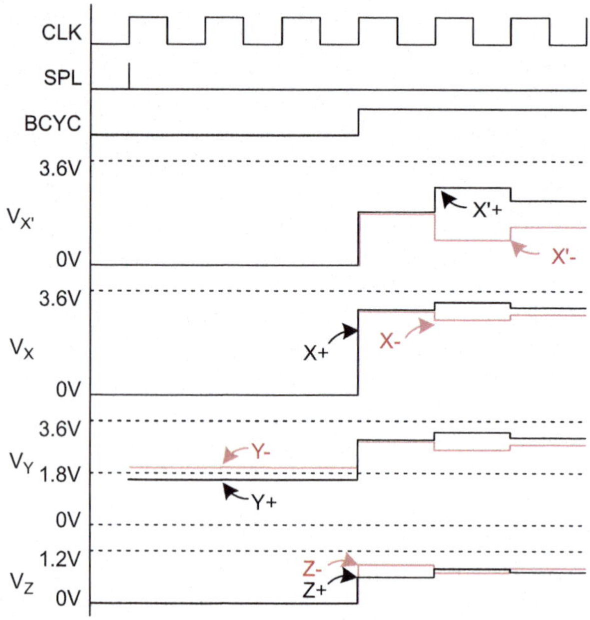

Fig. 11 Timing diagram of the ADC internal nodes: $V_{X'}$, V_X, V_Y, V_Z

is identical to that of a conventional SAR ADC. The comparator operates at 1.2 V (V_{1P2}) to reduce power, while the DAC operates at 3.3 V (V_{DAC}). The comparator is a conventional two-stage clocked comparator with 400 fF internal loading capacitor to enhance noise performance. Across all the phases, V_Z needs to stay between 0 and 1.2 V, and the common mode voltage can be tuned by adjusting the M code of the MSB DAC. The BDC can optionally run offset calibration. It operates with shorted inputs (SHRT = 1) and B_{OS} = 512 during the calibration, and its output is set as B_{OS} during normal operation. To accommodate multiple applications that require different resolutions, the BDC conversion can be oversampled with an oversampling rate (OSR) of 1–256. This approach repeats the entire conversion process OSR times and accumulates the output codes.

4 Measurement Results of Stand-Alone BDC Circuit

The BDC was fabricated in 180-nm CMOS technology and has an active area of 1.6 mm^2 (Fig. 12). The on-chip decoupling capacitor is made up of M1-M4 MOM and MIM, and its area is 0.48 mm^2. There is an RC-relaxation oscillator that generates an internal 17.2 kHz clock and a bus controller [18] that connects the chip to other chips in the overall sensor system. The input range was measured at different MSEL codes that set the DAC gain in the main DAC (Fig. 13). Across a

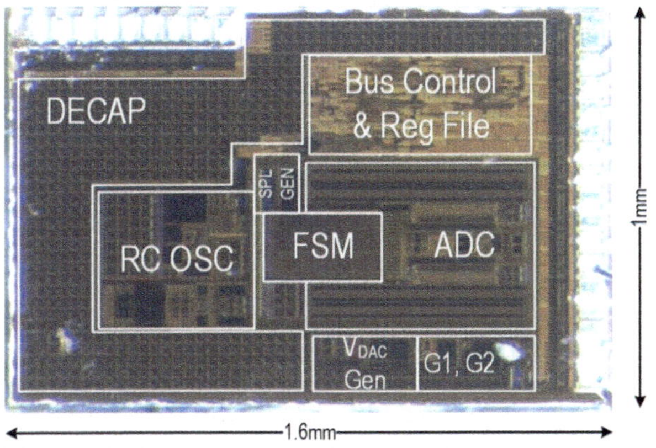

Fig. 12 BDC die photo

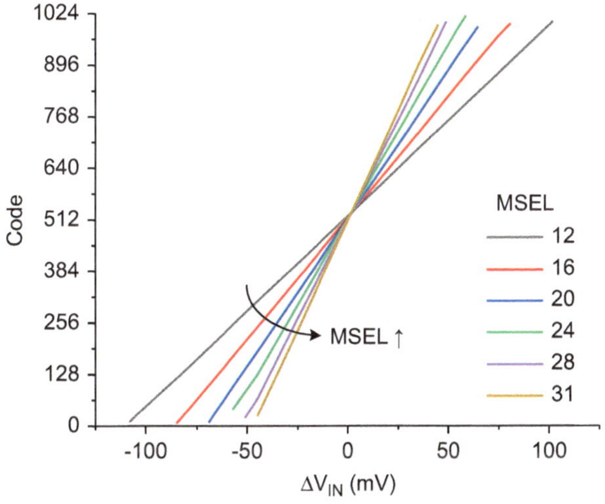

Fig. 13 Measured results of Code vs. V_{IN} at different MSEL

range of MSEL from 12 to 31, the input range changes from 45 to 110 mV with the slope changing accordingly, and the measured SNR is between 46 to 51 dB, as shown in Fig. 14. The BDC was then tested across 100–900 mmHg at 4, 3.8, and 3.6 V V_H (Fig. 15). The pressure sensitivity is 0.41 code/mmHg, and the pressure resolution is 2.2 mmHg, which is sufficient for many implantable applications. The code shift due to a V_H voltage shift from 4 to 3.6 V is 0.6, which can be readily calibrated out. Figure 16 shows the linearity error after code shift from V_H is calibrated and two-point calibration across pressure with 3.8 V V_H is applied. The maximum linearity error is 0.9 code. The total BDC conversion energy is 2.5 nJ at 1 ms conversion time, and its breakdown is shown in Fig. 17. Due to the duty-cycled

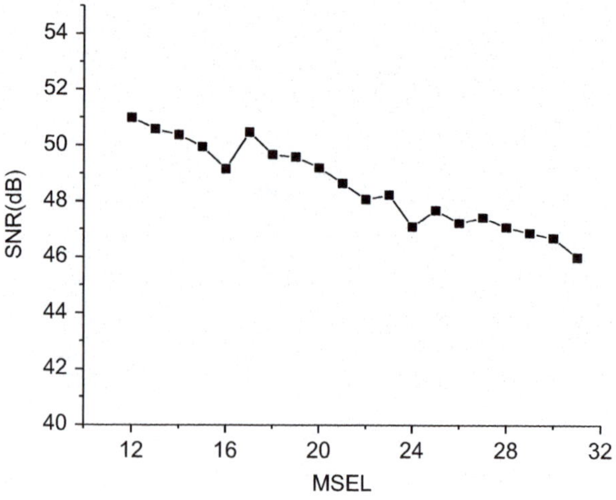

Fig. 14 Measured results of SNR vs. MSEL

Fig. 15 BDC pressure test results at 4.0, 3.8, and 3.6 V V_H

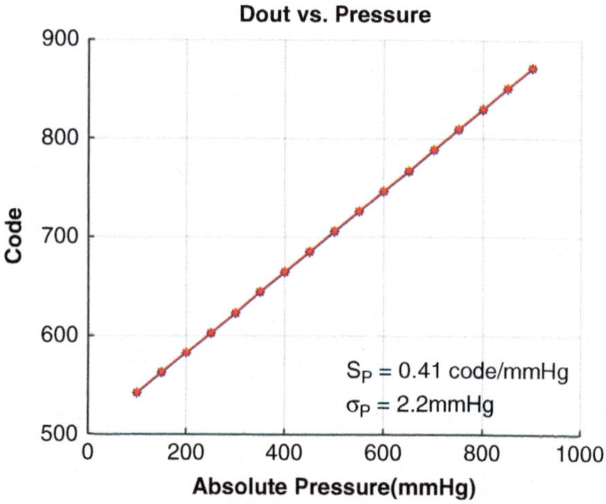

excitation, the excitation energy is greatly reduced such that it is similar to that of the other components. The BDC core conversion energy is 1.9 nJ when excluding the RC clock generator and bus controller. Figure 18 summarizes the BDC and also compares it with previous related BDC work. It draws 0.65 μA at 1.2 voltage domain and 0.52 μA at 3.6 voltage domain, which is significantly less than that obtained with prior designs and important for miniature sensor systems. Unlike other works where energy is dominated by the bridge excitation, the excitation energy of this work is just 20% of the total energy. That allows the overall energy to be the lowest among those listed in Fig. 18. To compare the BDC performance, two well-known ADC FOM metrics are redefined with energy/conversion including the

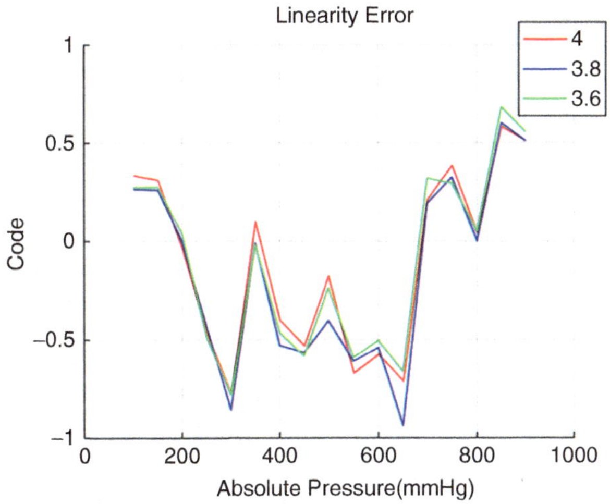

Fig. 16 BDC linearity error after code shift from V_H calibration

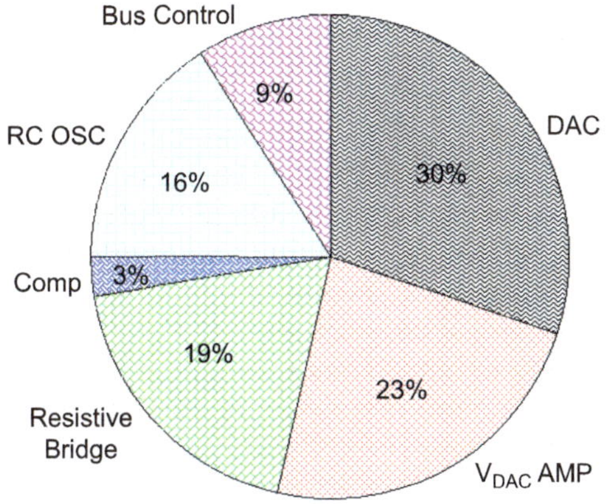

Fig. 17 BDC power breakdown

bridge excitation (Fig. 18). This work achieves the best reported FOMW in bridge interface circuits and also records a very high FOMS when compared with reported moderate SNR range bridge circuits.

5 System Integration for Barometric Range

The BDC was incorporated into an M-cube millimeter-size sensor system [19, 20], providing a pressure sensing capability of 0–900 mmHg. The M-cube sensor uses a stacked die composed of a MEMS pressure sensor, a battery, and 6 IC layers: radio, decap, processor, energy harvester [21], photovoltaic cells [22], and power management unit [23] (Fig. 19). The overall system dimensions are $3.9 \times 1.7 \times 1.9$ mm^3. The system is powered by two 8-uAh thin-film batteries

	This work	Grezaud, VLSI`17	Nguyen, Sens J`14	Jiang, ISSCC`17
Technology (nm)	180	180	90	180
Supply Voltage (V)	1.2, 3.6	1.8	1	1.8
Supply Current (μA)	0.65 @1.2V, 0.52 @3.6V	140	52	1200 @1.8V, 1500 @5V
Bridge Voltage (V)	3.6	1.8	1	5
Bridge Resistance (kΩ)	6	1	12	3.3
Conv. Time(μs)	1000	1000	96	500
Energy/Conv exclude Bridge (nJ/Conv)	2	61	1.88	1080
Energy/Conv include Bridge (nJ/Conv)	2.5	246	5	4870
+/-Input Range(mV)	68	16	12.8	10
SNR (dB)	49.2	59.0	44.1	95.5
FOMW[1] (pJ/c.s.)	10.6	337.9	38.3	100.0
FOMS[2] (dB)	132.2	122.1	124.1	145.6

[1]$\mathrm{FOMW} = \mathrm{E}_{\mathrm{CONV\_INC\_BRIDGE}}/2^{(\mathrm{SNR}-1.76)/6.02}$ [2]$\mathrm{FOMS} = \mathrm{SNR(dB)} + 10\log(1/(2\mathrm{E}_{\mathrm{CONV\_INC\_BRIDGE}}))$

Fig. 18 BDC performance summary and comparison

with 3.6–4.1 V output voltage, which is down-converted to 1.2 V and 0.6 V by the switched-capacitor power management unit [23]. The system includes 8-kB SRAM and an ARM Cortex-M0 processor, which controls the BDC operation. The MEMS pressure sensor [24] is on top of the entire stack with a pressure-sensitive top diaphragm. The four electrodes are directly wire-bonded to the proposed BDC chip.

The complete system was tested in a pressure chamber controlled by a pressure calibrator (Fig. 20). Figure 21 shows the serial bus clock and data waveform and V_H during system operation. As needed, the system wakes up from a sleep mode and enters an active mode by releasing power gates and isolation gates, turning on its RC clock and operating the BDC. After the BDC execution, it returns to the sleep mode by enabling the isolation gates and power gates in the reverse order of waking up. The data are saved in retentive SRAM and later read out with the radio. The system is completely functional in stand-alone wireless operation.

6 System Integration for Hydraulic Pressure Range

A slightly larger system for hydraulic pressure sensing was fabricated (Fig. 22). The system uses a pressure sensor [25] with a range of 0–10,000 PSI, powered by a 1-mAh coin cell battery. The other layers are stacked in the same order as described for the previous system. The system is hermetically sealed by black and clear epoxy. The black epoxy covers the IC layers, whereas the clear epoxy allows light on PV cells for recharging and optical communication. The overall system dimensions are $5 \times 5 \times 3\ \mathrm{mm}^3$.

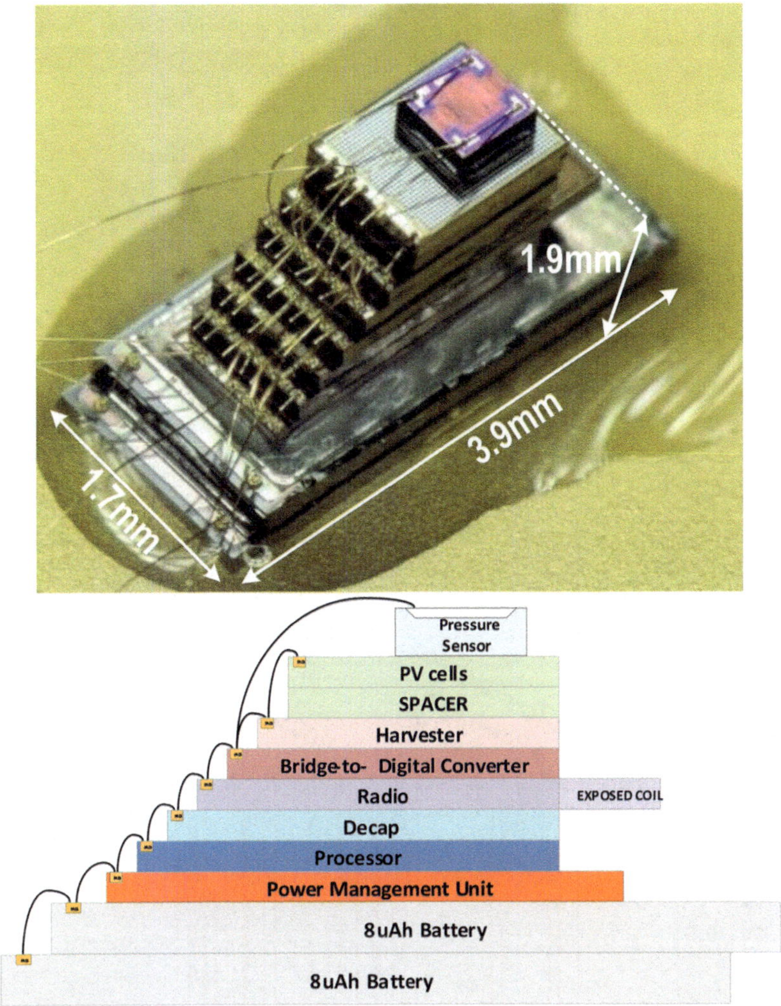

Fig. 19 Stacked-die system for barometric pressure sensing

The encapsulated system was initially programmed optically [26] and tested in a high-pressure chamber, as shown in Fig. 23. The system periodically recorded ambient hydraulic pressure in the SRAM according to a scenario similar to the barometric range system. After pressure testing, the system was unloaded from the chamber and was triggered to transmit the SRAM data wirelessly. Figure 24 shows the data collected by a software-defined radio receiver. The top shows three different sensors over time and the bottom shows one of those three relative to a reference gauge after linear calibration.

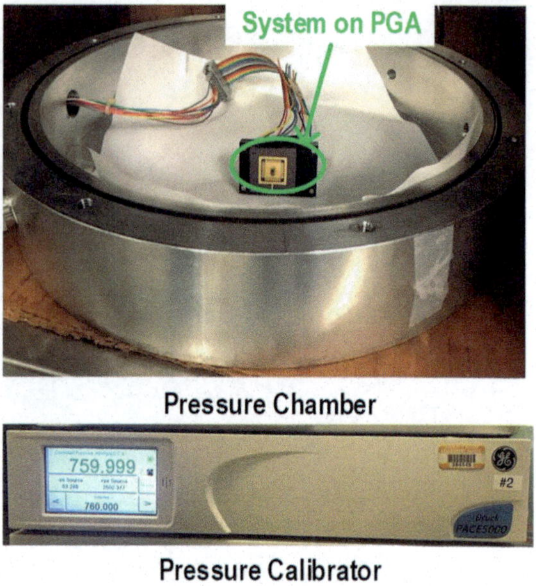

Fig. 20 System on PGA in barometric pressure chamber

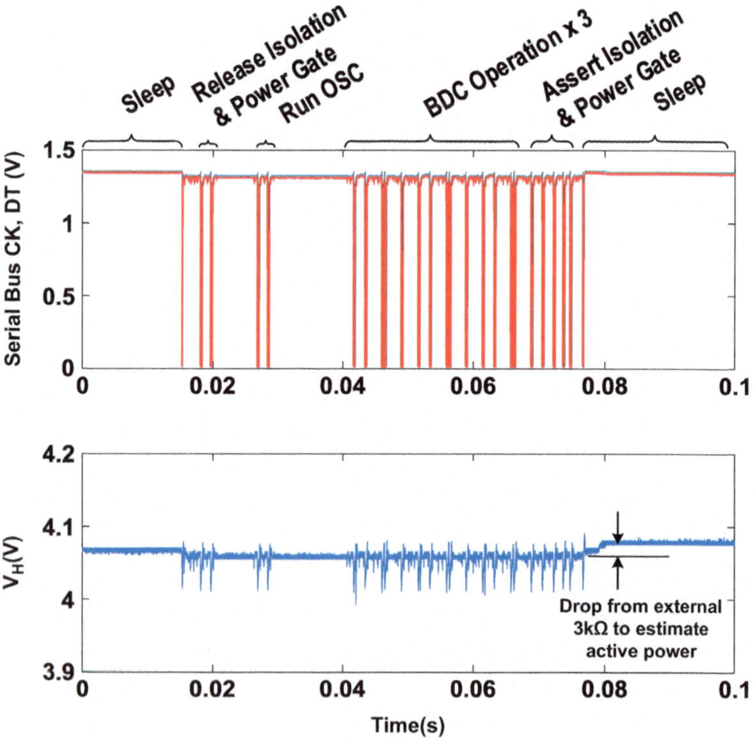

Fig. 21 Serial bus clock and data waveform and V_H during system operation

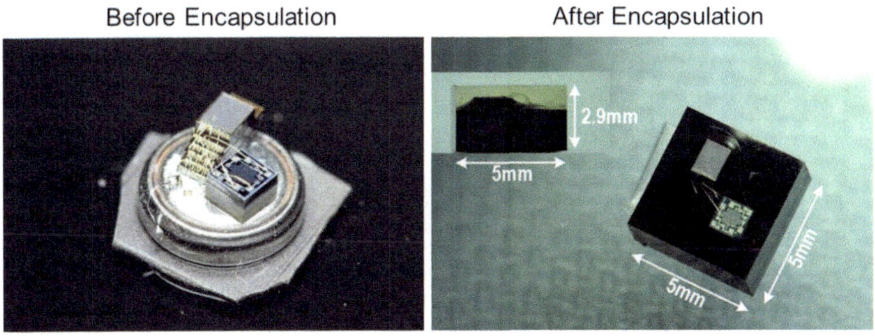

Fig. 22 Stacked-die system for hydraulic pressure sensing with and without encapsulation

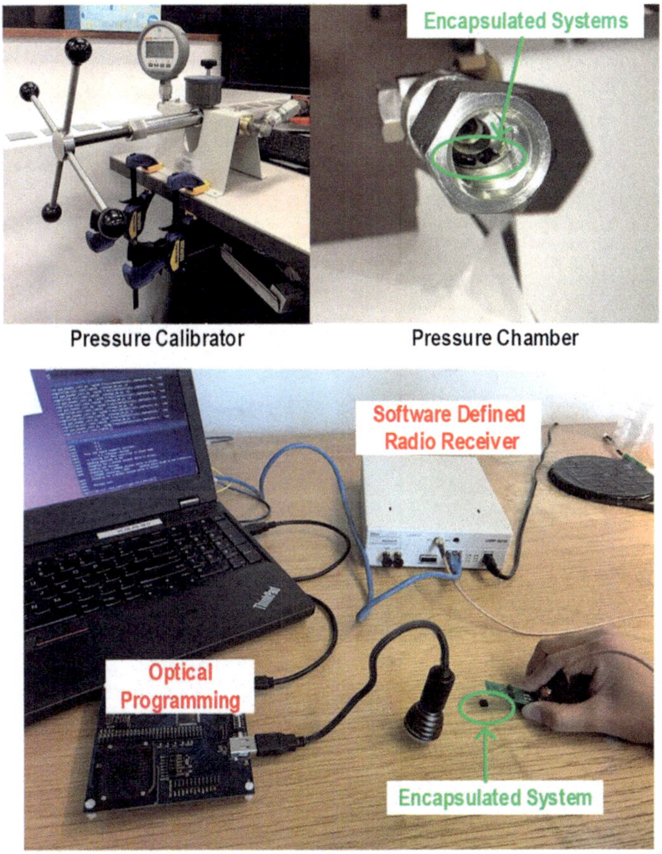

Fig. 23 Testing setup for the hydraulic pressure sensing system

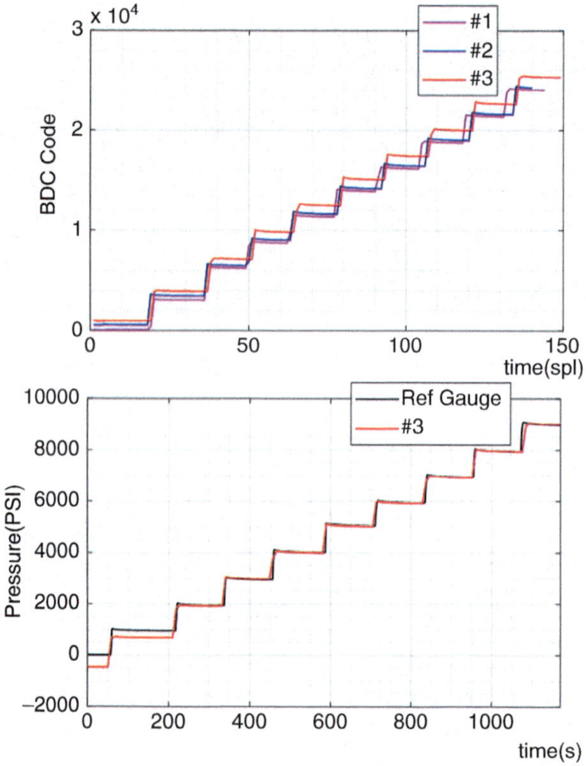

Fig. 24 Hydraulic pressure test results: three different sensors (top), one of them relative to a reference gauge after linear calibration (bottom)

7 Conclusions

In conclusion, we proposed a low-power, highly duty-cycled BDC that reduces excitation power by 6000× compared to conventional DC biasing. It consumes just 2.5 nJ per conversion and achieves 10.6 pJ/conversion-step FOMW. We also demonstrated a complete small form factor pressure sensing system including a pressure sensor, processor, memory, battery, power management unit, solar cell, and radio for different pressure applications.

References

1. Jiang H, Makinwa KAA, Nihtianov S. 9.8 An energy-efficient 3.7nV/$\sqrt{}$Hz bridge-readout IC with a stable bridge offset compensation scheme. In 2017 IEEE International Solid-State Circuits Conference (ISSCC), 2017, pp. 172–3.

2. Grezaud R, Sibeud L, Lepin F, Willemin J, Riou JC, Gomez B. A robust and versatile, -40°C to +180°C, 8Sps to 1kSps, multi power source wireless sensor system for aeronautic applications. In 2017 Symposium on VLSI Circuits, 2017, pp. C310–1.
3. Nguyen TT, Fernandes LAL, Hafliger P. An energy-efficient implantable transponder for biomedical piezo-resistance pressure sensors. IEEE Sens J. 2014;14(6):1836–43.
4. Donida A, et al. A circadian and cardiac intraocular pressure sensor for smart implantable lens. IEEE Trans Biomed Circuits Syst. 2015;9(6):777–89.
5. Zang Y, Zhang F, Di C, Zhu D. Advances of flexible pressure sensors toward artificial intelligence and health care applications. Mater Horiz. 2015;2(2):140–56.
6. Gubbi J, Buyya R, Marusic S, Palaniswami M. Internet of Things (IoT): a vision, architectural elements, and future directions. Future Gener Comput Syst. 2013;29(7):1645–60.
7. Oh S, et al. A dual-slope capacitance-to-digital converter integrated in an implantable pressure-sensing system. IEEE J Solid-State Circuits. 2015;50(7):1581–91.
8. Kim G, et al. A millimeter-scale wireless imaging system with continuous motion detection and energy harvesting. In 2014 Symposium on VLSI Circuits Digest of Technical Papers, 2014, pp. 1–2.
9. Blaauw D, et al.. IoT design space challenges: Circuits and systems. In 2014 Symposium on VLSI Technology (VLSI-Technology): Digest of Technical Papers, 2014, pp. 1–2.
10. EnerChip, Cymbet Corporation. [Online]. Available: http://www.cymbet.com/. Accessed: 13-Mar-2018.
11. Maruyama M, Taguchi S, Yamanoue M, Iizuka K. An analog front-end for a multifunction sensor employing a weak-inversion biasing technique with 26 nVrms, 25 aCrms, and 19 fArms input-referred noise. IEEE J Solid-State Circuits. 2016;51(10):2252–61.
12. Wu R, Chae Y, Huijsing JH, Makinwa KAA. A 20-b ±40-mV range read-out IC with 50-nV offset and 0.04% gain error for bridge transducers. IEEE J Solid-State Circuits. 2012;47(9):2152–63.
13. Jun J, Rhee C, Kim M, Kang J, Kim S. 19.7 A 21.8b sub-100μHz 1/f corner 2.4μV-offset programmable-gain read-out IC for bridge measurement systems. In 2018 IEEE International Solid-State Circuits Conference (ISSCC), 2018.
14. Oh S, et al.. 19.6 A 2.5nJ duty-cycled bridge-to-digital converter integrated in a 13mm2 pressure-sensing system. In 2018 IEEE International Solid-State Circuits Conference (ISSCC), 2018.
15. Bang S, Blaauw D, Sylvester D, Alioto M. Reconfigurable sleep transistor for GIDL reduction in ultra-low standby power systems. In Proceedings of the IEEE 2012 Custom Integrated Circuits Conference, 2012, pp. 1–4.
16. Jeong S, et al.. 21.6 A 12nW always-on acoustic sensing and object recognition microsystem using frequency-domain feature extraction and SVM classification. In 2017 IEEE International Solid-State Circuits Conference (ISSCC), 2017, pp. 362–3.
17. Ginsburg BP, Chandrakasan AP. 500-MS/s 5-bit ADC in 65-nm CMOS with split capacitor array DAC. IEEE J Solid-State Circuits. 2007;42(4):739–47.
18. Kuo YS, et al. MBus: A 17.5 pJ/bit/chip portable interconnect bus for millimeter-scale sensor systems with 8 nW standby power. In Proceedings of the IEEE 2014 Custom Integrated Circuits Conference, 2014, pp. 1–4.
19. Lee Y, et al. A modular 1 mm die-stacked sensing platform with low power I C inter-die communication and multi-modal energy harvesting. IEEE J Solid-State Circuits. 2013;48(1): 229–43.
20. Ghaed MH, et al. Circuits for a cubic-millimeter energy-autonomous wireless intraocular pressure monitor. IEEE Trans Circuits Syst Regul Pap. 2013;60(12):3152–62.
21. Jung W, et al. An ultra-low power fully integrated energy harvester based on self-oscillating switched-capacitor voltage doubler. IEEE J Solid-State Circuits. 2014;49(12):2800–11.
22. Teran AS, et al. AlGaAs photovoltaics for indoor energy harvesting in mm-scale wireless sensor nodes. IEEE Trans Electron Devices. 2015;62(7):2170–5.

23. Jung W, et al. 8.5 A 60%-efficiency 20nW-500μW tri-output fully integrated power management unit with environmental adaptation and load-proportional biasing for IoT systems. In 2016 IEEE International Solid-State Circuits Conference (ISSCC), 2016, pp. 154–5.
24. C39, TDK Europe – EPCOS. [Online]. Available: https://en.tdk.eu/pressure_sensor_elements. Accessed: 13-Mar-2018.
25. K-Series, Merit Sensor. [Online]. Available: https://meritsensor.com/. Accessed: 13-Mar-2018.
26. Lim W, Jang T, Lee I, Kim H-S, Sylvester D, Blaauw D. A 380pW dual mode optical wake-up receiver with ambient noise cancellation. In 2016 IEEE Symposium on VLSI Circuits (VLSI-Circuits), 2016, pp. 1–2.

Part II
Sensors for Mobile Devices

The second part of this book is dedicated to recent developments in the field of sensors for mobile devices. The first chapter provides a general overview of capacitive sensing systems, while the remainder discuss recent advances in microphones, gyroscopes, CO_2 sensors, and optical time-of-flight sensors.

Hans Klein et al. (Cypress Semiconductors) provide an overview of the technology behind capacitive sensing systems. Such systems have to detect small capacitance changes ranging from pico-Farads to atto-Farads, while also rejecting display and power-supply interference. Challenges and solutions for sensors, architectures, algorithms, and circuit technology are discussed.

Luca Sant et al. (Infineon) discuss the challenges involved in designing competitive MEMS microphones, which requires the co-design of the capacitive sensor, its package, and its readout circuitry. As an example, they present the design of a state-of-the-art digital microphone system that achieves 140 dB SPL full scale and an SNR of 67 dB at a 94 dB SPL reference level.

Zhichao Tan et al. (Analog Devices) present a readout ASIC for dual-axis (pitch and roll) MEMS gyroscopes. The ASIC also includes the sensor's high-voltage drive circuitry, digital filters, on-chip regulators, and a temperature sensor and occupies 7.3 mm^2 in a 0.18 μm CMOS technology. It consumes 7 mA from a 3 V supply. Over a 480 Hz signal bandwidth, the ASIC achieves a noise floor of $0.0032°/s/\sqrt{Hz}$ and a bias stability of 2.5°/h in a full-scale input range of 500°/s.

Burak Eminoglu and Bernhard Boser (UC Berkeley) describe an FM gyroscope, which, in contrast to vibratory gyroscopes, measures rate directly as a frequency variation. As a result, its scale factor can be accurately defined by a reference clock. Rate chopping is employed to reject drift, while symmetric and asymmetric readout modes enable a trade-off between long- and short-term errors without changing the transducer or its readout circuitry. When chopped at 10 Hz, a prototype achieves 40 ppm scale-factor accuracy, $1.5°/hr^{1.5}$ rate-random walk in symmetric mode, and $0.001°/s/\sqrt{Hz}$ Angle Random Walk (ARW) in asymmetric mode.

Zeyu Cai et al. (TU Delft) describe a CMOS-compatible CO_2 sensor. It consists of a suspended hot-wire transducer, whose heat loss is related to the CO_2 dependent thermal conductivity of air. Only a single extra etch step is required to realize the transducer in the tungsten via layer of a standard CMOS process. Together with its interface electronics, the resulting sensor achieves a state-of-the-art resolution of 94 ppm, while consuming 12 mJ per measurement.

Neale Dutton et al. (ST Microelectronics, University of Edinburgh) discuss the challenges involved in developing optical time-of-flight sensors based on the Time Correlated Single Photon Counting (TCSPC) technique. They describe a proof-of-concept sensor with a 10 GS/s folded-flash time-to-digital converter (TDC) and on-chip histogram generation. Fabricated in STMicroelectronics' 130 nm SPAD foundry process, the sensor consumes 178.1 pJ per photon at 899 M photon/s, while the TDC achieves state-of-the-art 0.48 pJ/S energy efficiency.

Advanced Capacitive Sensing for Mobile Devices

Hans W. Klein, O. Karpin, I. Kravets, I. Kolych, D. MacSweeney, R. Ogirko, D. O'Keefe, and P. Walsh

1 Introduction

Capacitive touch-sensing technology has existed for over 60 years, and with the progress of integrated semiconductors, it has become one of the dominant technologies in human–machine interfaces. Nowadays, most electronic interfaces utilize "Cap-Sense" technology in the form of buttons and sliders, arrays in touch pads or touch screens, and, more recently, fingerprint readers.

Indeed, Cap-Sense technology has very attractive attributes: it is low power, it is physically very thin and small, and it can be made very robust to real-world interferences. In addition, it can be made "smart"—allowing it to respond to changes in its use-case environment in sophisticated ways.

But with such a broad application space, challenges are abundant, and some of the most common ones will be presented in this chapter. They include dealing with a very large signal range, overwhelming noise levels, and spatial resolution.

In this chapter, we will start with simple buttons and show that they are not so simple after all. The techniques and solutions shown will lay the foundation for subsequently discussing more complex systems, like touchscreens and fingerprint readers.

H. W. Klein (✉) · O. Karpin · I. Kravets · I. Kolych · R. Ogirko
Cypress Semiconductor, San Jose, CA, USA
e-mail: Hans.Klein@cypress.com

D. MacSweeney · D. O'Keefe · P. Walsh
Cypress Semiconductor, Cork, Ireland

K. A. A. Makinwa et al. (eds.), *Low-Power Analog Techniques, Sensors for Mobile Devices, and Energy Efficient Amplifiers*,
https://doi.org/10.1007/978-3-319-97870-3_7

2 Buttons and Sliders: The pF Domain

2.1 The Basics of Capacitive Sensing

As the name suggests, capacitive sensing uses a capacitive element and an electronic read-out circuit to sense capacitance changes of the sensor. While there are many different shapes of capacitive sensors possible [2], the underlying concept is the same for all: detecting change in capacitance as an object approaches the sensor. The capacitance of a basic two-plate element can be calculated as:

$$C = \frac{\varepsilon_0 \, \varepsilon_r \, A}{d} \tag{1}$$

Here, the capacitance is a function of the permittivity of the dielectric material between two plates with area *A* and the distance *d* between them. An approaching object (such as a finger) changes the electric field between the electrodes, and thus the above parameters, thereby altering the capacitance. An illustration of a co-planar capacitive sensor is shown in Fig. 1.

A finger-touch of a simple capacitive "button" sensor creates a signal typically in the single-digit pF range and a variety of circuits exist to sense such change in capacitance. Simple solutions may be based on a time-delay circuit (using a comparator to detect the delay period) or an RC-oscillator to create a proportional frequency, as illustrated in Fig. 2. The capacitive sensing element is C_S.

As is apparent, the circuit complexity for the approaches shown is low, which generally translates into low cost and low power consumption, both very desirable. In practical applications, however, these basic solutions are generally not sufficiently robust against variations in temperature, component tolerances, power-supply variations, and various noise sources—to name a few factors. Furthermore, to make a product broadly applicable, it is advantageous to support a wide range of component values, especially for C_S.

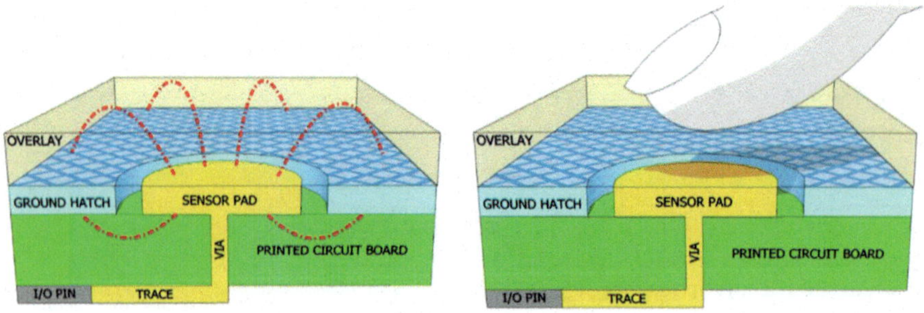

Fig. 1 Basic capacitive touch sensor with field lines (left) and an approaching finger (right) [4]

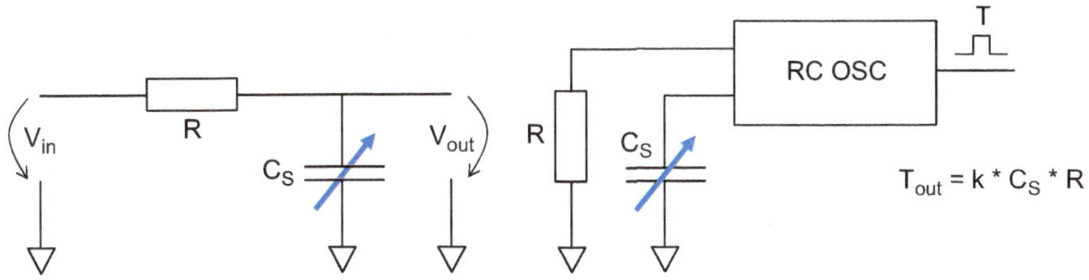

Fig. 2 RC delay circuit (left), RC-oscillator approach (right)

Fig. 3 Charge-based Cap-Sensing architecture

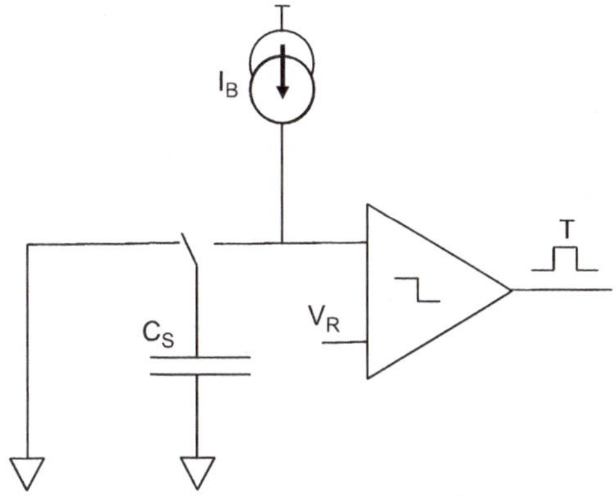

As an alternative to simple RC-based circuits, a charge-based architecture can be used; one example is shown in Fig. 3. Here, sense capacitor C_S is periodically discharged and then linearly recharged via a constant-current source.

The time required to reach a reference threshold V_R is proportional to the capacitor value C_S:

$$T_{out} = \frac{C_S V_R}{I_B} \tag{2}$$

where I_B from the current source may be internally derived from a reference voltage V_R and an internal or external resistor R:

$$I_B = k \frac{V_R}{R} \tag{3}$$

As capacitance C_S changes, T_{out} reacts proportionally. If the comparator output signal is used to reset sense capacitor C_S, a proportional output frequency can be generated just as easily.

Circuit elements such as switches, current sources, resistors, comparators, and a bandgap reference are all easy to implement in silicon. And more importantly, *all* these components can be designed to support a wide range of operating requirements, by giving them programmable characteristics. For example, current

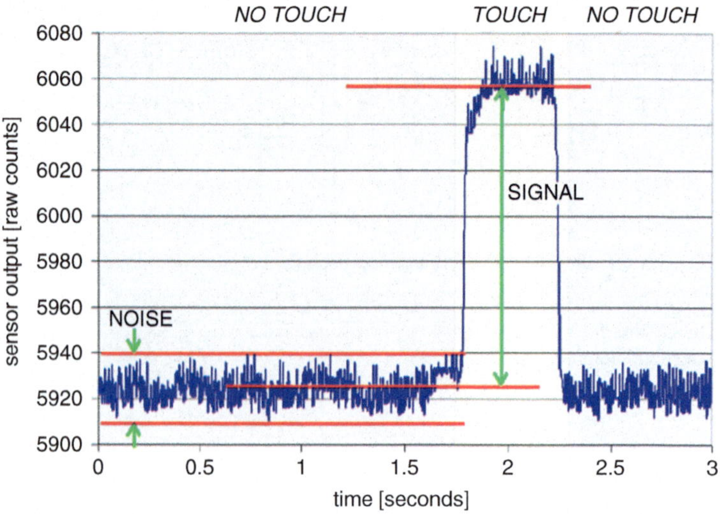

Fig. 4 Output signal from a Cap-Sense circuit

source I_B would in practice be a current-mode *D*/*A* converter, the reference voltage V_R would have multiple settings that may be derived from a bandgap or the power-supply, and the frequency may be selectable too.

One of the biggest challenges in real Cap-Sense applications is the presence of external noise. Especially for small signals (say, <1 pF), a sensing circuit becomes highly sensitive to noise coupling. An example for a touch event in the presence of noise is shown in Fig. 4.

The vertical axis in Fig. 4 is the "raw count" of the circuit, essentially a digital number representing the total capacitance measured. Without touch, the output value is referred to as "baseline capacitance" (5925 counts in the example). The noise contained in the signal amounts to about 30 counts. When touch occurs, the average count changes, in this case by 140, a little over 2% of baseline capacitance. To obtain robust detection (i.e., no false touches reported), the SNR should be 5:1. In the scenario shown above, this is nearly achieved. But in many applications, the SNR situation can be much more challenging.

An attractive option to combat often unknown noise is to narrow the channel's bandwidth. One concept for this is illustrated in Fig. 5.

The architecture shown here utilizes a passive large capacitor as a "low-cost integrator" ($C_{INT} >> C_S$). It serves as a "reservoir" for the charge subtracted by sensing capacitor C_S and the charge added back by the IDAC, which is gated by the comparator output. Of course, active integrator circuits are an option as well, but the large size of the integrator capacitor is a cost penalty for on-chip implementation.

A "single-shot" capacitance measuring method has limitations, in that it processes a short sample, corresponding to a wide bandwidth system response to noise. By contrast, integrating over a longer period of time (multiple samples) will result

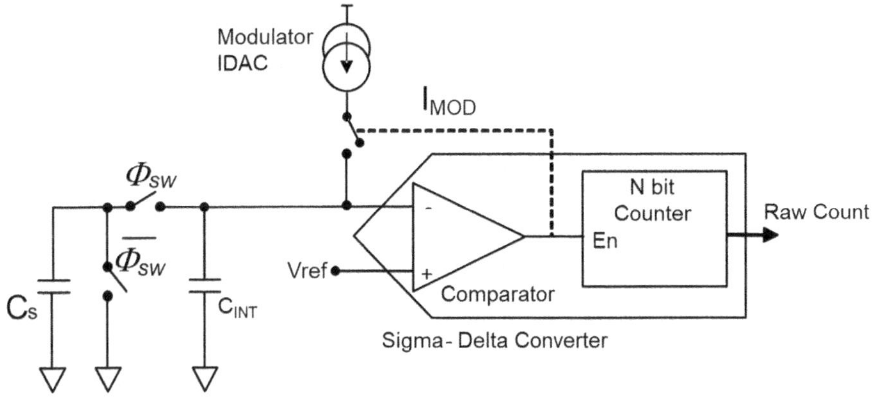

Fig. 5 Cap-Sensing utilizing an integrator and a Sigma-Delta feedback loop

in narrower bandwidth and can be achieved by simply adding up multiple digital samples in a counter.

One can use the simple architecture of Fig. 3 to achieve this. However, quantization errors occurring in each single-shot conversion do not average out. The architecture in Fig. 5 has an advantage here: As the IDAC reaches the rebalance point, current stops flowing, and then resuming after C_S subtracts the next charge packet, and so on. This continuous closed-loop operation can occur over an extended period of time, thereby essentially functioning as an integrator with a long-time window, with its desirable low-pass filter properties. Noise disturbances coupling into the signal path as well as individual quantization errors all average out, until the overall integration finally ends.

The only disturbance that will not get corrected occurs at the instant when the window closes and downstream counter stops accumulating. Thus, conversion errors due to noise and quantization will accumulate in C_{INT} and thus average out over the entire integration period.

The noise-transfer function NTF (for capacitively coupled noise onto the signal path), including the downstream digital accumulator, is shown in Fig. 6, below. Blue identifies the NTF for the accumulation approach, whereas the orange line corresponds to the simple "one-shot" approach.

For this scenario, AC-coupled noise (sine wave) is swept from DC to 1 MHz. The transfer functions reflect the converters behavior for a 300 kHz switching frequency. For the accumulation converter, the longer the integration time, the narrower the main lobe. Since noise can be injected into the touch-sensor only capacitively, there is a transfer-function "zero" at DC. The square-wave nature of the demodulation method (essentially polarity switches) explains that the spectrum repeats at uneven harmonics.

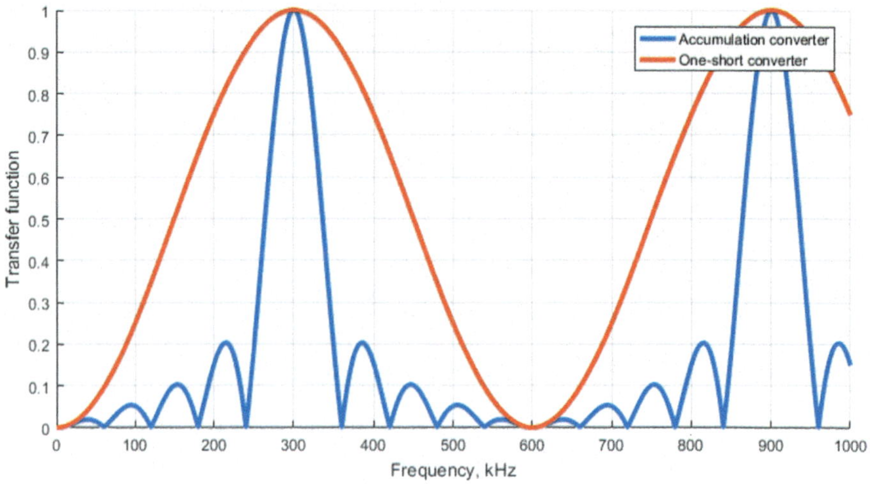

Fig. 6 Noise-transfer function for single-shot and accumulation conversions

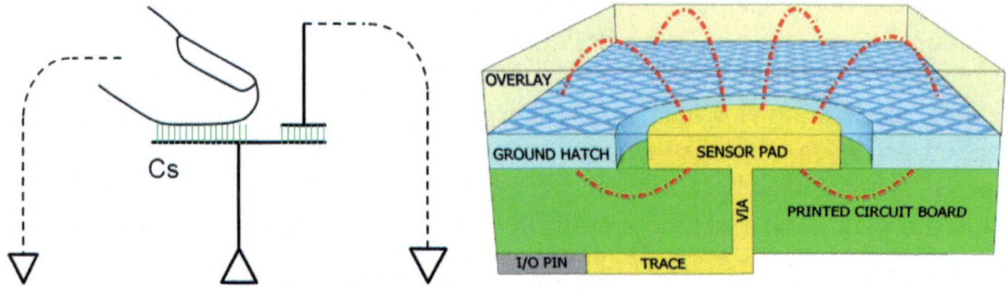

Fig. 7 Concept of "self-capacitance," where a finger adds field lines/charge

2.2 *Button and Slider Design*

One of the most popular applications of Cap-Sense technology is in as buttons. Two basic principles are used to sense the presence of a finger; they are based on sensing self- and mutual-capacitance. Sensing self-capacitance is illustrated in Fig. 7. There, we measure the change in capacitance from one plate to the sum of its environment (generally Ground or virtual Ground). A fingertip placed on the sensor pad will *increase* the element's total self-capacitance C_S to ground. The finger basically adds more field lines. This creates a "positive" signal.

In the case of mutual-capacitance, one of the electrodes is now driven with a stimulus (TX) signal, and the charge transferred through the sense capacitor is measured on the opposite plate, hence the term "mutual" capacitance. In the example shown in Fig. 8, the driver (TX) electrode is a ring surrounding the sensor pad (RX), similar to the layout in Fig. 7. Placing a fingertip near the capacitor will redirect field lines *away* from the signal path (and toward ground) and thus

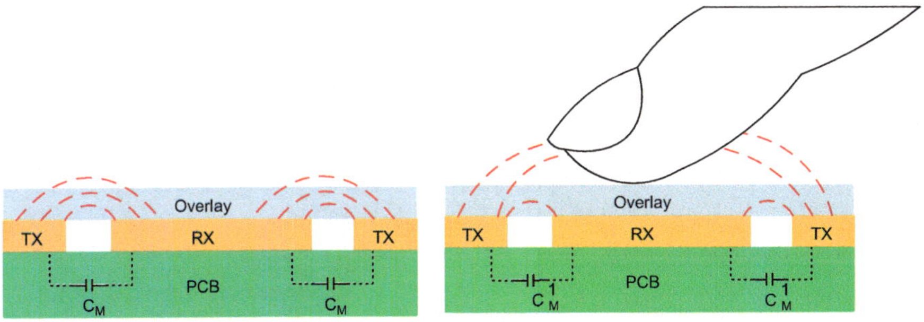

Fig. 8 Concept of "mutual-capacitance," where a finger subtracts charge from the mutual capacitance C_M

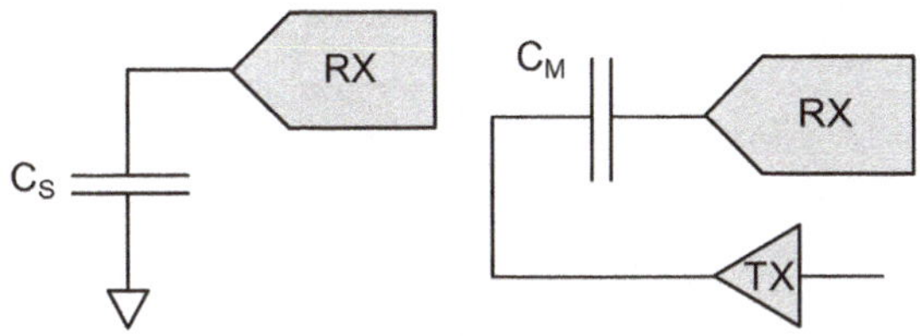

Fig. 9 System architecture for self-cap (left) and mutual-cap (right) sensing

reduce the mutual capacitance. This creates a "negative" signal. As a rule of thumb, a mutual-cap signal is smaller than a self-cap signal by 5 to 10×.

Both sensing methods have useful features and often both of them are used in actual applications. Fortunately, measuring self-capacitance and mutual-capacitance can be achieved with the same sense circuit. Simply applying a TX signal on one of the sensor electrodes and measuring the response on the other enables mutual-cap sensing. This is illustrated in Fig. 9.

Note that the mutual-cap approach always requires a "TX" signal which, for simplicity, is often simply a square wave, derived from a chip-internal clock. However, an extra pin is required for each capacitor driven in such manner. The benefit of this approach will become apparent later, when button *arrays* are discussed.

Another popular Cap-Sense application which expands on the single-button concept is a multi-button "slider" (e.g., useful for volume control or any other such proportional control function). One can think of them as a one-dimensional array of individual buttons, often self-cap based for simplicity. The sensing elements can be arranged in a straight fashion, angled, curved, or circular, as indicated in Fig. 10.

The figure also illustrates the change in signal magnitude across the various slider segments caused by finger touch. As the finger slides along the sensor array's segments, the signal magnitude for the approaching segment increases while it decreases on the departing one. Finger position and motion speed calculations are possible by interpolating across multiple segments. With sufficient SNR and resolution, achievable position accuracy is quite high, just fractions of a mm.

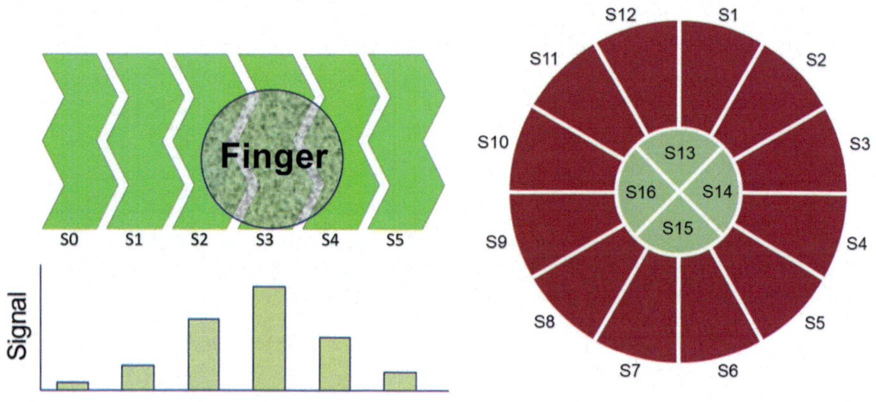

Fig. 10 Examples of linear and circular "slider" sensors

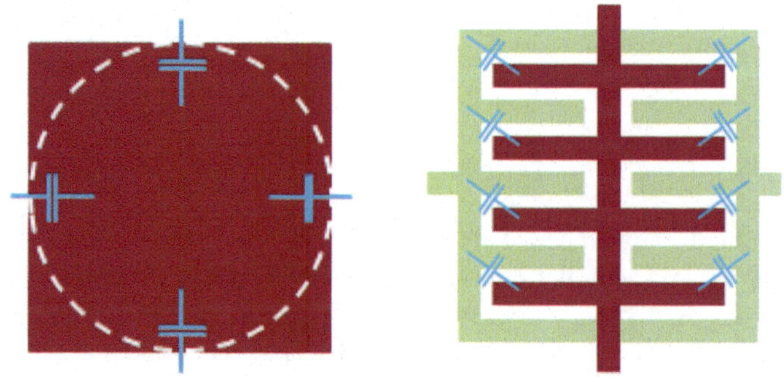

Fig. 11 Layout of Cap-Sense "buttons" optimized for self-capacitance (left) and mutual-capacitance (right)

2.3 Sensor Optimization

Given that SNR is one of the biggest challenges, the sensor should be optimized to maximize the useful signal. For that reason, sensors for self- and mutual-sensing are constructed differently.

For a self-cap type button sensor, one would maximize the area touched by the finger with minimal stray capacitance to its environment (generally Ground). Conversely, for a mutual-cap type button one would maximize the peripheral capacitance between the two plates in such way that an approaching finger could steal away as many field lines as possible. This is illustrated in Fig. 11.

Touch signals produced by the various capacitive sensors are roughly in the 0.1–1 pF range. Nearby noise sources can couple very strongly into the signal path and greatly disturb these small signals. This drives the need to optimize the sensor design, including layout, material choices, and layer thickness. Such sensor

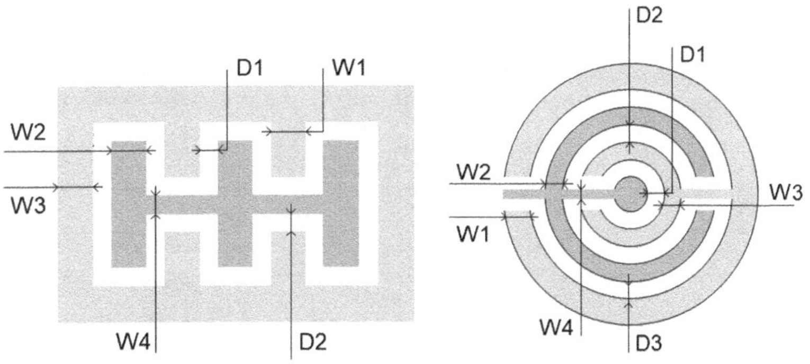

Fig. 12 Dimensions subject to optimization for two Cap-Sense button design examples

optimization can easily increase system SNR by 2–3×; and that is a critical factor toward making touch systems robust.

If the buttons are constructed by copper traces on the same PCB as the rest of the electronic circuit, the choice of material and their properties may not be on option. Even so, layout optimization is still critical and can be quite challenging. Figure 12 illustrates two different button designs with their various layout dimensions, demonstrating a rather complex optimization space.

In many cases, design experience provides good rules of thumb for layout optimization. However, as Cap-Sense structures are used not only in basic finger-touch applications but also liquid/solid-level detection, high-sensitivity proximity detection, and many others, the motivation for optimal sensor design is compelling.

To illustrate this, Fig. 13 shows the sensitivity exploration study of different square-layout sensor designs, all based on the same "stack-up" of layers of multiple materials. Each dot in the graph represents a particular sensor's response. The layouts illustrate how the sensors were modified (# of fingers, width, spacing). All these sensor designs have their respective pros and cons and clearly the choice is quite complex.

For example, the graph shows the mutual-capacitance change (dC_{m}) of a touched sensor vs. its baseline capacitance C_{m}, for various sensor shapes and dimensions. The "sweet spot" in that graph appears to be the top-left corner, where a large dC_{m} touch signal can be generated by a small C_{m} sensor (having a small baseline capacitance) which would suggest a clear winner. However, the structures shown also exhibit different sensitivity to power-supply coupling or injected-noise coupled to the routing path of the final system. Additional factors are introduced when the sense-electronics path is considered (e.g., noise gain) and the type of noise. For these reasons, structures other than the ones in the top-left corner of the graph may likely be a better choice. Quickly, the optimization challenge can be overwhelming.

To address this complex challenge, we created a front-to-end system simulator and coupled it to an AI system based on neural-networks, along with an appropriate learning algorithm to train it on ~100,000 data points.

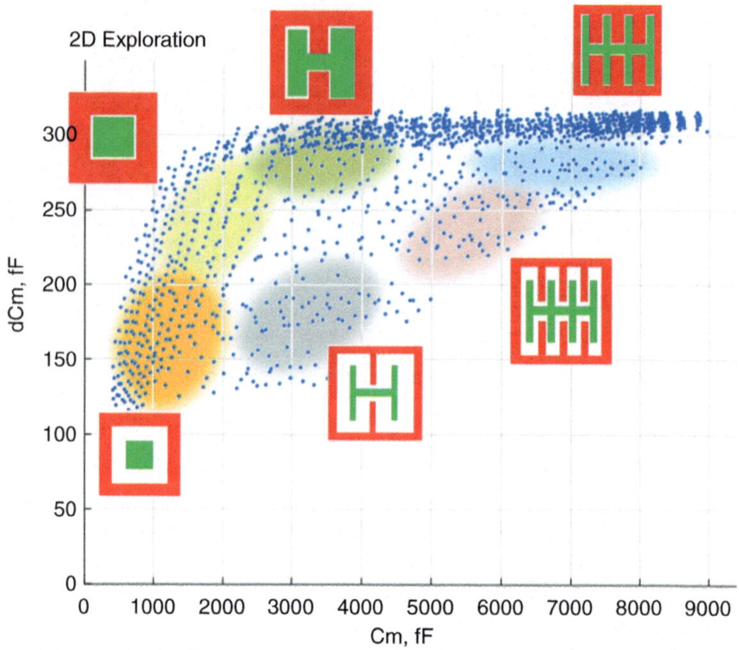

Fig. 13 Impact of layout and dimensions on the sensitivity for various mutual-cap sensor elements

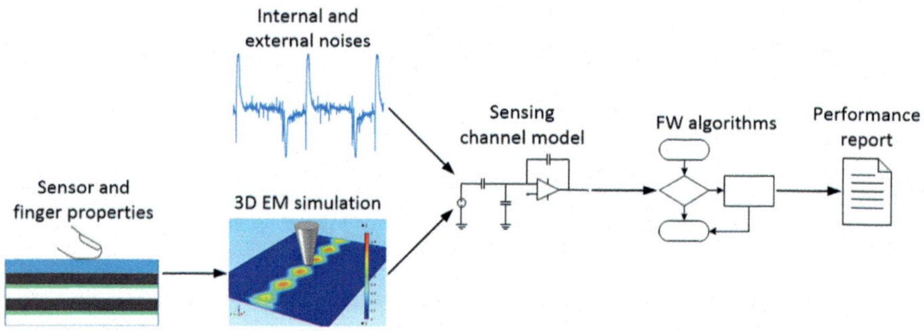

Fig. 14 Font-to-end Cap-Sense system-simulator architecture

The system-simulator architecture is illustrated below in Fig. 14. Multiple tools are integrated to achieve full front-to-end coverage: 3D simulations for the sensor's physical and electrical properties, frequency-domain and time-domain simulators for the signal path, and calculation of spacial (position, jitter) information.

The front-to-end system simulator comprehends all key parameters that determine the physical and electrical properties of the sensor in response to finger touch, noise-coupling effects from the environment, behavior of the sensing channel with its programmable parameters, subsequent algorithms for filtering, interpolating, motion vector extraction, and so forth, running on the embedded CPU of the

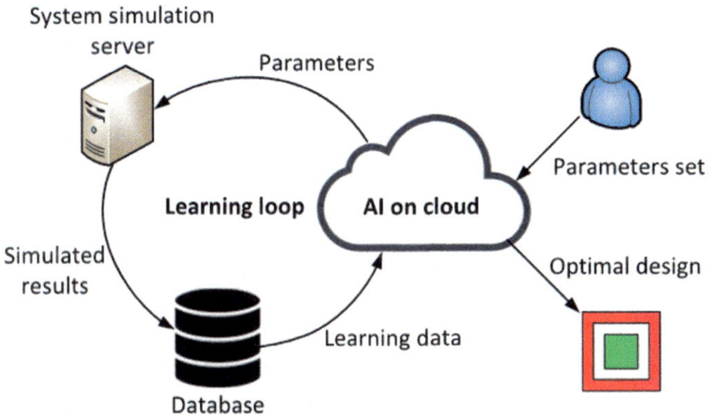

Fig. 15 System Simulation and an AI Learning Loop, utilizing cloud-based resources

Cap-Sense chip. A comprehensive performance report is generated including all the position and linearity errors, position jitter, drop outs, detection errors, and so forth.

With such system in place, it is possible to execute a large number of experiments in a reasonable time frame, using cloud-based resources. This allows to run a comprehensive set of design-of-experiments batch-producing a very large number of results which are then fed into the AI system's database for training. This general approach is shown in Fig. 15.

Upon completion of the system's training, a user can simply input various application constraints (input parameter sets). The neural network will then identify the best-suited solution: For a given chip, this normally means the best-suited sensor along with key operating ("tuning") parameters.

To identify a "near-optimum sensor design," such system can find a solution to a very complex optimization problem in a matter of seconds, rather than weeks previously. Of course, novel and the most-challenging cases still require highly skilled experts, but for most other cases, the system's suggested solution is very close to optimum and entirely sufficient.

One additional advantage of such AI-based approach is that new knowledge and new solutions identified by the experts can be added to the system over time, making it even more helpful to all nonexpert users.

3 Touchscreens: The fF Domain

Touchscreens can be considered a more sophisticated version of a Cap-Sense button array. Substantial complexity is introduced by the need for fine spacial resolution, often down to 0.1 mm. This necessitates size reduction of the sense elements (finer pitch), which in turn reduces the input signal. Interpolating the position of a finger covering multiple sensor elements accurately also demands high signal

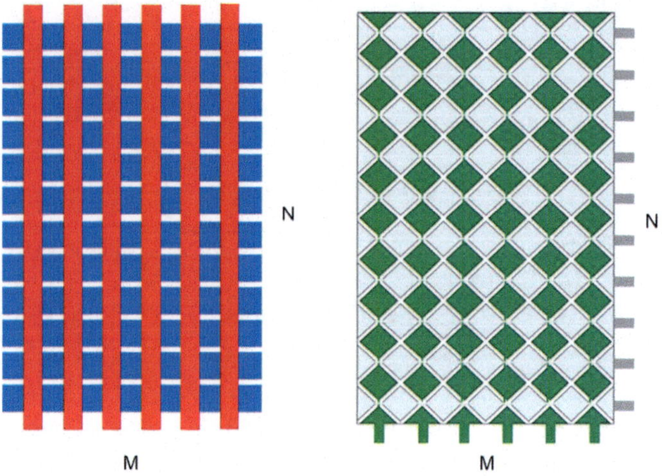

Fig. 16 Touchscreens layout examples: "Manhattan" (left) and "Diamonds" (right)

resolution (typically, 10–12 bits) for each scanned element. Furthermore, some high-end touchscreen controllers (TSCs) support the use of gloves and even fine-tip pencils. These bring the useable signal levels into the low fF range.

For example, a 10–15″ touchscreen on a tablet may need 1000–2000 sense elements with a 4–5 mm grid pitch. For such large arrays, only mutual-cap sensing is practical, allowing for profound reduction in wires and channels. For example, a two-layer grid of lines running in X and Y direction can be constructed, this is also referred to as "Manhattan" layout. One axis then serves as TX lines, the other as RX lines, with mutual-capacitance created between those RX and TX lines. A finger touch creates a signal typically around 1–100 fF.

"Diamond" structures are similarly popular, and they can be constructed much the same way or in a quasi-single layer with little "bridges" at all the intersections. Example layouts utilizing mutual-cap sensing are shown in Fig. 16.

For an example touchscreen with 30 by 50 sensor rows/columns, the total wire count for mutual-cap mode would be $N + M = 80$ compared to a nonpractical $N*M = 1500$ wires in self-cap mode.

In addition, only the mutual-cap mode allows for supporting multi-finger touch as each intersection of RX and TX lines can be individually measured. This is not practically achievable with self-capacitance methods.

3.1 Sense Channel Architecture

While an $N*M$ sensor structure is very compact and cost-friendly, all individual RX/TX intersections have to be measured, resulting in a large combined measurement time for the entire panel, corresponding to a low frame rate.

However, to support gesture detection like “flicks,” the need arises for scanning rapidly (e.g., 10 ms for a complete frame). A single Cap-Sense channel scanning across *N*M* intersections will not acquire the signals fast enough. A multichannel (parallel-scanning) approach is required which increases the measurement time per intersection proportionally without sacrificing frame rate.

Because of the many sense channels involved, it is also no longer practical to operate with an *external* integration capacitor for each channel. Rather, a fully integrated channel is required.

Unfortunately, incoming charge from a touchscreen can be quite large (such as from large noise spikes), saturating even the largest on-chip capacitors; but saturation must be avoided for downstream filtering to be effective. Thus, a signal attenuator is required, resulting in an architecture like in Fig. 17.

As screen size increases, so does the dynamic range requirement. After all, the finger signal remains the same, whether it is a small or large screen. Yet, external noise (e.g., from an LCD) couples into the sense path proportional to the length of the sensor strip, which grows with the size of the panel. Given the need for an attenuator at the channel input (largely due to noise reasons), the sought-after finger signal is attenuated as well. To maintain the same target SNR as before, the noise floor of the channel must now be reduced, and even after circuit optimization, this requirement often costs extra power. An example architecture for the attenuator is shown in Fig. 18 below, which is based on a transconductance amplifier with a programmable replica output stage.

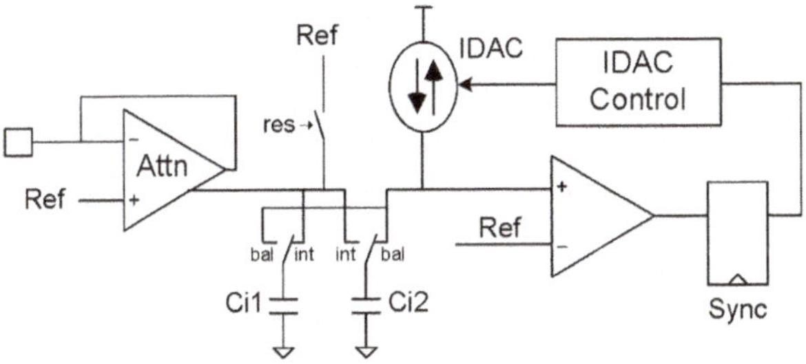

Fig. 17 Example architecture of a fully integrated sense channel for touchscreens

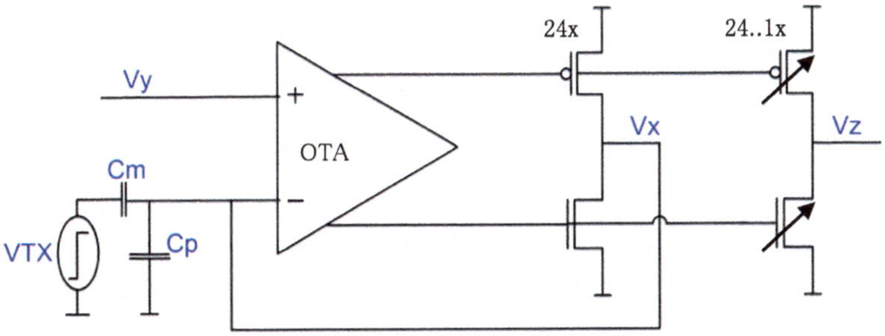

Fig. 18 Example architecture of a programmable active attenuator

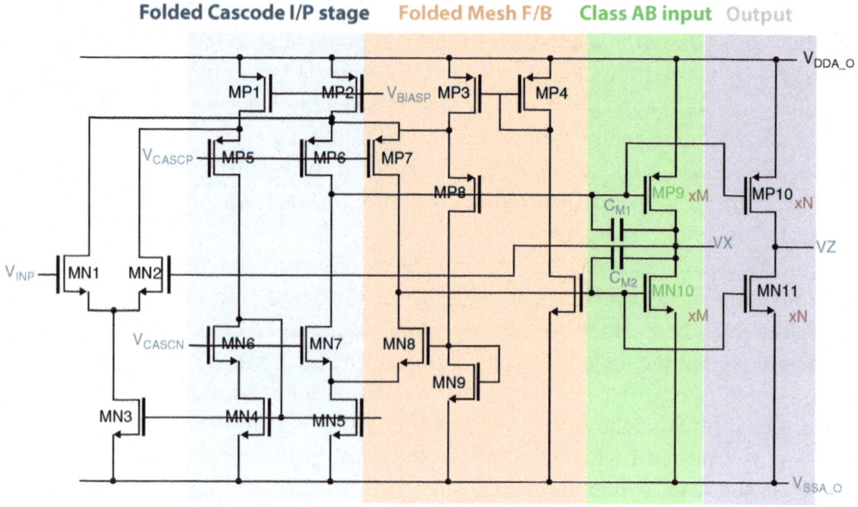

Fig. 19 Example circuit implementation for the programmable attenuator

For simplicity, V_{TX}, C_m, and C_p on the left represent the excitation signal source and one element of the sensor array, each having mutual and parasitic capacitance. The sense circuit's feedback loop creates a virtual ground node at the sensor's output, and the charge needed to keep that node steady is attenuated at the programmable replica output stage. The circuit can be implemented with an approach such as in Fig. 19.

Another interesting aspect of the overall architecture (Fig. 17) is that it features *two* integrator capacitors. This dual-integrator approach allows both positive (Q_{IN+}) and negative (Q_{IN-}) input signals to be accumulated in a sequential "ping-pong" fashion, produced by the positive and negative TX pulse transitions. After subtracting these positive and negative charge packets, such "full-wave" operation not only produces twice as much signal but also any common-mode content gets subtracted out:

$$Q_{out} = (Q_{IN+} + Q_{CM}) - (Q_{IN-} + Q_{CM}) = 2\,|Q_{IN}| \quad (4)$$

Common-mode signal Q_{CM} is the result of offsets in the attenuator and comparator and also from low-frequency (near DC) injected noise (e.g., from AC mains). To put all signal levels in perspective, the table below identifies the main contributors to the incoming signal:

The signal levels in Table 1 indicate the challenge: the finger signal is 20–40 dB below noise levels. The main noise sources include the LCD display (coupling across the entire area) and charger noise, discussed further in [3]. But additional noise sources exist and must be dealt with in many applications, such as finger/palm-coupled noise from CFL lamps, AC mains, and any nearby electrical objects.

Table 1 Typical signal and noise signals encountered in a touchscreen application

Item	Signal level	Comment
Baseline capacitance	~1000 fF	This is DC offset
Touch signal	10–100 fF	The "useful" signal
1 LSB (at 10b resolution)	10…100 aF	10b for good position accuracy
LCD noise	10–100 fF	Magnitude similar to finger signal
Charger noise	Up to 1000 fF	10× larger than finger signal
Channel self-noise	<1 fF	Allows high-sensitivity modes

Therefore, multiple methods to boost SNR must be applied. High TX frequencies (= large number of charge packets) would be very helpful, but touchscreen layers have limited bandwidth. This is due to the resistive nature of the ITO sense layers (kΩ range) and their self-capacitance, resulting in a panel signal-path bandwidth of generally <300 kHz, and even <100 kHz for large touchscreens. However, slowing down integration time (i.e., frame rate) to collect many charge samples is not acceptable, as was mentioned before.

One measure to boost the finger signal is the use of charge pumps for the TX drivers, and this can double or quadruple the finger signal. And while charge pumps add to chip size and power consumption, this brings back 6–12 dB in SNR. Yet more must be done.

A multichannel approach is another SNR-boosting technique. High-performance TSCs provide up to 60 channels, corresponding to the size of the targeted touch panels. The signal integration (measurement) time available for each intersection increases proportional to the channel count, correspondingly narrowing the noise bandwidth of the channels. A high channel count boosts the SNR enough to even allow for high-sensitivity modes, such as "Proximity" or "Hover" detection. There, a hand/finger is away from the surface by some distance, decreasing finger signals yet another factor of 10 or more.

All techniques combined can result in significant die size. Examples of fully featured 14-channel and 60-channel chips are shown in Fig. 20, where larger chips reach ~$20mm^2$ die size.

4 Fingerprint Readers: The aF Domain

One of the latest achievements in capacitive sensing with growing popularity is fingerprint (FP) scanning. While several sensing approaches exist in the market (such as thermal or optical imaging), capacitive sensing is attractive due to its low cost, low power, and compatibility with Silicon technology, including packaging. Only capacitive FP sensing is discussed here.

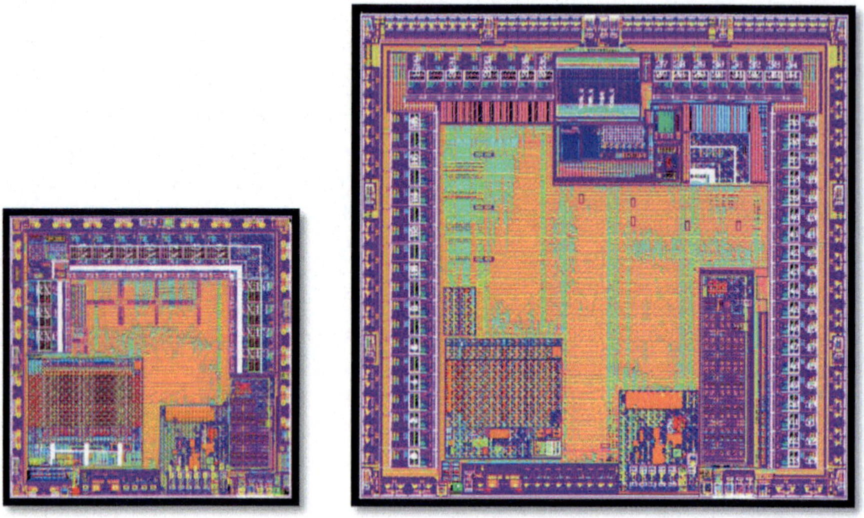

Fig. 20 Example layouts of a 14-channel and a 60-channel TSC (Cypress Semiconductor)

Fig. 21 Raw image from a FP scan

Figure 21 shows an image from an actual FP scan[1] without any image post-processing. Note the fine features where adjacent ridges may touch each other via thin "bridges" and other minutiae, such as skin pores. This level of detail is critical to robust detection by a FP matching algorithm. Robustness here means a low probability of "false rejection" of a valid FP image, and likewise a low probability of accepting a similar—but invalid—image ("false acceptance"). Thus, FP readers must deliver high spacial and signal resolution and still scan a complete image quickly.

[1]Source: Cypress Semiconductor.

4.1 Fingerprint Sensors

Capacitive FP sensors can be based on self-cap or mutual-cap methods. One can view such sensors as dramatically scaled-down versions of a touchscreen.

The periodicity of human-skin FP ridges is around 300–500 μm. Thus, the pitch of the RX/TX grid is driven by that dimension and the need to resolve minutiae, such as pores, bridges, and gaps. As a result, a typical sensor pitch size is 75 μm, which is a 75× reduction in each dimension compared to a touchscreen sensor. Given such >>2000× scale-down in area-per-sensor, each capacitive sensor element now operates in the fF..aF realm. Such tiny signals naturally present a new set of challenges. In addition, to cover an image size of $5 \times 5\ \text{mm}^2$, approximately 4000 "pixels" are required and proportionally more for larger sensor areas.

Sensors using the self-cap method could be constructed from an array of capacitive plates right on silicon (see Fig. 22 as an example[2]). The CMOS sense electronics can be placed underneath each plate ("pixel"). Basically, for a sensor with about 4000 sensor elements, such chip would comprise of 4000 sense-channel front ends.

One of the downsides of such approach is that for each different FP sensor, a different chip must be created. Furthermore, for the more-desirable "large-area" sensors (e.g., $10 \times 10\ \text{mm}^2$ or larger), chip cost grows proportionally, making a sensor-on-silicon approach commercially unattractive.

A more desirable alternative is enabled by a mutual-capacitance approach, utilizing a "Manhattan"-like sensor structure, similar in concept to that in touchscreens. Such an approach allows the sensor and the FP readout chip to remain *separate* items, connected by an $N + M$ bus (amounting to 130 connections for a $5 \times 5\ \text{mm}^2$ sensing area). That separation allows a variety of FP sensors to be supported by the same small FP chip. On the downside, mutual-cap sensing typically creates just ~10% of the signal compared to self-cap and that makes the SNR challenge that much greater.

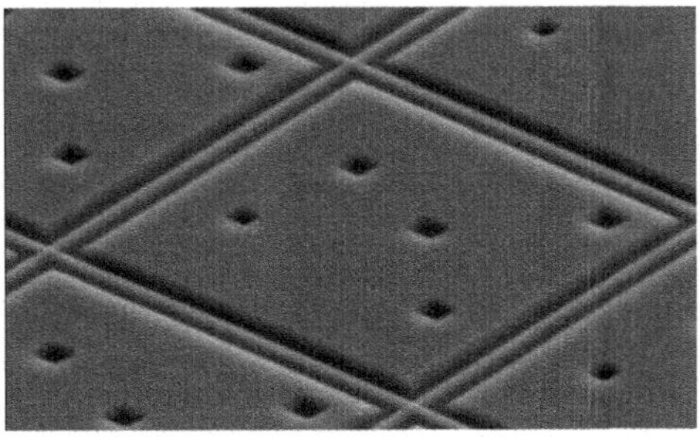

Fig. 22 Silicon-based FP sensor[2], showing the "sensing plates"

[2]© System Plus Consulting, France.

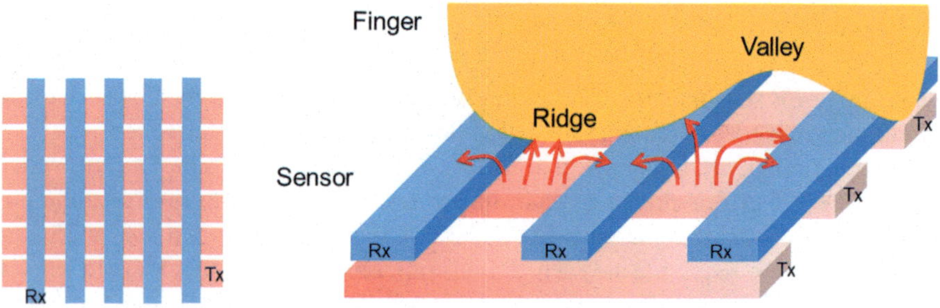

Fig. 23 Concept of a capacitive FP sensor based on mutual capacitance

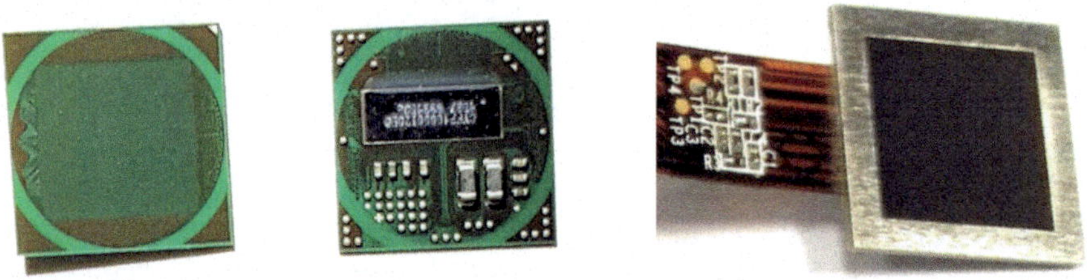

Fig. 24 A FP reader's sensing side (left), chip + component side (center), and complete module (right)

For the mutual-cap approach, a "Manhattan"-type FP sensor concept is shown in Fig. 23, with an enlarged segment on the right, illustrating the touch of a finger.

The "ridge" and "valley" labels reflect a human finger's skin topography. Indicated by the red arrows, electric-field lines are affected more strongly by a nearby ridge and less so by a more distant valley. This creates a difference in mutual capacitance at each of the sensor intersections (pixels).

A complete FP module, utilizing such sensor principle, is shown in Fig. 24. The left-most item shows the top-side of the sensor which is exposed to the finger (sensing side). In the middle, the bottom-side of the sensor is shown with the separate FP readout chip near the center, connecting to the sensor elements by $N + M$ vias. A few passive components (e.g., power-supply bypass caps) are also placed on that side. "Bumps" allow connection to a flex cable, for connecting the digital interface and power supply lines. The item on the right shows a fully integrated and coated "module," along with a metallized bezel—ready for easy connection and integration for a final product.

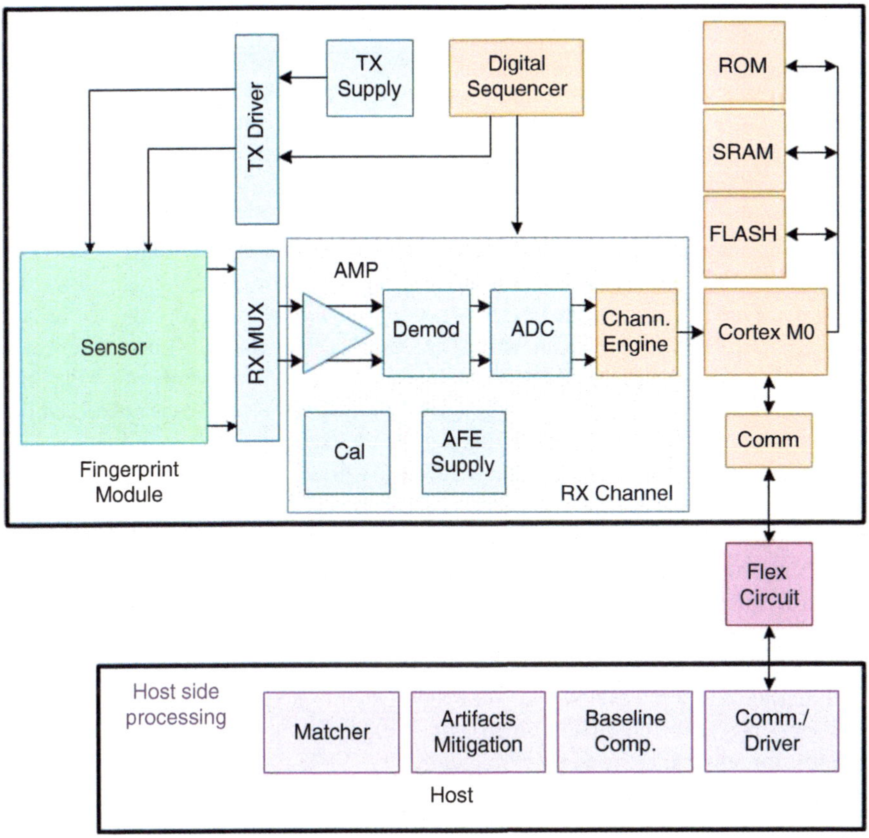

Fig. 25 A FP-Reader System Architecture: integrated FP module (top) and the Host (bottom), connected by a flex cable

4.2 Fingerprint-Reader System Architecture

A system architecture that can connect to an off-chip, separate FP sensor is illustrated in Fig. 25. The system comprises of a complete FP Module (top) and a Host (bottom), which could be a cell-phone's application processor.

The system is partitioned such that the FP module acquires the image and performs most of the image pre-processing and encryption, whereas the Host runs the matcher software and provides other system-level functions. Note, the FP image that comes off the module and is sent to the host should be encrypted, an increasingly important privacy and security requirement.

The FP Sensor is based on a Manhattan structure, thus $N + M$ wires (typically part of the package) connect the sensor array to the TX drivers and RX channels of the FP chip. In Fig. 25, everything to the right of the sensor (in green) is integrated in the FP reader chip. Details of the RX channel are elaborated next.

4.3 Fingerprint Sense-Channel

As was mentioned, the sensor elements are typically sized in the 75 × 75 μm^2 range. Because of such small size, the mutual-cap signals (dC_M) produced by ridges and valleys are about 1000 times smaller than touchscreen signals, falling into the 100aF range. For the matcher software to do a good correlation against a set of stored images, simple "black & white" FP images don't suffice. Therefore, high-quality FP matching requires a gray-scale image resolution of at least 8 bits. This leads to a self-noise requirement for the sensing channel <<10aF.

A complete FP image must be produced typically within <100 ms, so that the total system time (touch➔scan➔match➔report) is less than 200 ms, a generally accepted response time for human–machine interaction. Thousands of pixels must be scanned within this 100 ms budget, allowing just 20 μs or so per pixel. Unlike in touchscreen controllers, a multichannel parallel scan is not economical for a FP chip because a channel with extremely low noise floor is costly—both in power and area.

Fortunately, given the low resistivity of the FP sensor (metal) array, and the small capacitance of the sensor elements, it is possible to drive the sensor at much higher frequencies than touchscreens. In case of FP sensors, bandwidths can easily exceed 10 MHz. This allows for high TX frequencies resulting in charge packet accumulation at much higher rates than for touchscreens, an important factor for building up a usable pixel signal. As a side benefit, at such high frequencies, it is also possible to differentiate human-skin from rubber or most other materials an impostor might use to clone a FP image.

Thus, one approach to sensing FP signals is a single, high-performance channel, and then scan the entire image by multiplexing TX and RX lines. The corresponding channel architecture is shown in Fig. 26.

Key building blocks in the signal path are the low-noise amplifiers (LNA), followed by a chain of programmable-gain amplifiers (PGAs), quadrature demodulator, filter, and ADC. Typical for very small signal processing, the entire signal path is fully differential. For offset to not be amplified throughout the high gain path, the amplifiers are AC-coupled.

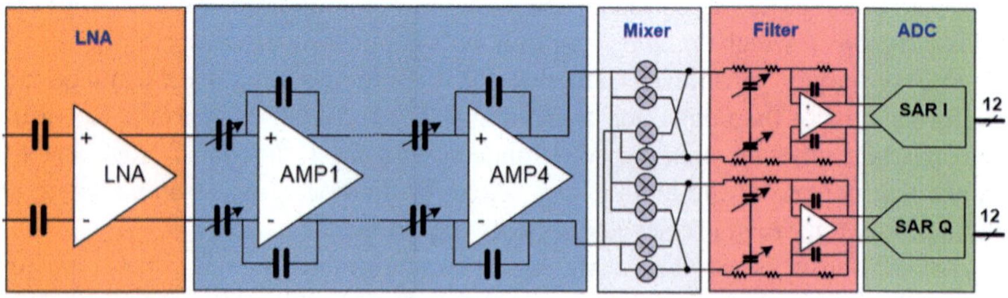

Fig. 26 Example architecture of a sense channel for capacitive FP sensing

This single RX channel is preceded by a wide "RX multiplexer," which connects the LNA to each of the sense lines in the sensor array. To achieve sufficient signal gain, the LNA is followed by multiple amplifiers with programmable-gain which accommodates a wide range of sensors and manufacturing tolerances. The overall signal-path provides a gain of up to 5000. After reaching a magnitude of a few hundred mV, the signal is demodulated, and many RX charge pulses are then averaged by a LPF before conversion by a 12bit ADC. A quadrature approach to demodulation removes the challenge of amplitude detection "at the right moment" for non-square wave signal shapes, and the I and Q results are easily combined into a single magnitude value by the downstream digital process.

4.4 The LNA

The LNA is perhaps the most critical building block in the RX path. It needs to have >250 MHz of gain-bandwidth, while containing its self-noise to the equivalent of <10 aF which corresponds to a noise floor of <10 nV/$\sqrt{}$Hz. Furthermore, there is significant common-mode signal riding on the small dC_m FP-signal, which requires CMRR around 60 dB in the first amplifier.

The ideal LNA should be simple and robust, with little or no need for feedback loops or compensation. A good approach to the problem is the circuit shown in Fig. 27. Other topologies are possible, such as in [1].

This circuit is based on a "g_mR" approach. The g_m of the input transistors MP_IN produces a signal current through load resistors R_1 thus setting the gain to g_m*R. The bias for the LNA is derived from a self-regulated bias circuit on

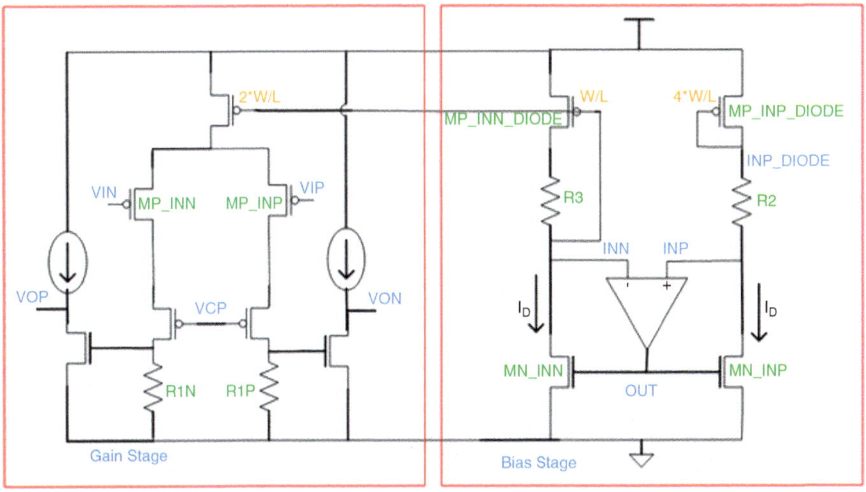

Fig. 27 Example of an efficient LNA Circuit

Table 2 Key performance specifications of the combined LNA cascade

2-stage Amp	Min	Max	Units
IDD	1.4	1.9	mA
3db_Freq	81	120	MHz
GBW	430	665	MHz
GAIN_9MEG	26.6	28.2	dB
GAIN_Tempco	7	20	mdB/degC
GAIN variation vs supply	0.2	4.2	mdB/mV
LINEARITY	−0.116	0.004	%
CMRR_9MHz	57	104	dB
PSRR_9MHZ	28	65	dB
ATT_9MHz	0.90	0.91	V/V
NOISE	7.4	9.0	nV/rtHz

the right and mirrored to the LNA's current sources on the left. SNR/power/area optimization resulted in an amplifier gain target of ~5, with a >50 MHz bandwidth (i.e., >250 MHz GBW) under worst case conditions, which is easily achievable in a 130 nm CMOS process. Two such LNA stage are cascaded for a combined gain of 25×. Table 2 summarizes some of the key performance specifications of the cascaded LNA.

This approach to the single-channel mutual-cap architecture has been proven to be very robust in production and has demonstrated its flexibility by successfully connecting to a large variety of FP sensors. Of course, for the actual chip implementation, additional functions are required, such as self-test, calibration sources, biasing, and an entire micro-controller subsystem with memory, encryption, etc., none of which are discussed here. The important point of this is that with large amounts of noise-generating digital circuitry surrounding the low-noise channel and its >140 input pins, extreme care must be taken to achieve highest levels of noise isolation for bias lines, supply lines, control signals for the RX muxes and other analog switches, and of course substrate isolation.

5 Conclusions

Cap-Sensing is a highly popular method in many types of products for implementing electronic buttons, sliders, touchscreens, and fingerprint readers. In this chapter, we have covered a broad range of sensors, architectures, circuits, and presented key challenges and proposed solutions.

While buttons and sliders produce signals in the low pF range, solution challenges are lowest-possible cost, low power, and robust system behavior. For touchscreens, the complexity of the sensor array increases, while the signal magnitudes available to receive channels are much smaller. In some applications, the signals drop into the low fF range. Finally, in FP readers, sensor complexity goes up again by another 100×, while area of sensing elements drops more than 5000 times, and to satisfy the required resolution, the signals of interest are now in the aF range.

Despite the many challenges associated with Cap-Sense technology, industry has found ways to expand its use beyond mobile devices. The technology is already ubiquitous in industrial equipment, home appliances, and many other consumer items, and it is also present in every modern cars. So although its large-scale introduction was only some 15 years ago, Cap-Sense technology might just stay with us forever.

Acknowledgments The authors would like to thank V. Bharathan, V. Bihday, O. Hoshtanar, O. Kapshii, A. Maharyta, and D. Starr for their various contributions to this chapter.

References

1. Steininger JM. Understanding wide-band MOS transistors. IEEE. 1990.
2. Baxter LK. Capacitive sensors: design and applications. Wiley; 1996.
3. Klein HW. Noise Immunity of Touchscreen Devices. Cypress White Paper, Feb 2013. www.cypress.com/file/120641/download. “Getting Started with CapSense”, 2017, Cypress Application Note AN64846.
4. Maltoni D, et al. Handbook of fingerprint recognition. Springer Professional Computing; 2003.

MEMS Microphones: Concept and Design for Mobile Applications

Luca Sant, Richard Gaggl, Elmar Bach, Cesare Buffa, Niccolo' De Milleri, Dietmar Sträussnigg, and Andreas Wiesbauer

1 Introduction

In the recent past, MEMS microphones have gained a significant market share in the area of consumer applications. Heavily driven by the need to achieve the highest possible audio quality when making phone calls in adverse conditions, such as windy weather and, in more recent times, by the need to correctly distinguish audio commands made to mobile devices placed in a very noisy ambient or at very large distances, specifications for MEMS microphone systems have become tougher. There is a clear trend toward significant increases both in the desired Signal to Noise Ratio (SNR) as well as in the Acoustic Overload Point (AOP, i.e., the sound pressure level at which the total harmonic distortion reaches 10%). Challenges like the miniaturization of the package size allowing placements in extremely thin bezels, as well as the request for adaptive power modes to limit the current consumption in battery powered devices contribute to an increasingly demanding system design. To keep up with the tight time schedules of the mobile phone market while guaranteeing the highest possible quality for the product, a structured approach in the development of both MEMS and readout electronics is mandatory. As an example, accurate modeling of the major electroacoustical effects in the sensor can highlight weaknesses in the early stages of its development, allowing modifications in a timely manner. It also enables a tailor-made design of the readout electronics leading to an optimum system performance. In the following sections, an insight to the proposed approach to structured development of a digital microphone from concept to implementation is given.

L. Sant (✉) · R. Gaggl · E. Bach · C. Buffa · N. De Milleri · D. Sträussnigg · A. Wiesbauer
RF & Sensors, Infineon Technologies Austria AG, Villach, Austria
e-mail: Luca.Sant@infineon.com

K. A. A. Makinwa et al. (eds.), *Low-Power Analog Techniques, Sensors for Mobile Devices, and Energy Efficient Amplifiers*,
https://doi.org/10.1007/978-3-319-97870-3_8

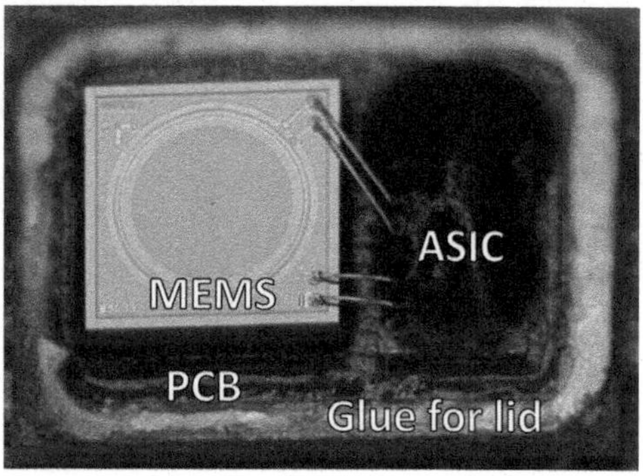

Fig. 1 Microphone system

2 System Overview

The micrograph of a microphone system where the lid has been removed is shown in Fig. 1. Both the readout Application Specific Integrated Circuit (ASIC) and the sensor are glued on a Printed Circuit Board (PCB) that provides mechanical support and electrical connectivity. The ASIC can be covered by an epoxy bubble to prevent environmental influences. A metallic lid that is glued to the PCB makes sure that the system is sealed and sound can only reach the sensor through the sound port opening. The acoustical response of the system can thus be tweaked by selecting different package sizes, sound port diameters, or positions.

3 Capacitive MEMS Microphones and Read-Out Methods

The sensor translates a variation of atmospheric pressure into an electrical quantity. A MEMS microphone generally consists of two conductive plates at a certain distance that form a parallel plate capacitor, as shown in the cross section in Fig. 2. One electrode, the back plate, is perforated but rigid enough not to move even when sound waves are applied. The perforation holes inside the back plate transmit any acoustical stimulus toward the second electrode, the membrane. The membrane is flexible and thus moves proportional to the amplitude and frequency of the impinging sound. The nature of sound can be described by a longitudinal wave, thus the readout of the transducer needs to sense the displacement of the membrane. The sound pressure varies the distances between the membrane and the back plate causing a change of capacitance that can be written as Eq. (1).

$$C\left(P_{\mathrm{S}}\right)=\frac{\varepsilon_0\cdot A}{x\left(P_{\mathrm{S}}\right)}=\frac{\varepsilon_0\cdot A}{x_0+\Delta x\left(P_{\mathrm{S}}\right)} \tag{1}$$

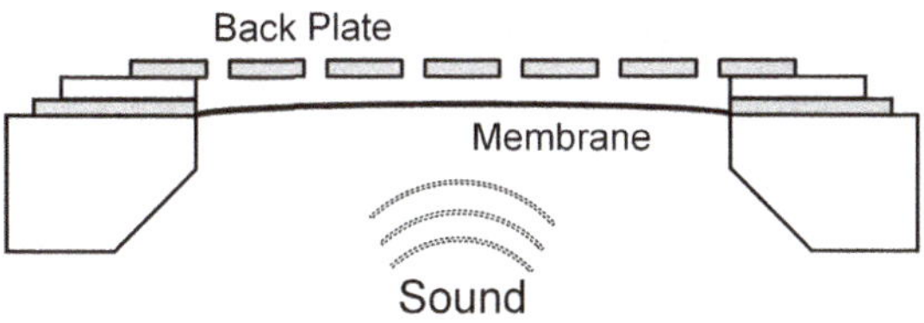

Fig. 2 Cross section of a single back plate MEMS

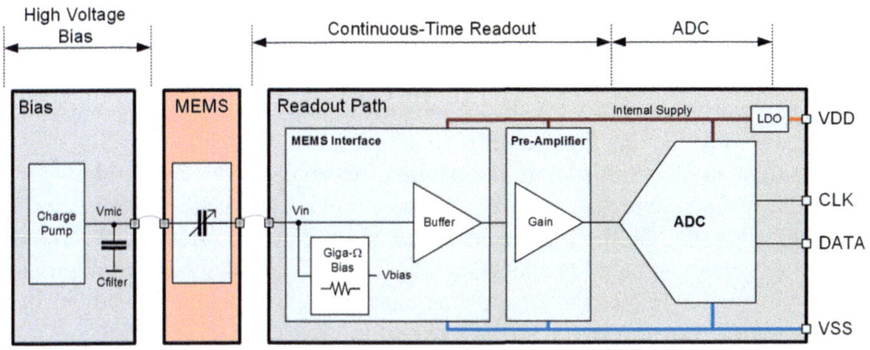

Fig. 3 Block diagram of a typical microphone system

where A is the area of the capacitor plates, ε_0 the vacuum permittivity, x_0 the gap between the plates with no acoustical perturbation, and $\Delta x(P_S)$ the displacement of the membrane when a sound pressure P_S is applied.

There are two possible ways to convert the sound-related capacitance variation into an electrical signal [1]. If a constant charge is applied to the MEMS capacitance by applying a high voltage biasing to the membrane and by isolating the back plate with a high impedance termination, the voltage change at the back plate electrode caused by any MEMS capacitance variations has a linear relation to the sound pressure level. If instead a constant voltage is applied and maintained across the MEMS capacitor, any acoustic perturbation will result in a charge difference that is this time inversely proportional to the displacement.

To achieve the highest possible linearity, the constant charge readout is the preferred solution. A typical block diagram of the interface electronics is shown in Fig. 3. The membrane bias voltage is generated by a charge pump as its value can be in the range of 10 V to achieve sufficient sensitivity. The back plate is terminated by high-ohmic impedance that sets also the correct input common mode for the preamplifier. Depending on the analog or digital nature of the system, a Delta-Sigma modulator converts the amplifier output into a pulse density-modulated data stream.

4 True Differential MEMS Microphone

The performance of the MEMS microphone suffers from the typical limitations of a single-ended system. Its most prominent imperfection is its limited linearity at medium to high signal levels. The request to handle higher acoustical levels at lowest possible distortion levels has led to the industrialization of a true-differential microphone [2] in dual back plate technology. Figure 4 presents a cross-sectional view with a bottom back plate as first signal electrode, a top back plate as second signal electrode, and a flexible membrane between the two electrodes. This dual capacitive construction provides two symmetrical 180° phase shifted signals when the membrane is moving stimulated by sound.

The advantages in terms of THD performance can clearly be noticed in Fig. 5a. Signal levels are expressed in dB Sound Pressure Level (dB SPL) with respect to the human hearing threshold at a sound pressure of 20 μPa corresponding to 0 dB SPL. The green and blue lines represent the THD of a single-ended configuration using only the top or bottom electrode for signal processing, while the red line uses the differential signal across both back plates. In the latter configuration, the THD is significantly reduced from 2% to less than 0.2% at 125 dB SPL. Also, the AOP is shifted by approximately 3 dB toward higher sound pressure levels reaching a distortion of 10% at an SPL of 136 dB. For both improvements, the underlying cause is a cancelation of even order harmonics in the differential output signal.

The benefits in terms of SNR might not appear obvious. At first glance, a second back plate with perforation holes adds an additional source of acoustical noise. On the other hand, the symmetric capacitive configuration can handle higher biasing voltage. Analytical investigations on the so-called pull-in voltage [3] show that a dual back plate MEMS sensor can allow 30% higher biasing voltages compared to a single-ended microphone. This increase in bias voltage can be leveraged to achieve higher sensitivity without any compromise on biasing stability. Figure 5b shows the measurement of sensitivity and SNR versus biasing voltage for a dual back plate microphone. As a result of the increased biasing voltage, a 4 dB higher sensitivity and a 1.5 dB higher signal-to-noise ratio can be obtained.

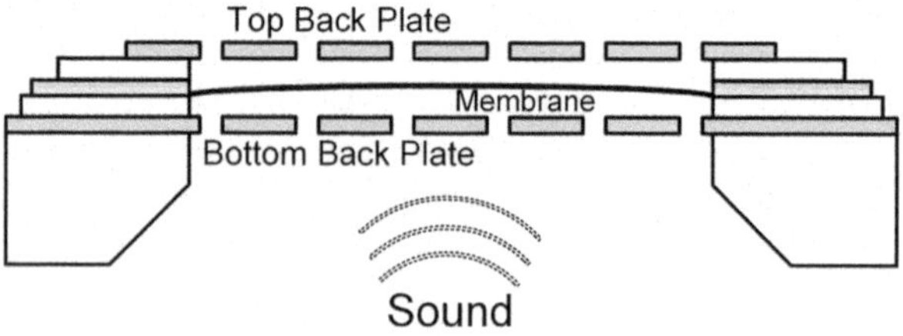

Fig. 4 Cross section of a dual back plate MEMS

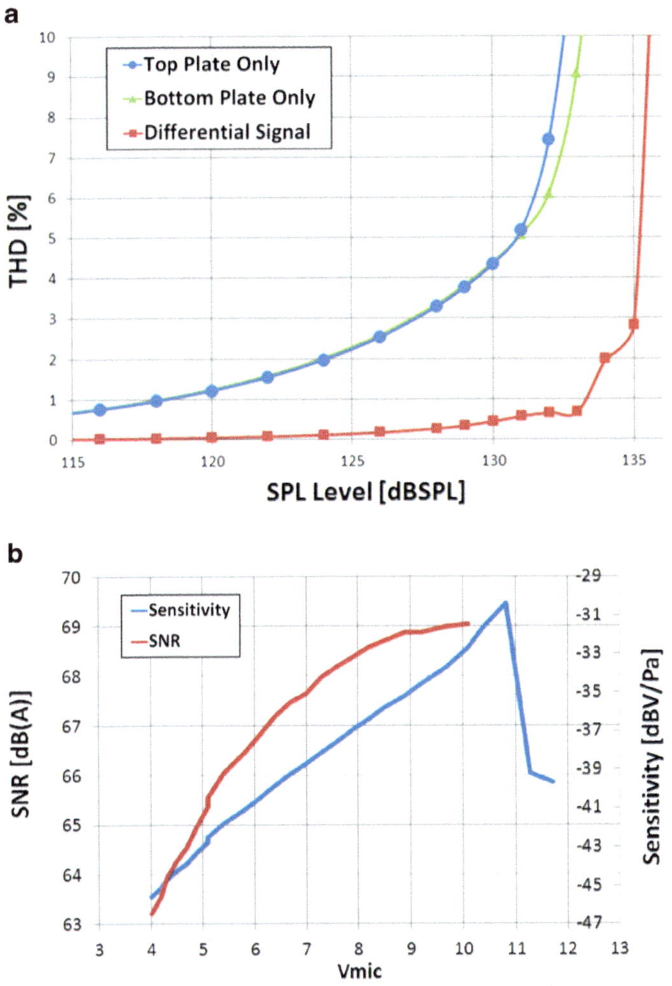

Fig. 5 (**a**) THD of single-ended versus differential MEMS. (**b**) SNR and sensitivity versus bias voltage across back plates

5 Modeling of a MEMS Microphone

MEMS and packaging modeling is an important step when designing MEMS sensors and systems. The models must take into account different physical effects in order to accurately predict the behavior of the physical systems within the frequency range of interest and in the needed application cases.

A first example of simple MEMS microphone model approximates the motion of the membrane and the stator with a piston behavior. This analogy allows the two electrodes of the variable capacitor, to be modeled as rigid flat plates that have only

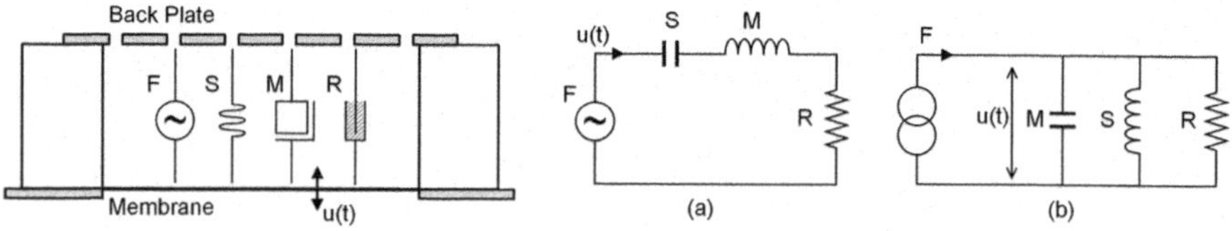

Fig. 6 Main electroacoustical analogies: (**a**) impedance and (**b**) mobility

one degree of freedom. The simple relation for the capacitance showed in Eq. (1) can be exploited to model its variation due to instantaneous pressure changes. The dynamic behavior of such a system follows the equation of motion given in Eq. (2) and can be conveniently approximated with a classic second-order system

$$[M] \cdot \frac{d^2U}{dt^2} + [D] \cdot \frac{dU}{dt} + [K] \cdot U = F_{\text{ext}}(t) + F_{\text{el}}(U, t) \qquad (2)$$

where U represents the motion matrix (i.e., the position), M is the membrane mass matrix, D is the damping matrix, K is the stiffness matrix, and F_{ext} and F_{el} are the acoustic and electrostatic forces applied to the structure.

By exploiting an electroacoustical analogy, a typical behavioral model that follows Eq. (2) is constituted by an RLC resonator tank. There are two main ways to build up the analogy between acoustical and electrical world, depending on the mapping of the two independent variables of the system: the impedance analogy or the mobility analogy, both shown in Fig. 6. In the first, the pressure is related to the voltage of the electrical system while the velocity is represented by the current, and as a consequence, the energy storage elements, that is, compliances and masses, are represented by capacitors and inductors, respectively, while the damping of the system is linked to electrical resistors. The second analogy is nothing more than the dual of the one described above. The choice between the two depends on the complexity as well as on the field of application. As a general concept, the impedance analogy keeps a direct relation between acoustical and electrical impedances while the mobility equivalence keeps the topology unchanged between the two domains.

The described model can be successfully used in almost all small signal applications (AC responses, noise analysis, etc.), but its accuracy tends to decrease when large signals are applied to the system (e.g., when analyzing linearity and distortion of the system). In those cases, more complex models need to be built, and usually a proper discretization of both the membrane and the stator is needed in order to better capture the different membrane oscillation modes. Considering that each node of the discrete system has three degrees of freedom as per Eq. (2), the complexity of such models dramatically increases. The reduction of the order of such a system can be achieved by different means; the most immediate one relies on the cylindrical symmetry of the structure, which allows the substitution of the classical 1-D nodes with an entire concentric annulus that moves as a whole.

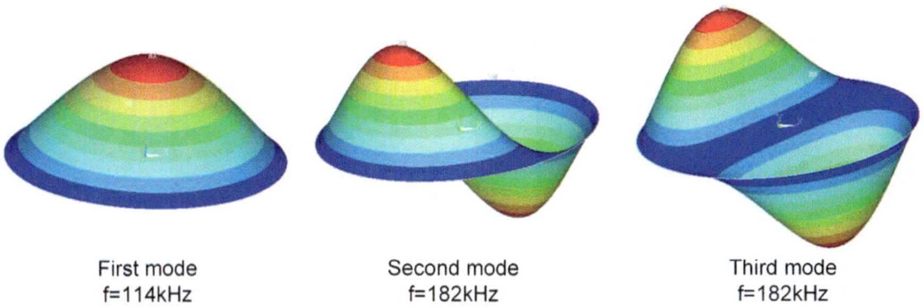

Fig. 7 Simulation of the modes of oscillation of a membrane

In addition, more sophisticated methods for order reduction can be applied; such techniques usually rely on the extraction of the eigenmodes of the system with higher energy content and the successive reduction of the required eigenfunctions to accurately describe the motion of the structure. Figure 7 shows the extraction of the first three modes of oscillation of a membrane obtained through simulation. Typically, the first mode is sufficient to accurately model the microphone at small pressures up to the clipping point of the microphone, while a combination of the first three (or more) modes is needed when describing the behavior of the diaphragm at high pressure levels that bring the device above the clipping threshold.

The convenience and accuracy of the mentioned models can be appreciated in Fig. 8, which shows some typical membrane motion half shape cross sections in the case of homogeneous applied pressure. The ideal stress-dominated (i.e., thin) diaphragm shows a parabolic relation with respect to the radial coordinate, and the structure-dominated case (i.e., a thick diaphragm) exhibits a biquadratic shape. For a direct comparison, the reduced order model shape of a real stress-dominated membrane is plotted. It shows some noticeable deviation from the analytical ideal cases, but significantly decreases the CPU time required to predict the linearity of a microphone system.

6 Modeling of Package Effects

As a first approximation, the microphone package can be modeled with simple lumped element networks that exploit mechanical acoustical to electrical analogies. These models are valid only if the frequencies of interest are low enough and the system dimensions are small enough. As a rule of thumb, the smallest wavelength of an input signal must be 10 times larger than the characteristic dimension of the modeled system. The typical audio band means a dimensional limit of roughly 1.7 mm. When particularly large packages need to be analyzed or high frequencies are involved, the modeling efforts have to include distributed elements approximations making use of the transmission line theory.

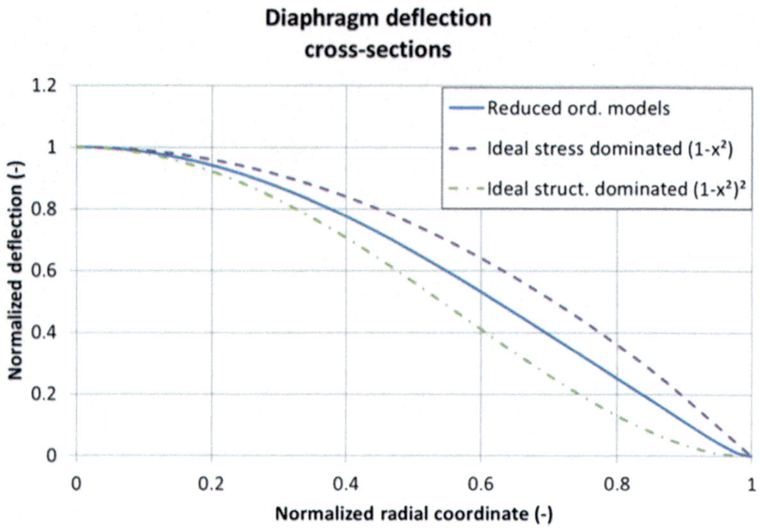

Fig. 8 Comparison between ideal and reduced order model of membrane motion

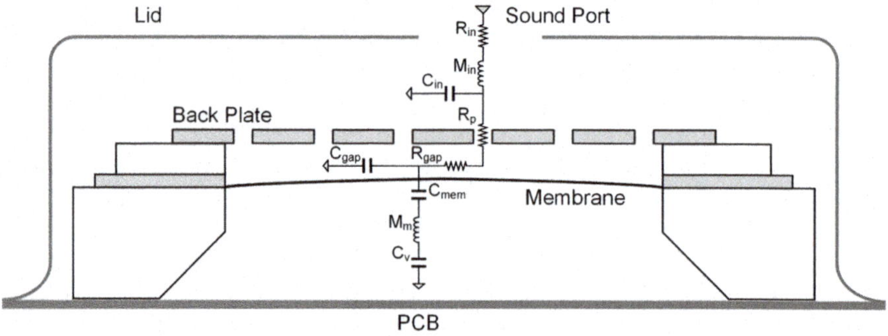

Fig. 9 Lumped element model of a microphone package

A typical lumped element model for a mic package is shown in Fig. 9. The system can ideally be divided into two main parts that interact with the MEMS device: the Helmholtz resonator, a second-order system formed by the sound port and the air in the front cavity and the air in the back cavity. Applying the motion analogy in the lumped element model of the package, the air masses can be represented as inductors and their compliances as capacitors while the damping is modeled with resistors. The frequency response of such a MEMS plus package system can be appreciated in Fig. 10.

It can be seen how the first resonance of the system is determined by the aforementioned Helmholtz resonator that becomes a decisive portion of the design. Usually, the resonance needs to be shifted above the audio band by properly dimensioning the structure, in order to avoid unwanted increase of in-band noise and distortion effects.

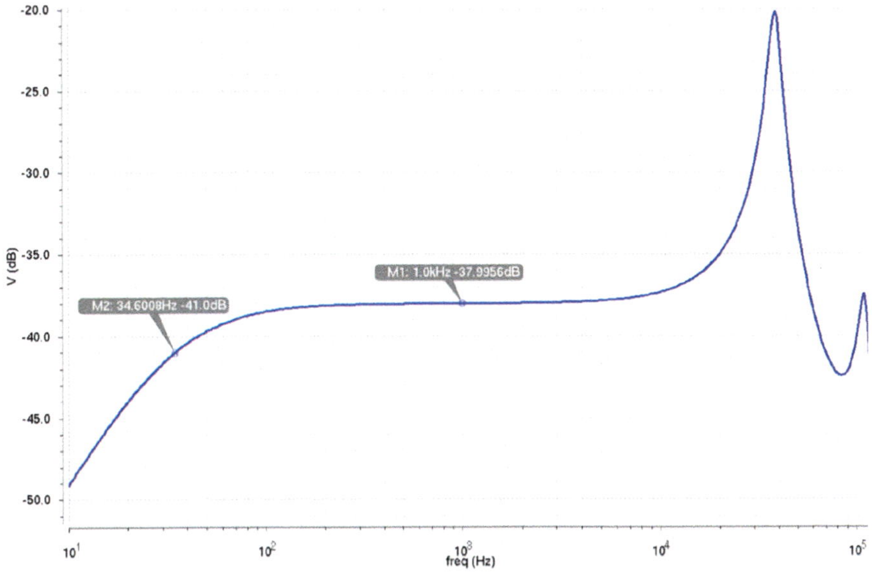

Fig. 10 – Typical frequency response of a microphone package

A further assumption that allows a considerable simplification of the package model is to assume that all the processes in the system are adiabatic (i.e., lossless). Applying the impedance analogy then allows us to model the cavities' compliances with capacitors that depend uniquely on the modeled volume of air and its compressibility, avoiding complex frequency-dependent acoustical impedances. The adiabatic approximation starts to fail when the package dimensions are shrunk; in such cases, more complex acoustical models are needed to account for the losses of the cavities and the nonidealities of the propagation of waves.

7 Design Example: A 140 dB SPL Digital Microphone with a 67 dB SNR

The chosen design example intends to highlight a selection of circuit design techniques that were needed to handle the demanding sound pressure requirements. Considering a typical sensitivity of −38 dBV/Pa at 94 dB SPL, differential signal swings up to 2.5 Vrms at 140 dB SPL must be processed by the readout electronics being supplied with 1.8 V ± 10% [4].

The architecture of the microphone system depicted in Fig. 11 is similar to the general block diagram already shown in Fig. 3. The MEMS biasing includes a high-voltage ripple optimized charge pump and an input buffer with high-ohmic termination. A third-order feed-forward 6b switched capacitor $\Delta\Sigma$ modulator with

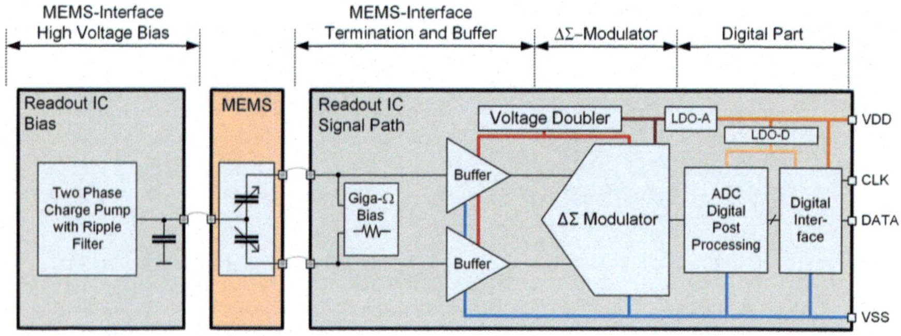

Fig. 11 Block diagram of proposed design example

high input impedance feeds the digital post processing that equalizes the MEMS resonance peak and filters the modulator quantization noise. A fifth-order digital noise shaper derives the single-bit pulse-density modulation (PDM) output. To avoid signal swing limitations, the MEMS interface buffer and parts of the $\Delta\Sigma$ modulator are supplied with a 2.7 V supply generated by an on-chip voltage doubler.

The charge pump can generate up to 14.5 V using a two-phase 10-stage Dickson structure and is designed without dedicated high-voltage transistors. Its output ripple is attenuated by a 10 Hz low-pass filter comprising PMOS diodes forming a larger than 1 GΩ resistor in series to a 10 pF metal capacitor.

Noise optimum PMOS source followers can be used to buffer the MEMS outputs, as the $\Delta\Sigma$ modulator has high input impedance. Two 150 GΩ termination resistors built as antiparallel branches of stacked PMOS diodes provide the DC operating point for the buffer. As the MEMS attenuates frequency components above 40 kHz, a weak 3 MHz cutoff anti-alias filter is added as a gm-C filter at the source follower output.

The readout IC's signal flow diagram is depicted in Fig. 12. Conventional $\Delta\Sigma$ modulators implement the subtraction of the feedback signal from the input signal in the current [5] or charge domain. This example shows a subtraction in the voltage domain enabling a high input impedance of the modulator using a low power concept for the DAC that acts as a floating voltage source. The modulator feedback controls the value of the floating voltage source V_{FB} based on the quantizer output. The additional ADC buffer reduces loading effects on the MEMS buffer and the floating voltage source. That enables the use of a Resistive-DAC (R-DAC) to generate the required voltage offset V_{FB} between the modulator input V_{SDM} and the loop filter input V_{LF}. To support large input signal swings, the R-DAC is operated from the 2.7 V supply. ADC buffer, loop filter, and quantizer can operate under a low supply voltage as the signal swing is already limited.

A draft schematic of the analog readout circuit is shown in Fig. 13. The buffered MEMS output is applied to the center tap of two parallel resistor strings that form the R-DAC together with two sets of switches and current sources. The switches are controlled by a 64b one-hot code generated by the quantizer. As the R-DAC

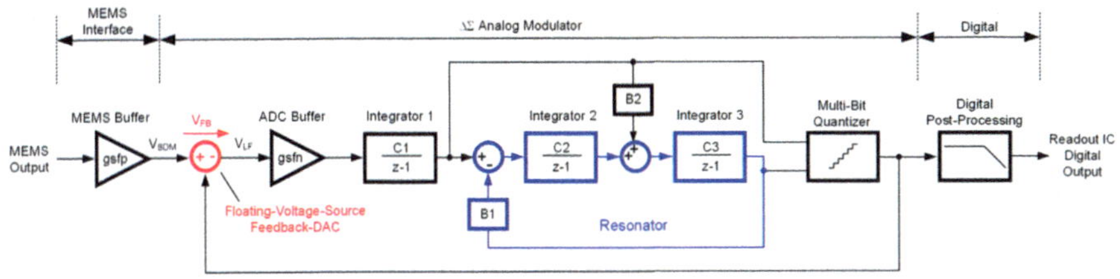

Fig. 12 Signal flow diagram of proposed design example

has a constant current consumption, a voltage doubler can be used to derive the 2.7 V supply. The loop filter is driven by a NMOS source follower to isolate the load of the first integrator from the R-DAC. Hence, the input impedance of the $\Delta\Sigma$ is determined only by the impedance of the R-DAC current sources and is independent of the input signal level over a wide range. The first integrator uses a telescopic gain-boosted OTA to obtain lowest possible noise with a 7.5 pF sampling capacitor. The second integrator is a standard folded-cascode design, while the third integrator uses a two-stage Miller OTA to drive the 2 Vpp-diff input range of the quantizer. The latter has been chosen to have an output of 6b, considering the trade-off between DR, quantization noise, unit resistance in the R-DAC, and dynamics inside the loop filter. The quantizer is designed as a tracking architecture comprising five comparators [6] to save area and power.

In terms of noise analysis, the R-DAC can be modeled as a programmable resistor between MEMS and ADC buffer. Hence, the DAC noise is a function of the signal level. At low signal levels, that is, in single-bit mode, only the inner taps of the R-DAC are used and the optimum THD and noise performance can be achieved. Approximately, 40% of the modulator's noise is due to the thermal noise of the R-DAC resistors and the ADC buffer. 1/f noise is a negligible part in the overall modulator noise due to application of correlated double sampling in the first integrator. At higher signal levels, multiple unit resistors are connected in series, increasing noise and degrading THD by INL limitations due to unit resistor mismatch. This is shown in Fig. 17 comparing the SNR and SNDR of the readout circuit to the microphone, showing that the electronics are not limiting the system performance. A reduced THD is acceptable for audio application as requirements relax with increasing SPL.

8 Digital Signal Processing

The third main block in the signal path (in addition to the analog front-end amplifier and to the ADC) is represented by the digital signal processing, whose main task is not only to convert the ADC multi-bit stream into a single-bit data stream (only one DATA pad is allowed due to area and package limitations), but also to shape

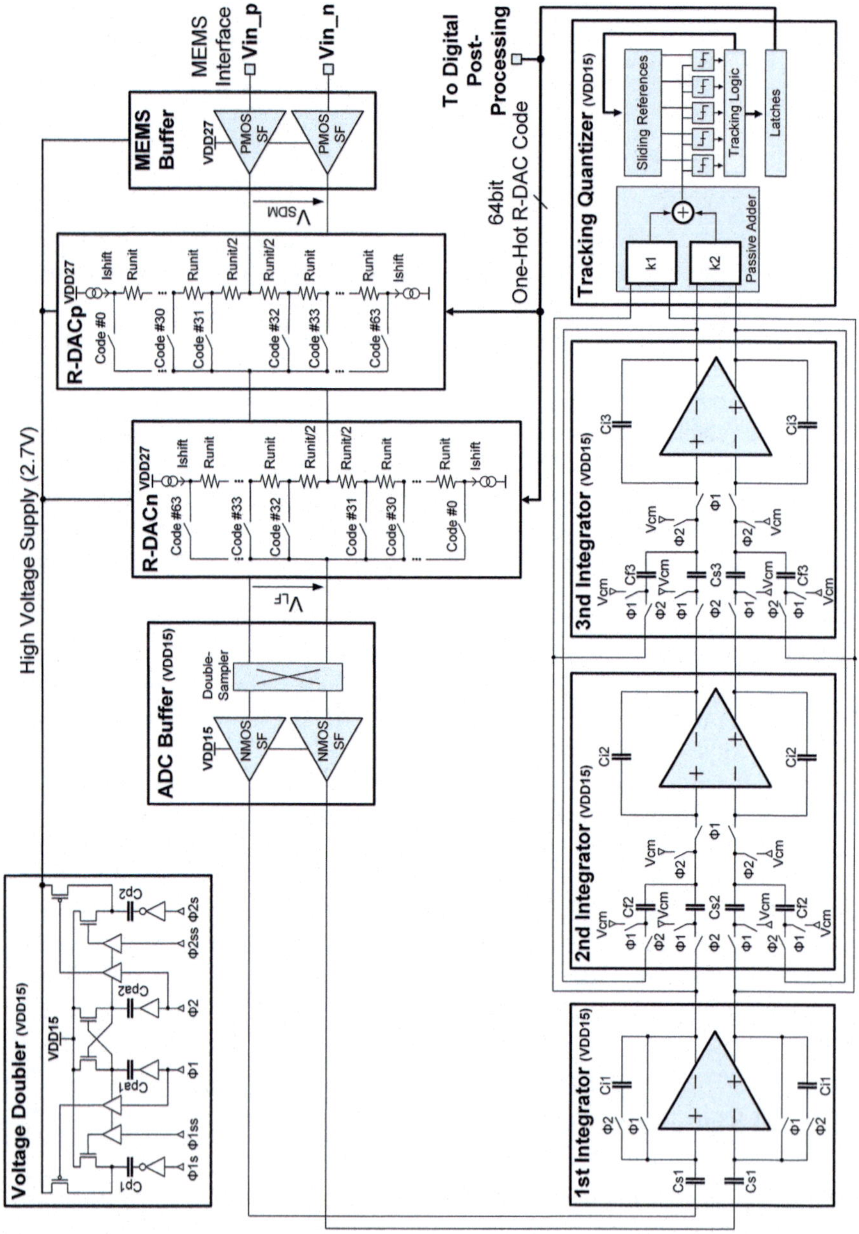

Fig. 13 Schematic of proposed design example

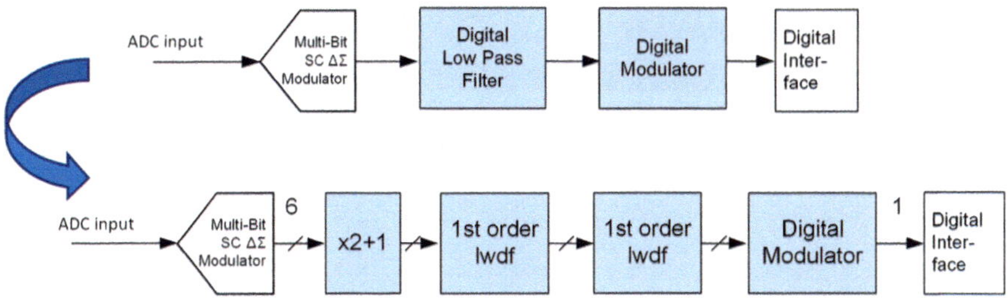

Fig. 14 Block diagram of the digital post-processing

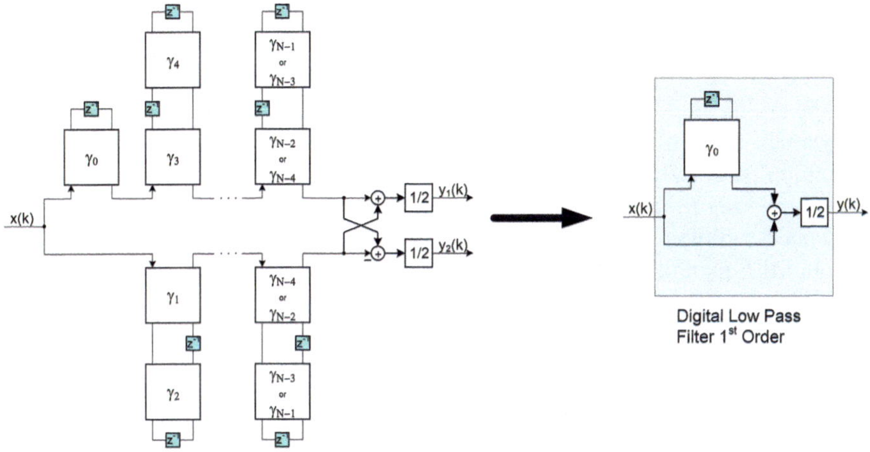

Fig. 15 Lattice Wave Digital Filter (LWDF): example of implementation of a low-pass filter cell

the frequency response of the microphone with SNR and bandwidth optimization. Figure 14 reports the block diagram of the overall digital filter chain which samples the signal at the output of the six-bit quantizer. A second-order programmable digital low-pass filter attenuates the mechanical resonance peak of the MEMS and improves SNR and a fifth order digital modulator implemented in cascade of resonators with feed-forward (CRFF) topology provides a single-bit PDM signal, typically required for these applications.

To implement the low-pass filter, the well-established wave digital filter (WDF) structure is used [7]. WDFs are highly efficient recursive IIR filters derived from a continuous time equivalent, and they are available in different types. The Lattice topology is used for this design as shown in Fig. 15. In general, WDFs have excellent stability properties under finite arithmetic conditions and show very low coefficient sensitivity. The number of coefficients is determined by the filter order; therefore, the low computational complexity enables area and power efficient implementations. In this design, two first-order low-pass filters are cascaded, because with lattice topology only odd order low pass-filters are available.

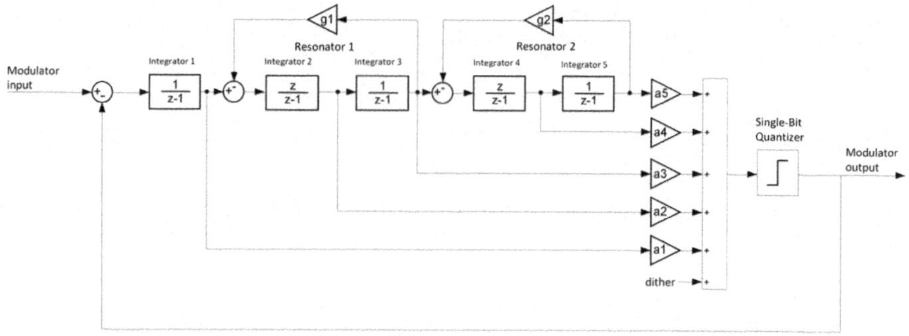

Fig. 16 Digital modulator: fifth-order CRFF topology with single-bit quantizer

The last block of the digital processing chain is a digital modulator, whose block diagram is shown in Fig. 16. This block is a modulator with two resonator poles, one integrator pole, and a one-bit quantizer. Integrator stages with optimized saturation levels are designed to meet high sound pressure level THD specifications. To avoid unwanted audible tones caused by the feedback structure of the modulator at the input of the quantizer, a pseudo random single-bit signal, which acts as a dither, is added.

9 Measurement Results

All system level measurements are performed in an anechoic box using a reference speaker to generate the acoustical stimuli driven by an R&S UPV audio analyzer. The characterization of the standalone readout circuit is performed using an APx525 audio analyzer. Two low-leakage capacitors of the same value of the MEMS rest capacitances are soldered at the circuit inputs to mimic the noise transfer function properly. The input signal is calibrated according to the target system sensitivity time window. Figure 17 reports SNR and SNDR measurements, while a 1 kHz sinusoidal input signal had been applied for ASIC only and for the fully system comprising the MEMS. A system SNR of 67 dB-A can be achieved with the proposed ASIC architecture and a MEMS SNR of about 74 dB-A.

On average, a sensitivity of −45.95 dBFS (nominal target is −46 dBFS) is reached over 120 samples with a system noise performance of 67.03 dB-A (Fig. 18). The signal to noise ratio is calculated according to

$$\mathrm{SNR} = \frac{\text{Signal power at 1Pa}}{A - \text{weighted noise power in } [20\ldots 20\text{k}]\,\text{Hz}} \tag{3}$$

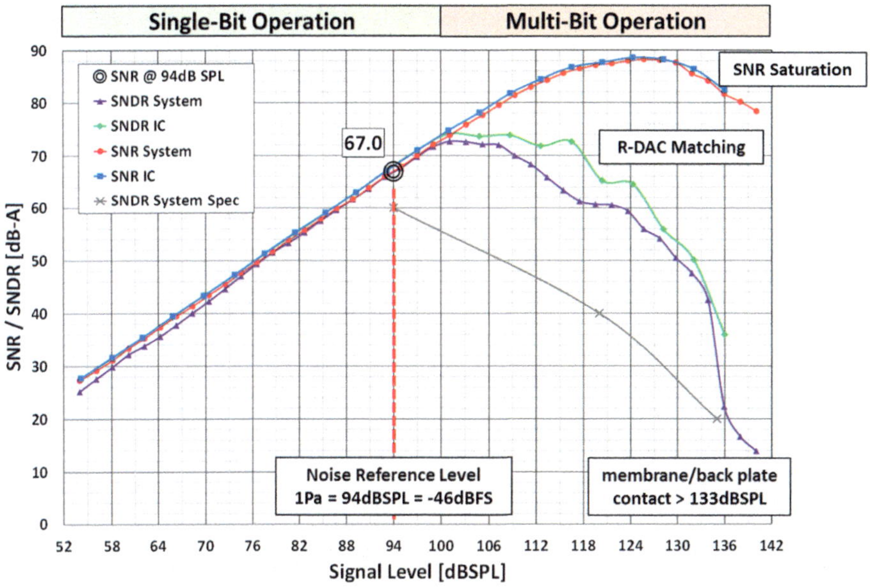

Fig. 17 Measurement results comparing SNR and SNDR of readout IC versus system

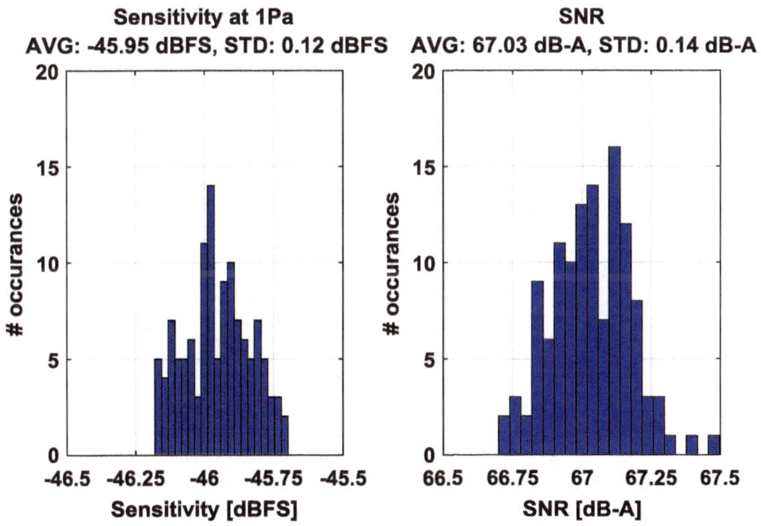

Fig. 18 Sensitivity and SNR distribution for 120 system samples measured with 1 kHz sinusoidal signal

Finally, Fig. 19 shows the THD + N plot of 120 samples with a negligible distortion at 94 dB SPL and a typical distortion of 1% at 130 dB SPL. The Acoustic Overload Point (AOP) of 10% THD is above 135 dB SPL and typical distortion at full scale of 140 dB SPL is less than 20%.

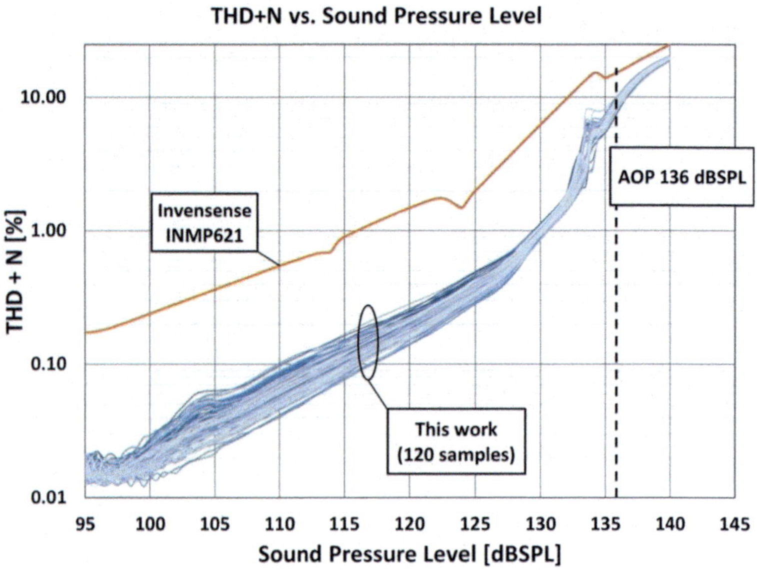

Fig. 19 THD + N versus sound pressure level of 120 system samples

10 Communication and Control Interfaces

The Latest MEMS microphones are equipped with a communication interface, which is used to program the memory registers (both volatile and nonvolatile) and to access the chip for testing and internal diagnostics. A few bytes of nonvolatile memory are typically integrated in microphones to support several audio full-scale and gain adjustments, trimming of the reference circuitry, and programming of the main charge pump to properly bias the sensor and compensate pull-in voltage production spread.

The main constraint for the communication interface is definitely represented by area (a lower complexity interface typically corresponds to a lower amount of gates to be synthetized) and pad limitations: standard acoustic packages for digital microphones have only five pins (VSS, VDD, CLK, L/R, and DATA pads). The DATA pad is used for the main data stream, transmitting a PDM signal from the microphone to the codec synchronous to the master clock CLK. The L/R pin (left/right) is used to enable stereo application by defining the time slot used for transmitting the PDM data stream on a shared DATA bus for two time-interleaved microphones. During normal operation, the L/R pin is connected either to GND or to VDD depending on the desired channel, but the same pin can be used in the calibration/debugging mode to establish a communication with the ASIC master logic using a one-wire protocol.

The one-wire interface is a full-duplex bidirectional programming, testing, and debug interface, allowing data rate up to about 500 kbit/s in a robust way through a

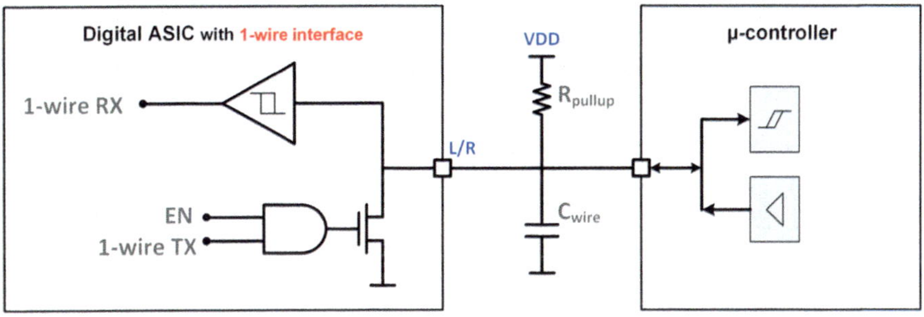

Fig. 20 – One-wire interface principle

Command	Command Code	Address	Data In/Out	Stop Bit
	C1:C0	A6:A0	D7:D0	STOP
READ	"01"	address<6:0>	write data<7:0>	'0'
WRITE	"10"	address<6:0>	read data<7:0>	'0'

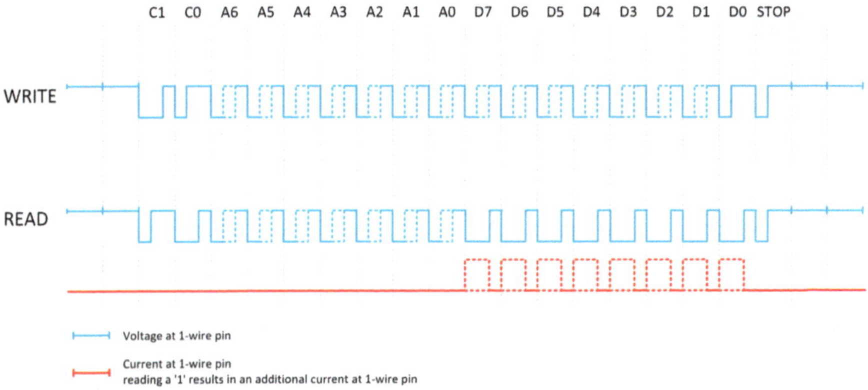

Fig. 21 One-wire interface: examples of transmission of a "1" and a "0"

single pin. It is meant to be available during the whole product lifetime; it features a low-voltage transmission (0 V–VDD) requiring an open-drain drive and it is based on a PWM modulation. In Fig. 20, a sketch of this interface concept is shown, where all four options to transmit a bit in both directions are visible.

The microphone RX channel (shared with the TX) senses the voltage and sends information by current impulses.

The bit period must stay constant during a read/write cycle and its duty cycle determines the transmission of a high or low signal: in order to communicate a "1," the duty cycle is 66% and in order to transmit a "0," the duty cycle is set to 33%. An example of a RX and TX communication is reported in Fig. 21: two bits are used to decode the command (read or write), seven bits dedicated to the address, and eight bits for the data; the communication is concluded with a 0-stop bit.

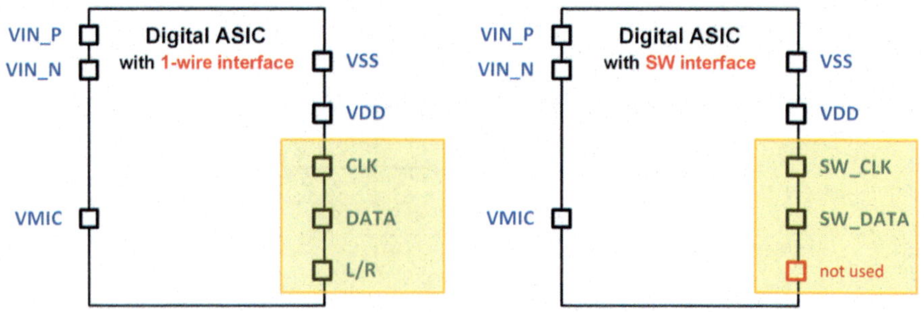

Fig. 22 One-wire and DATA/CLK versus MIPI SoundWire interface pinouts

Future microphone generations are expected to integrate the MIPI SoundWire[1] protocol, since this standard strives to be adopted as the main communication channel in mobile systems for audio applications. The SoundWire protocol is a digital interface whose specifications and versions are driven by MIPI Alliance; it supports bidirectional digital communication with the focus on being attractive for mobile audio systems and, therefore, facilitating low complexity and a minimum number of gates. It allows adding intelligence to audio peripherals, increasing the number of peripherals attached to a link and optimizing their implementations without compromising product cost, pin count, power consumption, software complexity, or key audio metrics. MIPI SoundWire provides built-in synchronization capabilities and optional multilane extensions, and it supports Pulse-code modulation (PCM) and PDM, multichannel data, and isochronous and asynchronous modes. Figure 22 compares the pinout of a current digital microphone where an auxiliary one-wire control interface is integrated together with the DATA/CLK lines with one where the SoundWire is adopted taking advantage of its low power, low latency, two-pin multidrop bus that allows for the transfer of multiple audio streams and embedded controls and commands. Thus, compared to the current solution, a saving of one pad is possible with the additional benefits listed earlier.

11 Conclusions

The silicon microphone market has been growing in recent years and faces new challenges, driven by the beamforming and wind suppression applications that require high SNR with an acoustic overload point of 135 dB SPL. To cope with such scenarios, a digital microphone based on a dual back plate MEMS has been fabricated using standard CMOS MEMS technologies. Measurements show that it achieves 67 dB SNR with an AOP of 136 dB SPL. The architecture which enables

[1] SoundWire is a copyright of MIPI Alliance (www.mipi.org).

Table 1 Digital microphones supporting high SPL: overview and comparison of specifications

Parameter	This work	InvenSense INMP621	ST Micro-electronics MP34DB02	Knowles SPH1668LM4H
Microphone type	Bottom Port	Bottom Port	Bottom Port	Bottom Port
Package size [mm]	4 × 3 × 1	4 × 3 × 1	4 × 3 × 1	3.5 × 2.65 × 0.98
Sound port diameter [mm]	0.35	0.5	0.25	0.325
Supply voltage [V]	1.8	1.8	1.8	1.8
Clock [MHz]	3.072	3.072	2.4	2.4
Sensitivity 1 kHz sinusoidal ® 94 dB SPL [dBFS]	−46	−46	−26	−29
Full-scale acoustic level (0 dBFS output) [dBSPL]	140	140	120	123
SNR A-weighted 20 Hz to 20 kHz [dB-A]	67	65	62.6	65.5
Dynamic range [dB]	113	111	88.6	94.5
Acoustic overload point (10% THD) [dBSPL]	136	133	120	122
THD 1 kHz sinusoidal @ 115 dB SPL [%]	<0.2	0.9	<5	–
THD 1 kHz sinusoidal @125 dB SPL[%]	<0.5	2	N.A.	N.A.
THD 1 kHz sinusoidal @ 140 dB SPL [%]	20	24[a]	N.A.	N.A.
Frequency response [Hz]	40–20 k	45–20 k	80[a]–20 k	80[a]–20 k
Current consumption [μA]	1200	1200	650	626

N.A.: not applicable
–: data not provided in datasheets
[a]Extrapolated data from charts available in datasheets

these specifications is based on a six-bit switched-capacitor ΔΣ modulator, with a voltage mode summing DAC combining low power and high DR requirements. The readout IC has been implemented in a standard 130 nm CMOS technology with 1.8 V supply. Table 1 lists other digital microphones based on MEMS technology, which are currently on the market, comparing their performance to that achieved in this work.

References

1. Malcovati P, et al. Interface circuits for MEMS microphones, Springer, Nyquist AD converters, sensor interfaces, and robustness. 2011. p. 149–74.
2. Füldner M, et al. Dual Back Plate Silicon MEMS Microphone: Balancing High Performance!. DAGA 2015: 41. Jahrestagung für Akustik, March 2015, Nürnberg.
3. Martin D. Design, fabrication, and characterization of a MEMS dual-backplate capacitive microphone, PhD thesis, University of Florida, 2007.

4. Bach E, et al. A 1.8V true-differential 140dB SPL full-scale standard CMOS MEMS digital microphone exhibiting 67dB SNR. In: 2017 IEEE International Solid-State Circuits Conference (ISSCC), San Francisco, CA, 2017, p. 166–7.
5. De Berti C, et al. A 106.7-dB DR, 390-μW CT 3rd-order ΣΔ modulator for MEMS microphones. In: Proceedings of ESSCIRC. 2015. p. 209–12.
6. Dörrer L, et al. A 3-mW 74-dB SNR 2-MHz continuous-time delta-sigma ADC with a tracking ADC quantizer in 0.13-μm CMOS. IEEE JSSC. 2005;40:2416–27.
7. Gazsi L. Explicit formulas for lattice wave digital filters. IEEE Trans Circuits Syst. 1985;32(1):68–88.

High-Performance Dual-Axis Gyroscope ASIC Design

Zhichao Tan, Khiem Nguyen, and Bill Clark

1 Introduction

Gyroscopes are widely used in motion sensing applications such as automotive safety, gaming, navigation, and so on [1–3]. High-performance gyroscopes are particularly important for new industry applications such as electronic stability control (ESC) systems, which improve vehicle safety [4, 5], or in optical image stabilization (OIS) systems that compensate for camera movement while a picture is taken [6, 7].

1.1 Sensor Physics

Sensors rely on the transduction of a desired measurand to a more easily measured and quantified electrical property [8, 9]. This may be a multistep process. In inertial MEMS, the final transduction step is most commonly capacitance. For example the displacement of a MEMS sensor will result in a change of capacitance and hence charge, and the voltage applied to a capacitor will result in an electrostatic force, which can also be springlike (a force that varies with displacement). As shown in Fig. 1, arrays of capacitors can be used to sense the displacements of, or apply forces the proof mass of a vibratory rate gyroscope.

The relevant capacitive transductions are below. Equation 1 expresses the change in capacitor charge resulting from either a change in voltage across the capacitor or from a change in capacitance in the presence of a constant voltage. The force in

Z. Tan (✉) · K. Nguyen · B. Clark
Analog Devices, Inc., Wilmington, MA, USA
e-mail: Zhichao.Tan@analog.com

K. A. A. Makinwa et al. (eds.), *Low-Power Analog Techniques, Sensors for Mobile Devices, and Energy Efficient Amplifiers*,
https://doi.org/10.1007/978-3-319-97870-3_9

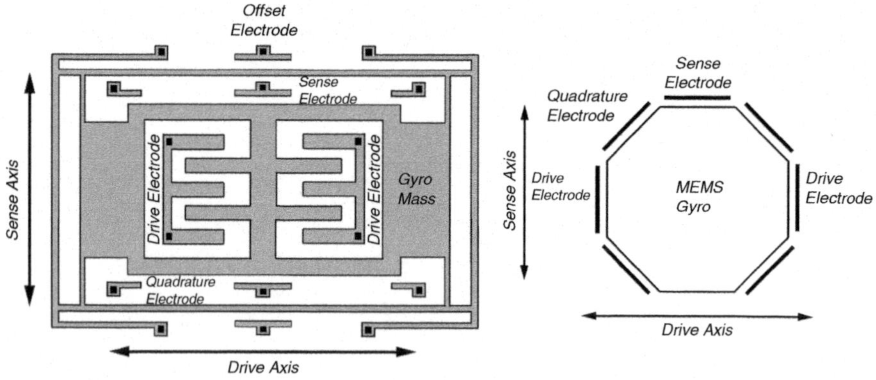

Fig. 1 MEMS gyroscope sensor transducer

the presence of a static voltage is also approximated below, showing both a constant force plus two springlike forces. A further approximation provides the linearized model of force electrostatically transduced in the presence of a large bias voltage V_b. Transduction of capacitance change is done to assess a change in sensor geometry due to the measurand, e.g. acceleration. Force transduction is generally used to counteract the measurand using a feedback to achieve force balance [3].

$$\Delta Q = C\Delta V + \boldsymbol{\Delta C V} \approx C\Delta V + \frac{\partial \boldsymbol{C_x}}{\partial \boldsymbol{x}} \boldsymbol{\Delta x V_b} + \ldots$$

$$F_x = \begin{cases} = \nabla\left[\frac{1}{2}C\left(x,y\right)V^2\right]_x \\ \approx -\left[\frac{\partial C}{\partial x} + \frac{\partial^2 C}{\partial x^2}\Delta x + \frac{\partial^2 C}{\partial x \delta y}\Delta y + \ldots\right]\frac{V^2}{2} \\ \approx -\left[\frac{\partial C}{\partial x} + \frac{\partial^2 C}{\partial x^2}\Delta x + \frac{\partial^2 C}{\partial x \delta y}\Delta y + \ldots\right]V_b V \end{cases} \quad (1)$$

In a vibratory rate gyroscope, the transductions above respond to Coriolis acceleration (Eq. 2) where a rotation rate Ω_z interacts with the amplitude and frequency of oscillation of a proof mass X_0 and ω_x, respectively. The input rotation rate is assumed to be about the Z-axis and a vibration or sustained oscillation is along the x-axis (or more correctly, in the x-mode) and the response to Coriolis acceleration is in the mutually orthogonal y-axis/mode.

$$\ddot{y}_{\text{Coriolis}} = 2X_0\omega_x\Omega_z \quad (2)$$

In an open-loop configuration, the deflection amplitude that results from the Coriolis acceleration Y_{Coriolis} is detected. The magnitude of the response, shown in Eq. (3), is also a function of the resonant frequency of the Coriolis mode ω_y. Two simplifications of the Coriolis response amplitude are included for cases where the

Table 1 Possible representations of input ratio rate based on transduction

Ω_z^{EST}/Ω_z	Resonator deflection transduction	Resonator force transduction
Coriolis deflection transduction	$\frac{2Q_y}{\omega_x}\frac{\partial x}{\partial C_x}\frac{\partial C_y}{\partial y}\frac{V_b}{V_b}$ or $\frac{1}{\mid\omega_y-\omega_x\mid}\frac{\partial x}{\partial C_x}\frac{\partial C_y}{\partial y}\frac{V_b}{V_b}$	$\frac{2Q_yQ_x}{M_x\omega_x^3C_{ref}}\frac{\partial C_x}{\partial x}\frac{\partial C_y}{\partial y}V_b^2$ or $\frac{2Q_yQ_x}{M_x\omega_x^2\mid\omega_y-\omega\mid C_{ref}}\frac{\partial C_x}{\partial x}\frac{\partial C_y}{\partial y}V_b^2$
Coriolis force transduction	$\frac{2M_y\omega_xC_{Ref}}{\frac{\partial C_x}{\partial x}\frac{\partial C_y}{\partial y}V_b^2}$	$\frac{2Q_x}{\omega_x}\frac{M_y}{M_x}\frac{\partial C_x}{\partial x}\frac{\partial y}{\partial C_y}\frac{V_b}{V_b}$

resonant frequencies are matched or near matched. In a closed-loop configuration, an electrostatic force is generated to balance the Coriolis acceleration and resulting inertial force scaled by the mass of the Coriolis mode M_y. For completeness, the magnitude of the force used to generate the driven oscillation is also included in Eq. (4).

$$Y_{\text{Coriolis}}\begin{cases} = \dfrac{2X_0\omega_x\Omega_z}{\sqrt{\left(\omega_y^2-\omega_x^2\right)^2+\left(\frac{\omega_y}{Q_y}\omega_x\right)^2}} \\ = 2X_0Q_y\frac{\Omega_z}{\omega_x} \quad \text{when } \omega_x=\omega_y \\ \approx X_0\frac{\Omega_z}{|\omega_y-\omega_x|} \quad \text{when } 1 \ll Q_{\text{eff}}=\frac{\omega_x}{2|\omega_y-\omega_x|} \ll Q_y \end{cases} \tag{3}$$

$$\begin{aligned} F_{Y_{\text{Coriolis}}} &= 2M_yX_0\omega_x\Omega_z \\ F_{X0} &= \frac{X_0M_x\omega_x^2}{Q_x} \end{aligned} \tag{4}$$

The open-loop deflection responses to rotation rate are summarized in Eq. (3). These can be transduced to charge using the relationship in Eq. (1). The closed-loop forces applied are captured in Eq. (4) and can also be transduced to voltage as described in Eq. (1). Given the transduced properties in the electrical domain identified earlier, there are fundamentally four options to quantify the rotation rate Ω_z. These are characterized in Table 1 as either displacement or force transduction for the Coriolis and resonator axes.

These relationships are related to the sensitivity of the sensor. Ideally, the resulting sensor will have a stable and predictable sensitivity. For best stability and predictability, ratios of similar parameter sets are preferred, as in the upper left and lower right of the table. The mass of the sense elements is arguably the most stable of the parameters shown, while the resonant frequencies of MEMS structures are also considered to be stable and easily measured. Typically, Q_x will vary significantly over temperature and time leading to instability. Mismatched elements as found in the lower left and upper right of Table 1 are less stable and this may be exacerbated by the $1/f$ noise-induced stability limit of the bias voltage V_b. Due to its simplicity, predictability, and stability, most commercially available gyroscopes use the frequency mismatch version of the upper left table entry.

In addition to simple sensitivity errors, there are a number of offset errors that can be identified in a vibratory rate gyroscope. Typically, the largest of these occurs when the resonator mode x is not orthogonal to the Coriolis sense capacitor (i.e., $\frac{\partial C_y}{\partial x} \neq 0$). This will result from an error in the manufacture of the Coriolis sense capacitor with respect to the manufacture of the spring/mass system, which defines the resonator mode. The error signal from this mechanism is proportional to the resonator displacement rather than resonator velocity and is therefore in quadrature with the Coriolis acceleration. This error can be detected in a manner similar to the rotation rate by demodulating the quadrature signal in addition to the in-phase signal. The relative signal magnitude resulting from this error mechanism is given by Eq. (5) in terms of both displacement and force equivalents. By design and with quality manufacture, the error source ($\frac{\partial C_y}{\partial x}$) is quite small, but the final quadrature error is still large, if only because the gain term of the ratio of resonant frequency to rotation rate is very large.

$$\begin{aligned}\frac{Y_{\text{Quadrature}}}{Y_{\Omega_z}} &= \frac{\partial C_y}{\partial x}\frac{\partial y}{\partial C_y}\frac{|\omega_y-\omega_x|}{\Omega_z}\\ \frac{F_{\text{Quadrature}}}{F_{Y\text{Coriolis}}} &= \frac{\partial C_y}{\partial x}\frac{\partial y}{\partial C_y}\frac{\omega_x}{2\Omega_z}\end{aligned} \tag{5}$$

The stability of this error term relies on the geometry of the sense capacitors and the geometry and the springs that define the resonator mode and can be reliably stable provided the geometry of the structure does not change (e.g., due to stress). Even if stable, the magnitude of this error source can be much larger than full-scale Coriolis signals and swamp the Coriolis signal path. The signal levels can be managed by injecting a counteracting charge into the signal path, for example. This generally involves dissimilar elements which, while reducing the signal content, can introduce additional sources of error to the signal path. In the system described below, an additional electrostatic trim is introduced.

A gyroscope sensing element along with electrostatic quadrature trim electrodes are shown in Fig. 1. Application of bias voltages on these electrodes results in precision cancellation of the quadrature error source. Electrostatic quadrature trim is used to accurately reorient the driven resonator mode to be orthogonal to the Coriolis sense electrode. This is accomplished using capacitors that have a deliberately large cross-axis component $\frac{\partial^2 C_Q}{\partial x \partial y}$. The force generated by the quadrature electrode is given by Eq. (6) along with the quadrature error we expect to cancel. Both terms are proportional to the displacement of the resonator mode in both magnitude and phase. By trimming or by continuously adjusting quadrature voltage V_Q in a feedback loop, this error trim can be eliminated in a phase accurate manner. Because this correction electrostatically manipulates a cross-axis spring, there will be subtle changes to the observed resonant frequencies of the two modes ω_x and ω_y, which are now aligned to the Coriolis sense capacitor.

$$F_Q = \frac{\partial C_Q}{\partial x \partial y} X_0 V_b V_Q = -\frac{\partial C_y}{\partial x}\frac{\partial y}{\partial C_y} X_0 M_x \omega_x^2 \tag{6}$$

1.2 Sensor Electronics

There are two primary signal-processing subsystems in a MEMS gyroscope readout ASIC, namely the drive and sense channels. The drive channel maintains a sustained oscillation of the proof mass (sensor disc) at resonance, thereby creating velocity, which combines with rotation rate to generate Coriolis acceleration. The sense channel senses the Coriolis acceleration centered at the resonant frequency, demodulates that to baseband to extract the rotation rate information, and finally converts that to either analog [10] or digital output [11–14]. The drive channel is usually implemented as a closed loop topology system that can sustain an oscillation of fixed amplitude. The sense channel can either be an open or closed loop system. In the case of a closed-loop system, electronic force feedback is used to null Coriolis acceleration [15, 16]. An open-loop topology, on the other hand, directly converts motion-induced capacitor changes to voltage or current for further processing. Open-loop sense channel topology is more popular due to its stability, robustness, and straightforward implementation [11–14].

This chapter describes a dual-axis (pitch and roll) MEMS vibratory gyroscope readout ASIC [17]. The readout ASIC includes complete signal chains of the drive channel, sense channels, and other signal-processing channels. An on-chip charge pump is also implemented to create a high-voltage supply for the sensor driving signal. The high voltage has two benefits: to drive the MEMS sensor disc with larger amplitudes and to yield larger displacement sensing signals in both the drive and sense channels, which lead to high signal to noise ratio (SNR). A regulator and temperature sensor are also integrated to give the analog building blocks better power supply rejection ratio (PSRR) and temperature monitoring capability.

This chapter is organized as follows: Sect. 2 describes the gyroscope ASIC design and some important circuit building block implementations. Sect. 3 presents measurement results. The chapter concludes in Sect. 4.

2 ASIC Design

Figure 2 shows a top-level diagram of the proposed gyroscope ASIC. The ASIC mainly consists of one drive channel, which drives the sensor at its resonant frequency, and two sense channels for pitch and roll sensing, which convert physical rotation information into digital output. The charge pump is built in to create gyroscope high drive voltage for high output signal level from the MEMS sensor. Detailed system and circuit design are presented below.

2.1 Coriolis Channel Design

Figure 3 shows a block diagram of the sense channel. There are two identical sense channels in the proposed ASIC for the pitch and roll axes. Each sense loop contains a transimpedance amplifier (TIA) to convert a displacement-induced current from the sensor to a voltage. To convert such small currents, the transimpedance needs to be in the hundreds of MΩ. This is achieved by using PMOS devices in weak inversion region, which saves lots of area compared with standard solution.

Figure 4 shows the diagram of the transimpedance amplifier used in the design. The AC signal gain is set to be $1/(sC_f)$, which is around 22 MΩ. The low-frequency components pass through the Gm cell, setting the DC bias for amplifier OTA1. To balance the DC handling capability and noise performance, a MOS pseudo-resistor current divider [18] is used as depicted in Fig. 4b. The advantage of this gm structure is high linearity and rail-to-rail swing due to negative feedback. Including the resistor Rin, the equivalent transconductance is:

$$g_{m,\text{dc}} = \frac{I_{\text{out}}}{V_{\text{in}}} = \frac{1}{R_{\text{in}} \frac{W_{\text{mp1}}}{W_{\text{mp2}}}} \tag{7}$$

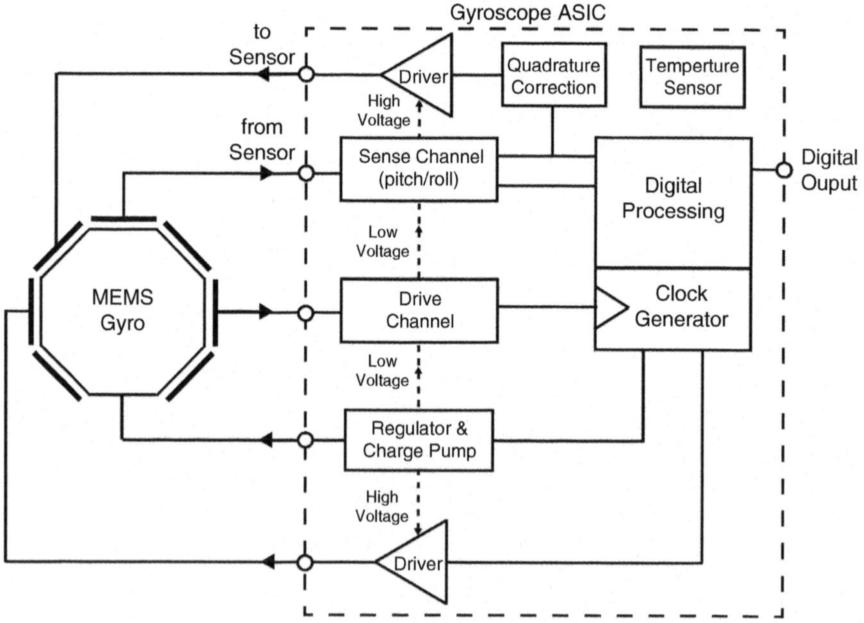

Fig. 2 Gyroscope ASIC top level diagram

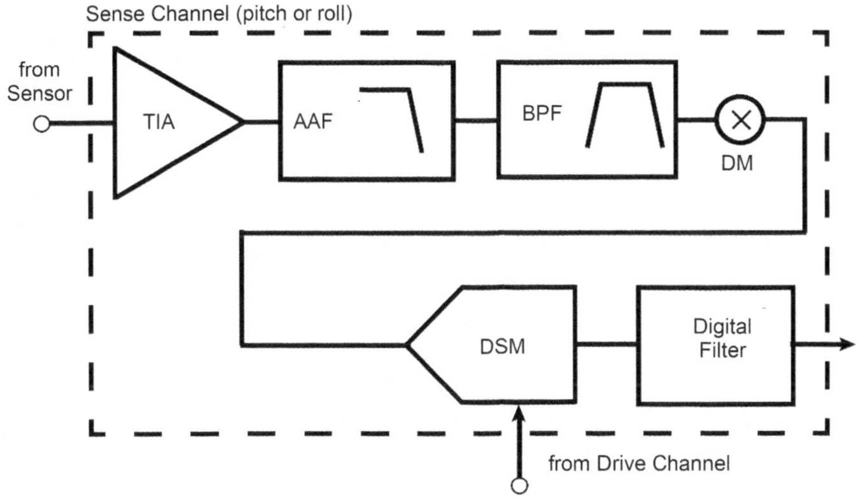

Fig. 3 Sense channel and drive channel detailed diagram

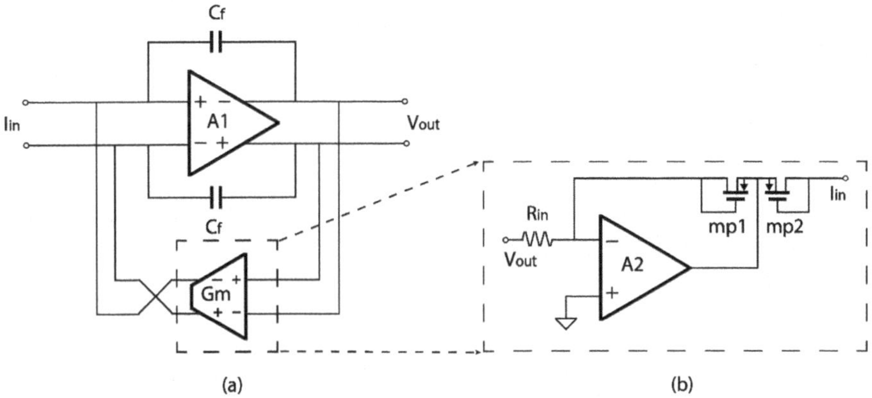

Fig. 4 Transimpedance amplifier (TIA) used in the design

If $R_{\text{in}} = 8\ \text{M}\Omega$, $W_{\text{mp1}}/W_{\text{mp2}} = 100$, $G_{m,\text{dc}} = 1/800\ \text{M}\Omega$. However, the noise from R_{in} is attenuated by $(1/100)^2$, with an equivalent R_{noise} of 80 GΩ. Thus, the noise is dominated by mp2, which operates in either subthreshold or drain-well diode mode, depending on the input low-frequency current direction. Choosing a long length ensures that its noise is not dominant. The capacitor C_f can be selected from either 100 fF or 200 fF, depending on the input range setting.

An anti-aliasing filter (AAF) follows the TIA to filter out the unwanted out-of-band high-frequency signals that might alias back in to baseband by the sampling action of the subsequent discrete-time bandpass filter (BPF). The anti-aliasing filter is a continuous-time, low-pass filter. Its cut-off frequency is determined by the pole formed by a resistor and a capacitor. Signal gain is also added at this stage to

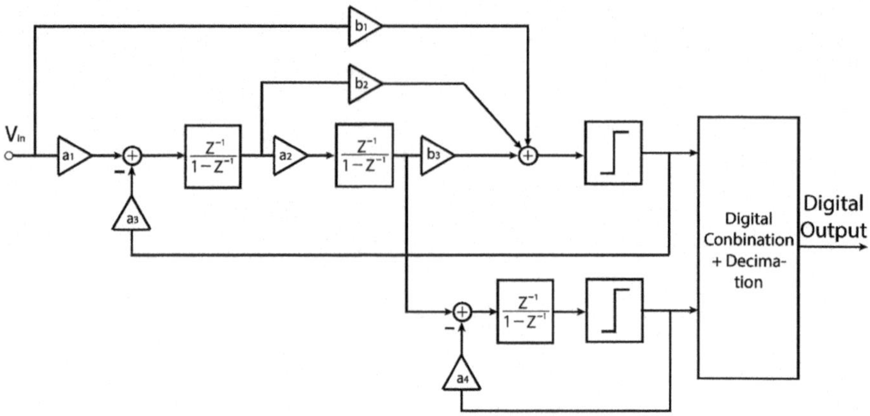

Fig. 5 2–1 MASH delta-sigma modulator

relax the switched-capacitor sampling noise (kT/C) requirement of the following bandpass filter.

The subsequent bandpass filter (BPF) is designed to filter out unwanted signals outside the band of interest close to the resonant frequency. Thus, the center frequency of the bandpass filter is at f_0 or approximately 72 kHz. The BPF is a four-pole switched-capacitor biquad circuit. The sampling frequency is 6.91 MHz, which is 96 times the center frequency. The gain of the BPF can be trimmed for different input range selections.

A demodulator (DM) is placed between the bandpass filter and the analog to digital converter (ADC) to demodulate the high-frequency input signal down to base band. Demodulation to baseband prior to the ADC eases the requirements for the ADC.

The delta-sigma ADC is widely used in sensing applications; it is particularly suitable for low speed and high-resolution applications such as humidity, temperature, and inertial sensors [14, 19, 20]. A switched-capacitor (SC) MASH 2–1 delta-sigma modulator is adopted in this design. Figure 5 shows the diagram of the 2–1 MASH delta-sigma modulator used. Both the second-order first stage and first-order second stage modulators have one-bit quantizers. The digital combination logic and digital filter are also integrated to process output of the delta-sigma modulator. The one-bit quantizer is adopted for its inherent linearity and also simplifies the backhand digital processing design [21].

To suppress the quantization noise down to the required level, third-order noise shaping is required. Compared with a single-loop design, the 2–1 MASH topology is chosen for its higher stability and larger input range. A folded cascode amplifier ensures sufficient DC gain of the integrator, which minimizes leakage of quantization noise of the first stage. The input of the second stage is taken from the output of the second integrator. This arrangement benefits from the fully feed-forward structure of the first stage, which ensures that the loop filter only processes

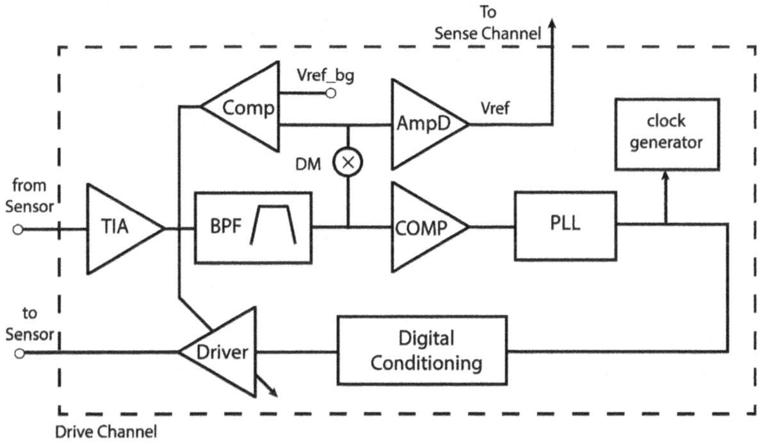

Fig. 6 Detailed drive loop diagram

the quantization noise [22]. The sampling clock for the ADC is 144 kHz, which is twice the resonant frequency. The oversampling ratio (OSR) is 128, which leads to an input bandwidth of 562.5 Hz.

2.2 Resonator Channel Design

Figure 6 shows the detailed diagram of the drive channel in the proposed ASIC. In the drive channel, an on-chip phase lock loop (PLL) produces a clock signal, which is level shifted to 31 V. This is sufficient to electrostatically actuate the sensor mass. The 31 V is generated on-chip by a high-voltage charge pump. The high-voltage clock signal then drives the MEMS sensor mass, whose motion is detected by a set of variable capacitors, which yields a displacement charge when biased at 31 V. That charge creates an output current from the sensor, which is sensed by the trans-impedance amplifier (TIA) in the drive channel. The TIA converts this charge into a sinusoidal voltage and passes it to a second-order bandpass filter to filter out unwanted out-of-band signal. The bandpass filtered output sinusoidal signal is then converted to a digital clock signal through a comparator, which has the same frequency as the drive stimulus, but a phase which is determined largely by the relative spacing separation between the drive frequency and the resonant frequency of the sensor. Using this signal as the reference to the same PLL, the driving clock signal from the PLL and the motion of the sensor form a loop which settles at the resonant frequency of structure at approximately 72 kHz (f_0). The PLL produces clocks for the entire ASIC, including both the analog and digital sections. The highest clock produced by PLL is around 27 MHz, which is used in the digital section.

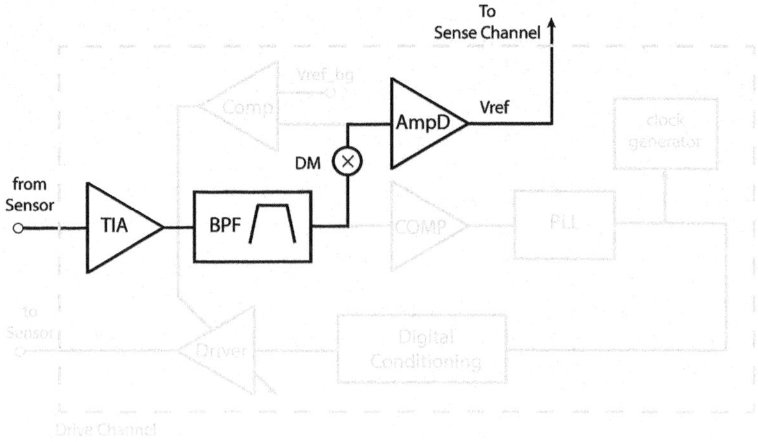

Fig. 7 ADC reference

The drive loop also provides the reference voltage of the on-chip analog to digital converter (ADC) in the sense channel. As shown in Fig. 7, the output of the bandpass filter is also picked up by a block called amplitude detector (AmpD) which captures the amplitude of the input sinusoidal signal and converts it to a DC voltage with a preset gain. This DC voltage will be used as the reference voltage of the ADC. Compared to using a static voltage reference for the ADC, using the sensor drive amplitude can improve the sensitivity accuracy of the gyroscope during transient events. This signal can also be used to monitor whether the resonator amplitude is within prescribed limits.

To maintain constant sensitivity of the gyroscope readout chain, the resonant amplitude of the sensor must be constantly monitored—either controlled or used to adjust the signal path gain. The monitoring can be done by checking the output value of amplitude detector as shown previously. The control can be done by comparing the bandpass filter output amplitude with a predefined reference voltage and further adjusting the gain of high-voltage driver, which drives the resonator as shown in Fig. 8. If the output of BPF is higher than the reference voltage, the gain of high-voltage driver will be adjusted to the small value to decrease the drive voltage, and a further decrease in resonate amplitude, the gain will be adjusted to higher value if output of BPF is smaller than the reference voltage. By doing so, the resonant amplitude of the sensor is controlled relatively constant which make sure gyroscope output have constant sensitivity against environment change or during resonator start-up.

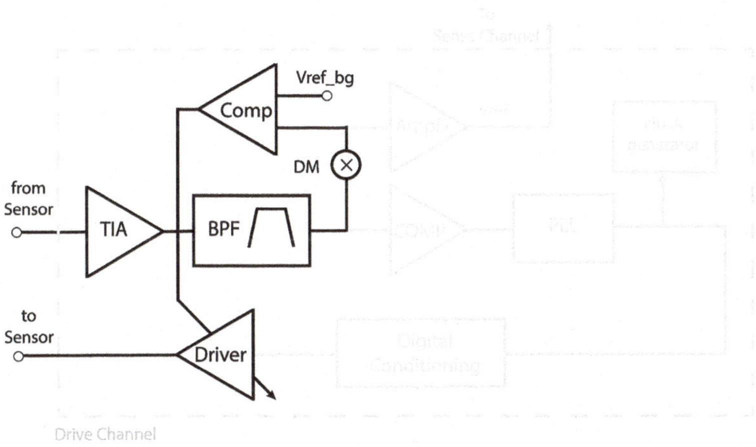

Fig. 8 Amplitude control loop

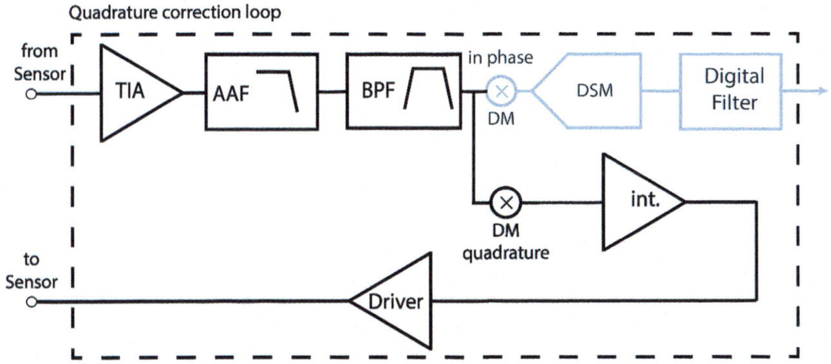

Fig. 9 Quadrature correction loop

2.3 Quadrature Correction Loop

The demodulation process yields the in-phase signal content at the resonator frequency, which contains the desired input rotation rate information. A quadrature demodulation of the same signal yields an error signal, quadrature error, resulting from nonideal motion of the sensor resonator with respect to the Coriolis sense detectors. This error signal can be many times larger than the desired in-phase signal. Thus, it is desirable to cancel this error at source. Shown in Fig. 9, a quadrature correction loop is included to track and correct the quadrature error. It samples the quadrature error at the output of the bandpass filter in the sense loop and accumulates it on an integrator. The integrator output is amplified by a high-voltage amplifier and fed back to the sensor to correct the quadrature error via quadrature correction electrodes on the MEMS sensor.

Fig. 10 Micrograph of the prototype gyroscope ASIC

3 Measurement Results

The proposed dual-axis gyroscope ASIC is fabricated in 0.18 μm CMOS IC technology. The ASIC includes all necessary function blocks to give the digital output of the input angular rate. Figure 10 shows a chip photo of the implemented ASIC, which occupies an area of 7.3 mm^2. The ASIC can work with a supply voltage from 2.7 V to 5 V. An integrated LDO supplies 1.8 V to all the major analog and digital building blocks. The whole ASIC draws 7 mA during measurement mode. For the measurement ranges of 500°/s and 2000°/s, the ASIC achieves noise floor of 0.0032°/s/$\sqrt{}$Hz and 0.0061°/s/$\sqrt{}$Hz, respectively, in an output signal bandwidth of 480 Hz.

Figure 11 shows the noise floor of the gyroscope ASIC at 2000°/s mode. Figure 12 shows the measured Root Allan Variance (RAV) of the proposed design. It achieves 2.5°/h stability in the 500°/s mode.

A performance summary and comparison with other recently published state-of-the-art works are presented in Table 2. The table includes dual- or three-axis digital output gyroscope ASIC which integrated complete signal chains such as drive channel, sense channel, high-voltage driver, and analog to digital converter. Our work achieves similar performance compared with [4] at lower power current consumption although [4] have three-axis outputs. Compared with other open-loop works, our work improves noise performance by more than 5× compared with [12, 13]. Sun et al. [10] has better energy efficiency which demonstrates low power potential of open sense channel topology.

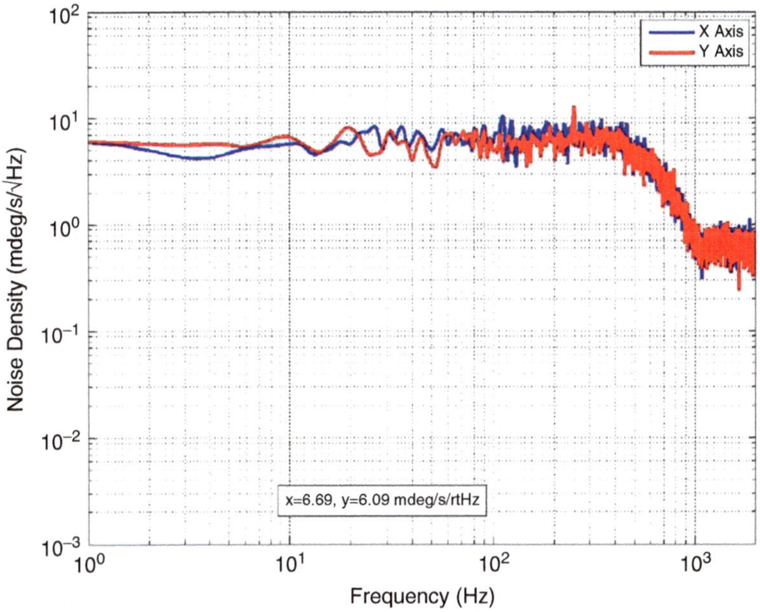

Fig. 11 Measured noise floor of the proposed gyroscope ASIC

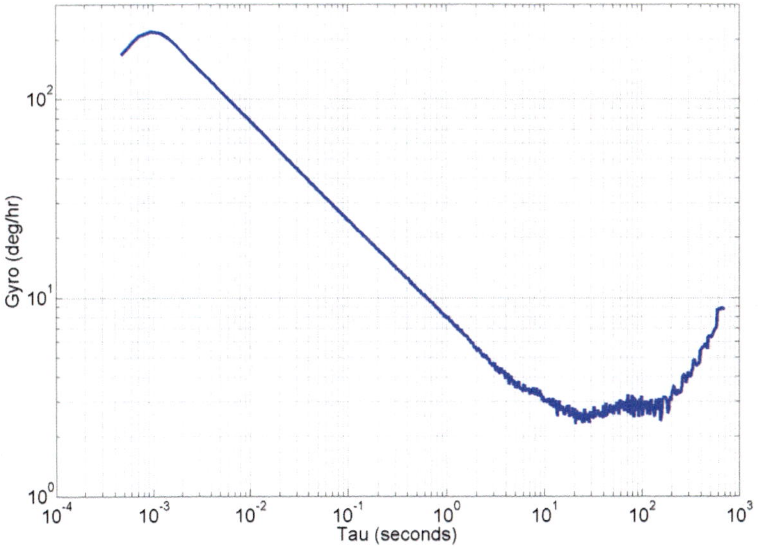

Fig. 12 Measured bias instability of propose gyroscope ASIC

Table 2 Performance summary and comparison with other state of the arts

	This work		JSSC16 [4]	ISSCC15 [10]	ISSCC11 [12]	JSSC11 [13]
Sense channel transduction	Displacement		Force	Displacement	Displacement	Displacement
Supply (V)	**3**	**3**	3	3	3	3
Current (mA)	**7**	**7**	8.8	0.85	6.1	2.2
No. of axis	**2**	**2**	3	3	3	2
Signal BW (Hz)	**480**	**480**	80	520	50	160
Full scale (°/s)	**500**	**2000**	630	2000	2000	1000
Noise floor (°/s/$\sqrt{}$Hz)	**0.0032**	**0.0061**	0.0038	0.007	0.03	0.028/0.032 (X/Y)
Bias instability (°/h)	**2.5**	**5.5**	1.2	N/A	N/A	18/22 (X/Y)

4 Conclusion

A dual-axis (pitch and roll) MEMS vibratory gyroscope readout ASIC has been presented. There are several signal channels surrounding the MEMS sensor. The sense channel converts a small output current from the sensor into voltage and passes it to an anti-aliasing filter and a bandpass filter. The output of the bandpass filter is demodulated to baseband and fed to an analog-to-digital converter for digitization. The drive channel is a closed-loop signal path around sensor, which drives the sensor at its resonant frequency, which is around 72 kHz. The motion of the sensor creates a velocity of the sensor disc to generate a Coriolis force during angular rotation. The design achieves noise floor of 0.0032°/s/$\sqrt{}$Hz and 0.0061°/s/$\sqrt{}$Hz in the full-scale input range of 500°/s and 2000°/s, respectively, over the signal bandwidth of 480 Hz. The bias instability is measured as 2.5°/h at input range of 500°/s. The whole ASIC consumes 7 mA from a single 3 V supply and occupies an area of 7.3 mm^2.

Acknowledgments The authors would like to thank their colleagues from the High-Performance Inertial sensor group at Analog Devices Inc. (both in Wilmington and Greensboro) for their help during design, layout, and chip evaluations.

References

1. Marek J. MEMS for automotive and consumer electronics. In: IEEE International Solid-State Circuits Conference (ISSCC) Digest of Technical Papers, San Francisco, 2010.
2. Yazdi N, Ayazi F, Najafi K. Micromachined inertial sensors. Proc IEEE. 1998;86(8):1640–59.
3. Clark WA. Micromachined vibratory rate gyroscopes, Dissertation, 1997.
4. Balachandran GK, Petkov VP, Mayer T, Blalslink T. A 3-axis gyroscope for electronic stability control with continuous self-test. IEEE J Solid State Circuits. 2016;50(1):177–86.

5. Sharma A, Zaman MF, Ayazi F. A Sub-0.2 hr bias drift micromechanical silicon gyroscope with automatic CMOS mode-matching. IEEE J Solid State Circuits. 2009;44(5):1593–608.
6. Masten MK. Inertially stabilized platforms for optical imaging systems. IEEE Control Syst. 2008;28(1):47–64.
7. Hilkert J. Inertially stabilized platform technology concepts and principles. IEEE Control Syst. 2008;28(1):26–46.
8. Meijer G. Smart sensor systems. Wiley; 2008.
9. Meijer G, Makinwa K, Pertijs M. Smart sensor systems: emerging technologies and applications. Wiley; 2014.
10. Sun H, Jia K, Liu X, Yan G, Hsu Y-W, Fox RM, Xie H. A CMOS-MEMS gyroscope interface circuit design with high gain and low temperature dependence. IEEE Sensors J. 2011;11(11):2740–8.
11. Ezekwe C, Geiger W, Ohms T. A 3-axis open-loop gyroscope with demodulation phase error correction. In: Proceedings of IEEE international solid-state circuits conference, San Francisco, 2015.
12. Prandi L, et al. A low-power 3-axis digital-output MEMS gyroscope with single drive and multiplexed angular rate readout. In: Solid-State Circuits Conference Digest of Technical Papers (ISSCC), San Francisco, 2011.
13. Aaltonen L, Kalanti A, Pulkkinen M, Paavola M, Kamarainen M, Halonen KAI. A 2.2 mA 4.3 mm ASIC for a 1000°/s 2 – Axis capacitive micro-gyroscope. IEEE J Solid State Circuits. 2011;46(7):1682–92.
14. Aaltonen L, Halonen KAI. Pseudo-continuous-time readout circuit for a 300°/s capacitive 2-axis micro-gyroscope. IEEE J Solid State Circuits. 2009;44(2):3609–20.
15. Chen F, Li X, Kraft M. Electromechanical sigma–delta modulators force feedback interfaces for capacitive MEMS inertial sensors: a review. IEEE Sensors J. 2016;16(17):6476–95.
16. Rombach S, Marx M, Nessler S, Dorigo DD, Maurer M, Manoli Y. An interface ASIC for MEMS vibratory gyroscopes with a power of 1.6 mW, 92 dB DR and 0.007°/s/ vHz noise floor over a 40 Hz band. IEEE J Solid State Circuits. 2016;51(8):1915–27.
17. Tan Z, Nguyen K, Yan J, Samuels H, Keating S, Crocker P, Clark B. A dual-axis MEMS vibratory gyroscope ASIC with 0.0061°/s/VHz noise floor over 480 Hz bandwidth. In: 2017 IEEE Asian Solid-State Circuits Conference (A-SSCC), Seoul, 2017.
18. Gozzini F, Ferrari G, Sampietro M. Linear transconductor with rail-to-rail input swing for very large time constant applications. Electron Lett. 2006;42(19):1069–70.
19. Tan Z, Daamen R, Humbert A, Ponomarev YV, Chae Y, Pertijs MAP. A 1.2-V 8.3-nJ CMOS humidity sensor for RFID applications. IEEE J Solid State Circuits. 2013;48(10):2469–77.
20. Souri K, Chae Y, Makinwa KAA. A CMOS temperature sensor with a voltage-calibrated inaccuracy of ±0.15°C from −55°C to 125°C. IEEE J Solid State Circuits. 2013;48(1):292–301.
21. Schreier R, Temes GC. Understanding delta-sigma data converters. Wiley-IEEE Press; 2004.
22. Silva J, Moon U, Steensgaard J, Temes G. Wideband low distortion delta-sigma ADC topology. Electron Lett. 2001;37(12):737–8.

Direct Frequency-To-Digital Gyroscopes with Low Drift and High Accuracy

Burak Eminoglu and Bernhard E. Boser

1 Introduction

Present implementations of MEMS gyroscopes measure rate indirectly by first converting it to a displacement [1, 2]. In this case, the scale factor is a complex function of the transducer and readout circuits. Changes of any of the underlying parameters result in measurement errors.

The solution presented here measures rate directly as frequency and converts it to a digital output by comparing it to a precision clock reference [3]. Figure 1 illustrates the principle. The transducer proof mass consists of two orthogonal resonators excited at their resonant frequencies f_o by two sustaining circuits. For a 90° phase shift in the displacements of the x- and y-channels, the motion of the proof mass follows a circular pattern. An observer in the rotating frame perceives a rate input as a shift of the observed oscillation frequency of the proof mass. The scale factor equals α_z, where α_z is the unitless transducer gain. It can be measured accurately with a frequency-to-digital converter with an explicit reference input f_{ref}.

2 Rate Chopping

The transducer resonance f_o appears as a huge offset in the output. Environmental variations preclude straightforward subtraction from the rate output. Instead, the direction of the circular path is altered periodically to modulate the sign of the rate sensitivity corresponding to rate being shifted to the modulation frequency. It is

B. Eminoglu (✉) · B. E. Boser
Electrical Engineering and Computer Sciences, UC Berkeley, Berkeley, CA, USA
e-mail: eminoglu@eecs.berkeley.edu

K. A. A. Makinwa et al. (eds.), *Low-Power Analog Techniques, Sensors for Mobile Devices, and Energy Efficient Amplifiers*,
https://doi.org/10.1007/978-3-319-97870-3_10

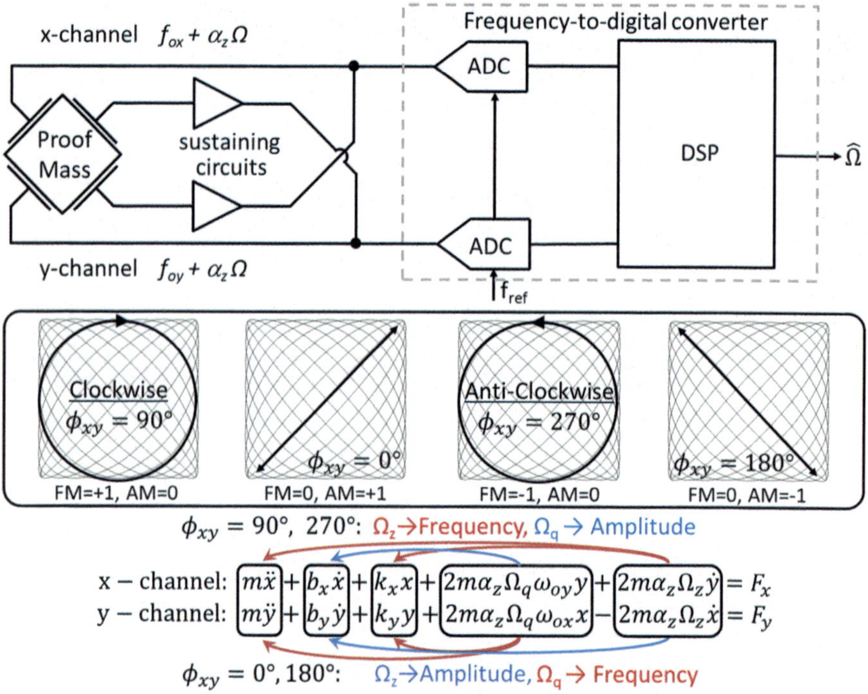

Fig. 1 Simplified block diagram and rate chopping for FM and AM channels

accomplished by deliberately mismatching the resonances f_{ox} and f_{oy} of the two axes by a small amount Δf (typically <100 Hz). Now, the relative phase Φ_{xy} of the x- and y-channels changes continuously, passing through 90° and 270°, corresponding to FM gains $+1$ and $-\ 1$. This is equivalent to chopper stabilization and rejects drift at frequencies below the modulation rate. At 0° and 180°, the rate modulates the amplitude rather than the frequency of the x- and y-displacements. The force equations illustrate that the rate signal Ω_z and the quadrature error Ω_q modulate the effective stiffness (k) and damping (b) terms with time-varying Φ_{xy} ($=2\pi\Delta f{\bullet}t$). Consequently, rate appears in the output both as a frequency shift (FM channel) and a change in the oscillation amplitude (AM channel). FM and AM signals are modulated at Δf with $\sin(\Phi_{xy})$ and $\cos(\Phi_{xy})$, respectively.

As in conventional AM gyroscopes, reducing the split Δf between the modes improves the ARW of the sensor [1, 4]. Since both modes are continuously driven, the split is observable and electrostatically tuned to 10 Hz in the prototype. The ability to accurately set the split frequency is an important advantage of FM over AM implementations and a consequence of both axes being driven.

3 Implementation

Figure 2 shows the readout circuits consisting of differential oscillators with amplitude control. It consists of a transcapacitance amplifier followed by a phase shifter, amplitude detector, and VGA. An active biasing circuit with long-channel transistors having less than 5 fA/rt-Hz current noise is used to provide the DC feedback in the front-end amplifier which has, in total, 17 fA/rt-Hz input-referred current noise. An SC peak detector is clocked at the zero crossings of the differentiator output to sample the oscillation amplitude. Unlike other options, this solution does not require a low-pass filter which would limit the measurement bandwidth (Fig. 6). The VGA ensures a stable oscillation amplitude and rejects the amplitude variations from the AM channel.

For testing, the circuit was connected to the symmetric quad mass gyroscope (QMG) (Fig. 3) with nominal $f_o = 24$ kHz and $Q = 100$ k. The frequency-to-digital conversion (FDC) is performed off-chip by digitizing two oscillator outputs and extracting the frequency with a software PLL. The output from the AM channel is obtained from the amplitude detector and also digitized off-chip.

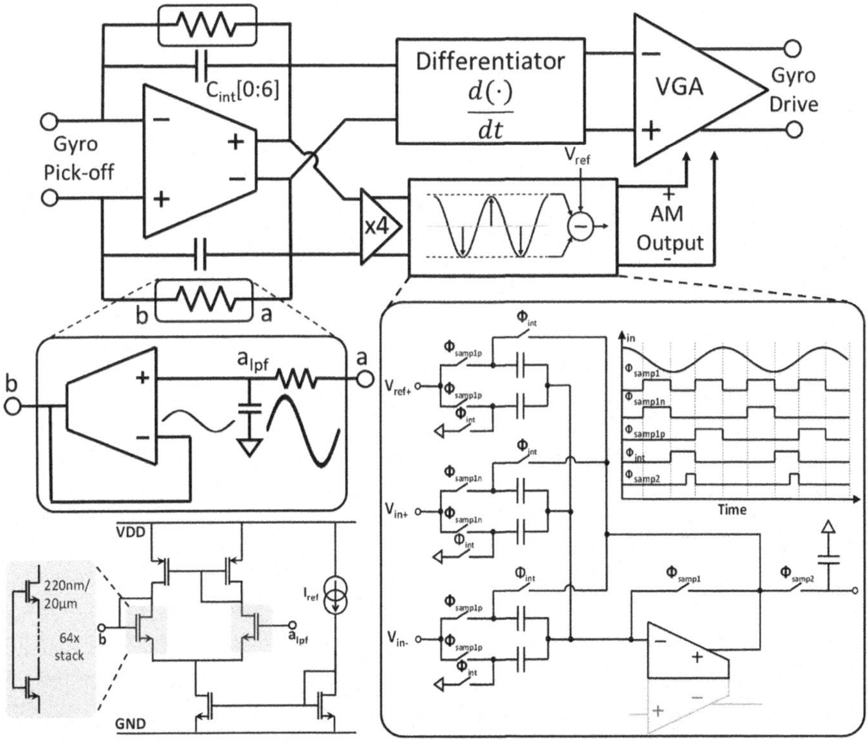

Fig. 2 Circuit schematic of a single oscillator with active biasing and SC amplitude controller

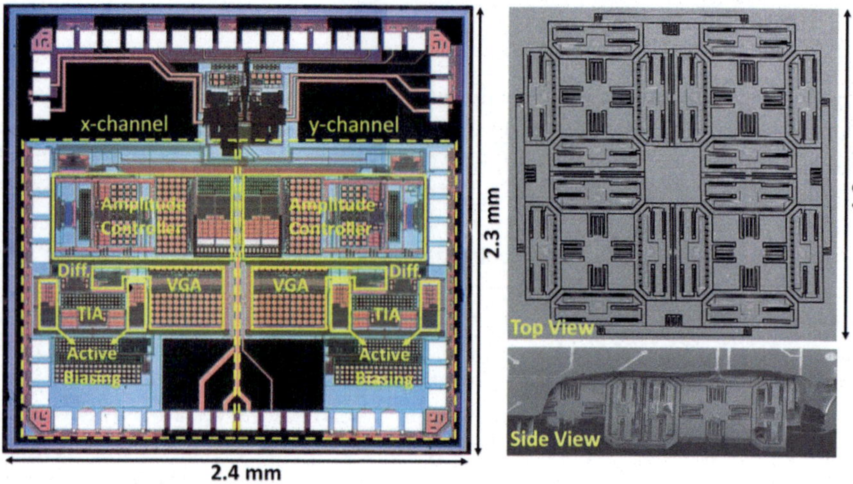

Fig. 3 Chip micrograph and SEM image of the MEMS gyroscope

4 Measurement Results

Figure 4 shows the measured linearity and scale factor stability. The sensitivity of conventional AM gyroscopes is determined by transducer bias, electrode gaps, oscillation amplitude, and VGA gain, which are difficult to control accurately. In contrast, FM sensitivity is set by an external reference clock and proportional to the slip factor α_z set by transducer geometry and the sum of the reciprocal velocity ratios ($v_x/v_y + v_y/v_x$). For best stability, the velocities are chosen to be equal, contributing only a 1 ppm error to the scale factor for as much as 1400 ppm velocity mismatch [5]. For this prototype, the AM and FM linearity over ±300 dps are 1830 ppm and 110 ppm, respectively. Measured over a 24-h period in an uncontrolled environment, the individual FM channels exhibit considerable fluctuations dominated by temperature variations. Summing the two outputs reduces this variation to ±150 ppm, a more than order of magnitude improvement over the AM performance. Also shown is a first-order compensated result which reduces the magnitude of the error to less than 40 ppm. The temperature of the sensor is obtained without extra circuitry from the FDC based on the transducer TCF of −30 ppm/°C.

Figure 5 shows the measured Allan variance for sensors operated at equal oscillation amplitudes and hence nearly identical velocities in both channels and with a deliberate amplitude mismatch. The mismatch increases the scale factor of the sensor, thereby reducing the noise contribution of the FDC. Note that the only change between the two measurements is a different setting in the amplitude controller. The possibility to dynamically adjust long- versus short-term stability without increased power dissipation is a unique feature of the FM gyroscope.

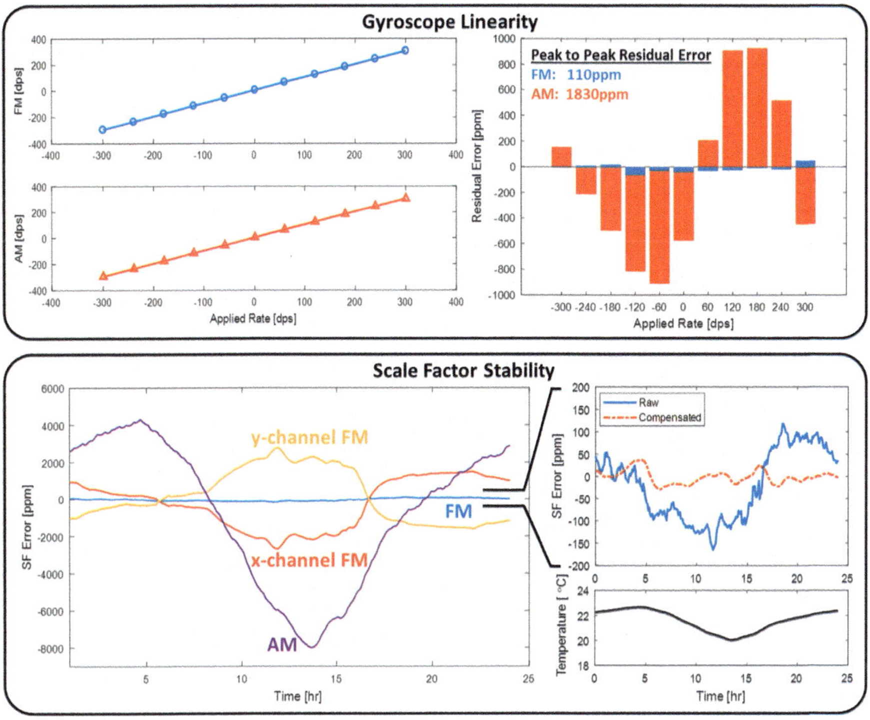

Fig. 4 Scale factor tests of FM and AM channels

These results were achieved with a transducer with parallel plate transduction, which due to its inherent nonlinearity results in noise folding, impairing the ARW. The 1 mdps/rt-Hz obtained with a transducer with comb-drive actuation confirms this hypothesis. Furthermore, comb drives enable to operate the transducer with a larger oscillation amplitude which minimizes the Brownian noise to less than 0.2 mdps/rt-Hz. Unfortunately, because this design has not been optimized, it exhibits poor long-term stability.

The noise is a function of Δf, and ARW of 1 mdps/rt-Hz at 10 Hz is achieved in the asymmetric mode, where total noise is dominated by the electronics. Below 5 Hz close-to-carrier phase noise dominates the ARW. Consequently, the mode-split of the gyroscope is tuned to 10 Hz with a servo loop with 20 ms settling time. Tuning accuracy is not critical, since the FDC extracts the instantaneous phase between the *x*- and *y*-axis motion for demodulation.

While reducing the mode-split is advantageous for noise, this also lowers the useful bandwidth since the input is chopped at this rate. Since the outputs from the FM and AM channels are in quadrature, the bandwidth of the sum of these outputs is limited only by the bandwidth of the amplitude controller. Figure 6 illustrates the summing process and the spectrum for a 25 Hz rate input. To show the effectiveness of the technique, these measurements have been performed with Δf tuned to 5 Hz.

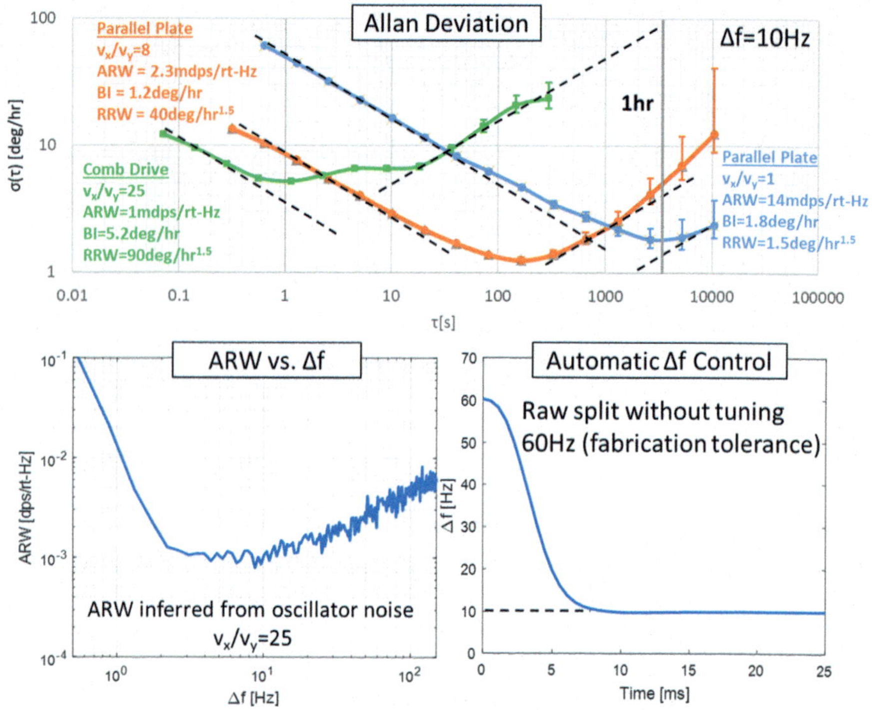

Fig. 5 Allan deviation of the FM channel, ARW versus mode split, and automatic mode split tuning

The tone at 20 Hz is due to transducer nonlinearity and can be reduced with an improved mechanical design. The image is the result of imperfect gain matching of the AM and FM channels. As expected, it disappears after trimming the AM scale factor.

5 Conclusion

Table 1 compares this result to solutions reported earlier. The FM gyro achieves competitive or better performance in all categories. Note that these results have been achieved without calibration. Not usually reported, but, a significant error source for applications such as navigation is scale factor accuracy. By relying on an explicit reference supplied in the prototype by an external (precision) clock, the FM gyro scale factor stability is better than the typical AM gyro accuracy. Further, significant advantages include the continuously tuned mode-split and the asymmetric mode of operation used to trade off long- and short-term stability without circuit changes.

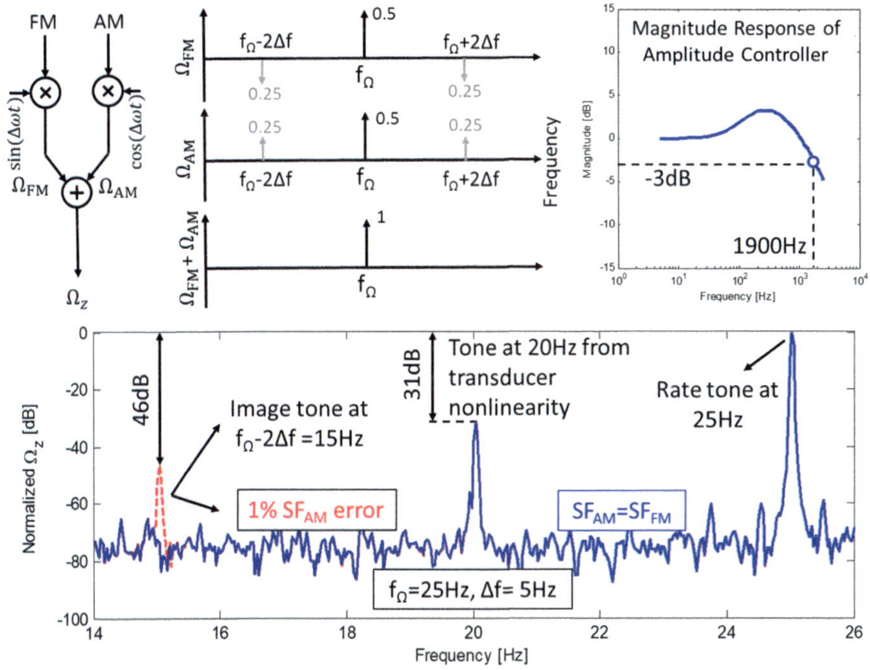

Fig. 6 Image rejection in combined FM and AM readout

Table 1 Performance summary and comparison table

	This Work	ISSCC'17 [1] Marx	ISSCC'15 [2] Ezekwe	ISSCC'08 [4] Ezekwe
ARW [dps/rt-Hz]	0.001[a]	0.0014	0.0049[b]	0.0028[b]
Bias stability [deg/h]	1.2[a]	0.9	n.a	n.a.
RRW [deg/h[a, c]]	1.5[d]	3.8	n.a.	n.a.
FS [dps]	1000[d,e]	800	2000	n.a.
Bandwidth [Hz]	1900[f]	50	520	50
Number of axes	1	1	3	1
Supply [V]	1.8	3.3	1.71–3.6	3.3
Power [mW]	0.45[c]	1.71	0.37/axis	1[g]
FoM[h] for ARW [dps[d]/Hz × W]	0.45n[a]	3.4n	8.9n	7.8n

(continued)

Table 1 (continued)

	This Work	ISSCC'17 [1] Marx	ISSCC'15 [2] Ezekwe	ISSCC'08 [4] Ezekwe
Read-out features	Simultaneous FM and AM	$\Delta\Sigma$ with tuned cont. time BPF	Open loop With HV	Closed loop
Bias stability methods	FM readout Symmetric transducer and readout	Manual quadrature tuning	Background phase error correction over temperature	n.a.
Mode split sensing	Direct readout of resonance frequencies	n.a. (initial tuning)	n.a.	Tone injection

[a]Asymmetric FM
[b]Rate noise density reported. ARW = Rate Noise Density/$\sqrt{2}$
[c]Power does not include off-chip ADCs and DSP
[d]Symmetric FM
[e]Circuit full-scale. Tested up to ±300dps (rate table limitation)
[f]Tested up to 25 Hz
[g]Drive electronics power not included
[h]FOM = Power × ARW^2 (per axis)

Acknowledgments The authors would like to thank Yu-Ching Yeh, Mithcell Kline, and Parsa Taheri for the transducer design and Yunhan Chen, Ian B. Flader, Dongsuk D. Shin, and Professor Thomas W. Kenny at Stanford University for the MEMS fabrication. Authors acknowledge the support of this project by DARPA under the PASCAL program and thank the TSMC University Shuttle Program for CMOS chip fabrication.

References

1. Marx M, et al. A 27μW 0.06mm^2 background resonance frequency tuning circuit based on noise observation for a 1.71mW CT-$\Delta\Sigma$ MEMS gyroscope readout system with 0.9°/h Bias instability. In: ISSCC Digest of Technical Papers. 2017. p. 163–4.
2. Ezekwe C, et al. A 3-Axis open-loop gyroscope with demodulation phase error correction. In: ISSCC Digest of Technical Papers. 2015. p. 478–9.
3. Izyumin I, et al. A 7ppm, 6°/hr frequency output MEMS gyroscope. In: Digest of MEMS. 2015. p. 33–6.
4. Ezekwe C, Boser B. A mode-matching $\Delta\Sigma$ closed-loop vibratory gyroscope readout interface with a 0.004°/s/$\sqrt{\text{Hz}}$ noise floor over a 50Hz band. In: ISSCC Digest of Technical Papers. 2008. p. 580–1.
5. Eminoglu B, et al. Comparison of long-term stability of AM vs. FM gyroscopes. In: Digest of MEMS. 2016. p. 954–7.

CMOS-Compatible Carbon Dioxide Sensors

Zeyu Cai, Robert van Veldhoven, Hilco Suy, Ger de Graaf, Kofi A. A. Makinwa, and Michiel Pertijs

1 Introduction

CO_2 sensors can provide important information for indoor air-quality monitoring, given that CO_2 concentration in the air is highly correlated with the occupancy of a building [1]. As this increases, human-related pollutants such as bacteria, molds, and volatile organic compounds also increase, meaning a heightened risk for the health of the occupants. On the other hand, the more people there are in a building, the higher the CO_2 concentration will be as a by-product of human respiration, implying that CO_2 concentration can be used as an indicator for air quality. To do so, concentrations of up to 2500 ppm need to be measured with a resolution of better than 200 ppm [1]. Currently available sensors that meet these requirements are optical sensors based on nondispersive infrared absorption (NDIR) [2–4]. Despite their merits of being accurate and selective, optical CO_2 sensors also have several considerable downsides. They require a cavity or tube for optical waveguiding, making them relatively bulky and expensive. Moreover, the infrared source typically consumes a significant amount of power [2–4].

Z. Cai (✉)
Delft University of Technology, Delft, The Netherlands

NXP Semiconductors, Eindhoven, The Netherlands
e-mail: zeyu.cai@nxp.com

R. van Veldhoven
NXP Semiconductors, Eindhoven, The Netherlands

H. Suy
ams AG, Eindhoven, The Netherlands

G. de Graaf · K. A. A. Makinwa · M. Pertijs
Delft University of Technology, Delft, The Netherlands

K. A. A. Makinwa et al. (eds.), *Low-Power Analog Techniques, Sensors for Mobile Devices, and Energy Efficient Amplifiers*,
https://doi.org/10.1007/978-3-319-97870-3_11

Alternatively, CO_2 concentration can be determined indirectly by measuring the thermal conductivity (TC) of the air [5–8]. This approach exploits the fact that the TC of CO_2 is lower than that of the other constituents of air, and thus the thermal conductivity of the air is CO_2 dependent. Therefore, when a resistive transducer is suspended in ambient air and heated up, its temperature rise is CO_2 dependent. This temperature rise can be detected as a change of the transducer's electrical resistance. Since the required transducer can be realized with minimum post-processing in standard CMOS technology, this approach has important cost and miniaturization advantages compared to NDIR-based sensors. However, as the difference in the thermal conductivities of the air and CO_2 is very small, tiny changes in resistance need to be sensed (e.g., 1.5 $\mu\Omega$ change per ppm change in CO_2), making the measurement extremely demanding [5].

To perform accurate measurement, a proper reference must be chosen. Furthermore, the dynamic range (DR) requirement of the readout circuit must be significantly reduced to achieve energy-efficient measurements. This chapter presents two different designs, implemented in the amplitude domain and time domain, respectively. In the first approach [5], the steady-state temperature rise and power dissipation of the transducer are measured relative to those of a capped reference transducer. The sensor achieves a CO_2 resolution of about 200 ppm (1 σ) in a measurement time of 30 s. In the time-domain approach [9], the CO_2-dependent thermal time constant is measured using a low-noise PD$\Delta\Sigma$M. This sensor achieves a CO_2 resolution of 94 ppm while dissipating only 12 mJ per measurement, which is best in class in both resolution and energy consumption for CO_2 sensors in CMOS technology.

This chapter is organized as follows: The analysis, design, and measurement results of the amplitude-domain approach will be introduced in Sect. 2. The operating principle and design of the time-domain approach are explained in Sect. 3, together with its measurement results. The pros and cons of these two approaches are analyzed in Sect. 4 and the chapter ends with conclusions in Sect. 5.

2 CO_2 Sensor Based on Amplitude-Domain TC Measurement

2.1 Operating Principle

2.1.1 Thermal Resistance Measurement Using a Resistive Transducer

Figure 1a shows a hot-wire resistive transducer realized in a VIA layer in the metal stack of a standard CMOS process. When power is dissipated in such a transducer, it loses heat through two main paths, both of which can be modeled as thermal resistances, as shown in Fig. 1b: one to the surrounding air (R_{th_air}) and the other to the substrate through the anchor points (R_{th_sub}). The temperature rise (ΔT) of the transducer relative to the ambient temperature ($T_{amb.}$) caused by the power

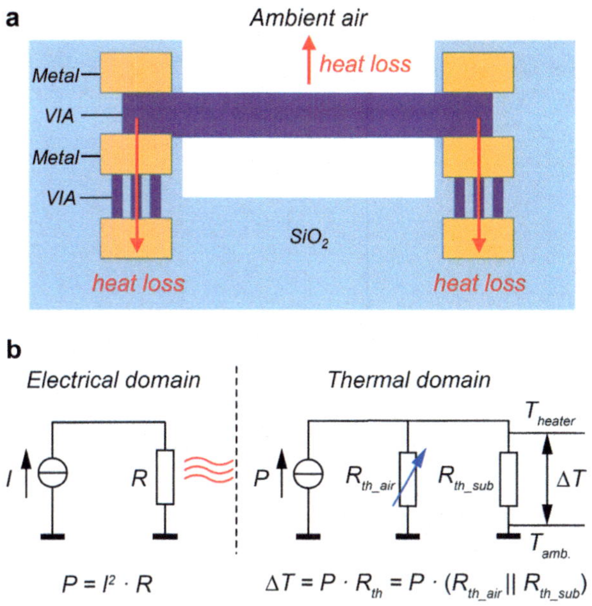

Fig. 1 Amplitude-domain TC sensing principle: (**a**) cross-sectional view of a suspended hot-wire resistive transducer and its heat loss paths; (**b**) the equivalent model in both electrical and thermal domains [5] (reproduced with permission)

dissipated in the transducer (P) is directly proportional to the parallel combination R_{th} of $R_{\mathrm{th\_air}}$ and $R_{\mathrm{th\_sub}}$:

$$\Delta T = P \cdot R_{\mathrm{th}} = P \cdot \left(R_{\mathrm{th\_air}} \parallel R_{\mathrm{th\_sub}}\right) \tag{1}$$

Since the TC of CO_2 is lower than that of the other constituents of air, $R_{\mathrm{th\_air}}$ is CO_2 dependent, and so ΔT can be used to determine the CO_2 concentration in air. To maximize sensitivity, heat loss to the substrate must be minimized, which is usually achieved by employing suspended transducers [10, 11], realized in our case by a single release etch step.

While ΔT in (1) can be measured using a dedicated temperature sensor (e.g., thermopiles; [7]), the use of the electrical resistance of the heater to measure its temperature greatly simplifies the fabrication process, allowing a single resistive transducer to be used as a heater, as shown in Fig. 1a [12].

To a good approximation, the resistance of the tungsten transducer is a linear function of temperature:

$$R = R_0 \cdot (1 + \alpha \cdot (T - T_0)) \tag{2}$$

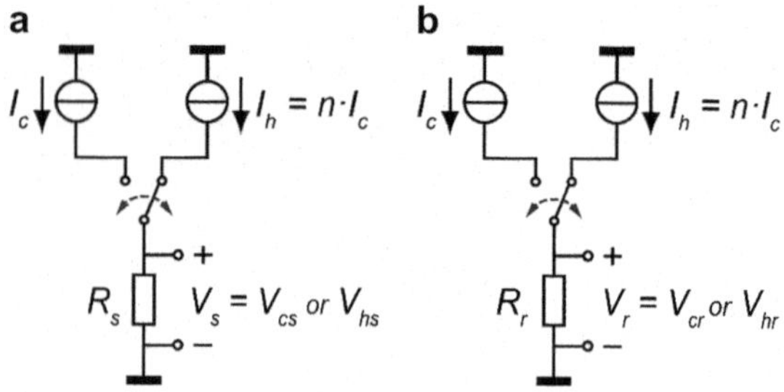

Fig. 2 (**a**) CO_2-sensitive and (**b**) reference transducer alternately biased at two different current levels [5] (reproduced with permission)

where R_0 is the nominal electrical resistance of the transducer at room temperature T_0 and α its temperature coefficient. For our transducers, the nominal resistance R_0 and temperature coefficient α are 110 Ω and 0.0017/K, respectively. The nominal resistance R_0 is set by the aspect ratio of the resistor and was designed to allow sufficient power to be dissipated in the resistor at the available voltage headroom.

To generate ΔT, the transducer is alternately biased at a low current I_c and a high current I_h (Fig. 2a) corresponding to a "cold" and a "hot" state. The power dissipation in (1) then becomes the difference in power dissipation between these two states (ΔP), and R_{th} becomes:

$$R_{th} = \frac{\Delta T}{\Delta P} = \frac{T_h - T_c}{P_h - P_c} = \frac{R_h - R_c}{R_0 \alpha \left(I_h^2 R_h - I_c^2 R_c \right)} \tag{3}$$

In order to accurately measure R_{th}, both ΔT and ΔP need to be accurately measured. The nominal R_{th} of our tungsten transducer (i.e., at 400 ppm CO_2 and 25 °C) is about 53,500 K/W. A change of 200 ppm CO_2 results in about 80 ppm change in the thermal resistance of air. Taking account of the substrate heat loss, a 200 ppm change in CO_2 only corresponds to a relative change of R_{th} of about 50 ppm. This implies that the power levels and temperature measurement should be stable to within ±25 ppm, making the measurement very challenging.

2.1.2 Ratiometric Thermal Resistance Measurement

Measuring the thermal resistance of a CO_2-sensitive transducer relative to that of a (CO_2 insensitive) reference transducer, rather than measuring the absolute thermal resistance of one transducer, greatly relaxes the power stability and temperature measurement requirements, as the absolute accuracy requirement is replaced by a matching requirement of the CO_2-sensitive and reference transducers. The reference transducer is biased in the same way as the sensitive transducer (Fig. 2b). Their thermal-resistance ratio can be derived from (3) and expressed as a multiplication of two ratios: the ratio of the temperature difference of the sensitive transducer (ΔT_s)

and the temperature difference of the reference transducer (ΔT_r) and the ratio of their power differences ($\Delta P_r/\Delta P_s$):

$$\frac{R_{ths}}{R_{thr}} = \left(\frac{\Delta T_s}{\Delta T_r}\right)\left(\frac{\Delta P_r}{\Delta P_s}\right) = \left(\frac{R_{hs}-R_{cs}}{R_{hr}-R_{cr}}\right)\left(\frac{n^2 I_c^2 R_{hr}-I_c^2 R_{cr}}{n^2 I_c^2 R_{hs}-I_c^2 R_{cs}}\right)$$
$$= \left(\frac{V_{hs}-nV_{cs}}{V_{hr}-nV_{cr}}\right)\left(\frac{nV_{hr}-V_{cr}}{nV_{hs}-V_{cs}}\right) \tag{4}$$

where $n = I_h/I_c$, $V_{hs} = n \cdot I_c \cdot R_{hs}$, $V_{cs} = I_c \cdot R_{cs}$, $V_{hr} = n \cdot I_c \cdot R_{hr}$, $V_{cr} = I_c \cdot R_{cr}$, and the transducers are assumed to have identical R_0 and α, which therefore cancel out. The last term in (4) shows that the thermal-resistance ratio can be written as a product of two voltage-difference ratios, which in this work are digitized sequentially by a dual-mode switched-capacitor incremental $\Delta\Sigma$ ADC and multiplied in the digital backend.

2.1.3 Ratiometric Readout with Transducer Pairs

With the sequential readout of hot and cold states, the voltage drop across the transducers varies significantly between the two states. Consequently, to avoid clipping, the following ADC must have a large dynamic range. To relax the dynamic range, a pair of CO_2-sensitive transducers (R_{s1}, R_{s2}) and a pair of reference transducers (R_{r1}, R_{r2}) are employed (Fig. 3).

Package-level sealing ensures that both reference transducers are isolated from the ambient air. In each pair, the transducers are alternately biased at I_c and $I_h = n \cdot I_c$, generating simultaneously "hot" and "cold" voltages for both sensitive (V_{hs}, V_{cs})

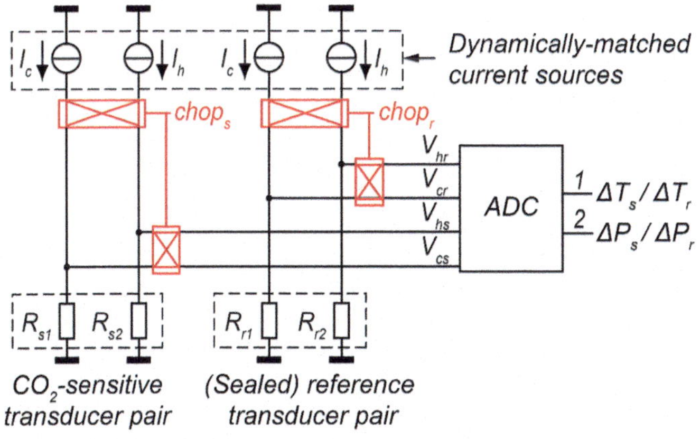

Fig. 3 Block diagram of the ratiometric thermal-conductivity sensor readout with transducer pairs for baseline-resistance cancellation [5] (reproduced with permission)

and reference transducers (V_{hr}, V_{cr}). The current ratio n is chosen to optimize the signal-to-noise ratio (SNR). For a given power consumed in biasing the transducers, a smaller n gives smaller signal amplitude, while a larger n reduces the current in the "cold" state, thereby increasing the noise level associated with that state. Therefore, SNR degrades for small and large values of n and an optimum can be found for which the SNR is maximized. A parametric simulation of our design shows that this optimum is reached at a ratio of $n = 5$. Mismatches are averaged out, as the transducers in each pair are periodically swapped by the chopper switches around them. The "hot" and "cold" voltages are simultaneously sampled by scaled switched-capacitor circuits and merged together (detailed in the next subsection).

2.2 *Design and Implementation*

2.2.1 Charge-Balancing Incremental ΔΣ Modulator

To obtain a CO_2-sensing resolution of the order of 200 ppm (corresponding to 50 ppm or 14.3 bits resolution in the R_{th} ratio), we digitize each of the voltage ratios in (4) with a resolution better than 15.3 bits (equivalent to 100 ppm CO_2). An incremental delta-sigma ADC is suitable, as CO_2 concentration tends to change relatively slowly. A charge-balancing ΔΣ modulator is used that operates in two modes: temperature mode and power mode. First, in temperature mode, it produces a bitstream *bs* proportional to the first voltage ratio in (4), which equals the temperature-difference ratio. Then, in power mode, it produces a bitstream proportional to the second voltage ratio in (4), which equals the power-difference ratio. These bitstreams are decimated by an off-chip decimation filter and the results are multiplied in the digital domain to obtain the thermal-resistance ratio.

Figure 4 shows the switched-capacitor implementation of the modulator. It consists of a switched-capacitor integrator, with four parallel input branches connecting to the four transducers and a clocked comparator. The sample and hold circuit uses a double-sampling scheme, which doubles the signal amplitude compared to the single-sampling scheme used in [12]. The integrator employs a gain-boosted folded-cascode OTA with a unity-gain bandwidth of about 2.5 MHz and a nominal DC gain of 140 dB, to make sure that the settling is accurate and to prevent integrator leakage from limiting the resolution [13]. The comparator is a latched comparator using a preamplifier to reduce the kick-back effect of the positive feedback latch [14]. To minimize charge injection, minimum size switches ($W/L = 0.8\ \mu m/0.16\ \mu m$) are used. In order to obtain accurate CO_2 measurements, accurate matching is required for the capacitors, especially for the current sources.

2.2.2 Dynamically Matched Current Sources and Current Trimming

Errors in the 1:n bias current ratio lead to errors in the measured R_{th} ratio. According to system-level simulations, in order to reduce the resulting error in the measured

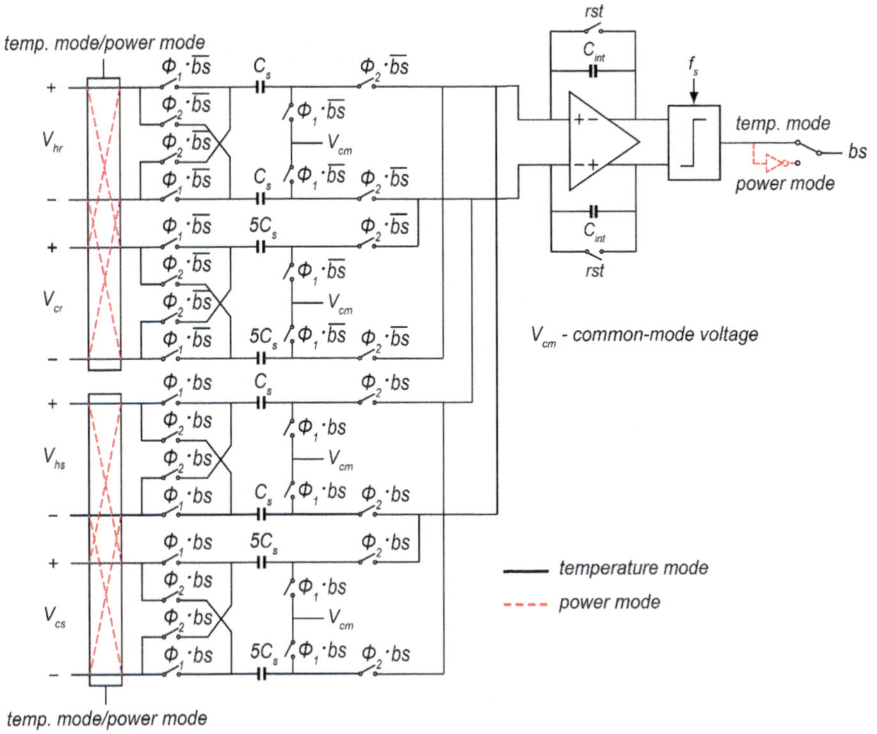

Fig. 4 Simplified circuit diagram of the switched-capacitor delta-sigma modulator in both temperature and power modes [5] (reproduced with permission)

CO_2 concentration to less than 200 ppm by a one-point offset trim, the error in current ratio should be less than 0.06%. To achieve this, dynamic element matching (DEM) is applied (Fig. 5).

Each transducer is associated with a set of five unit-current sources, each of which can be connected to the transducer through a switch. These switches are digitally controlled according to the DEM timing diagram shown in Fig. 5. When one sensitive transducer is biased by all five unit current sources (in the hot state), the other sensitive transducer is sequentially biased by one unit current source (in the cold state), generating an accurate average current ratio of five. The same biasing approach is applied to the reference transducers. It should be noted that the current-domain chopping (indicated in Fig. 3) is also implemented by this switching scheme (switching between I_c and $5I_c$).

In principle, the periodic chopping of the transducers modulates the errors due to transducer mismatch to an AC signal. However, it may still be necessary to trim the initial mismatch, so that the ripple caused by the mismatch at the output of the first integrator will not overload the $\Delta\Sigma$ modulator. As shown in Fig. 6, the two currents I_{OUT1} and I_{OUT2}, biasing the sensitive transducers R_{s1} and R_{s2}, are generated using

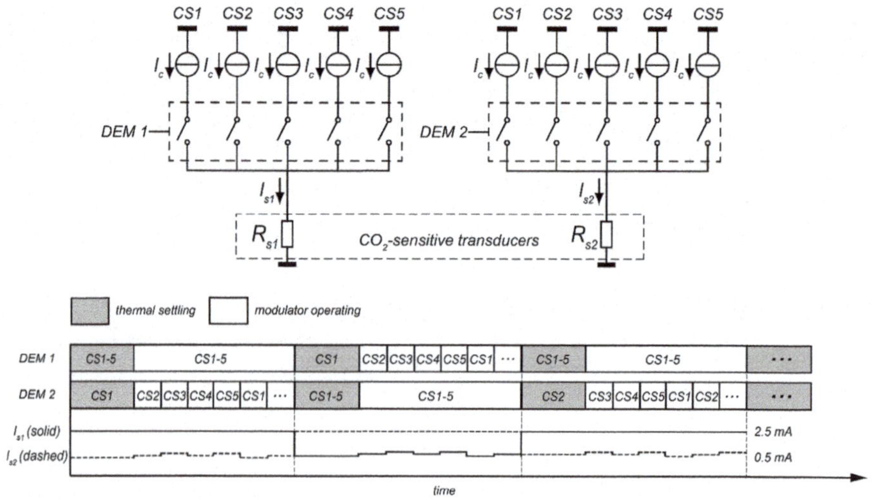

Fig. 5 Dynamically matched current sources and associated timing diagram (same algorithm applies to the current sources for the reference transducers) [5] (reproduced with permission)

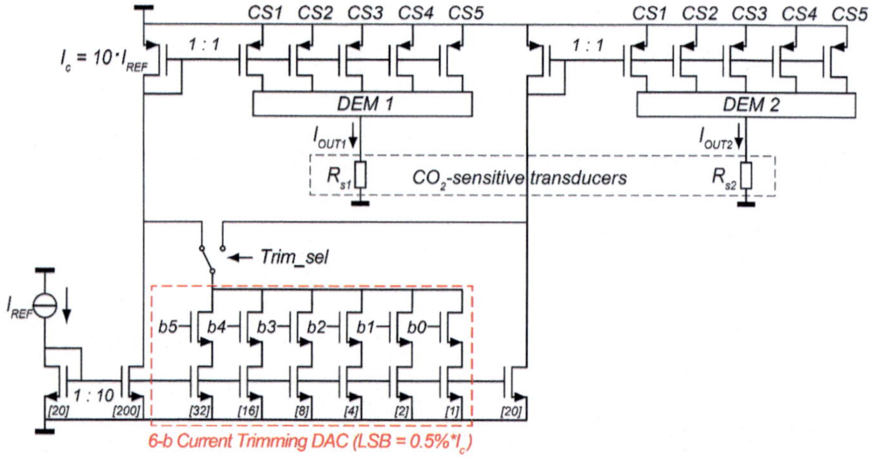

Fig. 6 Circuit diagram of the current sources with a 6-bit current trimming DAC (LSB current = 0.5% × I_c; the current sources as well as the current trimming DAC for the reference transducers are identical, not shown here; cascode transistors omitted for simplicity) [5] (reproduced with permission)

PMOS current mirrors (an identical circuit, not shown, is used for the reference transducers). A 6-bit binary-weighted current DAC (LSB current = 0.5% · I_c), embedded in the current source circuit, is used to trim the input current of these current mirrors, thus effectively compensating for the resistance mismatch between R_{s1} and R_{s2}. Note that the current ratio between I_{OUT1} and I_{OUT2} does not need to be accurate, as long as both I_{OUT1} and I_{OUT2} provide an accurate 1:n current ratio sequentially to the individual transducers, which is ensured by the current-source DEM (Fig. 5). In this way, the measured thermal-resistance ratio is a ratio between the averaged thermal resistances of the two sensitive transducers and that of the two reference transducers.

2.3 Measurement Results

The readout circuit as well as the tungsten-wire transducers have been designed and fabricated in 0.16 μm CMOS technology [5]. Figure 7 shows the layout plot and micrographs of the integrated readout circuit and one of the transducers. The active die area of the circuit equals 0.7 mm^2, of which 0.37 mm^2 is occupied by the current sources and 0.33 mm^2 by the switched-capacitor $\Delta\Sigma$ modulator. The transducers are on another chip for flexibility, fabricated using the same CMOS process followed by an etch step to release the wires.

To demonstrate the insensitivity of the ratiometric measurement to the absolute current and power levels, Fig. 8 shows the measured temperature, power, and

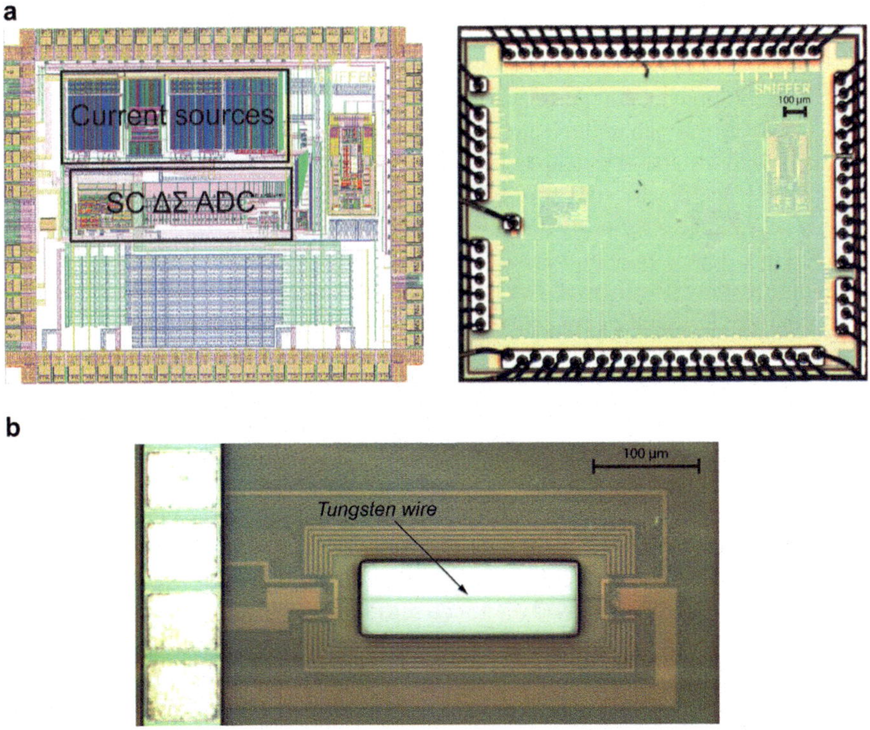

Fig. 7 (**a**) Layout plot and micrograph of the integrated readout circuit and (**b**) micrograph of a CMOS-compatible tungsten-wire transducer [5] (reproduced with permission)

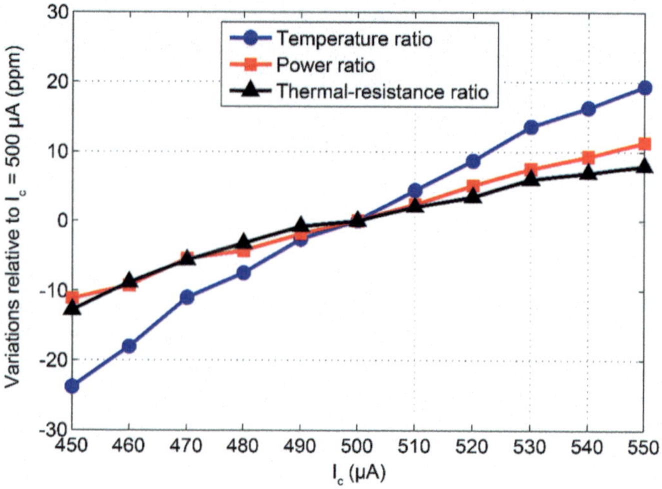

Fig. 8 Variations in temperature, power, and thermal-resistance ratios between the sensitive and reference transducers as a function of the bias current at "cold" state (I_c) [5] (reproduced with permission)

thermal-resistance ratios as a function of I_c (the bias current at the "cold" state). For a $\pm 10\%$ change in I_c, the power ratio only changes by about ± 10 ppm. For our CO_2 sensor, a 200 ppm change in CO_2 will result in about a 50 ppm change in thermal-resistance ratio, which requires the errors of power ratio to be within 25 ppm. Thus, the measured results indicate that the ratiometric measurement effectively alleviates the dependence on the stability of the power dissipation. The measured thermal-resistance ratio also varies by about ± 10 ppm, which could be due to the secondary temperature dependence of the thermal conductivity of air.

Figure 9 shows the thermal-resistance ratio of the tungsten transducers measured using the single-sampling readout circuit described in [12], while the CO_2 concentration was changed stepwise from 500 ppm to 9000 ppm. Like other TC-based CO_2 sensors [5–7], the readings of the sensor are affected by variations in ambient conditions, which need to be compensated for in a final product. In our experiment, ambient temperature, humidity, and pressure sensors were applied to facilitate cross-sensitivity compensation. To lower the noise level, a measurement time of 70 s was used for this experiment, at which the measured resolution is equivalent to 228 ppm CO_2 (1σ). The thermal-resistance ratio measured using the readout circuit has a sensitivity of 0.29 ppm/ppm CO_2, and shows good correlation with the CO_2 level measured using an NDIR-based reference sensor.

The improvement of the double-sampling readout circuit (Fig. 4) has also been validated with discrete heaters Figaro TGS-8100 [15]. These discrete devices have similar electrical and thermal properties to the tungsten transducers and are more readily sealed manually at the package level, so they are used here as a substitute for the tungsten transducers. The results are shown in Fig. 10. Four different CO_2 levels (500 ppm, 2500 ppm, 4500 ppm, 9000 ppm) and a baseline (pure

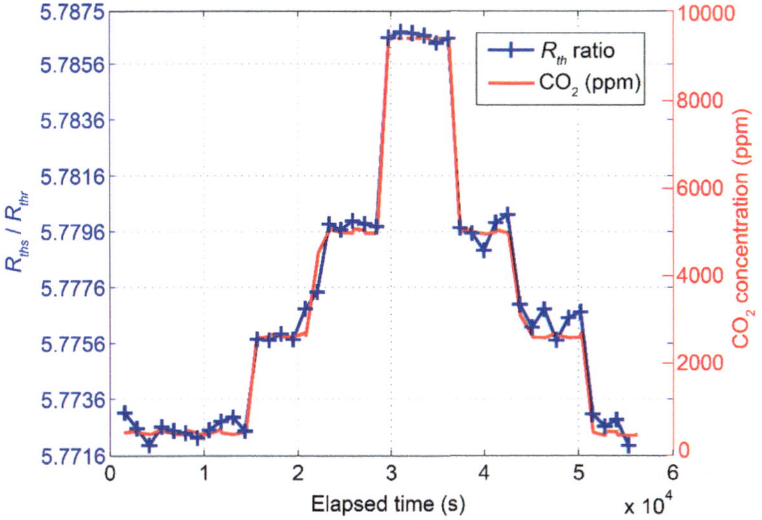

Fig. 9 Thermal-resistance ratio measured using the tungsten-wire transducers in combination with the single-sampling readout circuit [12], for stepwise changing CO_2 concentration with compensation for temperature, humidity, and pressure cross-sensitivity, along with CO_2 concentration measured using an accurate reference NDIR sensor [2, 5] (reproduced with permission)

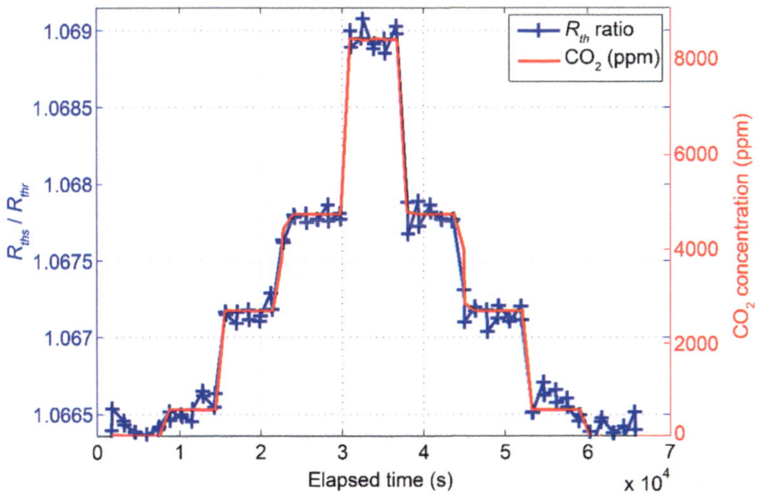

Fig. 10 Thermal-resistance ratio measured using the double-sampling readout circuit (Fig. 4) in combination with Figaro TGS 8100 transducers, with compensation for temperature, humidity, and pressure cross-sensitivity, for stepwise changing CO_2 concentration, along with CO_2 concentration measured using an accurate reference NDIR sensor [2, 5] (reproduced with permission)

air) were used in this experiment. The thermal-resistance ratio measured has a sensitivity of 0.27 ppm/ppm CO_2, and shows good correlation with the reference CO_2 measurements. In these experiments, a measurement time of 30 s was used.

3 CO_2 Sensor Based on Time-Domain TC Measurement

3.1 Operating Principle

3.1.1 Time-Domain TC Measurement

An alternative to measuring the steady-state temperature rise of a hot-wire is to characterize its thermal time constant τ_{th}, which is the product of the wire's ambient thermal resistance (R_{th}) and its thermal capacitance (C_{th}) [6, 7, 16]. When the wire is driven with a current I_d pulsed at a frequency f_{drive} and is thus periodically heated, its temperature transients are delayed relative to the driving pulses. The delay is determined by the thermal time constant τ_{th}, which in turn depends on the TC of the surrounding air (Fig. 11).

Such a TC sensor can be modeled as a first-order low-pass filter. Using a fixed driving frequency will then result in phase-delayed temperature transients relative to the driving pulses, from which τ_{th} can be derived. The optimal driving frequency equals the filter's pole frequency, i.e., $1/2\pi\tau_{th}$, at which the sensitivity of the phase shift to the changes of τ_{th} is maximized. For our devices, $\tau_{th} \approx 17$ µs, leading to an optimal f_{drive} around 9–10 kHz.

Earlier TC sensors based on transient measurements use separate resistive heaters and temperature sensors, either thermistors [6] or thermopiles [17], which are

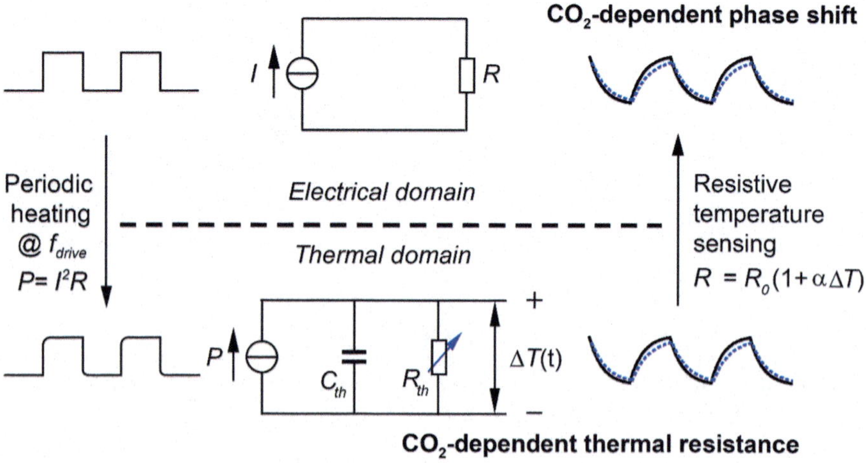

Fig. 11 Transient thermal-resistance (thermal delay) measurement principle [9] (reproduced with permission)

mounted together on a thermally isolated membrane. This separates the temperature transients from the electrical transients and thus simplifies the readout at the cost of fabrication complexity and hence cost. Since resistive transducers can be used as both a heater and a temperature sensor, the heating and sensing functions can, in principle, be combined in a single resistor, provided an appropriate readout scheme is devised. This will challenge the design of the readout circuit but will greatly reduce the fabrication cost.

In earlier work, both a sine wave [6] and a square wave [17] were used to drive the heater. The benefit of sinusoidal driving is that it contains only a fundamental frequency without harmonics, and thus the readout circuit can directly measure the phase shift of the temperature signal by filtering and zero-crossing detection [6]. In contrast, a square-wave driving signal will generate a sinusoidal fundamental signal and a series of odd-order harmonics. The phase shift of the fundamental can be detected using synchronous detection [17]. In terms of circuit implementation, a square-wave excitation is much easier to generate than a sine wave. The phase-domain delta-sigma modulator (PDΔΣM) enables the use of square-wave excitation of the sensor [18, 19], which greatly simplifies the design of the driving source.

3.1.2 Phase-Domain Delta-Sigma Modulator

The phase shift ϕ_{sig} of the transducer's temperature when it is driven at f_{drive} can be found by coherent detection, that is, by multiplying with a reference signal at the same frequency f_{drive} with phase ϕ_{ref}, as illustrated in Fig. 12a. Assuming sinusoidal

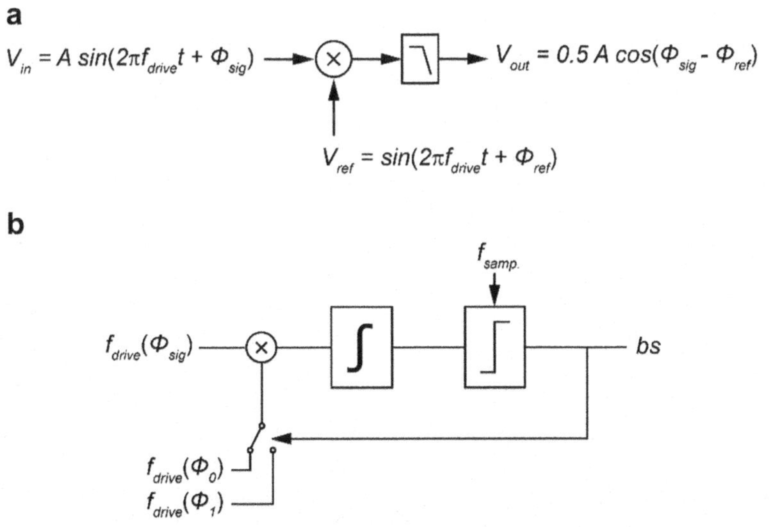

Fig. 12 (**a**) Phase detection by means of coherent detection. (**b**) Phase detection using a delta-sigma feedback loop [9] (reproduced with permission)

signals for simplicity, the result is a DC component proportional to the cosine of the phase difference and a component at $2f_{drive}$ that can be removed by a low-pass filter:

$$\begin{aligned} &A \cdot \sin\left(2\pi f_{drive}t + \phi_{sig}\right) \cdot \sin\left(2\pi f_{drive}t + \phi_{ref}\right) = \\ &0.5 \cdot A \cdot \left[\cos\left(\phi_{sig} - \phi_{ref}\right) - \cos\left(4\pi f_{drive}t + \phi_{sig} + \phi_{ref}\right)\right] \end{aligned} \tag{5}$$

As shown in Fig. 12b, the coherent detector can be embedded in a delta-sigma ($\Delta\Sigma$) loop, where an integrator serves as a low-pass filter and feedback is applied in the phase domain, by toggling ϕ_{ref} between two phase references ϕ_0 and ϕ_1 depending on the bit-stream output *bs* [18, 19]. The feedback loop, on average, nulls the input of the integrator and thus ensures that the average phase reference tracks the phase of the input signal, which can therefore be derived from the average value of the bit-stream.

From (5), it can be seen that in order to allow the $\Delta\Sigma$ modulator to track ϕ_{sig}, the reference phase ϕ_{ref} should toggle between two values such that the term $\cos(\phi_{sig} - \phi_{ref})$ presents inverse polarity. This implies that the two reference phases (ϕ_0 and ϕ_1) should be located on both sides of the 90° phase shift of the input signal. The resolution with which the phase shift can be determined depends on the number of clock cycles N that the $\Delta\Sigma$ modulator is operated per measurement, that is, the oversampling ratio (OSR), and equals $(\phi_1 - \phi_0)/N$ for a first-order $\Delta\Sigma$ modulator [19].

Simulation shows that the phase shift induced by a 1 ppm change in CO_2 concentration is roughly 7 μ°. This can be used to estimate the required OSR to arrive at a desired CO_2 resolution. For example, for a full scale $\phi_0 - \phi_1 = 4°$, the required OSR for a quantization step equivalent to 100 ppm CO_2 is about 6000.

3.2 *Design and Implementation*

3.2.1 Dynamic Range Requirement Reduction

While the voltage across the transducer in Fig. 11 contains temperature information, its sensitivity to temperature will change with the current level. To mitigate this, an additional sense current I_s, switched at a much faster rate f_{sense}, produces a modulated voltage proportional to $R(t)$ with a fixed sensitivity to temperature, independent of the drive current (Fig. 13a).

To make it easier to detect this voltage in the presence of the large voltage transients at f_{drive} (about 300 mV peak-to-peak), a pair of transducers are heated simultaneously by pulsed currents I_d (=2 mA) and read out differentially by out-of-phase sense currents I_s (=0.5 mA), switched at $f_{sense} = 15 \times f_{drive}$ (Fig. 13b). Thus, the signal at f_{drive} is converted into a common-mode signal and can be rejected by the differential readout circuit. Each transducer is also biased by an additional

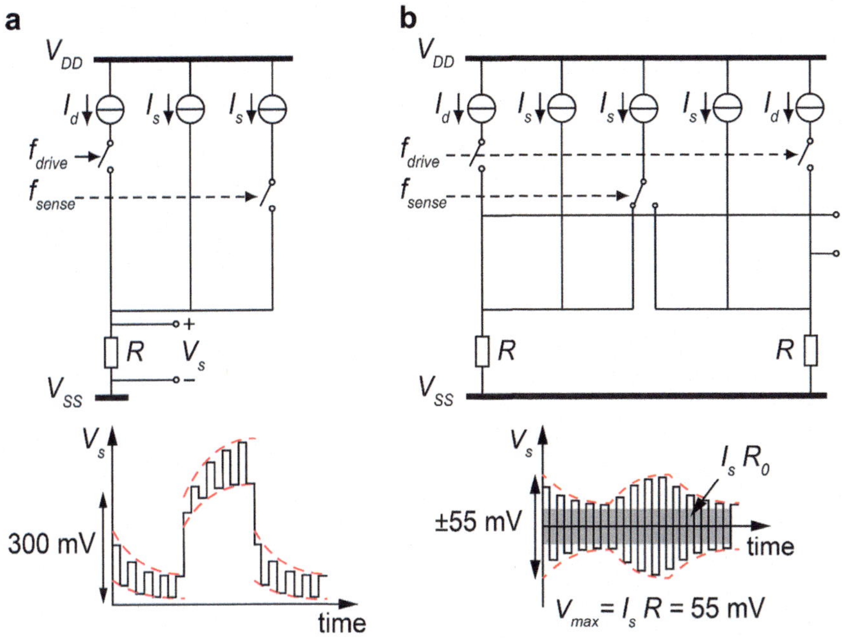

Fig. 13 Sensing the temperature-induced resistance changes using (**a**) current modulation and (**b**) differential sensing

constant sense current I_s (=0.5 mA) to provide a voltage signal to be sensed when I_d is switched off.

An odd ratio of 15 between the f_{sense} and f_{drive} is chosen here to prevent errors due to the down conversion of harmonics of the drive signal. Due to mismatch of the drive signals, a fraction of the common-mode drive signal will be converted to a differential-mode signal. If f_{sense} was an even multiple of f_{drive}, the odd harmonics of this differential-mode signal would be down converted to f_{drive} by the chopper demodulation at f_{sense}, and would then be detected by the PDΔΣM, affecting the decimated results. As f_{sense} is chosen to be an odd multiple of f_{drive}, the down-converted harmonics end up at DC and are rejected by the PDΔΣM.

Even with this arrangement, a large dynamic range is still required, since the temperature-induced resistance increase ($\Delta R \approx 3\ \Omega$) is small compared to the baseline resistance ($R_0 = 110\ \Omega$), while the variation in ΔR as a result of variations in CO_2 concentration is even smaller (about 1.5 μΩ per ppm CO_2). To cancel the voltage steps associated with R_0, two poly resistors R_p (= R_0) are connected in series with the transducers and the sense currents are routed such that the additional voltage drop $I_s R_p$ cancels out $I_s R_0$ (Fig. 14). The remaining differential signal V_s is ideally equal to $I_s\,\Delta R$ and reflects the transient temperature change, which is about 1.5 mV, 200× smaller than the initial 300 mV transients.

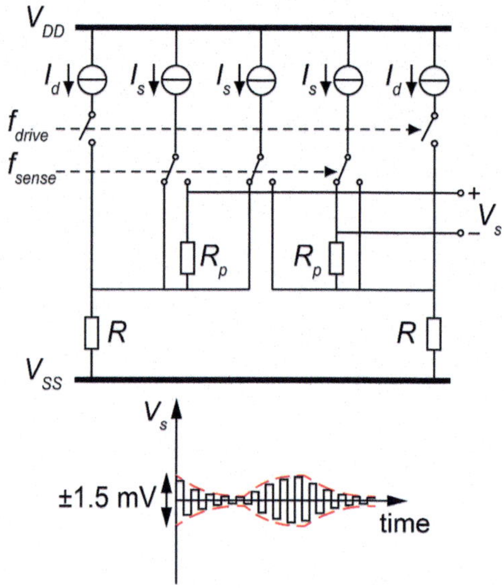

Fig. 14 Baseline resistance compensation using a pair of poly resistors

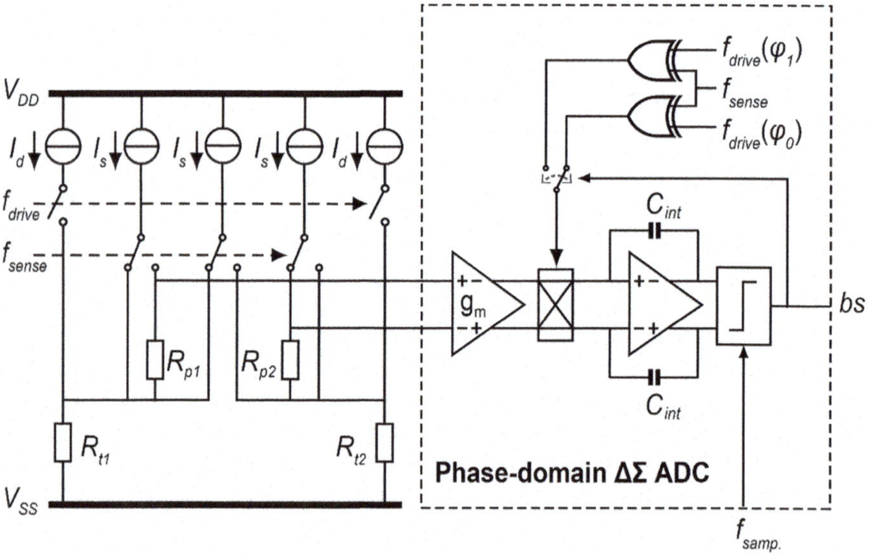

Fig. 15 Circuit diagram of the proposed readout circuit [9] (reproduced with permission)

3.2.2 Phase-Domain Delta-Sigma Modulator

The phase shift of the temperature-related differential signal V_s ($\approx I_s\,\Delta R$) is digitized by a low-noise phase-domain $\Delta\Sigma$ modulator similar to that described in [20]. As shown in Fig. 15, before demodulation by f_{sense}, a low-noise transconductor g_m converts the differential voltage V_s into a current. This current passes through a chopper switch, which serves the dual purpose of demodulation by f_{sense} (i.e., down converting the desired phase information at $f_{sense} + f_{drive}$ to f_{drive}), and multiplication with the phase-shifted versions of f_{drive} as a function of the bit-stream (as in Fig. 12b). This combination is realized by multiplying the phase-shifted versions of f_{drive} with f_{sense} by means of XOR gates. The resulting demodulated current is proportional to the phase difference between $V_s(t)$ and the selected phase reference. This difference is integrated on capacitors C_{int} of an active integrator and quantized using a clocked comparator to form a $\Delta\Sigma$ loop, which nulls the input of the integrator and thus ensures that the average phase reference tracks the phase of $V_s(t)$, which can therefore be derived from the average value of the bit-stream.

To ensure that the noise from the transconductor is lower than that from the transducer and its bias circuit, the g_m of the transconductor should be at least 400 μS. The transconductance of the g_m stage is about 560 μS. The sampling frequency is chosen to be the same as f_{drive}. Both f_{drive} and f_{sense}, including the feedback signals at f_{drive} with reference phases ϕ_0 and ϕ_1, are derived from a single off-chip master clock. The capacitor C_{int} in the integrator is 50 pF.

3.3 Measurement Results

Both the transducers and the readout circuit have been implemented in the same 0.16 μm CMOS technology (Fig. 16), with an active area of 0.3 mm^2 and 3.14 mm^2, respectively [9]. For flexibility, they have been realized on separate chips and connected at the PCB level. The modulator's control signals were generated using

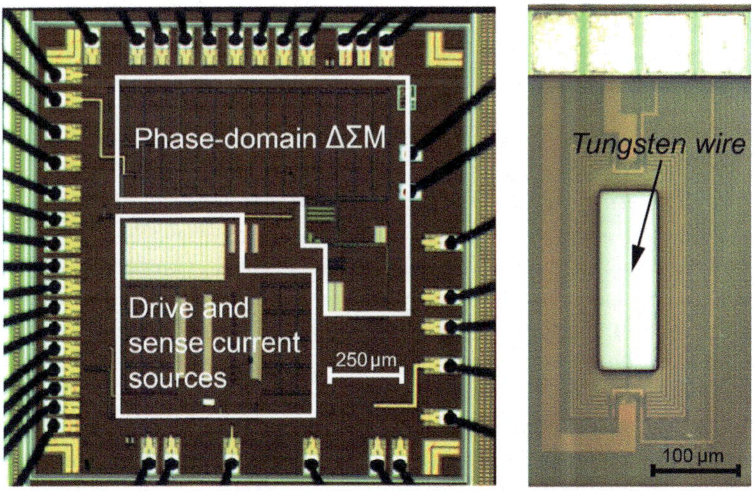

Fig. 16 Micrograph of the readout circuit and the transducer [9] (reproduced with permission)

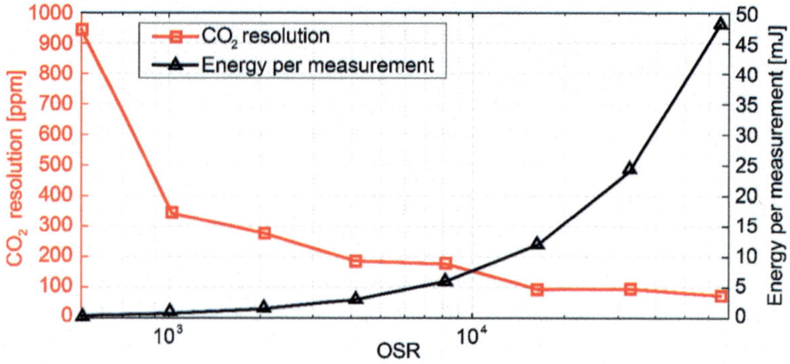

Fig. 17 Measured resolution (standard deviation of 20 consecutive measurements) and energy per measurement as a function of OSR [9] (reproduced with permission)

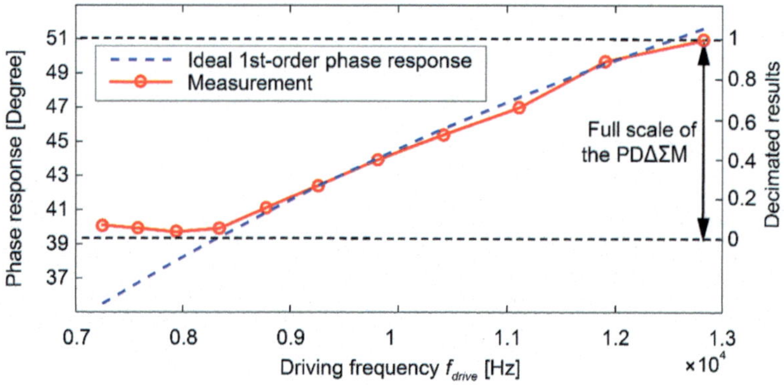

Fig. 18 Measured phase shift as a function of the drive frequency [9] (reproduced with permission)

an FPGA. The readout circuit consumes 6.8 mW from a 1.8 V supply, 6.3 mW of which is dissipated in the transducers.

Figure 17 shows the measured resolution at different oversampling ratios (OSR). A resolution equivalent to 94 ppm CO_2 is reached at an OSR of 16,384, which corresponds to a measurement time of 1.8 s and an energy consumption of 12 mJ.

The thermal delay or equivalently the measured phase shift, caused by the thermal resistance and thermal capacitance, should present a first-order behavior as a function of the driving frequency, like a first-order electrical low-pass filter. This is confirmed by measurements shown in Fig. 18. The measured phase shift as a function of the drive frequency shows a good agreement with the ideal first-order behavior associated with the hot-wire's thermal time constant (measured using a larger full scale $\phi_0 - \phi_1 = 12°$ for clarity).

To measure its CO_2 response, the sensor was placed in a sealed box along with an NDIR reference CO_2 sensor [2]. As before, ambient temperature, humidity, and pressure sensors were placed in the box to facilitate cross-sensitivity compensation.

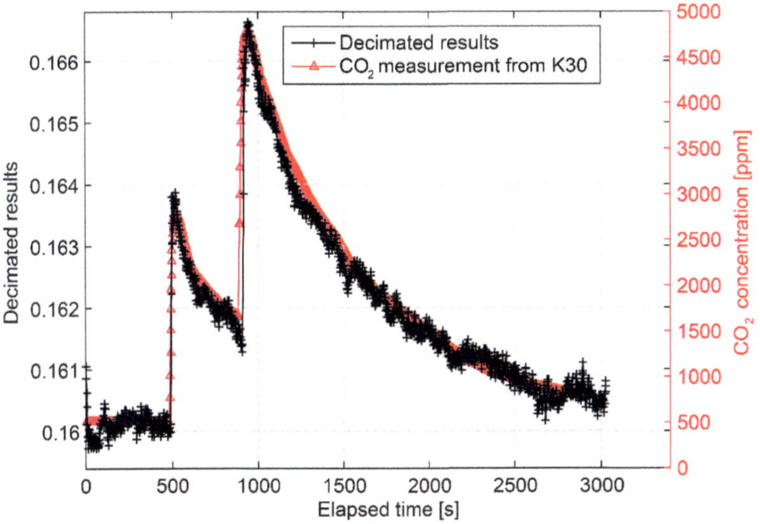

Fig. 19 Transient CO_2 response of the CO_2 sensor and an NDIR-based reference sensor K30 [2, 9] (reproduced with permission)

Figure 19 shows the good agreement between the readings of our sensor and the CO_2 concentration measured by the reference sensor.

4 Comparison and Benchmarking

Table 1 summarizes the performance of the amplitude-domain and time-domain TC-based CO_2 sensors and compares it with prior works. The two sensors were implemented in the same CMOS technology and have similar chip areas. In both cases, most of the power is dissipated in the transducers for heating. The time-domain sensor consumes about half the average power of the amplitude-domain sensor due to the circumvention of the reference transducers. By moving the design from the amplitude domain to the time domain and using a PD$\Delta\Sigma$M based on a continuous-time g_m-C integrator, the noise performance and the energy efficiency of the readout circuit have been greatly improved. The sensor based on time-domain TC measurement achieves a CO_2 resolution of 94 ppm while dissipating only 12 mJ per measurement, representing a significant improvement in energy efficiency compared to previously reported CO_2 sensors.

The attained results indicate that TC-based CO_2 sensors realized CMOS technology are a promising candidate for air-quality monitoring, provided they are cointegrated with appropriate sensors to compensate for the cross-sensitivity of TC measurement to temperature, humidity, and pressure. The resulting advantages in cost (>10×), volume (>100×), and energy consumption are significant, making them a promising candidate for CO_2 sensing in cost- and energy-constrained applications.

Table 1 Performance summary of the amplitude-domain and time-domain TC-based CO_2 sensors and benchmarking

Parameter	Time domain	Amplitude domain	[6]	[2]	[21]
Method	TC	TC	TC	NDIR	NDIR
Technology	CMOS (0.16 μm)	CMOS (0.16 μm)	SOI MEMS	Module	SOI MEMS
On-chip readout	Y	Y	N	N	N
Area (sensor)	0.3 mm^2	0.6 mm^2	16 mm^2	–	[a]0.3 mm^2
Area (readout)	3 mm^2	3 mm^2	–	–	–
Supply voltage	1.8 V	1.8 V	–	5–14 V	–
Power consumption	6.8 mW	11.2 mW	3 mW	200 mW	200 mW
Meas. time	1.8 s	30 s	60 s	2 s	2.4 s
CO_2 resolution	94 ppm	202 ppm	456 ppm	20 ppm	250 ppm
Energy/meas.	12 mJ	336 mJ	180 mJ	400 mJ	480 mJ

[a]Area of the IR emitter only, excluding 80 mm light tube and an infrared detector

5 Conclusion

In this chapter, CMOS-compatible CO_2 sensors based on thermal-conductivity measurement have been presented. They detect the CO_2-dependent thermal conductivity of ambient air by measuring the heat loss of a suspended hot-wire transducer, realized in the VIA layer of a standard CMOS process to reduce fabrication cost and complexity. Two sensor designs have been presented, based on different readout approaches. The first approach is amplitude-domain TC sensing, which measures the steady-state temperature and power of a resistive transducer using a dual-mode incremental $\Delta\Sigma$ ADC. The second approach is time-domain TC sensing, which measures the phase shift of the temperature transients of a periodically heated transducer using a phase-domain $\Delta\Sigma$ ADC. Both approaches have been validated and proven to be capable of measuring CO_2 concentration with a resolution suitable for indoor air-quality monitoring. The time-domain approach achieves best-in-class energy efficiency. These results make the design a promising alternative to NDIR-based CO_2 sensors, especially in cost- and energy-constrained applications.

Acknowledgments This work was in part supported by NXP Semiconductors, The Netherlands, and in part by ams AG, The Netherlands. The authors want to thank Lukasz Pakula and Zu-yao Chang for their technical support.

References

1. Emmerich SJ, Persily AK. Literature review on CO_2-based demand-controlled ventilation. ASHRAE Trans. 1997;103:229–43.
2. SenseAir K30 datasheet, SenseAir [Online]. Available: http://www.senseair.com/.
3. SGX Sensortech IR11BD datasheet, SGX Sensortech [Online]. Available: http://www.sgxsensortech.com/.

4. Frodl R, Tille T. A high-precision NDIR CO_2 gas sensor for automotive applications. IEEE Sensors J. 2006;6(6):1697–705.
5. Cai Z, et al. A ratiometric readout circuit for thermal-conductivity-based resistive CO_2 sensors. IEEE J Solid-State Circuits. 2016;51(10):2463–74.
6. Kliche K, et al. Sensor for thermal gas analysis based on micromachined silicon-microwires. IEEE Sensors J. 2013;13(7):2626–35.
7. Kliche K, et al. Sensor for gas analysis based on thermal conductivity, specific heat capacity and thermal diffusivity. In: Proceedings of IEEE international conference on MEMS. 2011 p. 1189–92.
8. XEN-5310 datasheet. Xensor Integration [Online]. Available: http://www.xensor.nl/.
9. Cai Z, et al. A phase-domain readout circuit for a CMOS-compatible hot-wire CO_2 sensor. IEEE J. Solid-State Circuits. (in press) doi: 10.1109/JSSC.2018.2866374
10. Simon I, Arndt M. Thermal and gas-sensing properties of a micromachined thermal conductivity sensor for the detection of hydrogen in automotive applications. Sens. Actuators A: Phys. 2002;97–98:104–8.
11. Ali SZ, et al. Tungsten-based SOI microhotplates for smart gas sensors. J Microelectromech Syst. 2008;17:1408–17.
12. Cai Z, et al. An integrated carbon dioxide sensor based on ratiometric thermal-conductivity measurement. In: Proceedings of IEEE international conference on solid-state sensors, actuators and microsystems (Transducers '15). 2015. p. 622–5.
13. Bult K, Geelen GJGM. A fast-settling CMOS op amp for SC circuits with 90-dB DC gain. IEEE J Solid-State Circuits. 1990;25(6):1379–84.
14. Fiedler H, et al. A 5-bit building block for 20 MHz A/D converters. IEEE J Solid-State Circuits. 1981;16(3):151–5.
15. FIGARO TGS8100 datasheet (rev06), FIGARO [Online]. Available: http://www.figaro.co.jp/.
16. Mahdavifar A, et al. Transient thermal response of micro-thermal conductivity detector (μTCD) for the identification of gas mixtures: an ultra-fast and low power method. Microsyst Nanoeng. 2015;1:15025.
17. van Vroonhoven C, de Graaf G, Makinwa KAA. Phase readout of thermal conductivity-based gas sensors. In: Proceedings of IEEE International Workshop on Advances in Sensors and Interfaces (IWASI). 2011. p. 199–202.
18. van Vroonhoven CPL, Makinwa KAA. A thermal-diffusivity-based temperature sensor with an untrimmed inaccuracy of ±0.5°C (3σ) from –40 to 105°C. In: Digest ISSCC. 2008. p. 576–7.
19. Kashmiri M, Xia S, Makinwa KAA. A temperature-to-digital converter based on an optimized electrothermal filter. IEEE J Solid-State Circuits. 2009;44(7):2026–35.
20. Kashmiri SM, Souri K, Makinwa KAA. A scaled thermal-diffusivity-based 16 MHz frequency reference in 0.16 μm CMOS. IEEE J Solid-State Circuits. 2012;47(7):1535–45.
21. Vincent TA, Gardner JW. A low cost MEMS based NDIR system for the monitoring of carbon dioxide in breath analysis at ppm levels. Sens Actuators B Chem. 2016;236:954–64.

Time of Flight Imaging and Sensing for Mobile Applications

Neale A. W. Dutton, Tarek Al Abbas, Francescopaulo Mattioli Della Rocca, Neil Finlayson, Bruce Rae, and Robert K. Henderson

1 Introduction

Time correlated single photon counting (TCSPC) systems are found in a wide range of optical sensing applications from medical, scientific, aerial mapping, space and defense to consumer and automotive [1]. In medical, positron emission tomography (PET) and scientific fluorescence lifetime imaging microscopy (FLIM) systems operate in controlled environments where key parameters are timing precision and accuracy (jitter), whereas mobile and consumer (PC, laptop) time of flight (TOF) sensors and light detection and ranging (LIDAR) sensors for automotive operate in uncontrolled environments where measurement rate and system power become important considerations under wide variation of background light and temperature.

The integration in a CMOS IC of both single photon avalanche diode (SPAD)-based photo sensing (with intrinsic picosecond temporal resolution) and dedicated timing and processing electronics provides an all-in-one compact solution for optical distance measurement with millimeter accuracy [2]. For mobile applications, energy efficient measurement and data conversion are highly significant. This chapter describes the design of an energy efficient TCSPC optical sensor with

N. A. W. Dutton (✉) · B. Rae
Imaging Sub-Group, STMicroelectronics, Edinburgh, UK
e-mail: neale.dutton@st.com

F. M. D. Rocca
Imaging Sub-Group, STMicroelectronics, Edinburgh, UK

School of Engineering, University of Edinburgh, Edinburgh, UK

T. Al Abbas · N. Finlayson · R. K. Henderson
School of Engineering, University of Edinburgh, Edinburgh, UK

K. A. A. Makinwa et al. (eds.), *Low-Power Analog Techniques, Sensors for Mobile Devices, and Energy Efficient Amplifiers*,
https://doi.org/10.1007/978-3-319-97870-3_12

10 GS/s conversion rate, designed to minimize time-domain distortion and energy efficiency in particular.

We report a previously published sensor [3–5], providing here a comprehensive overview and new circuit-level detail. This chapter presents here a revised version of the original sensor with a slightly lower conversion rate with improved PLL linearity, revised pixel routing for better timing performance, and characterization results are provided for this new device. This chapter is organized as follows: Sect. 2 gives an introduction to TOF sensing and TCSPC sensors. In Sect. 3, circuit design of a 10 GS/s time converter (TC) is described and an energy efficient time-domain signal processing block creating a histogram on-chip is shown. Sect. 4 provides measurement results, followed by discussion and conclusion.

2 Background

As illustrated in Fig. 1, a TOF sensor performing optical distance sensing and imaging based on the TCSPC method has four main parts [1], where this chapter focuses on the first three:

1. Single photon detection and signal routing
2. Event-driven measurement of photon arrival time
3. Arrival time data collection and processing
4. Synchronous laser transmission and control

The optical distance measurement to a target is calculated by the following equation:

$$d = \frac{\text{TOF}}{2}.c \tag{1}$$

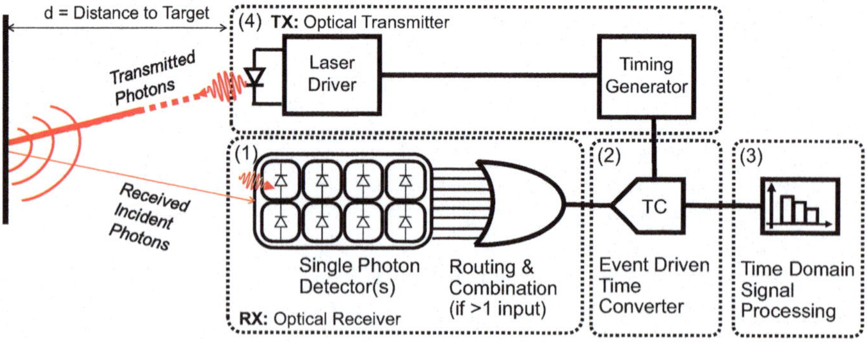

Fig. 1 Time correlated single photon counting (TCSPC) system with SPAD-based receiver and synchronized laser-based optical transmitter in four key parts: (1) detection, (2) time measurement, (3) data processing, and (4) synchronous laser transmission

Table 1 TOF system temporal resolution compared to distance resolution

Time resolution of TOF system	TOF distance resolution
6.67 ns	1 m
667 ps	10 cm
66.7 ps	1 cm
6.67 ps	1 mm

Where d is the round-trip distance travelled by an optical laser pulse to a target and back (hence, the factor 2), TOF is the round-trip time of the laser pulse and c is the speed of light in a vacuum at 3×10^8 ms^{-1}. From Table 1, it can be seen that for millimeter to centimeter resolution, as are required for consumer distance measurement, the time resolution of the TOF system must be in picoseconds. It is important to note that most TOF systems are based on oversampling; therefore, this is not the time converter resolution but the oversampled output resolution.

2.1 *Motivation*

Scientific FLIM, space, defense, and aerial mapping LIDAR TCSPC systems consist of many discrete components, primarily pulsed laser, detector, and processing electronics. Discrete high gain single-photon-sensitive photodetectors are used such as an avalanche photodiode (APD), a single photon avalanche diode (SPAD), or a photomultiplier tube (PMT) matched to the wavelength of interest [1, 6]. The measurement function is accomplished by either a front-end circuit (high-speed amplifier, filter, and/or comparator), followed by event-driven time conversion circuit based on either time to analogue (TAC) or time to digital converter (TDC) generating a timestamp per detected photon.

Additionally, a PC or embedded system is employed for timestamp data collection and post-processing (commonly histogram generation). Such setups are physically large with high overall cost (>$10 k). CMOS integration of the majority of the TCSPC system onto one chip brings down size and cost by orders of magnitude suitable for mobile TOF applications [2]. To understand the challenges of TCSPC sensor design for uncontrolled conditions, first we examine the input signal. The input to a TCSPC system has three distinct properties:

1. Asynchronous with respect of the system operation, that is, it can arrive at any time.
2. Discrete in time, for example, a spiking signal or rising edge trigger.
3. Poissonian in event rate and have multiple events per synchronous cycle, that is, at high input event rates, the variance is proportional to the mean event rate.

This input signal is comprised of two constituent parts: correlated (from the transmitter source) and uncorrelated (from another asynchronous source). In TOF and LIDAR, the correlated received signal from the return TOF is proportional to the inverse square of the distance of the target. In contrast to ADCs operating in the voltage domain converting a *single* input voltage, for TCSPC systems operating in the event-driven time domain, there are *multiple* events (both correlated and uncorrelated) from received photon events across the temporal dynamic range. At far distance, there can be single returned photons, and in contrast at close distance there can easily be $>10^9$ photons. Furthermore, in uncontrolled environments, there is background ambient light or dark count rate (generated by thermal or by defectivity and not by incoming photons) which are uncorrelated received signals manifesting as a white noise floor for the data processing and can be greater in average event rate than the correlated signal.

As shown in Fig. 2a, photon arrivals are a Poissonian process, the received correlated and uncorrelated received signals both have shot noise approximately $\sigma = \sqrt{N}$ for a high number of N events. The greater the number of events in the correlated signal, the lower the overall contribution of the shot noise in the signal processing to resolve the object distance. Therefore, the maximization of conversion rate is important because each and every photon contributes to the optical distance measurement. The target of this work is to overcome the limits on optical TCSPC systems on photon processing rate set by all three components: the event-driven signal routing, the time converter and the histogram generation DSP. This is achieved by increasing the event-driven signal rate, the conversion rate, and the histogram generation rate through a multi-event time to digital converter (TDC) with parallelized direct-to-histogram output as shown in Fig. 2b, c.

2.2 *Pile-Up Distortion*

Pile-up distortion is the degradation of time-domain information due to insufficient temporal bandwidth in any or all parts of a TCSPC system [7]. For example, in a time-domain data converter with asynchronous event-driven input, if any input trigger events occur before the conversion has finished from a previous data sample, then following events are missed. This time-domain signal clipping is due to the converter dead time and is an example of pile-up distortion [8]. An illustration is shown in Fig. 3.

There are two primary solutions for pile-up distortion minimization. The first applies to controlled condition environments (such as scientific applications), the input signal event rate must be controlled to less than a 10th of the rate of the laser transmitter and uncorrelated background signal must be suppressed to maintain linearity [1]. On the other hand, to a first order, for uncontrolled conditions in mobile applications, the primary choice is to increase the rate of the time conversion; the more frequent the conversion, the more input events that can be captured, hence the lower the probability of missing data conversions and distorting the measurement. However, greater conversion rate induces higher power consumption. Therefore, the energy efficiency of routing logic and time-domain data conversion is an important factor to manage the power consumption of high input event rate systems operating in real-world conditions.

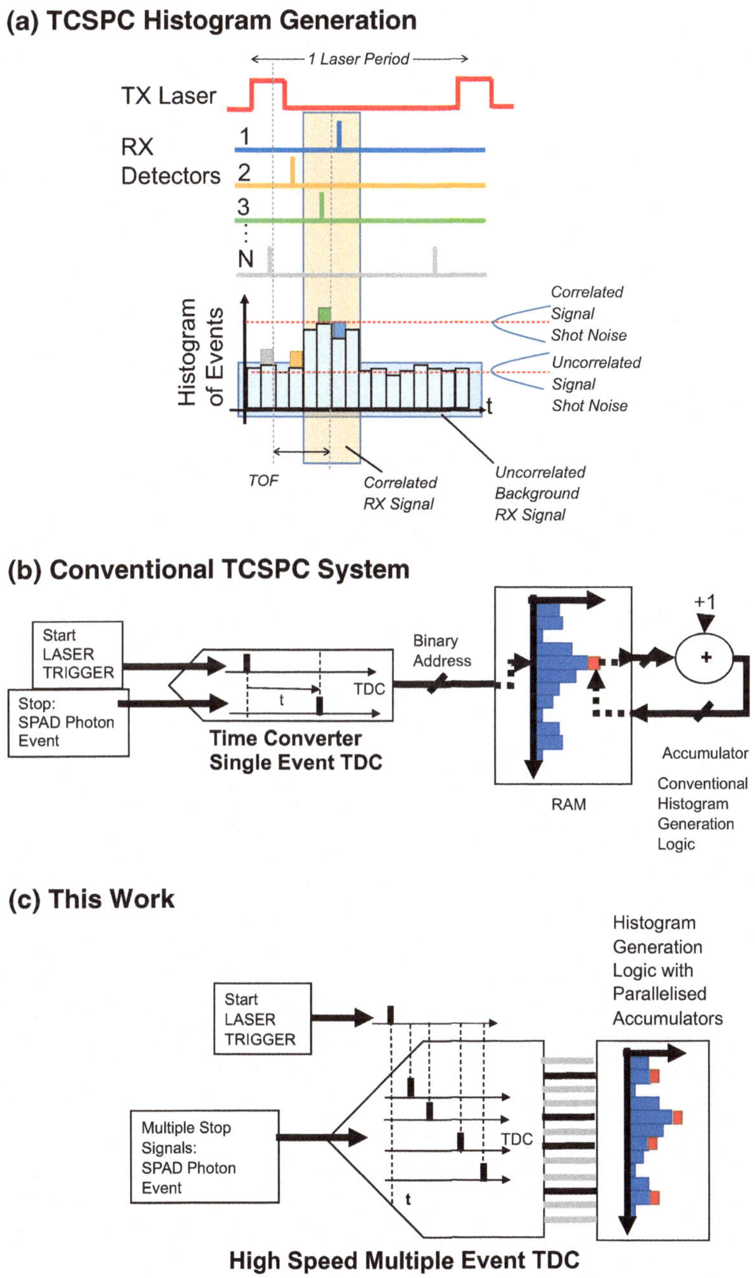

Fig. 2 (**a**) Conceptual diagram of TCSPC histogram generation. (**b**) Conventional TCSPC system with single channel time converter and RAM-based histogram generation. (**c**) This work: multi-event time converter with parallelized histogram generation

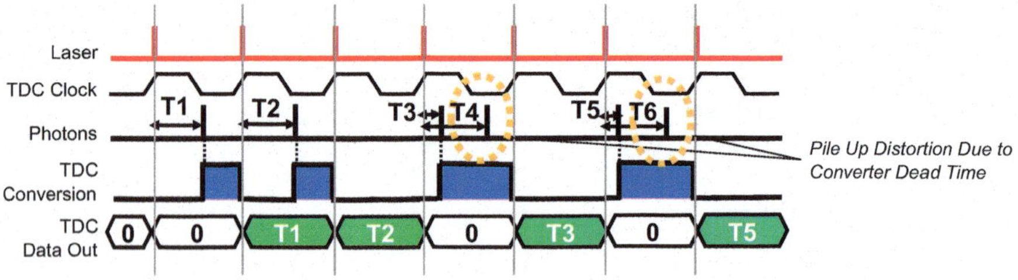

Fig. 3 Timing diagram showing the source of pile-up distortion in time-based data converters due to converter dead time

3 Sensor Design

This section describes the design of the three key parts of the TCSPC system optimized for high event rate. The first part describes the SPAD pixel and the pulse combination logic. The second provides detail of the multiple-event TDC. The third part details the parallelized histogram generation.

3.1 SPAD Input and Combination Logic Design

The SPAD pixel provides timing event inputs to the system. The SPAD diode device is a class of APD operating in "Geiger" avalanche mode (G-APD) [2]. The three regions of photodiode operation are illustrated in the gain to reverse bias plot in Fig. 4a: integration, avalanche, and "Geiger" single photon avalanche mode. The SPAD is a reverse biased PN junction biased and operated above its breakdown voltage (V_{BD}) by an excess bias (V_{EB}) shown by the red dot on the I-V plot in Fig. 4b. An electron-hole pair generated by impact ionization of a photon within the PN junction may trigger a current avalanche in the active region of the device. The time duration of the avalanche and recharge is known as the SPAD dead time (indicated in Fig. 4d), as the detector has a reduced sensitivity to incoming photons. It is controllable by the recharge resistance (or g_m) and is in the order of nanoseconds. Figure 4c shows the simplified SPAD pixel circuit diagram. To reduce the event rate of the SPAD pixel, a positive edge-triggered toggle flip-flop is placed on the output to half the output rate providing a reduction in switching activity to the next stage of the system which is a dynamic power saving technique. Encoding asynchronous events on both rising and falling edges provides double the maximum rate for the same fcv^2 dynamic power (over rising edge only). The pixel timing diagram is shown in Fig. 4d.

Figure 5 illustrates three methods of routing and combining detected SPAD events from multiple SPAD pixels through a logic tree to a single output channel. Figure 5a illustrates the simplest technique, consisting of an OR tree. A problem with this method is that simultaneous pulses from different inputs merge during

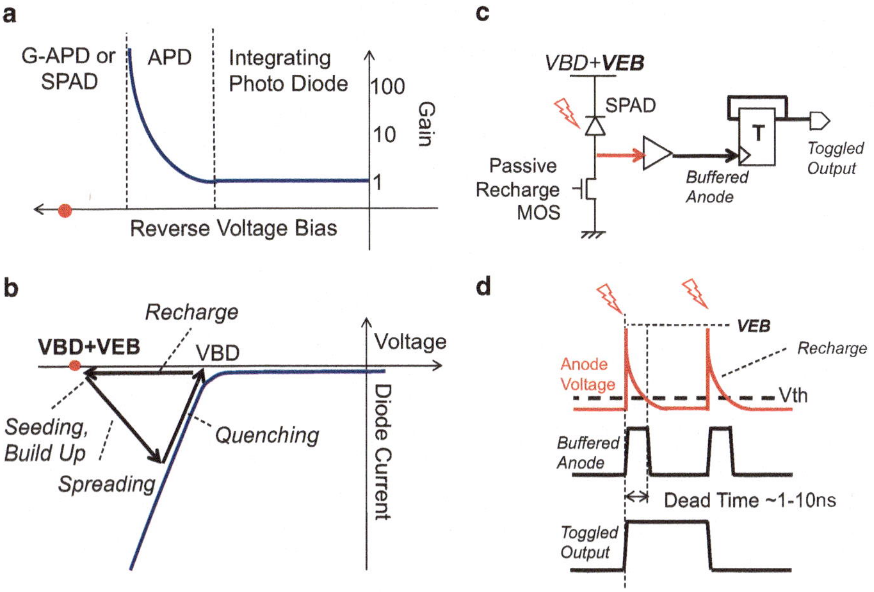

Fig. 4 (**a**) Gain versus reverse voltage bias for different photodetectors. (**b**) The I-V plot for a SPAD showing the four regions of operation during an avalanche event. (**c**) Passive quench and recharge SPAD pixel with toggled output. (**d**) Timing diagram of the SPAD pixel

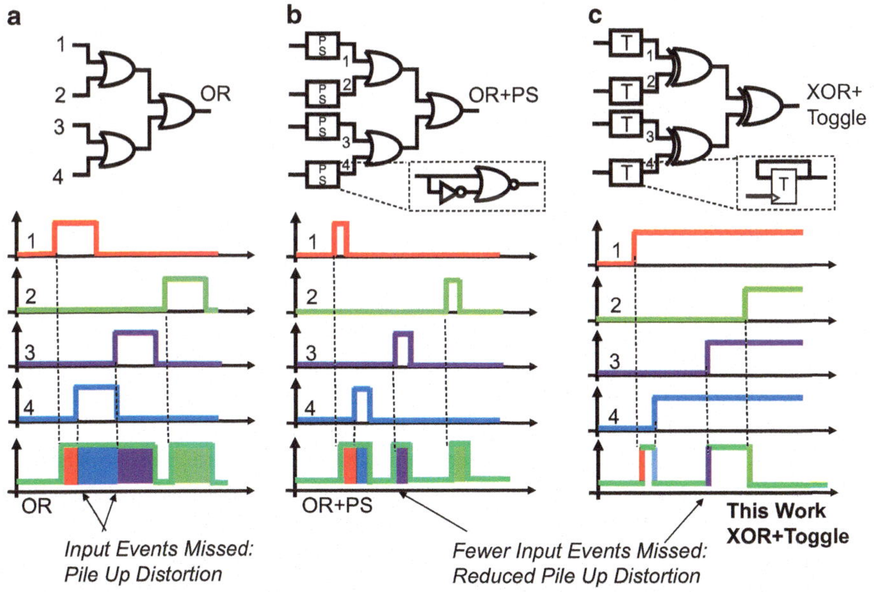

Fig. 5 (**a**) OR tree with pile-up distortion indicated. (**b**) OR tree with monostable pulse shaper input reduces pile-up distortion. (**c**) XOR tree with toggle flip-flop input reduces the rate of the output

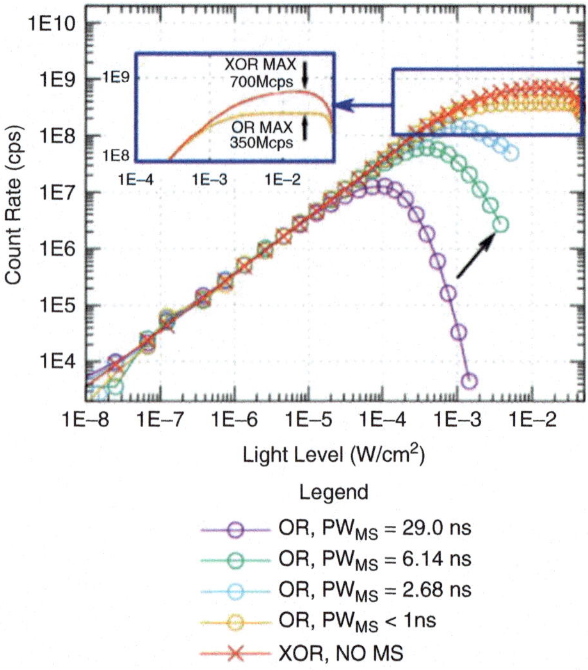

Fig. 6 Measured comparison of OR tree to XOR tree using data from [10]. The plot shows the count rates of the XOR tree (cross) and the OR tree with different pulse widths PW_{MS} set by the monostable circuits (circles)

any overlaps in their dead time. This limits the maximum rate of output events to a rate proportional to the reciprocal of the dead time. By adding a pulse-shortening monostable [10], photon arrivals are still represented by the rising edges but have reduced duration output pulses (Fig. 5b). This reduces the chance of pulse coalescence and increases the maximum rate of output events by the ratio of SPAD dead time to monostable time. In our recent work [5, 9, 10] and in this work, we implement an asynchronous dual data rate (DDR) encoding scheme which uses both rising and falling edges to represent a detected photon event to be fed to the data converter. Several toggled outputs from SPAD pixels are combined by an XOR combination tree. This is a power saving measure compared to the OR tree as the XOR tree output rate is reduced by up to two (in the ideal case with separated input pulses). Yet, if two events close in time (within a gate delay) try to propagate through the same XOR logic gate, the events will cancel out resulting in loss of data and pile-up distortion, but this only occurs when the mean input arrival rate approaches the inverse of a gate delay (i.e., at very high frequency and very high photon rates). Figure 6 is generated from the measurement data from [10], showing a comparative measurement of the OR tree with varying monostable pulse width to the XOR tree showing the double maximum rate. The interested reader is directed to our previous work on comparison between XOR and OR trees [9, 10].

3.2 Comparison of Time Converters

In this section, we first examine different architectures of time converter (TC) circuits in respect to their conversion rate: the time to amplitude converter (TAC) [11], the flash delay line TDC (DL-TDC) [12], and the gated ring oscillator TDC (GRO-TDC) [13], in comparison to the TDC described in this work. All data converters, including TDCs, naturally have a conversion dead time following the input sampling phase (of course, this can be mitigated by pipelining or time interleaving). The majority of TAC or TDC circuits for TCSPC have converter dead time, limiting the system to one photonic event per laser excitation and time conversion cycle [1]. This motivates the design of the multi-event TDC capable of converting multiple events per laser cycle to decrease significantly the pile-up distortion contribution from the time converter.

TAC converters are based on an event-driven analogue ramp function and an ADC (Fig. 7a) [11, 14]. The longer the duration of the ramp, the greater the temporal dynamic range of the converter. The combination of the ramp time and the ADC resolution sets the temporal resolution. The ramp is activated for one photon and must be reset, so limits the maximum conversion rate. Flash delay line and GRO-TDCs are two similar architectures of TDC using the propagation delay time of a logic gate (inverter or buffer) to provide a timing window forming the Least Significant Bit (LSB) of the time to data conversion [15]. As illustrated in Fig. 7b, c, an open-loop delay line or closed-loop ring of delay gates is employed, where each delay element is tapped to a sampling flip-flop. A variant of the flash TDC applies a

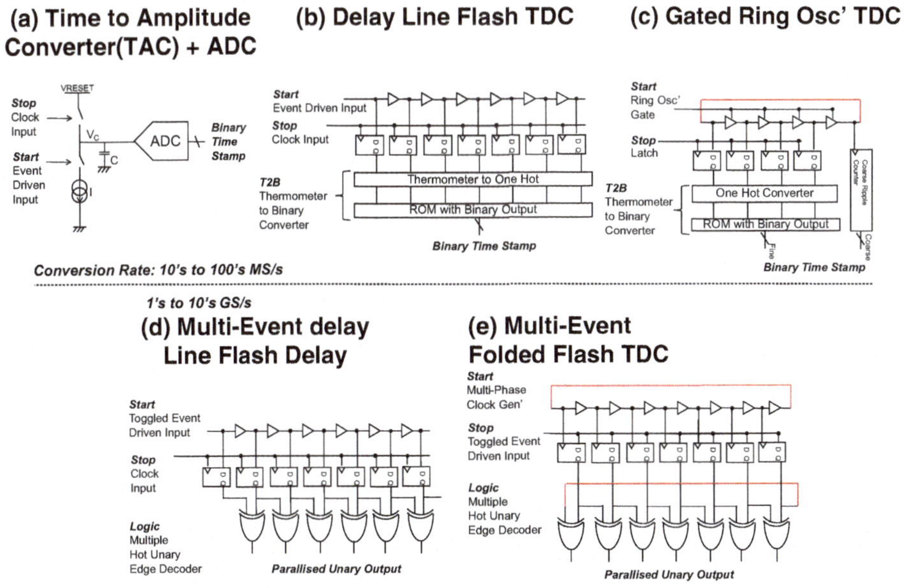

Fig. 7 Time converter circuits: (**a**) TAC and ADC, (**b**) flash delay line TDC, (**c**) GRO-TDC (**d**) multi-event DL-TDC, (**e**) this work: multi-event folded flash TDC

Table 2 Comparison of high conversion rate time converters for TCSPC applications with the rate-limiting circuit identified

Architecture	Reference	Output	Rate-limiting circuit	Maximum published conversion rate
TAC + ADC	[11]	Binary timestamp	Single event ramp circuit	16 MS/s
GRO-TDC	[13]	Binary timestamp	Ring-oscillator based counter	12.5 MS/s
Time-interleaved GRO-TDC	[8]			100 MS/s
Flash DL-TDC	[12]	Binary timestamp	Thermometer to binary converter	300 MS/s
Vernier flash DL-TDC	[17]	Binary timestamp	Thermometer to binary converter	500 MS/s
Multi-event flash DL-TDC	[16]	Parallelized unary	Gate delay	6.2 GS/s
Multi-event folded flash TDC	This work, [3–5]	Parallelized unary	Gate delay	14 GS/s

delay line to both clock and data. The conversion rate of the flash and Vernier flash TDCs and the GRO-TDC is inversely proportional to the "Stop" input frequency and thus is traded off against temporal dynamic range. Time interleaving GRO-TDC's increases the conversion rate at the cost of area and power, and similar to time-interleaved ADCs needs channel to channel calibration. All three converters (TAC, DL-TDC, and GRO-TDC) output a binary value based on the time of arrival of the event also known as a timestamp. This step is power-hungry, as one photon must be encoded by multiple binary bits, then be pipelined to the data processing. Instead, a power-saving and rate-increasing step is to encode a photon by the transmission of a single bit and the use of a parallelized unary data output. A logical high from the logic decoder indicates the arrival of a photon and the position of it in the logic decoder parallel output denotes the time conversion value. As a comparison, in our previous work on a FPGA-based multiple event delay line flash TDC (Fig. 7d), with multiple unary outputs, it achieved 6.2 GS/s [16] measured conversion rate over a conventional binary version at 300 MS/s [12] using a similar reference clock (316–300 MHz, respectively) on the same Xilinx Virtex 5 FPGA architecture. Table 2 presents a comparison of these architectures for TCSPC.

3.3 Folded Flash TDC Design

The multi-event folded flash TDC operation is described in this section. As drawn in Fig. 8, a multiphase PLL creates "*N*" multiple clocks (bus "C" in the figure), which are each attached to a front-end D-type flip-flop (DFF) that samples the output of

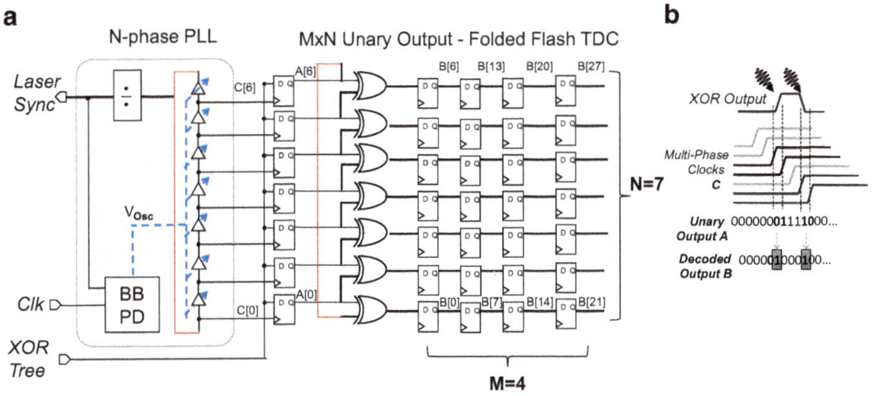

Fig. 8 Folded flash TDC (**a**) circuit diagram for $N = 7$ PLL clocks and $M = 4$ data pipeline generating 28 ($M \times N$) unary outputs and (**b**) timing diagram

the XOR-tree. The difference between two successive positive clock edges creates a sampling window in time. The output of each front-end flip-flop is XOR'd with the output from the next clock phase to detect a logical change in the SPAD array output representing a photon arrival in the timing window. As the multiphase VCO loops the final stage back to the first, the final clock C[6] precedes the next cycle of C[0]. To match, the final data output A[6] is folded back to be compared with the first A[0]. This technique creates no converter dead zone and permits continuous operation.

By inspection, the conversion rate of the circuit is the reciprocal of the sampling window. However, there is a downside as the temporal dynamic range is limited to a single period of one clock phase. There are two approaches to extend the dynamic range by increasing the parameter N or M, shown in the figure. The first is to increase the number "N" of VCO stages (clock phases) and "N" matching front-end flip-flops. This has the advantage that it also reduces the clock rate of each clock phase. Second, a pipelining shift register is attached to the output of the logic decoder. Each shift register stage increases the dynamic range by one clock period. As shown in Fig. 8, adding a four-stage shift register to the decoder outputs extends the dynamic range to four clock periods, taking it from an 7b unary output (M of 1) to a 28b unary output (M of 4). The subsequent histogram generation block is connected to the unary "B" outputs creating an $N \times M$ width histogram.

3.4 *On Chip Histogram Generation*

In TCSPC systems, a common first step of data processing is to collect the TDC binary timestamp data output codes into a histogram. One technique is to employ a RAM with an address per histogram bin, the respective memory location data

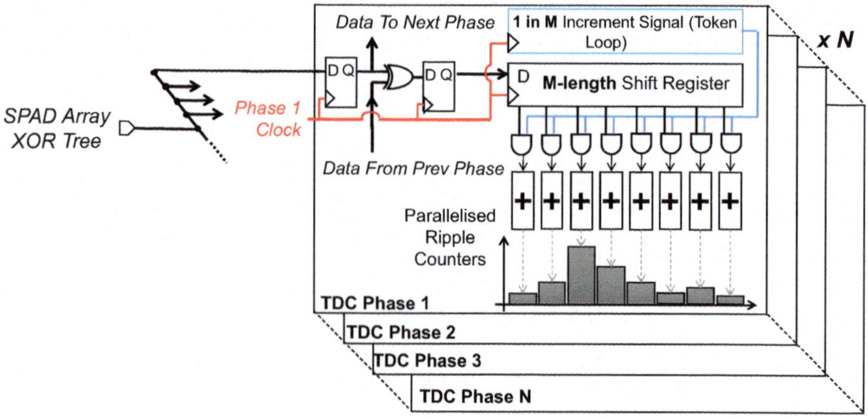

Fig. 9 Direct histogram generation on chip. Histogram is *M* by *N* bins, where each bin depth is set by the number of ripple counter stages

value containing the integrated count for that bin [1]. This conventional three-step histogram generation process is:

1. RAM address look up and obtain previous histogram bin count.
2. Increment the count by one.
3. Write the new count value back to the RAM.

This process may be optimized so the data read and write are pipelined to function in a single clock cycle [18]. The alternative approach in this work, which is also a single clock cycle operation, is to add simply a ripple counter to the output of each unary output from the TDC shift register. This results in the direct creation of an $M \times N$ bin histogram directly on-chip as displayed in Fig. 9. A counter increment signal is created by a copy of the shift register with a single looping token bit shown in blue. The period of the looping token bit is therefore reciprocal to the length of the TDC shift registers. For an M-bit shift register, the increment signal is high for 1 in M clock periods. This on-chip histogram processing uses the parallelized interface to attain multi-Gb/s operation with each parallel shift register and ripple counter block running at the period of the single clock.

The TDC is operated for an exposure time to build up the histogram, and following this, the integrated values of the bank of counters are read sequentially off-chip. This on-chip integration of the TDC Gb/s output into a parallel counter bank is both a data compression and the first stage of digital signal processing. For TCSPC systems based on multiple discrete ICs, this would also be of benefit as it significantly reduce the I/O data rate and associated power consumption.

4 Sensor Implementation

A proof-of-concept sensor was designed in ST Microelectronics 130 nm FE/90 nm BE 1P4M SPAD-optimized foundry process. Figure 10a provides a system overview consisting of a 32 × 32 SPAD array, 1024-to-1 channel XOR tree, 33-phase PLL, matching 33-phase multi-event folded flash TDC, 33 blocks of 8 length shift register, and 264 bin histogram using 16b ripple counters. Figure 10b shows the IC photomicrograph measuring 1.7 × 2.4 mm. The schematic of a SPAD pixel is shown consisting of a passively quenched SPAD, an 8 T input buffer, a 17 T toggle flip-flop (TFF), and a 6 T SRAM (not shown) which can activate or disable the pixel. As displayed in the layout image in Fig. 10c, the pixel is implemented with a p-well to deep n-well SPAD at 21 × 21 μm^2 pitch and 43% fill factor. The sensor region of interest (RoI) is programmed by writing to each pixel SRAM allowing both a RoI to be selected and high dark noise pixels to be disconnected from the TDC input. The output of each SPAD pixel TFF in the array is combined into a single output through a timing-balanced cascaded XOR H-tree, consisting of a column-wise five-stage cascaded XOR-tree using 10 T single-ended XORs arranged between pixels in a vertically flattened tree with centrally tapped output.

Both the TFF and the single-ended XOR are compact in area, as they must fit in the pixel array (too high area consumption would reduce the fill factor of the pixel further), yet as the output is based on a single-ended inverter the PMOS to NMOS mismatch will contribute a randomized pixel to pixel timing offset. The toggle flip-flop rise and fall time is an important parameter to correct in order to have a low systematic spread in the final histogram. A simulation extracting the delay from input clock-pin trigger to Q pin output is performed. As shown in Fig. 11, before optimization of rise and fall times, the simulated difference in the mean value is 809 ps/390 ps, respectively, under Monte Carlo extracted simulation. After rise and fall alignment by simply balancing the two inverters triggering the clock and clock bar internal signals of the flip-flop, this improves to 8 ps and the standard deviations of both decrease from ~25 ps to ~10 ps. Moreover, the absolute delay is reduced.

For the second half of the XOR tree outside the array, the area constraint is lifted so a differential structure is employed. A single-ended to differential buffer and edge aligner is placed on each column output connecting into the pseudo-differential horizontal five-stage XOR tree. The full 10-stage XOR tree is designed to have balanced timing from each pixel to the TDC input. The last XOR-tree output is toggled when any activated pixel in the array flips its state from receiving a SPAD event. Combining the pixel and TFF extracted simulation with the full 10-stage XOR tree, the overall worst case min/max simulated spread is 213 ps. Figure 12 shows the 1024 to 1 XOR tree in two halves. The first five stages are shown in Fig. 12a with the in-column XOR tree. The second in Fig. 12b shows the final five stages based on a pseudo-differential XOR, edge aligner input, and pseudo-differential output to the TDC front-end flip-flops.

The TDC has 33 parallel flip-flops with common data and data bar input from the SPAD array, and each flop has an individual clock phase input generated from

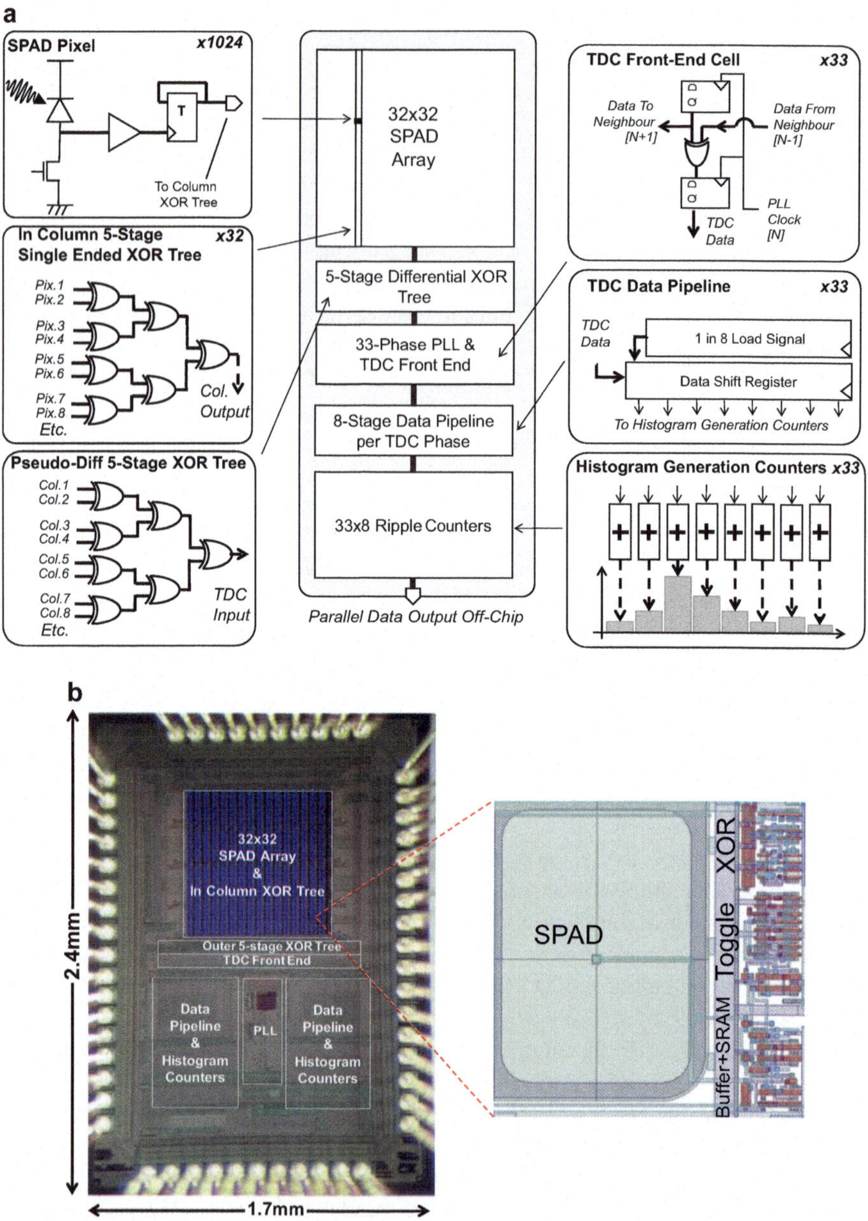

Fig. 10 (**a**) Block diagram of the TCSPC sensor with detail of each main block reproduced from [3]. (**b**) IC photomicrograph of the proof-of-concept sensor with 1.7 mm width and 2.4 mm height. (**c**) Pixel layout showing SPAD diode, buffer, SRAM, toggle, and XOR

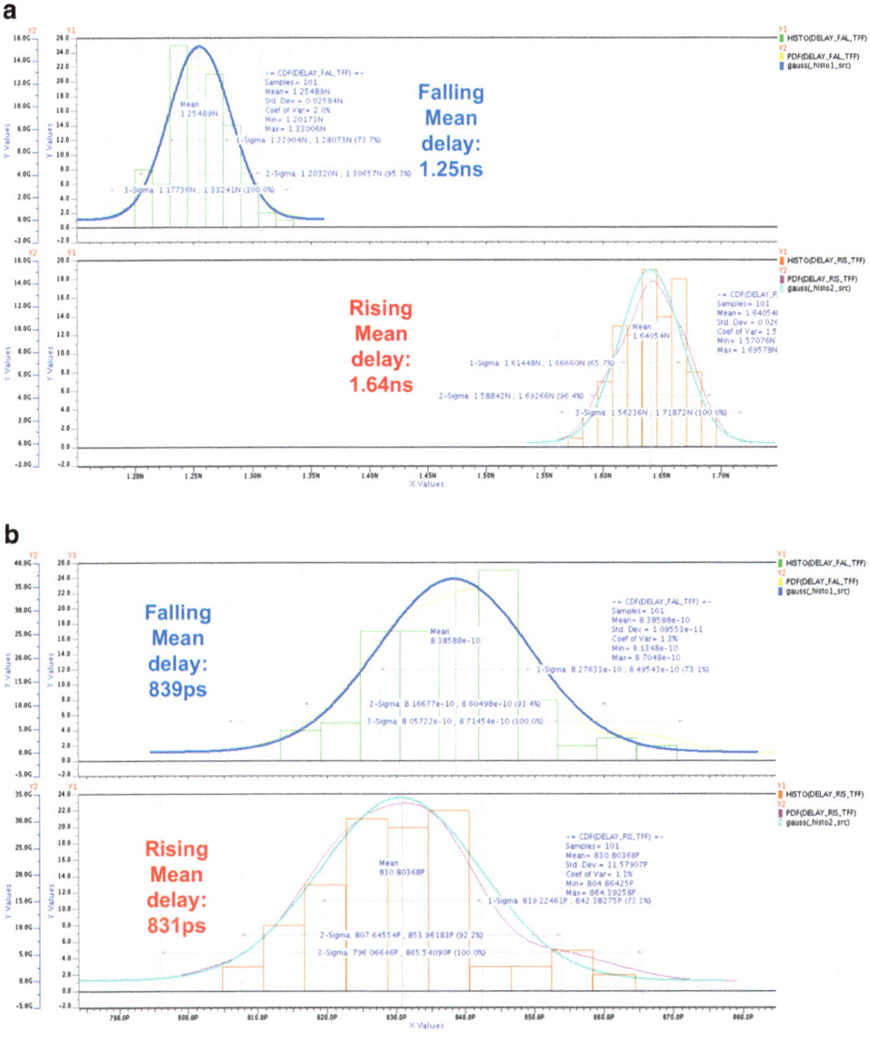

Fig. 11 Toggle flip-flop output delay from clock trigger to Q output: without and with balancing of rise and fall times. (**a**) Without rise/fall balancing. (**b**) With rise/fall balancing

the PLL. The flip-flop is a differential input, single-ended output based on a sense amplifier. The circuit is shown in Fig. 13 indicating the differential input stage (essentially a historical design of a StrongArm latch) and cross-coupled output stage. The output of each flip-flop is passed through an XOR acting as a multiple hot code edge detector. If the sampled state of the SPAD-array output changes from the previous sample (either low to high or high to low), this indicates the arrival of a SPAD event within the time window between the two clock edges, and this is decoded by the XOR edge detector which outputs a logic high in this condition.

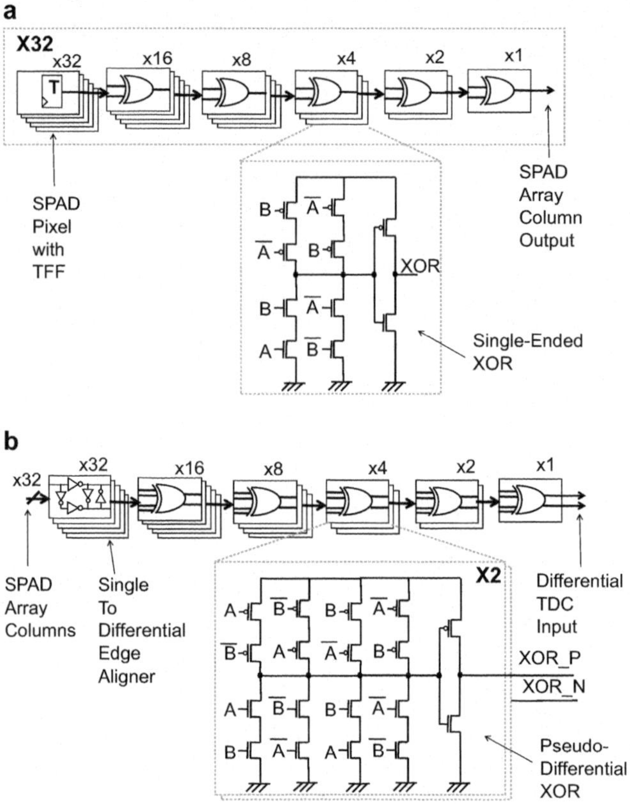

Fig. 12 1024 to 1 asynchronous combining logic XOR tree: (**a**) First half of the XOR tree: in-column five stage XOR tree and there are 32 instances of this block and (**b**) the second half of the XOR tree: 32 to 1 pseudo-differential XOR with single to pseudo-differential edge aligner

A multiphase PLL is designed to generate the clock phases for the TDC. Table 3 shows the trade-off of the number of VCO stages "N" based on a 100 ps time difference between clock phases. A lower number of stages creates a high frequency per phase. The trade-off is area (of both VCO and TDC) versus clock frequency (Table 3). A lower number of stages "N" in the PLL increases the number of pipeline stages "M" for the same size of histogram. The VCO is designed with a ring of 33 inverters, each with a minimum of 70 ps gate delay in simulation (maximum VCO control voltage), providing a maximum 432 MHz PLL phase frequency and typical 100 ps based on a 9.469 MHz reference clock and 1/32 divider ratio. The PLL reference clock input derives from an off-chip source, typically an FPGA or laser trigger. A selectable PLL feedback integer divider provides a range of frequencies for the oscillator or laser trigger.

For each clock phase, an eight-stage ($M = 8$) data pipeline creates a 264-bin histogram ($N \times M = 33 \times 8 = 264$), where each bin is a 16b ripple counter. A TDC

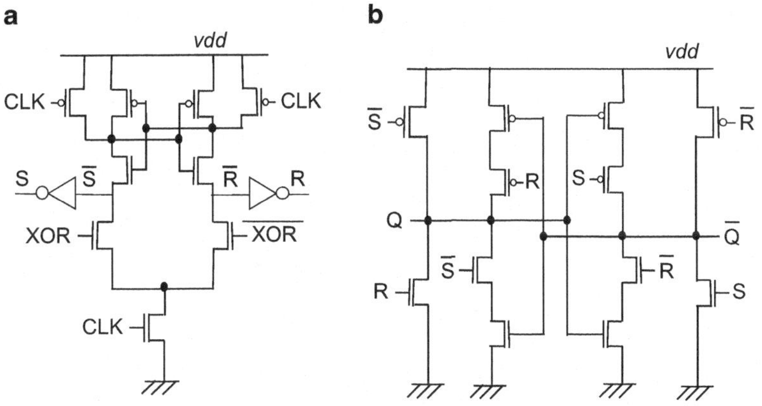

Fig. 13 Flip-flop based on sense amplifier with (**a**) differential input stage and (**b**) output stage

Table 3 Design parameters for sizing the number of VCO inverter stages "*N*" based on 100 ps per VCO stage

'N' VCO inverter stages	3	9	17	*33*
Min. Clock phases period (ns) based on 100 ps per VCO stage	0.3	0.9	1.7	*3.3*
Clock Freq (MHz) per VCO phase	3333	1111	588	*303*

gating signal is generated by a controlling FPGA, which permits an integration of multiple laser repetitions to build up a TCSPC histogram. Exposure times can be as short as a single histogram cycle to capture rapid single-shot transient optical events. In TCSPC mode, the sensor is typically operated for many thousand cycles to build a histogram from sparse photons stimulated by the synchronous laser source. At the end of an exposure cycle, the off-chip data readout is via a 16b parallel bus and one ripple counter is read per clock cycle. The off-chip readout operates at a maximum of 50 MHz or 800 Mb/s transferring a histogram (528 bytes) in a minimum of 5.3 μs to the FPGA for subsequent data transfer to PC.

5 Measurement Results

5.1 Electrical Test Signal

To confirm the bin resolution and integrated jitter, the TDC and histogram generation logic are first tested electrically, employing a dual-channel LeCroy Wavestation 3082 signal generator. The first generator channel is used to generate the PLL reference clock and the second channel generates a synchronous electrical test signal with variable duty cycle. The reference clock is set at 9.475 MHz and the 1/32 feedback divider sets the PLL VCO frequency at 303.2 MHz. Corresponding to a

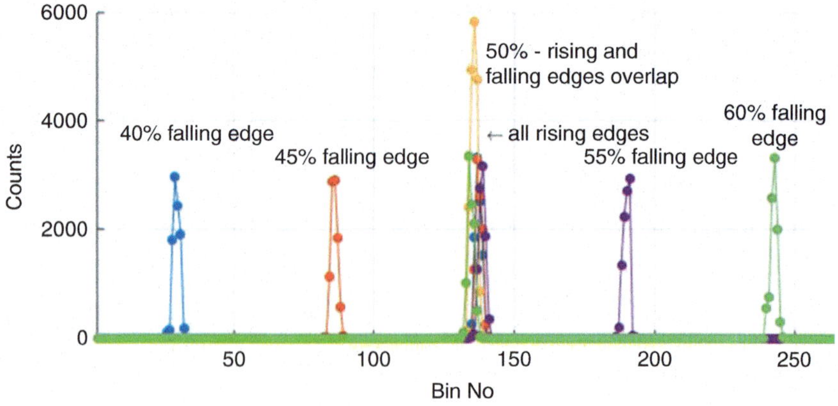

Fig. 14 Synchronous TDC electrical input with duty cycle variation from 40% to 60% duty cycle in 5% increments

VCO period of 3.3 ns and TDC dynamic range of 26.38 ns, each histogram bin has a width of approximately 100 ps. The TDC's input is connected to an electrical test signal with a variable duty cycle from 40% to 60% in 5% increments at 9.475 MHz, such that four TDC conversions are performed per cycle of the TDC. The signal generator positive edge is fixed and synchronously aligned with the PLL reference clock, whereas the falling edge is varied. Rising edges are captured on the first TDC cycle and falling edges on the third TDC cycle. Figure 14 shows histograms corresponding to rising and falling edge positions for the five different duty cycle settings. The rising edge for all settings remains static at the middle bin, while the falling edge appears offset by the duty cycle delay deviation from 50%. Peak counts at exactly 50% duty cycle condition are a factor of two greater than for other duty cycle values as rising and falling edges coincide. For example, for 45% duty cycle (orange plot), there are 53 bins between the rising edge and falling edge for a 5.28 ns time difference, which is approximately 100 ps per bin. A Gaussian fit on each of the histogram peaks yields an average FWHM 287 ps, subtracting in quadrature signal generator jitter (the oscilloscope measured) leaves a 103 ps FWHM mean integrated electrical TDC jitter, which is acceptably in the order of a single histogram bin.

5.2 Linearity and Count Rate

An uncorrelated light source (with variable power) is used to characterize the count rate of the SPAD and the XOR tree. Figure 15a shows saturation limits for a single passively quenched SPAD, as the SPAD recharge MOS voltage (VQ) increases, the dead time decreases resulting in a higher maximum count rate and saturation level. For VQ = 1.6 V, the maximum count rate is around 100 M count/s; this is equivalent to a dead time around 3.5 ns. Figure 15b shows the saturation limit of the

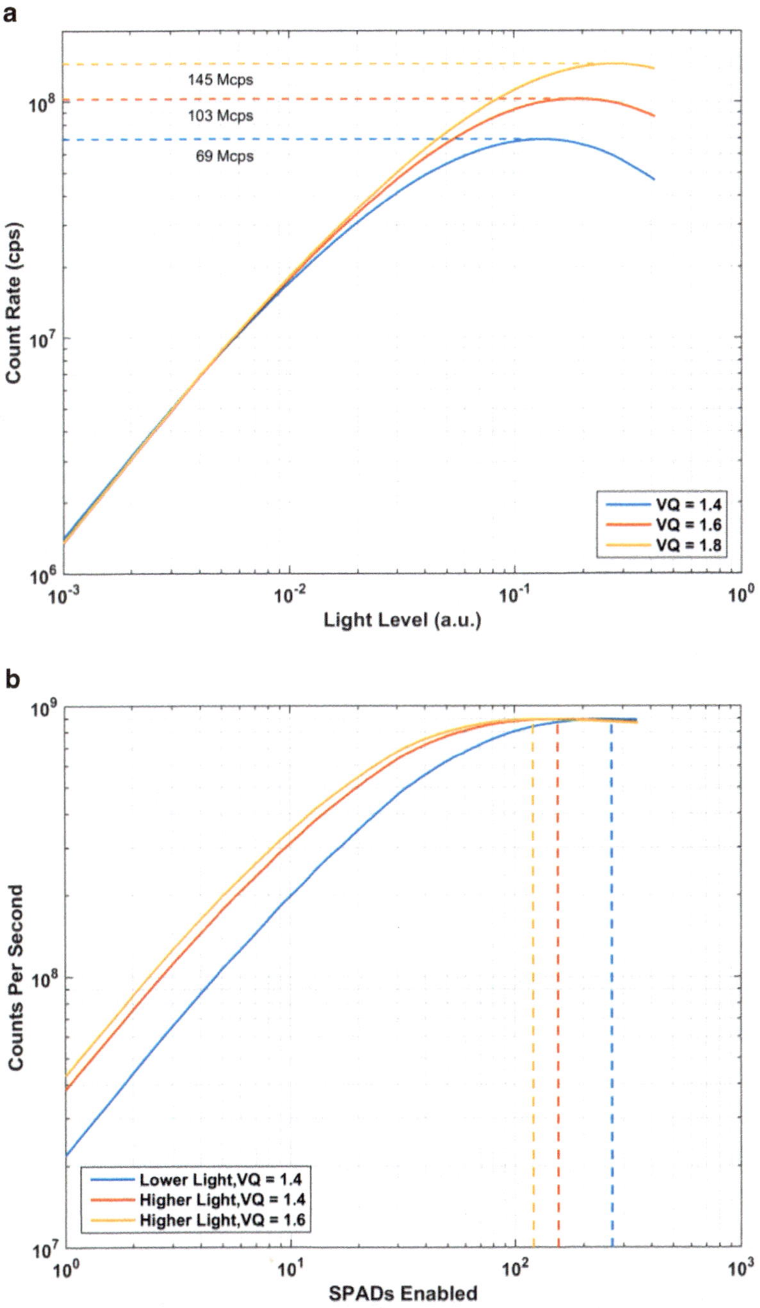

Fig. 15 (**a**) Saturation limit of a single SPAD versus recharge MOS voltage (VQ) using uncorrelated light source. (**b**) Saturation limit of the XOR tree versus number of SPADs activated in the Region of Interest. (**c**) TDC nonlinearity from optical statistical code density test using uncorrelated light source. DNL Max/Min: +0.53LSB/−0.47LSB (**d**) TDC INL Max/Min: +0.50LSB/−0.17LSB. (**e**) Layout view of the VCO indicating source of DNL

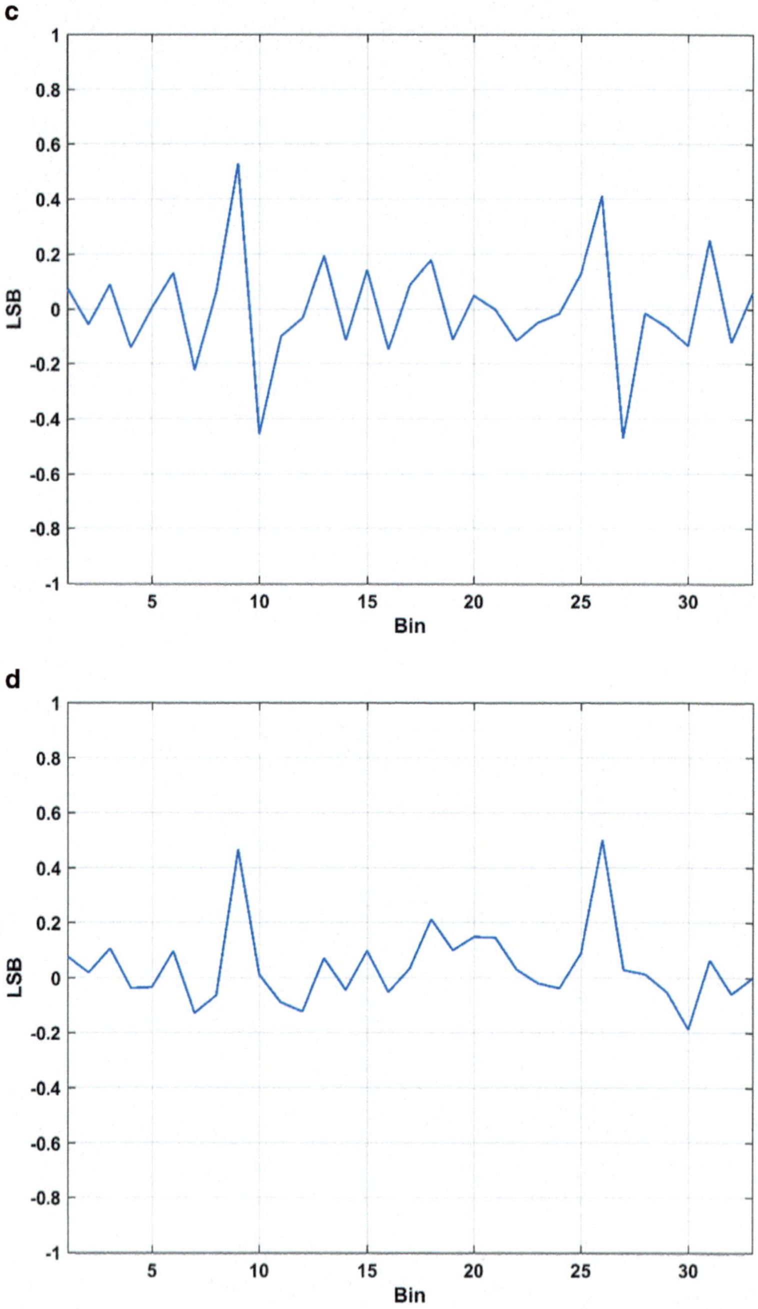

Fig. 15 (continued)

Fig. 15 (continued)

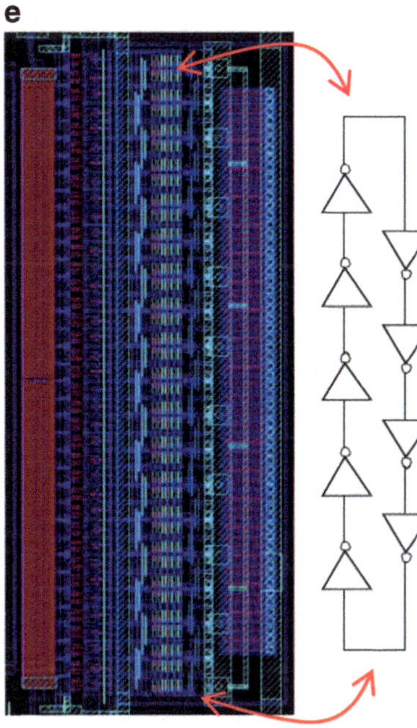

XOR tree combining 32 SPADs as 900 M counts/s. Dotted lines indicate the number of SPADs needed to saturate the XOR channel. At higher ambient conditions, this number is influenced by the SPAD recharge MOS voltage (VQ), as the SPADs are operating close to their saturation limit and experience different count rates. The optical statistical code density test is performed to estimate the TDC linearity [19]. Continuous light from an LED biased with a dc source provides uncorrelated SPAD trigger events or white temporal noise to the TDC. The Differential and Integral Nonlinearity (DNL/INL) across the 33-phase TDC front end is characterized and shown in Fig. 15c, d, respectively. The worst-case measured DNL is +0.53/−0.47 LSB and INL is +0.5/−0.19 LSB. The two spikes, at bins 9 and 26, are systematic and are attributed to the VCO layout (Fig. 15e), comprising two interleaved banks of inverters and specifically the two routes at both ends connecting the two banks. Using the sensor for TCSPC applications, the nonlinearity may be compensated by scaling the individual bin values by the inverse of the DNL [14].

5.3 TCSPC

To characterize the TCSPC accuracy of the sensor and to confirm the dynamic range, the PLL is configured as in the previous experiment and the TDC input is multiplexed back to the SPAD array. The laser synchronization trigger connects to a Stanford DG645 delay generator with minimum 5 ps time-step resolution to perform a time sweep. This in turn connects to a Hamamatsu PLP10 driver with 443 nm output head with 70 ps FWHM quoted electrical integrated jitter. Both IC and laser face a fixed-distance white diffuser with no optical lensing. Figure 16a shows the

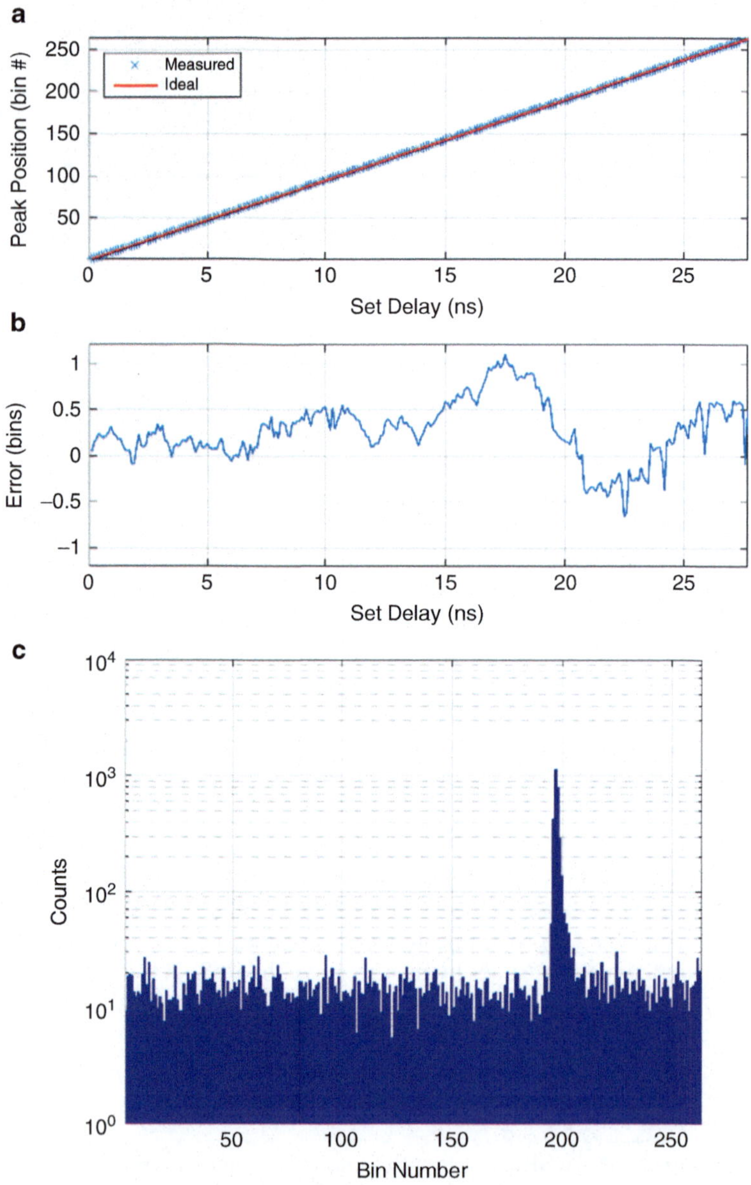

Fig. 16 (**a**) Pulsed laser TCSPC dynamic range sweep. (**b**) TCSPC error: the calculated center histogram peak position against delay generator setting. (**c**) Typical optical TCSPC histogram (log scale) from the sensor with 231 ps FWHM with a single SPAD enabled

incremental delay sweep performed capturing one histogram per 100 ps delay step with 20 ms exposure time. The centroid of the histogram peak is calculated by the weighted mean using the center of mass method (CMM) [8]. The calculated average bin position is plotted against absolute delay in Fig. 16a and the error of calculated peak position to absolute delay in Fig. 16b, indicating a TCSPC precision (σ) of 34 ps and accuracy of +116 ps, −70 ps across the 27.78 ns full-scale dynamic range. No dead zone is evident in the dynamic range, confirming that the folded flash TDC architecture has its last histogram bin contiguous with the first bin. A typical histogram from one 10 ms exposure is shown in Fig. 16c, and a Gaussian-fitted curve indicates a FWHM of 231 ps with a single SPAD enabled. Subtracting the external sources of jitter and the SPAD jitter of ~200 ps [20] in quadrature yields a similar result to the electrical test of 103 ps.

5.4 Multiple Photons per Laser Cycle

To demonstrate the capability of the TCSPC sensor to capture multiple photons per laser excitation cycle, a laser repetition rate of 72 MHz is selected, corresponding to pulse intervals of 13.89 ns or 2 laser pulses per one histogram cycle. This is achieved by locking the TDC PLL to a 9 MHz clock from channel 1 of a Keysight 33,250 A function generator while using channel 2 to trigger the laser. This lowers the resolution per bin to 105 ps and the dynamic range to 27.78 ns. Figure 17a shows 2000 successive single shot histograms (arrayed vertically) captured with a single 30 ns exposure (limited by the FPGA experimental controller) where each dot represents a single time correlated photon. Figure 17b displays one example histogram of one of the exposures demonstrating four photons captured. No TDC dead time is evident as two photons are captured in neighboring bins. The upper graph in Fig. 17c is a summation of the 2000 single exposures, in effect a 60 μs exposure histogram, showing the dual peaks from the lasers. A second experiment is performed with a second laser added to the experimental setup, adding an incremental delay to this laser with respect to the first. Figure 17d shows the successively captured histograms as a 3D plot, with histogram bin on the X-axis, histogram frequency on the Z-axis, and successive histograms from the incremental delay on the Y-axis. As the lasers cross each other in the histograms, there is no pile-up distortion evident that would be revealed as a reduction in the peak intensity of the two pulses when closely spaced in time.

5.5 Moving Targets and Sensor Bandwidth

Like high-speed image sensors in machine vision or automotive applications capturing moving targets, TCSPC sensors capturing fast moving subjects require fast acquisition and readout rates. The Nyquist sampling rate of image sensors is

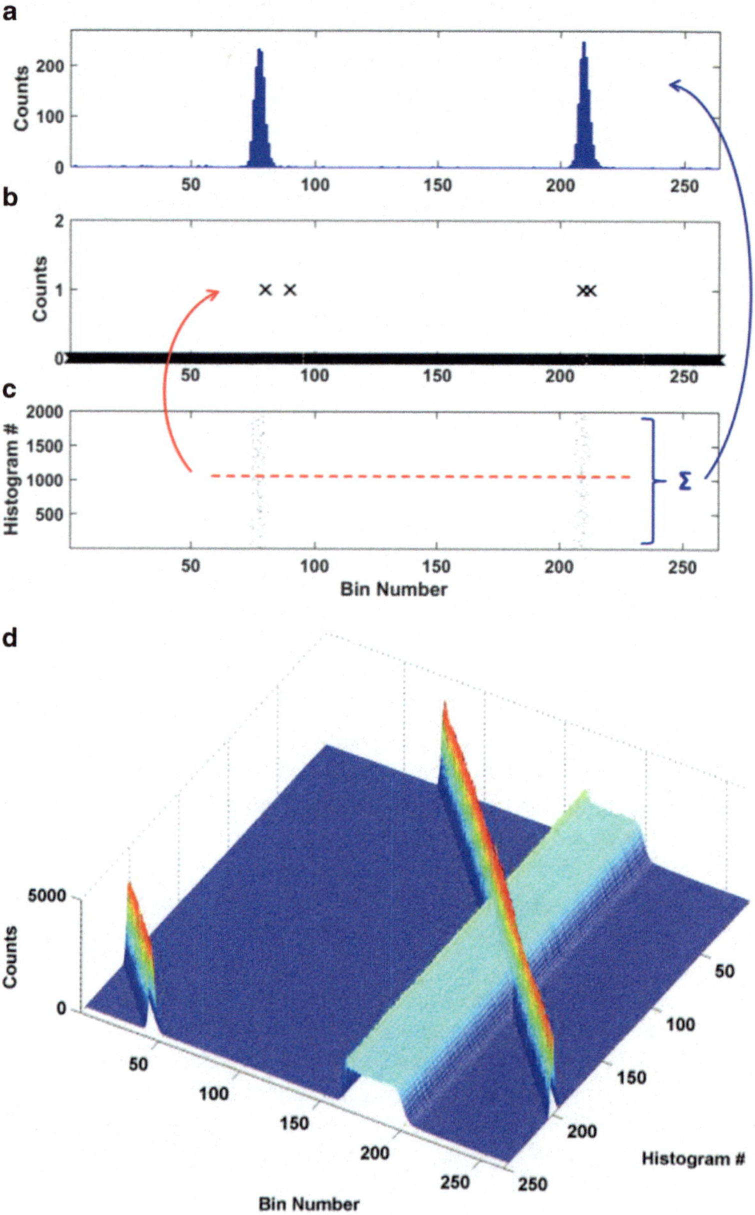

Fig. 17 (**a**) A histogram formed by summing 2000 single-shot 30 ns exposures with two laser pulses (short pulse width). (**b**) One example single-shot 30 ns exposure showing four multiple photons captured where no TDC dead time is evident as two photons are captured in neighboring bins. (**c**) 2000 successive single-shot histograms. (**d**) Incremental delay sweep of one laser (short pulse width) against a fixed delay of a second laser (long pulse width). The 3D plot has histogram bin on the x-axis, histogram frequency or count on the z-axis, and successive histograms on the y-axis

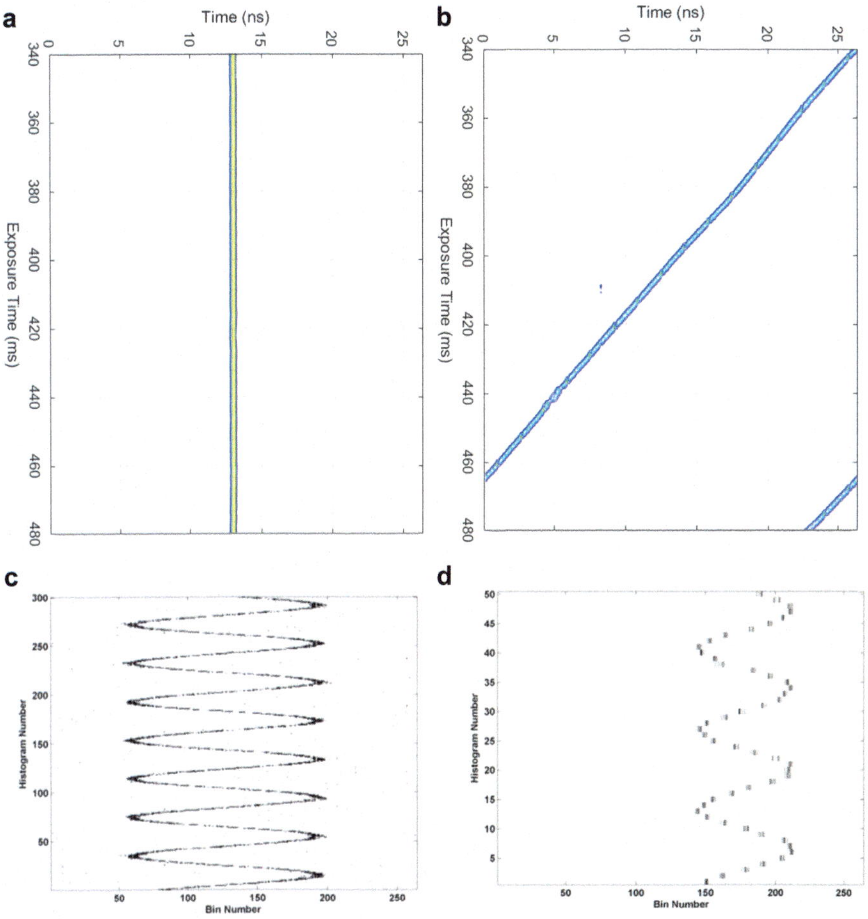

Fig. 18 For all four plots: *X*-axis is histogram position, *Y*-axis is time series of histograms. (**a**) Static target 1 ms exposure time per histogram. (**b**) Emulation of approaching target 1 ms exposure time per histogram. (**c**) Emulation of target moving backwards and forwards 200 μs exposure per histogram. (**d**) 20 μs exposure per histogram

simply the reciprocal of two frame periods. However, with the event-driven nature of single photon sensing, the trade-off is the exposure time needed for the target application to achieve the requisite number of captured photons per frame (or histogram, in this case) against the desired frame rate for high sampling rate.

The speed measurement capability of this sensor is evaluated in [4]. This speed measurement is emulated using a static target and a detuned laser synchronization pulse using two outputs of a Keysight 33,250 A arbitrary waveform generator used to generate a 9 MHz PLL reference clock for the sensor PLL, while a second output is frequency modulated around 9 MHz and used to trigger a Hamamatsu PLP10 443 nm laser. Figure 18a shows the time series of histograms for a static target and

Table 4 Power consumption breakdown of the sensor and figure of merit calculation

System block	Measured power (mW)
TDC pipeline and histogram counters at 10 GS/s	144
I/O pads at 50 MHz	10
Second stage XOR tree (stages 6–10)	10
TDC front end at 10 GS/s (including multiple edge detector)	2
PLL	2.4
Column-wise first stage XOR tree (stages 1–5)	2.4
Total	*170.6 mW*
TDC FOM1 = (PLL + TDC FE power) / (sample rate)	*0.48 pJ / S*
System FOM2 = (Total power w/o IO) / (max photon rate)	*178 pJ / photon*

0 Hz detuning, whereas Fig. 18b shows -1 Hz detuning to emulate an approaching target at 16 ms^{-1}.

A sensor bandwidth experiment is performed to test these limits of moving target detection, with two examples shown in Fig. 18c, d again displayed as a sequential continuous time series of captured histograms. Figure 18a shows a 100 Hz mean, frequency deviation ±9 Hz and exposure time 200 μs. The final plot in Fig. 18d is of a 1 kHz input with 400 Hz deviation and exposure time of 20 μs. At this point, the readout dead time of 5.3 μs imposes gaps between samples and the waveform is more discretized. The laser can now be seen within a histogram to be streaking across bins at rates determined by the sine wave rate of change. Capture of this type of signal in a TOF context represents high rates of velocity (at km/s) and acceleration (at km/s^2) [4]. The maximum sampling rate of the sensor is measured at 188 k histograms per second, limited by the 5.3 μs readout time.

5.6 *Power Consumption*

Table 4 details the power consumption of each constituent part of the sensor, giving a total of 170.6 mW consumption at 10 GS/s with conversion rates of 1 GPhoton/s recorded limited by the output rate of the XOR tree. This equates to a Figure of Merit (FOM1) of 0.48 pJ per TDC sample (S) or TCSPC timestamp considering only the TDC front end and PLL, and 16.1 pJ/S for the whole optical sensor not including I/O and 178.8 pJ per photon (FOM2) with power measured at 899.1 M photons per second.

6 Conclusion

The proof-of-concept sensor is evaluated in a side by side comparison in Table 5. This work achieves the highest single channel TDC conversion rate and highest of any CMOS sensor ASIC implementation for TCPSC. The single shot time resolution of 100 ps and full-scale range of 26.8 ns matches the requirements

Table 5 CMOS time domain sensor comparison table

	This work	[21]	[8]	[17]	[22]	[23]
TC architecture	Folded flash TDC	Col flash TDC	Interleaved GRO-TDC	Interleaved Vernier TDC	Oversampling SRO-TDC	Column parallel GRO-TDC
Application	TCSPC	TCSPC	TCSPC	TCSPC	ADPLL	TCSPC
Interleaved TCs per channel	1	4[a]	16	2	1	1
Parallel channels	1	96	1	1	1	512
Data processing on chip	Histogram	Histogram	Fluorescence lifetime	–	Mean time difference	Histogram
Tech.	130 nm	180 nm	130 nm	130 nm	90 nm	130 nm
Supply	1.2 V	1.8 V	1.2 V	1.3 V	1 V	1.2 V
Single shot resolution	100 ps	208 ps	52 ps	31 ps	156 ps	50 ps
Dyn. Range	26.8 ns	853 ns	3.6 μs	2 ns	2–840 ns	3.2 μs
External cal. Needed	No	No	Yes	Yes	No	Yes
System conv. Rate	10 GS/s	6 GS/s[b]	100 MS/s	500 MS/s	750 MS/s	16.5 GS/s (in Histogramming mode) 194 MS/s (in TCSPC mode)
TDC conv. Rate		62.5 MS/s	12.5 MS/s	250 MS/s		32.2 MS/s
Sensor bandwidth	30 kHz	60 Hz	–	–	–	–
Off-chip sensor rate	188 k histogram / s	30 FPS	60 k	500 M Ph/s	–	6.06 M histogram /s (194 MS/s in TCSPC mode)
TC power	TDC 2 mW + PLL 2.4 mW	–	1.8 mW	1 mW	2 mW	1.8 mW
TC FOM1[a]	0.48 pJ / S	–	18.0 nJ / S	2.00 nJ / S	2.67 nJ / S	36.0 nJ / S

[a]FOM1 = Power/Conversion rate = J/Time stamp

of optical distance measurements for mobile and is adequate for many other applications. The speed increase of TCSPC to high sample rates permits high numbers of photons to be time resolved per second by reducing pile-up distortion and enables tracking of high velocity objects.

Acknowledgments Salvatore Gnecchi and Luca Parmesan contributed to the design of the sensor. Oscar Almer contributed to the FPGA controller and measurement evaluation system.

Technical discussions with Pascal Mellot, Bruce Rae, Graeme Storm, Andrew Holmes, Lindsay Grant, Sara Pellegrini, and J. Kevin Moore have been influential in this research.

We are grateful to ST Crolles for silicon fabrication and ST for PhD student support for Francescopaulo Mattioli Della Rocca.

Tare Al Abbas acknowledges funding from The University of Edinburgh and PROTEUS project (http://proteus.ac.uk EPSRC grant number EP/K03197X/1).

References

1. Becker W. Advanced time-correlated single-photon counting techniques. Berlin/Heidelberg/New York: Springer; 2005.
2. ST VL531LX TOF Sensor: http://www.st.com/content/st_com/en/products/imaging-and-photonics-solutions/proximity-sensors/vl5311x.html.
3. Dutton NAW, et al. A time-correlated single-photon-counting sensor with 14 GS/S histogramming time-to-digital converter. In: IEEE international solid-state circuits conference (ISSCC) digest of technical papers, 2015.
4. Finlayson N, Al Abbas T, Mattioli Della Rocca F, Almer O, Gnecchi S, Dutton NAW, Henderson RK. Hypervelocity time-of-flight characterisation of a 14GS/s histogramming CMOS SPAD sensor. In: Proceedings of SPIE 10111, quantum sensing and nano electronics and photonics XIV, 101112Z, 27 Jan 2017.
5. Dutton NAW, Al Abbas T, et al. A CMOS SPAD sensor with a multi-event folded flash time-to-digital converter for ultra-fast optical transient capture. IEEE Sens J. 2018;18(8):3163–73.
6. Charbon E. Single-photon imaging in complementary metal oxide semiconductor processes. Philos Trans R Soc A. 2014;372:1–31.
7. Arlt J, et al. A study of pile-up in integrated time-correlated single photon counting systems. Rev Sci Instrum. 2013;84(10):103–5.
8. Tyndall D, et al. A high-throughput time-resolved mini-silicon photomultiplier with embedded fluorescence lifetime estimation in 0.13 μm CMOS. IEEE Trans Biomed Circuits Syst. 2012;6(6):562–70.
9. Gnecchi S, et al. Digital silicon photomultipliers with OR/XOR pulse combining techniques. IEEE Trans Electron Devices. 2016;63(3):1105–10.
10. Gnecchi S, et al. A comparative analysis between OR-based and XOR-based digital silicon photomultipliers for PET. In: Proceedings of IEEE nuclear science symposium, 2015.
11. Crotti M, Rech I, Ghioni M. Four channel, 40 ps resolution, fully integrated time-to-amplitude converter for time-resolved photon counting. IEEE J Solid-State Circuits. 2012;47(3):699–708.
12. Favi C, Charbon E. A 17ps time-to-digital converter implemented in 65nm FPGA technology. In: Proceeding of the ACM/SIGDA international symposium on field programmable gate arrays, 2009, p. 113.
13. Richardson J, et al. A 32x32 50ps resolution 10 bit time to digital converter array in 130nm CMOS for time correlated imaging. In: Custom integrated circuits conference 2009. CICC '09. IEEE, 2009, p. 77–80.

14. Kalisz J. Review of methods for time interval measurements with picosecond resolution. Merologia. 2004;41:17–32.
15. Roberts G, Ali-Bakhshian M. A brief introduction to time-to-digital and digital-to-time converters. IEEE Trans Circuits Syst II Exp Briefs. 2010;57(3):153–7.
16. Dutton N, et al. Multiple-event direct to histogram TDC in 65nm FPGA technology. In: Proceedings of IEEE PRIME conference, 2014.
17. Yousif AS, et al. A fine resolution TDC architecture for next generation PET imaging. IEEE Trans Nucl Sci. 2007;54(5):1574–82.
18. Haraszti TP. CMOS memory circuits. Norwell: Kluwer; 2000.
19. Doernberg J, Lee H-S, Hodges DA. Full-speed testing of A/D converters. IEEE J Solid-State Circuits. 1984;SSC-19(6):820–7.
20. Richardson JA, Grant LA, Henderson RK. Low dark count single-photon avalanche diode structure compatible with standard nanometer scale CMOS technology. IEEE Photon Technol Lett. 2009;21(14):1020–2.
21. Niclass C, Soga M, Matsubara H, Kato S, Kagami M. A 100-m range 10-frame/s 340x96-pixel time-of-flight depth sensor in 0.18-μm CMOS. IEEE J Solid-State Circuits. 2013;48(2):559–72.
22. Elshazly A, Rao S, Young B, Hanumolu PK. A noise-shaping time-to-digital converter using switched-ring oscillators—analysis, design, and measurement techniques. IEEE J Solid-State Circuits. 2014;49(5):1184–1197.
23. Erdogan AT, Walker R, Finlayson N, Krstajić N, Williams GOS, Henderson RK. A 16.5 Giga events/s 1024 $\times$ 8 SPAD line sensor with per-pixel zoomable 50ps-6.4ns/bin histogramming TDC. In: Proceedings of VLSI symposia, 2017.

Part III
Energy Efficient Amplifiers and Drivers

The third part of this book is dedicated to recent developments in the field of Energy Efficient Amplifiers and Drivers. Some papers deal with theoretical limitations and circuit solutions, while others address problems at the system level.

The first paper from Klaas Bult (Delft University) presents a comprehensive method of estimating the power dissipation of residue amplifiers. This method is then used to analyze the power efficiency of some recently published residue amplifiers. The most power efficient topologies share the same core circuits and mainly differ in how these are driven by the input signal. Finally, an overview is given of these topologies, ranked according to their power efficiency.

The second paper from Youngcheol Chae (Yonsei University) discusses the design and biasing of energy-efficient inverter-based amplifiers. The paper discusses recent developments and presents some examples of state-of-the-art designs.

The next two papers deal with class-D amplifiers, typically adopted for their energy efficiency.

In the third paper, Marco Berkhout (NXP Semiconductors) gives an overview of innovative class-D architectures and how they balance efficiency, EMI, and application cost.

In the fourth paper, Mark McCloy-Stevens (Cirrus Logic) presents a digital Class D amplifier architecture that combines open-loop and closed-loop configurations to provide high performance over the full signal range. At low signal levels, low noise and power is achieved with open-loop digital operation. At larger signal levels, a closed-loop digital Class D mode is used to deliver low THD and high PSRR with minimal analog circuitry.

In the fifth paper, Lorenzo Crespi (Synaptics) analyzes system and circuit solutions for improving efficiency in microphone audio interfaces. The main specifications of typical microphone interfaces are illustrated to exhibit the advances in their development toward the maximization of their efficiency in each block (preamplifiers and ADCs).

In the sixth paper, Khaled Khalaf (Imec) deals with efficiency improvement in digital transmitter implementations for millimeter wave wireless communication systems. Despite having a higher bandwidth baseband and more complex digital processing, high efficiency front-ends in digital polar architectures are closer to show a power consumption advantage in phased arrays, where the front-end contribution dominates.

High-Efficiency Residue Amplifiers

Klaas Bult, Md. Shakil Akter, and Rohan Sehgal

1 Introduction

All ADC architectures with higher SNR need some form of amplification to suppress noise from comparators and/or back-end circuitry. In a pipelined ADC [1] (Fig. 1), residue amplifiers serve this purpose. In each stage, first a number of bits (n) are resolved (usually by comparators). In the digital domain, a coarse representation is formed of the input signal. A DAC converts this signal back into the analog domain, and a residue signal is formed by subtracting the coarse signal from the input signal. This residue signal needs to be converted into the digital domain by the subsequent stages, but as is shown in Fig. 1 (bottom), the amplitude of the residue signal is much lower than the amplitude of the original input signal. It is the task of the residue amplifier to amplify the residue signal to the maximum signal level that can be handled by the subsequent stages.

As the residue amplifiers take up a considerable part of the power consumption budget of the ADC, it is critical to design them with the highest possible power efficiency. The goal of this paper is to derive expressions that provide us with a tool to critically examine and compare different amplifier topologies and determine which is the most power efficient one. Subsequently, we will provide an overview of the most recently published advances in residue amplifier design and examine them using this tool.

In this paper, we will first establish the requirements for a residue amplifier (Sect. 2). Next, an estimate will be made of the power dissipation of a straight-

K. Bult (✉)
Analog Design Consult B.V., Bosch en Duin, Netherlands
e-mail: klaas.bult@icloud.com

M. S. Akter · R. Sehgal
Broadcom Netherlands B.V., Bunnik, Netherlands

K. A. A. Makinwa et al. (eds.), *Low-Power Analog Techniques, Sensors for Mobile Devices, and Energy Efficient Amplifiers*,
https://doi.org/10.1007/978-3-319-97870-3_13

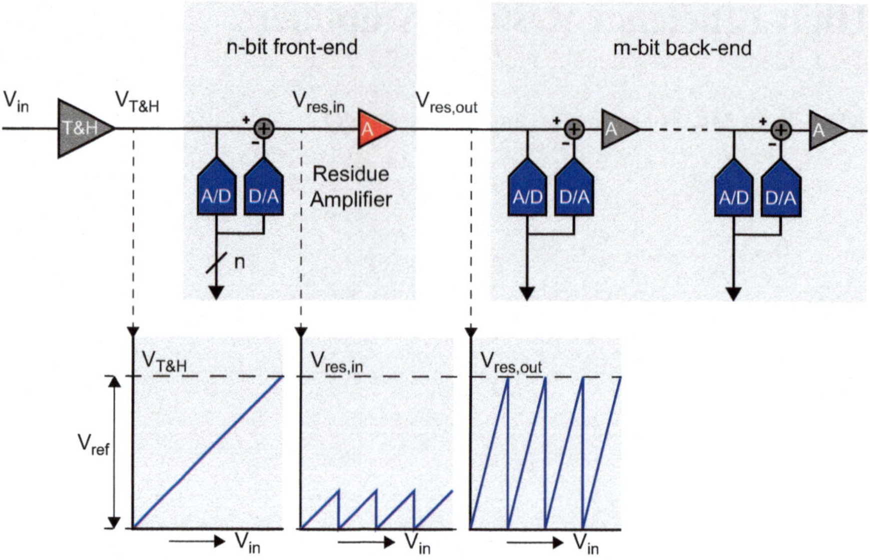

Fig. 1 A generic pipelined ADC with *n*-bits resolved in the first stage

forward classic solution [1, 2] (Sect. 3). Using this estimate, a definition of power efficiency will be derived (Sect. 4). Subsequently, factors in the expression for power efficiency will be analyzed that are determined by the chosen circuit solution, and directions will be presented for better, more efficient solutions (Sects. 4.2, 4.3, 4.4, and 4.5). Among these directions will be the required amount of settling for a residue amplifier, as this depends on the circuit architecture chosen. In Sect. 5, incomplete settling will be analyzed, together with its effect on power efficiency. The results from this analysis will be used to derive the result of power efficiency for ideal integrating amplifiers (Sect. 6). Dynamic amplifiers are an embodiment of the integrating amplifiers and will be analyzed in Sect. 7. Next, in Sect. 8, the core task of a residue amplifier will be analyzed, resulting in a core circuit that can be considered the minimum required circuitry. It will be shown that all of the recently published advances in residue amplifier design make use of this core circuit. All of these proposed solutions have their own type of overhead, resulting in different power efficiencies. Section 9 concludes this paper with an overview of the power efficiencies of the various topologies discussed.

2 Residue Amplifier Requirements

As the residue amplifier operates after the T&H circuit (Fig. 1), it operates in the *discrete-time domain* at a fixed clock rate F_{clk}, and settling as a response to an

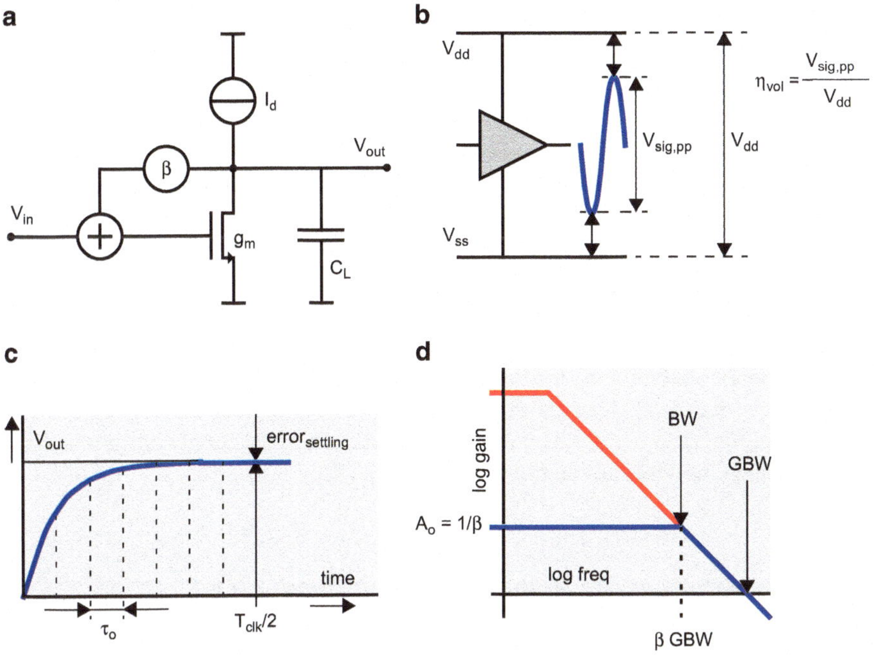

Fig. 2 A generic amplifier in a feedback situation (**a**), with (**b**) the signal swing indicated, (**c**) the step response, and (**d**) the Bode diagram

Table 1 Residue amplifier requirements

Parameter	Requirement	Example: $n = 3$, $m = 9$, $V_{ref} = 0.8$ [V]
A_{eff}	2^n	8
Accuracy A_{eff}	$<<2^{-m}$	<<0.19 [%]
$V_{in,noise}^2$	$<<V_{ref}/2^{(n+m)}$	<<0.19 [mV_{rms}]
$V_{in,max}$	$V_{ref}/2^n$	100 [mV]

input step is the normal mode of operation (Fig. 2c). The required *effective gain* A_{eff} is usually equal to 2^n, where n equals the number of resolved bits in that stage. The required *accuracy* with which this gain has to be achieved is dependent on the number of bits that still need to be resolved after this stage.

The input-referred *Noise* requirements are equal to the input-referred noise requirements of the entire ADC, as there is no gain between the ADC input and the first residue amplifier. The linearity requirements are part of the accuracy requirements, meaning that the required gain has to be achieved including finite settling and nonlinearity effects. The maximum input swing is equal to the reference voltage V_{ref} divided by a factor 2^n. An overview of these parameters is given in Table 1.

3 High-Gain OpAmp with Feedback

The classic [1, 2] implementation for a residue amplifier is a high-gain OpAmp with feedback. This ensures accurate gain and robustness against PVT spread. As such it is a self-contained solution that does not require any calibration. Figure 2 shows a model of such an amplifier (Fig. 2a) with its Bode diagram (Fig. 2d). In the next sections, we will derive a generic expression for the power dissipation of a high-gain amplifier employed in a feedback situation, as depicted in Fig. 2a.

To determine the device current required for the amplifier in Fig. 2a, we will follow the procedure as outlined in [3]: first the size of the load capacitance will be determined from the noise requirements (Sect. 3.1), and next the required device current will be determined from the settling requirements while driving this load capacitance (Sect. 3.2). Then the supply current will be determined from an analysis of the amplifier architecture (Sect. 3.3). Multiplication of the supply current with the supply voltage will lead to the power dissipation (Sect. 3.5).

3.1 Derivation of the Required Load Capacitance C_L

Consider the amplifier in Fig. 2. The input-referred noise power density is given by:

$$\mathrm{NPD_{input}} = 4kT/g_{\mathrm{m}} \tag{1}$$

The 3-dB bandwidth can be written as:

$$\mathrm{BW_{-3dB}} = g_{\mathrm{m}}/\left(2\pi A_{\mathrm{o}} C_{\mathrm{L}}\right), \tag{2}$$

with A_{o} being the closed-loop gain at low frequencies and C_{L} the load capacitance. The effective noise bandwidth equals:

$$\mathrm{NBW_{eff}} = \mathrm{BW_{-3dB}}\,(\pi/2) \tag{3}$$

Combining (1) and (3) allows us to determine the total input-referred integrated noise power:

$$V_{\mathrm{in,noise}}{}^{2} = kT/\left(A_{\mathrm{o}} C_{\mathrm{L}}\right) \tag{4}$$

Now that we have determined how much noise the circuit is producing, we need to determine how much noise the system can handle. To that extent we will start with the maximum signal swing that the circuit can handle, $V_{\mathrm{sig,pp}}$, and divide that by the required dynamic range. Defining the voltage efficiency η_{vol} as the ratio between the maximum peak-to-peak signal voltage and the supply voltage (see Fig. 2b):

$$\eta_{vol} = V_{sig,pp}/V_{dd} \tag{5}$$

we can express the maximum signal as:

$$V_{sig,pp} = \eta_{vol} V_{dd} \tag{6}$$

Since noise is usually expressed in V_{rms}, we also need to express our signal swing in V_{rms}:

$$V_{sig,rms} = \eta_{vol} V_{dd}/\mathrm{sqrt}(8) \tag{7}$$

To find the maximum allowable noise power, we need to take the square of (7) and divide by the dynamic range squared to obtain [3]:

$$V_{in,noise}^2 = V_{dd}^2 \eta_{vol^2} / \left(8\ \mathrm{DR}^2\right) \tag{8}$$

Combining (4) and (8), we can derive a (minimum) value for the load capacitance:

$$C_L = 8kT\ \mathrm{DR}^2 / \left(A_o\ V_{dd}^2\ \eta_{vol}{}^2\right) \tag{9}$$

3.2 Derivation of the Required Device Current I_d

Next, we need to determine how the amplifier has to drive this load capacitance. As depicted in Fig. 2c, the amplifier is required to settle to a certain accuracy as a response to an input-step voltage, within half the clock period, $T_{clk}/2$, with:

$$T_{clk} = 1/F_{clk}, \tag{10}$$

where F_{clk} is the clock frequency at which the circuit is operating. The relative settling error can be expressed as:

$$\mathrm{error}_{settl.rel.} = \exp\left(-N_\tau\right) \tag{11}$$

with:

$$N_\tau = T_{clk}/2\tau_o \tag{12}$$

the number of time constants allowed for settling and to the settling time constant of the amplifier. The allowable settling error is part of the error budget, including noise, mismatch, and nonlinearity. If we, for a moment, disregard other sources of error and finite settling would be the only cause of errors, one could say:

$$\mathrm{error}_{\mathrm{sett.rel}} < 1/\mathrm{DR}. \tag{13}$$

Here we speak of "fully settled" if condition (13) is met, but in practice (13) has to be met with a significant margin.

The time constant τ_{o} may be derived from inspection of the Bode diagram, so from Fig. 2d we may conclude:

$$\tau_{\mathrm{o}} = C_{\mathrm{L}}/\left(\beta g_{\mathrm{m}}\right) \tag{14}$$

From the same figure, we can also see that the low-frequency closed-loop gain equals:

$$A_{\mathrm{o}} = 1/\beta \tag{15}$$

From Eqs. (10), (12), (14), and (15), we can determine what the minimum value is of the transconductance g_{m}:

$$g_{\mathrm{m}} = 2N_{\tau}F_{\mathrm{clk}}A_{\mathrm{o}}C_{\mathrm{L}} \tag{16}$$

To be able to deduce the required current from the g_{m}, we need to establish a link between the device current I_{d} and the transconductance g_{m}. By defining a parameter V_{gt} as shown in [3]:

$$V_{\mathrm{gt}} = 2I_{\mathrm{d}}/g_{\mathrm{m}} \tag{17}$$

we may write in strong inversion:

$$V_{\mathrm{gt}} = V_{\mathrm{gs}} - V_{\mathrm{th}}, \tag{18}$$

also known as the overdrive voltage and in weak inversion:

$$V_{\mathrm{gt}} = 2nkT/q \tag{19}$$

which at room temperature is approximately equal to 80 [mV]. In this equation n is the weak-inversion slope [4], modeling the body effect.

Parameter V_{gt} is now our link between device current I_{d} and transconductance g_{m}. According to (17), the highest $g_{\mathrm{m}}/I_{\mathrm{d}}$ ratio is obtained for the lowest V_{gt}, which of course means biasing in weak inversion. Since in weak inversion the highest possible g_m/I_d ratio is obtained, we may define a transconductance efficiency as the ratio between the real g_m/I_d ratio and the g_m/I_d ratio in weak inversion:

$$\eta_{gm} = (g_m/I_d)/(g_m/I_d)_{WI} = 2nkT/(qV_{gt}) \tag{20}$$

This allows us to write:

$$I_d = g_m (nkT/q)/\eta_{gm} \tag{21}$$

Combining (16) and (21), we are now able to derive the required device current for the amplifier in Fig. 2a:

$$I_d = N_\tau F_{clk} A_o C_L (nkT/q)/\eta_{gm} \tag{22}$$

In Sect. 3.1 we found an expression for the load capacitance C_L (9). By inserting (9) into (22), we obtain:

$$I_d = 8kTN_\tau F_{clk} DR^2 (nkT/q)/\left(V_{dd}^2 \eta_{gm} {\eta_{vol}}^2\right) \tag{23}$$

This is the minimum required current in the MOS device of Fig. 2a to satisfy both the noise requirement (8) and settling with N_τ time constants, achieving a relative settling accuracy given by (11).

3.3 *Current Efficiency η_{cur} and Supply Current I_{dd}*

The current derived in (22) and (23) describes the current of a single device, i.e., g_m in Fig. 2a. Of course in a real circuit, there are more branches carrying current. For instance, in a folded-cascode amplifier [5] as depicted in Fig. 3, usually the currents in the transistors M1, M2, M5, and M6 are chosen equal. Hence, the current drawn from the supply would be, in this case, 4× larger than the device current of M1. We now define the current efficiency η_{cur} as the ratio between the current (I_d) required to obtain a certain g_m in the circuit of Fig. 2a and the total current drawn from the supply (I_{dd}) in the circuit of Fig. 3 to obtain the same g_m:

$$\eta_{cur} = I_d/I_{dd} \tag{24}$$

Using this definition, the current efficiency of the circuit in Fig. 3 would be $\eta_{cur} = 25\%$. Combining (20) and (22) allows us to express the total current drawn from the supply as:

$$I_{dd} = N_\tau F_{clk} A_o C_L (nkT/q)/(\eta_{cur}\eta_{gm}) \tag{25}$$

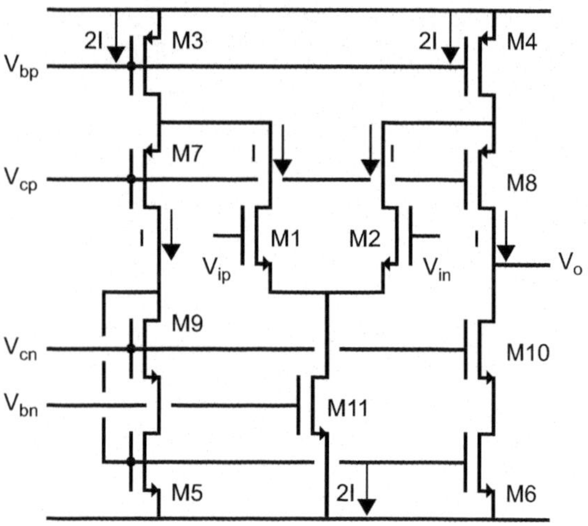

Fig. 3 Folded-cascode implementation of the amplifier of Fig. 2

3.4 Noise Excess Factor NEF

In the derivation of the size of the required load capacitance C_L, only the noise of the transistor(s) that embodies the g_m (Fig. 2) is taken into account. Of course, in a real amplifier, like depicted in Fig. 3, more transistors contribute noise. In particular, M1, M2, M3, M4, M5, and M6 contribute to noise considerably. The noise contribution of transistors M7–M11 usually is negligible. As transistors M1 and M2 embody the g_m, their noise contributions are unavoidable, but the noise from transistors M3–M6 is additional, in excess of the noise of M1–M2. We now define the noise excess factor (NEF) as follows:

$$\mathrm{NEF} = \frac{\text{Total Noise Power}}{\text{Noise Power } g_m \text{ transistor (s)}} \tag{26}$$

The excess noise has to be taken into account when calculating the size of the load capacitance C_L, as derived earlier in Eq. (9). In fact, the size of the load capacitor has to be increased by a factor NEF to accommodate the excess noise:

$$C_L = 8kT\ \mathrm{NEF}\ \mathrm{DR}^2 / \left(A_o V_{dd}^2 \eta_{vo1}{}^2\right) \tag{27}$$

3.5 Derivation of the Amplifier Power Dissipation

We can now insert (27) in (25) to obtain the supply current of a generic single-stage amplifier:

$$I_{\mathrm{dd}} = \frac{8kT\ \mathrm{NEF}\ N_\tau F_{\mathrm{clk}}\,(nkT/q)\,\mathrm{DR}^2}{\left(V_{\mathrm{dd}}{}^2\eta_{\mathrm{gm}}\eta_{\mathrm{cur}}\eta_{\mathrm{vol}}{}^2\right)} \tag{28}$$

Multiplying this result by the supply voltage yields an expression for the total power dissipation of a generic single-stage amplifier:

$$P = \frac{8kT\ \mathrm{NEF}\ N_\tau F_{\mathrm{clk}}\,(nkT/q)\,\mathrm{DR}^2}{\left(V_{\mathrm{dd}}\eta_{\mathrm{gm}}\eta_{\mathrm{cur}}\eta_{\mathrm{vol}}{}^2\right)} \tag{29}$$

This expression is valid for a generic single-stage amplifier, operating in a discrete-time environment, settling to a relative accuracy equal to $\exp(-N_\tau)$, and exhibiting a dynamic range DR.

4 Power Efficiency

Now that we have a generic expression for power dissipation (29), we can analyze the various parameters from this expression and derive a definition of power efficiency.

Expression (29) can be split into a system-dependent power, a circuit-dependent factor, and a supply-dependent factor in the following way:

$$P = P_{\mathrm{system}} \times F_{\mathrm{circuit}}/F_{\mathrm{supply}} \tag{30}$$

with:

$$P_{\mathrm{system}} = 32kT\ F_{\mathrm{clk}}\mathrm{DR}^2, \tag{31}$$

$$F_{\mathrm{circuit}} = 0.5\mathrm{NEF}\ N_\tau / \left(\eta_{\mathrm{gm}}\eta_{\mathrm{cur}}\eta_{\mathrm{vol}}{}^2\right) \tag{32}$$

and:

$$F_{\mathrm{supply}} = V_{\mathrm{dd}}/\,(nkT/q) \tag{33}$$

P_{system} is, as expected, a power dissipation (measured in Watts), fully determined by system specifications. F_{circuit} is a dimensionless factor, fully determined by circuit implementation choices. The factor "0.5" in (32) is chosen such that the best possible F_{circuit} equal 1. Why this is necessary will become clear in Sect. 6. F_{supply} can be seen as the supply voltage measured in the number of times (nkT/q) fits in it. Factor F_{circuit} requires more analysis, as it can give directions to power efficient circuit design. In the following we will discuss each factor in expression (32).

4.1 Noise Excess Factor NEF

This factor is increased by all noise-producing elements that contribute to the total integrated noise power of the circuit, with the exception of the device(s) that embody the transconductance g_m of the amplifier. A low NEF can be obtained by avoiding the use of current sources, load resistors, or switched-capacitor loads (although the latter may be hard in an ADC environment). A good example is a push-pull stage [6, 7] where both NMOS and PMOS are used actively as part of the effective transconductance g_m and no active current sources are necessary.

4.2 Number of Settling Time Constants N_τ

The settling accuracy is increased as more time constants are given to the amplifier to settle. However, for a fixed clock frequency, that translates to a lower time constant τ_o, therefore increasing the bandwidth of the circuit. This in turn increases the power dissipation proportionally. If a way can be found to reduce N_τ, the power dissipation will reduce accordingly. As will be seen in Sect. 6, for ideal integrating amplifiers, the effective N_τ (see Sect. 5.5) can drop to as low as 2, thus allowing for a very significant improvement of the power efficiency.

4.3 The g_m-Efficiency η_{gm}

Low power dissipation is achieved if the highest possible transconductance g_m is obtained for the smallest possible current I_d. This of course means the highest possible g_m-efficiency η_{gm}. As Eq. (20) shows, this is achieved for the lowest possible value of V_{gt}. The lowest possible value of V_{gt} is obtained in weak inversion and is approximately 80 mV at room temperature. This reflects the well-known notion that, if possible, one should operate devices in weak inversion. However, if the required speed is approaching the F_t of the technology, a larger V_{gt} has to be chosen. As a result, the power efficiency is affected negatively. Because for all lower speeds V_{gt} can remain at 80 mV and only when pushing the technology, V_{gt} has to be increased, the curve depicting power efficiency versus speed, Fig. 4, resembles a hockey-stick shape, and the effect is called the hockey-stick effect.

4.4 Current Efficiency η_{cur}

As is clear from (32), the current efficiency η_{cur} should be as high as possible. This means that when possible, current mirrors or folded-cascode structures (which

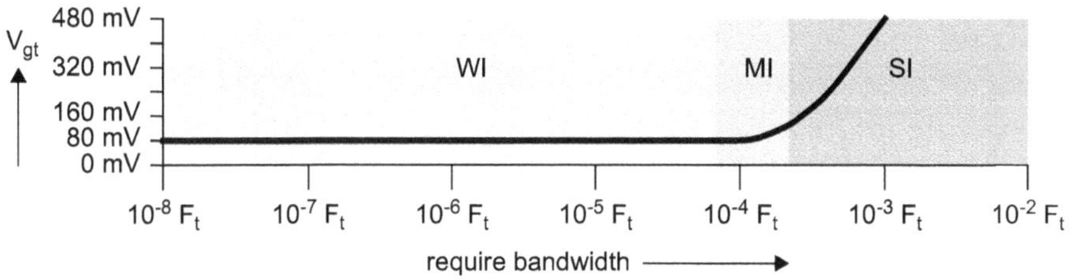

Fig. 4 The hockey-stick curve: V_{gt} versus the required bandwidth relative to the technology F_{t}

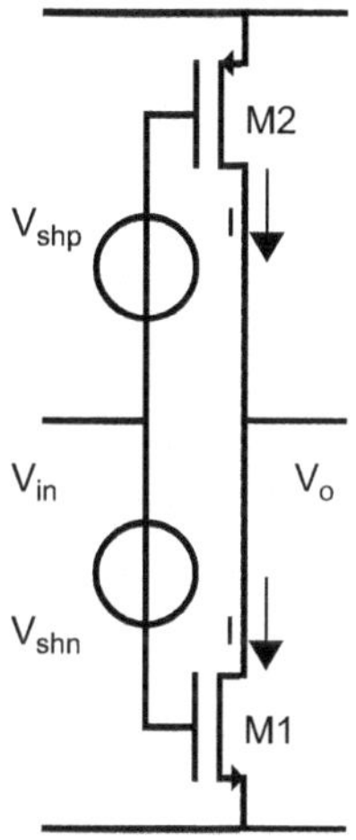

Fig. 5 Example of current sharing for increased current efficiency, with $\eta_{\text{cur}} = 200\%$

introduce additional current branches) should be avoided. As is generally known, telescopic topologies exhibit higher efficiency than, for instance, a folded-cascode structure.

Current sharing (i.e., an NMOS and a PMOS transistor in the same branch that are both actively driven, as depicted in Fig. 5) increases the current efficiency by approximately a factor of 2 and has a major (positive) effect on the power efficiency. This is because the same current flows through the NMOS as well as the PMOS transistor (so only one current is drawn from the supply), whereas the effective transconductance is the sum of the NMOS and PMOS transconductances: $g_{\text{m,eff}} = g_{\text{mp}} + g_{\text{mn}}$. However, stacking of devices can reduce the maximum output swing, which in turn reduces the voltage efficiency. Due to current sharing, the current efficiency potentially could reach values of 100% for differential circuits (and even 200% for single-ended circuits).

However, circuits using a differential pair with a tail current often exhibit a current efficiency of only 50%, as the input voltage is divided equally over the two devices, and as a result produce an output current of $0.5V_{\text{in}}g_{\text{m}}$ instead of the full $V_{\text{in}}g_{\text{m}}$. In the example of the circuit depicted in Fig. 6, both M1 and M2 carry a current equal to I_{d}, and in total $2I_{\text{d}}$ is drawn from the supply. But both M1 and M2

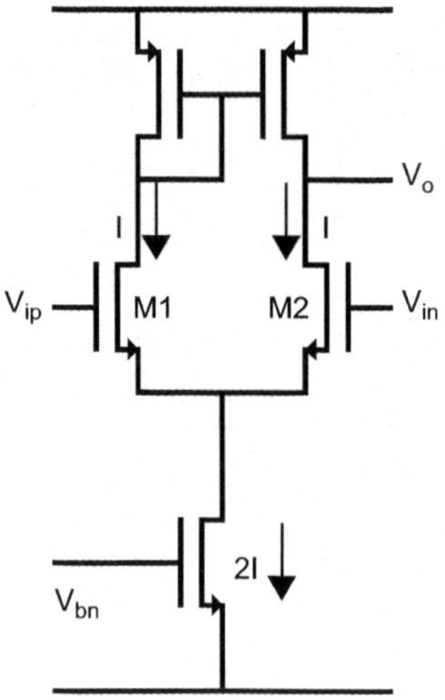

Fig. 6 Differential pair with $\eta_{cur} = 50\%$

only see half of the input voltage across their individual V_{gs}'s. As the drain currents are subtracted at the output, the effective transconductance of the entire circuit is equal to the g_m of a single device, and, as we have just seen, the current drawn from the supply is twice the individual current, hence $\eta_{cur} = 50\%$. However, if the entire circuit is differential and hence, also a differential output is used, the voltage efficiency (see following section) is doubled, and the product of current efficiency and voltage efficiency remains constant.

4.5 Voltage Efficiency η_{vol}

A high voltage efficiency requires a high output swing and hence benefits from the absence of cascodes. Also differential pairs in the same branch as the output node (like in Fig. 6) negatively affect the voltage efficiency. In contrast, a common-source NMOS and a common-source PMOS, with their individual sources connected to the supplies (as, for instance, shown in Fig. 5), deliver excellent voltage efficiency.

4.6 Definition of Power Efficiency

Equations (30)–(33) give a good insight in to the power dissipation of a generic single-stage amplifier operating in a discrete-time environment using feedback. A good definition of power efficiency should include the effects of the choices made in the circuit implementation of the amplifier but should not include any influence from system specifications. As such, the parameter $F_{circuit}$ (32)is an excellent definition of power efficiency in itself. It is good to point out that this definition of power efficiency is similar to the Schreier figure of merit (FoMS):

$$\text{FoMS} = 20 \log \left(P / \left(\text{DR}^2 F_{\text{clk}} \right) \right) \tag{34}$$

Applying (34) to the expression for power dissipation found in (30)–(33) yields:

$$\text{FoMS} = 20 \log \left(32kT F_{\text{circuit}} / F_{\text{sup ply}} \right) \tag{35}$$

Comparing (35) to (32) reveals that F_{circuit} and FoMS differ on three points: the Schreier FoM is using dB's (and F_{circuit} is on a linear scale), the Schreier FoM has a factor $32kT$ included in its expression, and the supply factor F_{supply} is included in the Schreier FoM. The factor $32kT$ converts the Schreier FoM into the dB's of a power (in Joules) per clock cycle, whereas F_{circuit} is dimensionless.

5 Incomplete Settling [8]

As was discussed in Sect. 4.2, the number of settling time constants N_τ has a significant influence on the power dissipation and hence on the power efficiency of an amplifier. According to Eq. (11), a 77-dB accuracy would require $N_\tau = 9$, whereas only $N_\tau = 3$ would be required if 26-dB accuracy would be sufficient. If we let go of the notion of obtaining gain accuracy by settling completely, we could possibly reduce the power dissipation by a significant amount ($3\times$ in the above example). In the past 12 years, a large amount of work has been done on the idea of incomplete settling [8, 9, 17, 21, 29].

5.1 The Idea of Incomplete Settling

Figure 7 shows a set of curves that depict an amplifier's response to an input-step voltage versus time as a fraction of the allowed settling time. Settling starts at $t = 0$ and in experiment stops at $t = 1$. Each curve has a different time constant (as indicated in the figure) and hence, after a fixed amount of time ($t = 1$), varying settling accuracy. The number of time constants settling realized at $t = 1$ is indicated in the figure and varies down from 7 to 1. As discussed above, a significant amount of power savings could be achieved by reducing N_τ. However, by observing Fig. 7, we see that reducing N_τ not only causes a reduction in gain accuracy, but the gain achieved at $t = 1$ is also dropping significantly.

Figure 8 shows the same set of curves but with the difference that the feedback factor β is adapted for each individual curve such that at $t = 1$ the exact gain is achieved for all curves. This shows that it is indeed possible to achieve the power

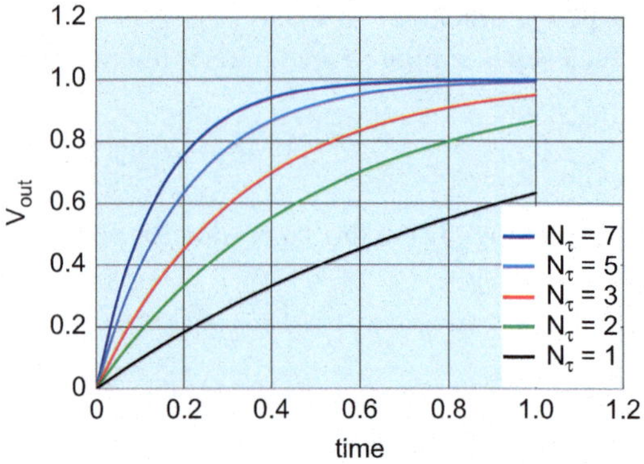

Fig. 7 Incomplete settling versus time as a fraction of the allowed time. Gain loss is visible, especially for $N_\tau = 3$ or less

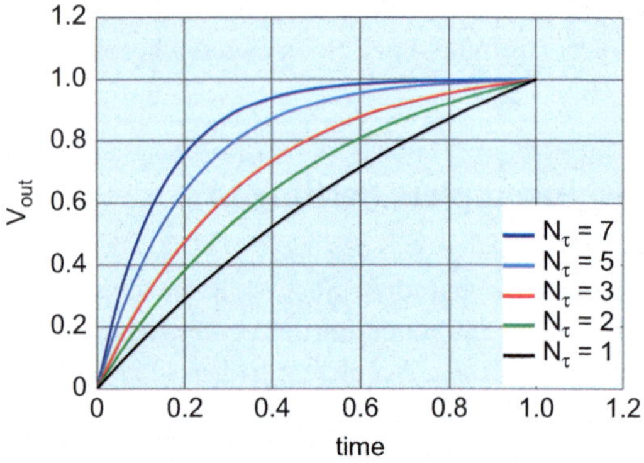

Fig. 8 Incomplete settling with compensated gain loss

savings together with the desired gain. However, some form of calibration will be required to obtain the correct β to produce an accurate gain.

If we were to push this idea to the limit, i.e., making N_τ increasingly smaller and simultaneously adapting β to obtain the correct gain, the required β would become extremely small, approaching an open-loop solution, as depicted in Fig. 9. In Fig. 10 this circuit is used for even smaller values of N_τ, and the gain $g_m R_L$ is adapted such that at $t = 1$ the correct gain is achieved. By doing so we left the path of feedback completely and are operating open-loop. We now have to rely on some form of calibration to achieve the desired accuracy. Nevertheless, significant power savings could be obtained by using such small values of N_τ. In the limit the amplifier would become a pure integrator with infinite output impedance R_L and hence infinite dc-gain $g_m R_L$.

5.2 *Incomplete Settling and Noise*

In the analysis below, we will assume, for simplicity reasons, the noise parameter [8] $\gamma = 1$ and $g_m R_L >> 1$. In [8] Iroaga and Murmann analyzed the noise behavior

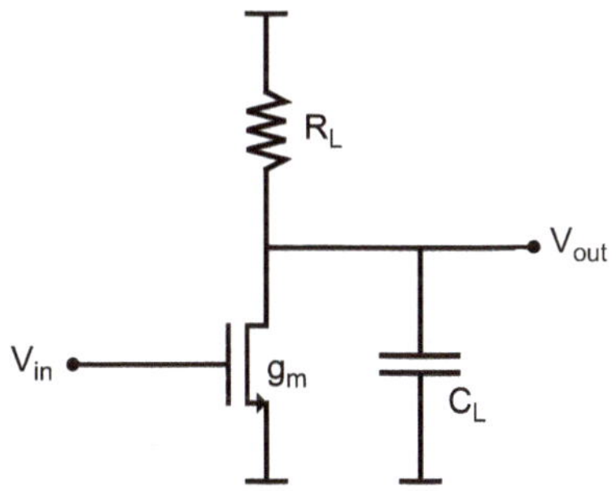

Fig. 9 Open-loop amplifier

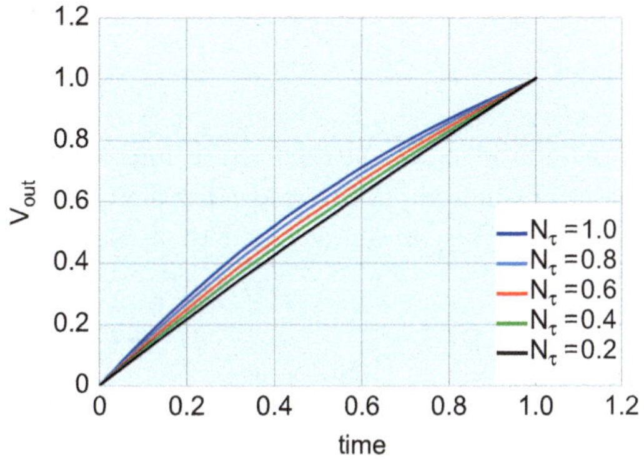

Fig. 10 Severe incomplete settling going toward pure integration

of the circuit in Fig. 9 in case of (severe) incomplete settling. In their analysis the settling was stopped at $t = t_\mathrm{s}$. They found the following result for output-referred noise power at that moment:

$$P_{\mathrm{noise-output}}\,(t_\mathrm{s}) = (kT/C_\mathrm{L})\, g_\mathrm{m} R_\mathrm{L}\,(1 - \exp\,(-2N_\tau)) \tag{36}$$

with N_τ given by:

$$N_\tau = t_\mathrm{s}/\tau_0 \tag{37}$$

and:

$$\tau_0 = R_\mathrm{L} C_\mathrm{L} \tag{38}$$

Based on the above time constants, the effective gain achieved at $t = t_\mathrm{s}$ is:

$$A_\mathrm{eff} = A_\mathrm{o}\,(1 - \exp\,(-N_\tau)) \tag{39}$$

with:

$$A_\mathrm{o} = g_\mathrm{m} R_\mathrm{L} \tag{40}$$

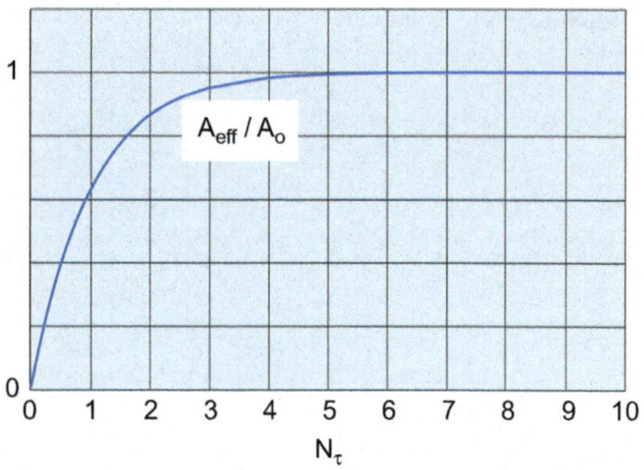

Fig. 11 The normalized gain $A_{\text{eff}}/A_{\text{o}}$ versus the number of settling time constants N_{τ}

Figure 11 shows the relative settling of the gain, $A_{\text{eff}}/A_{\text{o}}$, versus N_{τ}. Equation (36) describes the output-referred integrated noise power. To derive the input-referred integrated noise power, we have to divide (36) by the square of the gain in (39) and obtain:

$$P_{\text{noise-input}}(t_{\text{s}}) = \frac{(kT/C_{\text{L}})\,(1-\exp(-2N_{\tau}))}{A_{\text{o}}(1-\exp(-N_{\tau}))^{2}} \tag{41}$$

For N_{τ} approaching infinity (fully settled), expression (41) shows the steady-state input-referred integrated noise power:

$$P_{\text{noise-input,SS}} = kT/\,(A_{\text{o}}C_{\text{L}}) \tag{42}$$

This allows us to rewrite (41) as:

$$P_{\text{noise-input}}(t_{\text{s}}) = P_{\text{noise-input,SS}}F_{\text{settling}}(N_{\tau}) \tag{43}$$

where:

$$F_{\text{settling}}(N_{\tau}) = \frac{(1-\exp(-2N_{\tau}))}{(1-\exp(-N_{\tau}))^{2}}, \tag{44}$$

is the increase in noise power relative to the steady-state input-referred noise power, as a function of the number of settling time constants N_{τ}. This function is shown in Fig. 12 and shows the relative increase in noise power as a result of incomplete settling.

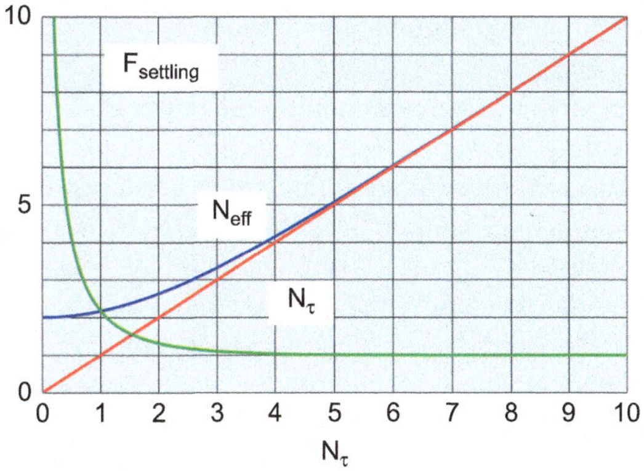

Fig. 12 The settling factor $F_{settling}$ and the effective number of settling time constants N_{eff} versus the number of settling time constants N_τ

5.3 *Effective Noise Bandwidth* NBW_{eff}

The increase in total integrated input-referred noise for N_τ approaching zero is a result of an increase of the effective noise bandwidth due to incomplete settling. This can be seen by rewriting Eq. (41) as follows:

$$P_{noise-input}\left(t_s\right) = \frac{4kT}{g_m}\frac{1}{4\tau_o}F_{settling}\left(N_\tau\right) \quad (45)$$

or:

$$P_{noise-input}\left(t_s\right) = \mathrm{NPD}_{input}\mathrm{NBW}_{eff} \quad (46)$$

with:

$$\mathrm{NPD}_{input} = 4kT/g_m \quad (47)$$

being the input-referred noise power density and:

$$\mathrm{NBW}_{eff} = \mathrm{NBW}_{SS}F_{settling}\left(N_\tau\right) \quad (48)$$

the effective noise bandwidth, with:

$$\mathrm{NBW}_{SS} = 1/\left(4\tau_o\right) \quad (49)$$

the steady-state effective noise bandwidth. The last factor in (48) indicates how the effective noise bandwidth will change from its steady-state value ($1/(4\tau_o)$) to a value approaching infinity, if N_τ is reduced from a large value (fully settling) to a value approaching 0 (ideal integration). This is depicted in Fig. 12 ($F_{settling}$).

5.4 Incomplete Settling and Power Dissipation

In Sect. 3.4 an expression for the power dissipation was derived, based on complete settling. As we have seen in the previous section, incomplete settling increases the input-referred total integrated noise power. Equation (43) shows that a factor $F_{settling}(N_T)$ needs to be included in the expression for $P_{noise\text{-}input}$, where $F_{settling}(N_T)$ is described by (44). The effect of this increase in input-referred noise power is that, in order to achieve the same overall dynamic range, the load capacitance C_L has to be increased by the same factor. In turn, the effect on overall power dissipation is the same: the device current as well as the power dissipation will increase by the factor $F_{settling}(N_T)$. This leads to the following expression for the power dissipation:

$$P = P_{system} \times F_{circuit} \times F_{settling}/F_{sup\,ply} \tag{50}$$

with P_{system}, $F_{circuit}$, $F_{settling}$, and F_{supply} given by (31), (32), (44), and (33), respectively.

5.5 Effective Settling Parameter N_{eff} [29]

The discussion on incomplete settling (Sect. 4.1) started with the idea that reducing the amount of settling time could significantly reduce the power dissipation. In fact, seeing that the power dissipation is proportional to N_τ may lead to the thought that the power dissipation could become negligible altogether, if we were to reduce N_τ to values close to zero. However, our analysis in the previous section revealed that due to incomplete settling, an additional factor has to be added to the expression for the power dissipation: $F_{settling}$, as given by (44). This means that, instead of being proportional to N_τ, the power dissipation is proportional to $N_\tau \times F_{settling}$. By defining a parameter N_{eff} in the following way:

$$\begin{aligned} N_{eff}\,(N_\tau) &= N_\tau\, F_{settling} \\ &= N_\tau \frac{(1-\exp(-2N_\tau))}{(1-\exp(-N_\tau))^2}, \end{aligned} \tag{51}$$

we may simply replace N_τ in Eq. (29) by N_{eff} (as defined by (54)) to obtain an expression for power dissipation that is valid also in case of incomplete settling:

$$P = \frac{8kT\ \mathrm{NEF}\ N_{eff} F_{clk}\,(nkT/q)\,\mathrm{DR}^2}{\left(V_{dd}\eta_{gm}\eta_{cur}\eta_{vol}{}^2\right)} \tag{52}$$

Since the validity of this expression is independent of the amount of settling, we can find the associated expressions for ideal integrating amplifiers, by calculating the limit value for N_τ going to 0.

Figure 12 shows a curve depicting N_{eff} as a function of N_τ. As can be seen from this plot, for larger values of N_τ, $N_{\mathrm{eff}} \sim N_\tau$, but for lower values of N_τ, especially for N_τ going to 0, N_{eff} goes to 2. This is the result of the increase in effective noise bandwidth, as is indicated by Eq. (48).

6 Ideal Integrating Amplifiers

Integrating amplifiers, often referred to as dynamic amplifiers, have become very popular in the past 10 years [10–16]. In this section we will derive expressions for all relevant parameters (gain, effective noise bandwidth, input-referred integrated noise power, power dissipation, and power efficiency) by calculating the limit values of the expressions found in the previous section. The implications of these results will be discussed here, whereas the circuit implementations will be discussed in Sect. 7.

6.1 Effective Gain A_{eff}

Equations (37)–(40) combined yield the following expression for the effective gain with incomplete settling:

$$A_{\mathrm{eff}}\left(t_{\mathrm{s}}\right) = \frac{g_{\mathrm{m}} t_{\mathrm{s}}}{C_{\mathrm{L}} N_\tau}\left(1 - \exp\left(-N_\tau\right)\right) \tag{53}$$

The limit value for this expression for N_τ going to 0 yields the expression for effective gain for an integrating amplifier:

$$A_{\mathrm{int}}\left(t_{\mathrm{s}}\right) = \frac{g_{\mathrm{m}} t_{\mathrm{s}}}{C_{\mathrm{L}}} \tag{54}$$

The ratio $g_{\mathrm{m}}/C_{\mathrm{L}}$ is equal to a single-stage amplifier gain-bandwidth product (GBW) and is usually equal to or close to the unity-gain frequency ω_{unity}. The GBW is showing what gain can be achieved in combination with a certain bandwidth. Similarly, in the case of an integrating amplifier, the associated time constant τ_{unity} is also setting the measure of what gain can be achieved: if the integration time t_{s} equals τ_{unity}, the effective gain is also unity. Gain can only be achieved if the integration time is larger than τ_{unity}.

6.2 Effective Noise Bandwidth NBW_{eff} [29]

Combining Eqs. (44), (48), and (49) results in the following expression for the effective noise bandwidth:

$$\mathrm{NBW}_{\mathrm{eff}} = N_\tau / (4t_s) \frac{(1 - \exp(-2N_\tau))}{(1 - \exp(-N_\tau))^2} \tag{55}$$

The limit value for this expression for N_τ going to 0 yields the expression for effective noise bandwidth for an integrating amplifier:

$$\mathrm{NBW}_{\mathrm{int}} = 1/(2t_s) \tag{56}$$

From this equation it is indeed clear that the effective noise bandwidth goes to infinity if the integration time goes to zero.

6.3 Input-Referred Integrated Noise Power

To calculate the input-referred total integrated noise power for an ideal integrating amplifier, we start with using (37), (40), and (51) in Eq. (41) and obtain:

$$P_{\mathrm{noise-input}}(t_s) = \frac{kT}{g_m t_s} N_{\mathrm{eff}} \tag{57}$$

As we have seen in Sect. 5.5, the limit value of N_{eff} for N_τ going to 0 equals 2, which results in the following expression for the integrated input-referred noise power:

$$P_{\mathrm{noise-input}}(t_s) = \frac{kT}{g_m} \frac{2}{t_s} \tag{58}$$

To get to a more recognizable expression, we use (54) in (58), which results in:

$$P_{\mathrm{noise-input,int}} = 2kT/(A_{\mathrm{int}} C_L) \tag{59}$$

This expression very much resembles the expression (42) for $P_{\mathrm{noise-input,SS}}$. Both Eqs. (42) and (59) express the integrated input-referred noise power as being proportional to $(kT/A_{\mathrm{eff}}C_L)$, where in steady-state $A_{\mathrm{eff}} = A_o$ and for the ideal integrating amplifier $A_{\mathrm{eff}} = A_{\mathrm{int}}$. However, an interesting difference is that (59) shows an additional factor of 2, compared to (42).

6.4 Power Dissipation and Power Efficiency

The power dissipation for an amplifier with incomplete settling is given by (51). Using (51) we find:

$$N_{\text{eff}}(0) = 2 \tag{60}$$

and as a result we find the following expression for power dissipation for an ideal integrating amplifier:

$$P = \frac{16kT \text{ NEF } F_{\text{clk}} \, (nkT/q) \, \text{DR}^2}{\left(V_{\text{dd}}\eta_{\text{gm}}\eta_{\text{cur}}\eta_{\text{vo1}}{}^2\right)} \tag{61}$$

6.5 Overview for Ideal Integrating Amplifiers

In the previous section, we have found expressions for ideally integrating amplifiers. Table 2 gives an overview of the expressions for various parameters found for the fully settled versus ideal integrating amplifiers.

The initial idea to investigate incomplete settling (and ideal integrating amplifiers) was to see how far we could reduce power dissipation by allowing less settling. It is clear from the results we have obtained, especially shown in the row 5 of Table 2, that indeed the power dissipation is lowest for ideally integrating amplifiers. This is also visualized in Fig. 12 by the curve labeled "N_{eff}." As described by Eq. (51), the power dissipation is proportional to N_{eff} (with the two extremes, "fully settling" and "integrating," shown in Table 2). So as Fig. 12 shows, less settling reduces the power dissipation until the effects starts to saturate around $N_\tau = 2$ and is, in the extreme, limited to a value of 2P1 (see Table 2).

As opposed to fully settling approaches, where accuracy is obtained through feedback and complete settling, integrating approaches do require calibration to achieve the required accuracy. Although calibration is not the focus of this paper, it can be shown that such a gain calibration can be achieved with negligible power [16, 28]. As an example, if the gain-error detection is performed in the digital

Table 2 Fully settling compared to integration

Parameter	Fully settling	Integrating
$A_{\text{eff}} =$	$g_{\text{m}}R_{\text{L}}$	$g_{\text{m}}t_{\text{int}}/C_{\text{L}}$
$\text{NBW}_{\text{eff}} =$	$1/(4\tau_{\text{o}})$	$1/(2t_{\text{int}})$
N_{eff}	N_τ	2
$P_{\text{noise-inp}} =$	$kT/(A_{\text{o}}C_{\text{L}})$	$2kT/(A_{\text{int}}C_{\text{L}})$
P	$N_\tau P_1$	$2P_1$
$P_1 =$	$\frac{16kT \text{ NEF } F_{\text{clk}}(nkT/q)\text{DR}^2}{\left(V_{\text{dd}}\eta_{\text{gm}}\eta_{\text{cur}}\eta_{\text{vo1}}{}^2\right)}$	

domain at a (low) subsample rate and the gain-error correction is done in the analog domain (by controlling the bias current using a small DAC), the total effect on power dissipation can be very small indeed [16].

7 Dynamic Amplifiers [10–16]

The past 20 years of data converter design have witnessed great emphasis on power efficiency. Since residue amplifiers are a significant percentage of the overall power budget, especially in the higher SNR range, a lot of work has been done on power-efficient residue amplifiers. Dynamic amplifiers [10–16] have been a very significant step forward in this respect.

7.1 *Basic Architecture*

Figure 13 shows a circuit diagram of a dynamic amplifier [10]. Resistive loads are completely omitted and are replaced by switches (S1 and S2). These switches short the output voltages V_{op} and V_{on} to the positive supply until $t = 0$. At $t = 0$ the switches open and the tail current switches on. As of that moment, both output nodes (V_{op} and V_{on}) start discharging, as is depicted in Fig. 14. The side that has the largest current (with the highest input voltage) will discharge faster, gradually producing a differential output voltage. At $t = t_{int}$ the tail current is switched off and the integration process stops.

7.2 *Gain Limitations*

If, for simplicity, we can assume the output impedance of the differential pair in Fig. 13 to be infinite, the circuit behaves as a true integrator, and Eq. (54) is a good

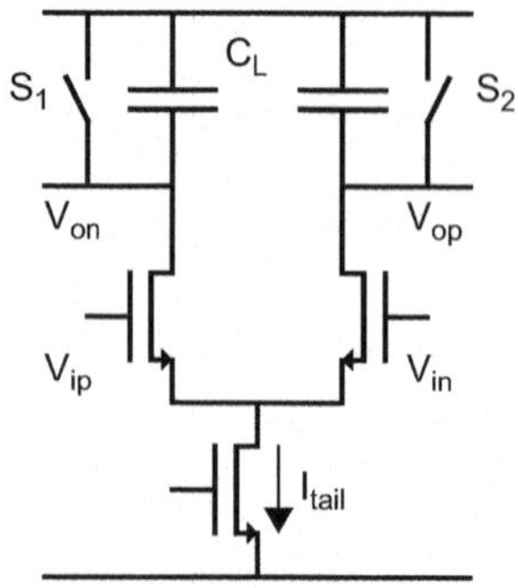

Fig. 13 Dynamic amplifier

Fig. 14 Output signals versus time

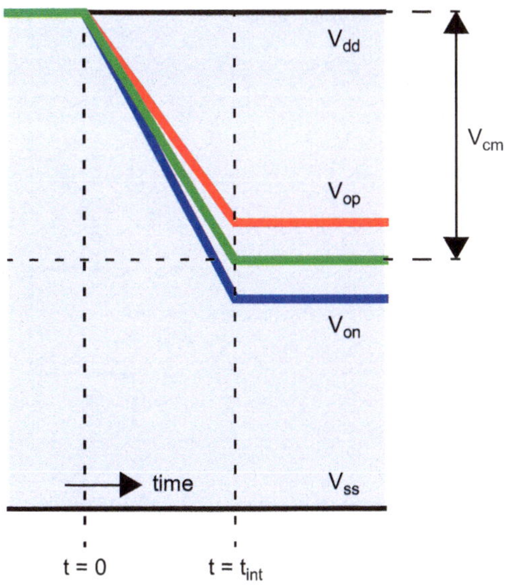

approximation of the gain. Using Eq. (17) to substitute g_m, we may write:

$$A_{int} = \left(2I_d/V_{gt}\right) t_{int}/C_L \tag{62}$$

with V_{gt} given by (18) and (19). By using $2I_d = I_{tail}$, we obtain:

$$A_{int} = \left(I_{tail}/V_{gt}\right) t_{int}/C_L \tag{63}$$

Figure 14 also shows the common-mode voltage. The amount the common-mode voltage (measured with respect to V_{dd}) comes down is dependent on the tail current and the load capacitance in the following way:

$$V_{cm}\left(t_{int}\right) = I_{tail} t_{int}/\left(2C_L\right) \tag{64}$$

Note that there is no dependence on the input signal in (64). Comparing (63) and (64), we can see that a large part of (63) actually describes the common-mode voltage V_{cm}. Hence we may also write:

$$A_{int} = 2V_{cm}/V_{gt} \tag{65}$$

This poses a serious limitation on the maximum achievable gain for this circuit. By observing Fig. 8a, b, we can estimate that making V_{cm} larger than half V_{dd} may be difficult, as the input devices will no longer be saturated. On the other hand, V_{gt} cannot be made lower than 80 [mV] at room temperature (Eq. 19). Let's take some numbers to exemplify the issue: if we assume a supply voltage of 1.0 [V],

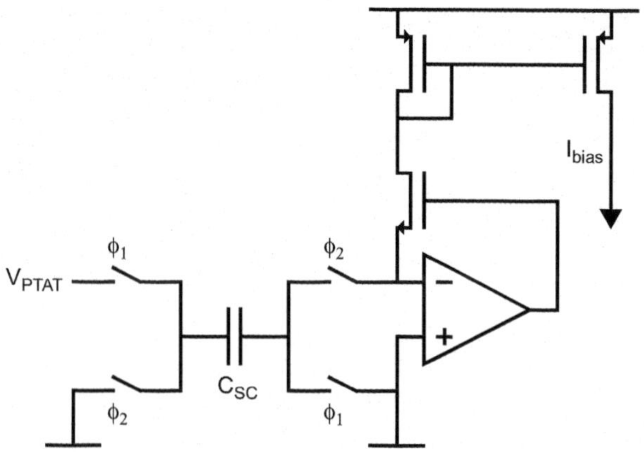

Fig. 15 Switched-capacitor biasing

the maximum gain that could theoretically be obtained is 2 × 0.5/0.08 = 12.5, and in practice it is even lower, as we need to reserve headroom for PVT variations. Achieving more gain (without changing the architecture) would require a larger V_{cm} drop.

7.3 *Gain Accuracy*

Inspecting the expression for gain (63) reveals what is required to control the gain in a robust way. There are various ways to control the gain, two of which will be discussed in the following.

Biasing Using a Switched-Capacitor Circuit Figure 15 shows a switched-capacitor bias generator. During ϕ_1, capacitor C_{SC} is being charged to a voltage V_{PTAT}, and during ϕ_2 the capacitor is discharged into the virtual ground of the amplifier. This produces an average current of:

$$I_{bias} = V_{PTAT} F_{clk} C_{SC} \tag{66}$$

Using this bias current as the tail current in Eq. (64) results in the following expression for the gain:

$$A_{int} = F_{clk} t_{int} \left(V_{PTAT} / V_{gt} \right) \left(C_{SC} / C_L \right) \tag{67}$$

Assuming $t_{int} = 1/2F_{clk}$ (half the clock period), the device is biased in weak inversion, such that $V_{gt} = 2nkT/q$ (Eq. 19) and $V_{PTAT} = \alpha T$ the expression for A_{int} becomes:

$$A_{int} = 2 \left(\alpha q / (nk) \right) \left(C_{SC} / C_L \right) \tag{68}$$

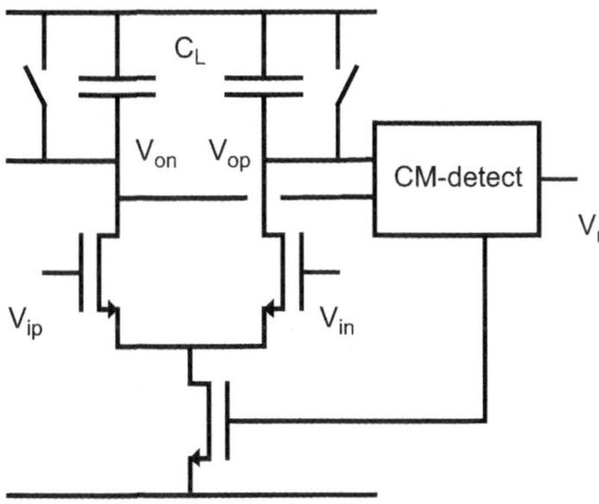

Fig. 16 Dynamic amplifier with a CM-mode detect circuit to fix the effective gain [12]

Equation (68) consists of constants and a ratio of capacitors which can be implemented with reasonable accuracy. Do note, however, that the weak-inversion slope factor n is dependent on technology. The accuracy of this approach is limited by the spread in the factor α (of the V_{PTAT} source), the spread in n, and the matching of the capacitors.

Controlling t_{int} by Forcing a Fixed V_{cm} [12] Figure 16 shows another way to control the gain. From (65) we can read that if we control V_{cm}, we can control the gain. In the circuit of Fig. 16, this is done by comparing V_{cm} to a fixed reference voltage V_{ref} and simply switching off the tail current I_{tail} [12] when V_{cm} reaches V_{ref}. At that moment the integration stops and the gain is fixed. In this approach, the accuracy is limited because of a slight dependence of the V_{cm} on the input voltage (which gives rise to distortion) and the way the V_{cm} is detected from V_{op} and V_{on}.

The above two methods to control the gain under PVT variations are, generally speaking (especially in the higher SNDR range), not enough, and calibration will be required. However, even though calibration may be used, the above described methods to control the gain are still needed to bring the initial gain (at the beginning of calibration) close enough to the optimal value, such that the system at least works reasonably and the calibration loop converges. If the starting point of the gain is too far off from the ideal value, the ADC will not be able to provide the calibration logic with the correct information.

7.4 Improved Gain Architectures

In Sect. 7.2 it was discussed how dynamic amplifiers are limited in maximum gain by the drop in common-mode voltage. As is also clear from (66), more gain would require more headroom for the common-voltage drop. Here we will discuss two circuits that increase the maximum achievable gain.

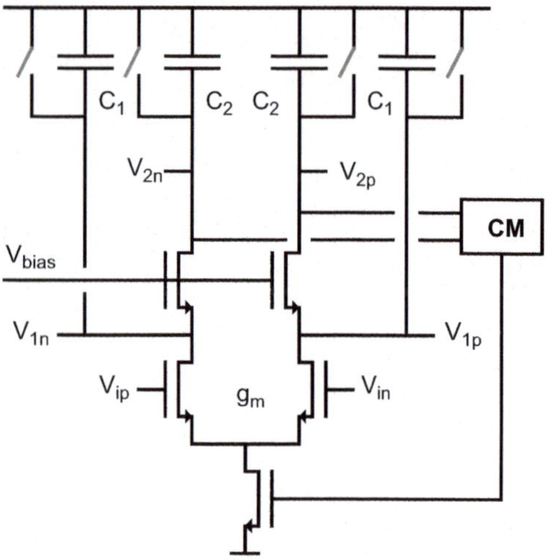

Fig. 17 Extended headroom through cascoding [18]

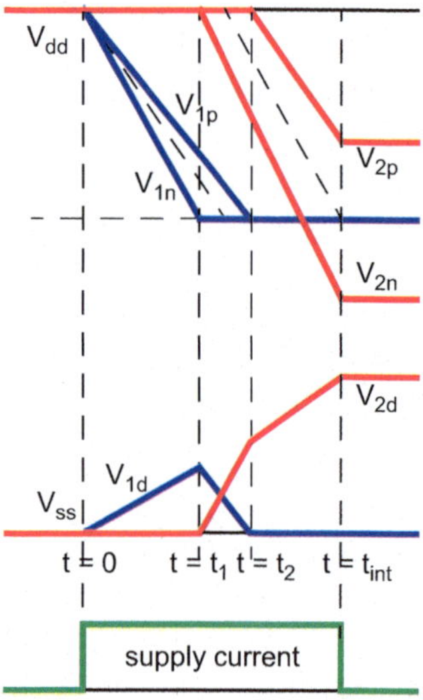

Fig. 18 Signals in the circuit of Fig. 17 [18]

Extended Headroom Through Cascoding (Fig. 10) [18]
In a classic amplifier, higher gain can be obtained through cascoding. The circuit in Fig. 10 [18] shows that a similar technique can also be used in a dynamic amplifier. The operation of the circuit can be explained as follows. During reset, nodes V_{1n}, V_{1p}, V_{2n}, and V_{2p} are all reset to V_{dd}, as shown in Fig. 18. At $t = 0$, the tail current is switched on and the differential pair becomes active. Since the nodes V_{1n} and V_{1p} are still high (close to V_{dd}), the cascodes are not yet ON and no current arrives at nodes V_{2n} and V_{2p}.

As a result, node V_{2n} and V_{2p} are not yet moving. So only nodes V_{1n} and V_{1p} are moving down.

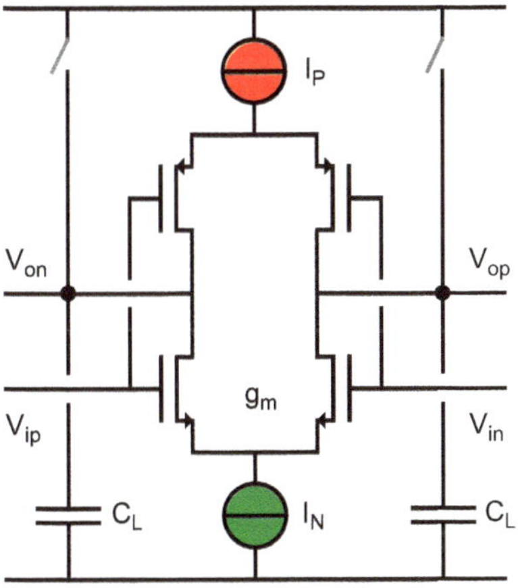

Fig. 19 Up and down integration to extend the common-mode range and hence increase the gain [19]

Let's assume V_{ip} is higher than V_{in}. That means that $V_{1\text{n}}$ is moving down faster than $V_{1\text{p}}$ (as depicted in Fig. 18). As soon as $V_{1\text{n}}$ is low enough ($t = t_1$), the first cascode will turn on, and as of that moment, the current will flow through the cascode and arrive at $V_{2\text{n}}$. As of that moment, $V_{2\text{n}}$ starts moving down.

At the moment that $V_{1\text{p}}$ is also low enough ($t = t_2$), also the second cascode is turning on and $V_{1\text{p}}$ stops moving as well. This happens at the same voltage level as where $V_{1\text{n}}$ stopped and reduces the differential voltage, $V_{1\text{d}} = V_{1\text{p}} - V_{1\text{n}}$, to zero.

However, as $V_{2\text{n}}$ was already moving, the differential voltage $V_{2\text{d}} = V_{2\text{p}} - V_{2\text{n}}$ has already been building up and after $t = t_2$ continues to do so, due to the differential current, until it is stopped by the shutting down of the tail current (at $t = t_{\text{int}}$). In the circuit shown in Fig. 17, this is done automatically by a CM-detect circuit, as discussed in Sect. 7.3.

For each set of nodes ($V_{1\text{p}}$ and $V_{1\text{n}}$ and $V_{2\text{p}}$ and $V_{2\text{n}}$), the gain that is being built up is proportional to the common-mode voltage drop, as described by (65). As both sets of nodes can drop approximately the same V_{cm}, the total gain built up on nodes $V_{2\text{p}}$ and $V_{2\text{n}}$ is approximately twice the value of the circuit of Fig. 13.

Up and Down Integration

Another way [19] to increase the maximum possible gain is shown in Fig. 19. The idea is pretty straightforward. As the common-mode voltage drops if an NMOS differential pair is used to drive the load capacitors, a second differential pair is employed, in this case a PMOS pair, to raise the common-mode level in exactly the same way, extending the integration activity over a longer period of time. As is shown in Fig. 20, the NMOS and PMOS differential pair work one after the other, causing the common-mode level to go down and up again.

The possible increase in gain is not exactly a factor of 2 but somewhat less. This is due to the fact that the second differential pair does not start with the output voltages at V_{ss}-level but at a significantly higher level (see Fig. 20). This causes the second part of the gain to be significantly smaller than the first part. Table 3 lists

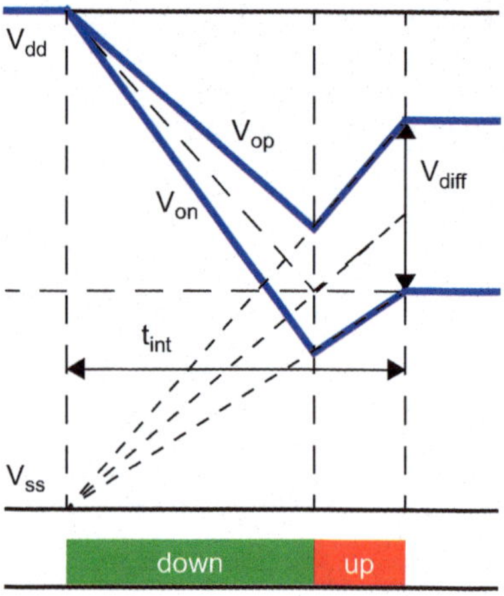

Fig. 20 Output signals in the circuit of Fig. 19

Table 3 Achievable gain

Integration type	Theoretical A_o	Practical A_o
	$V_{dd} = 1.0$	$V_{dd} = 0.9$
	Temp = 2 7 °C	Temp = 125 °C
Single (Fig. 13)	16.5	9.1
Up-down (Fig. 19)	24.5	10.3
Cascode (Fig. 17)	30	15.2

an example of achievable gains for the three different topologies, assuming weak-inversion operation. The second column shows typical conditions, whereas the third column shows corner conditions.

8 High-Efficiency Topologies

After the discussion on power efficiency and integrating amplifiers, in this section we will discuss the most efficient architectures published to date. All of these architectures share a common "core" circuit, which will be discussed in the following.

8.1 The Core Circuit

In Sect. 4, we discussed power efficiency and analyzed the expression for power dissipation to find ways to reduce the power dissipation. The factor F_{circuit} (30) shows which parameters are important: NEF, N_τ, V_{gt}, η_{cur}, and η_{vol}. To obtain NEF $= 1$ and $\eta_{\text{cur}} = 100\%$, all transistors driving the load capacitance need to be driven actively and share their bias current (push-pull). To reduce the amount of

Fig. 21 The "core" circuit

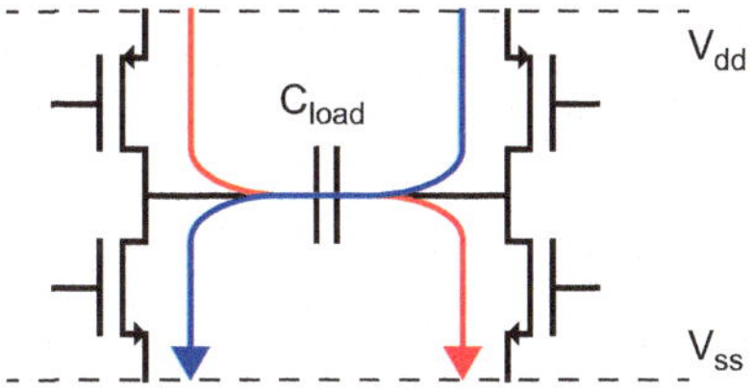

energy lost on settling as much as possible, we need to use integrating architectures, which results in $N_{\mathrm{eff}} = 2$. For the lowest possible V_{gt}, we need to operate the devices in weak inversion, and to obtain the maximum voltage efficiency η_{vol}, we need to make sure that the sources of the driving devices are biased around the level of the supplies (NMOS close to or at V_{ss} and PMOS close to or at V_{dd}). All these considerations together result in a "core" circuit, as depicted in Fig. 21, that is the "minimum" circuit required to drive the load capacitor. The circuitry around it that drives the gates of the four devices should be as simple as possible and drain negligible power. However, all topologies will have some form of "overhead." In the following sections, we will discuss various architectures that try to minimize this overhead as much as possible.

8.2 Complementary Dynamic Amplifier [24]

The first example of a high-efficiency topology is shown in Fig. 22. This complementary dynamic amplifier [24] shows the core circuit in transistors M1–M4. Transistors M5–M8 can be seen as overhead with respect to the core circuit. In this case they do not add any additional current drawn from the supply. As the circuit is differential and the two complementary differential pairs both add to the transconductance g_{m}, the current efficiency of this topology is $\eta_{\mathrm{cur}} = 100\%$. But M5–M8 together with the series switches do reduce the headroom for output swing and that affects the voltage efficiency negatively. So, in this case, the overhead in cost with respect to the core circuit is in the reduction of the voltage efficiency. Nevertheless, this amplifier is a very power-efficient topology. Table 4 shows an overview of a number of amplifier topologies and lists estimates for N_{eff}, NEF, T_{cur}, η_{vol}, and F_{circuit}. As can be seen from this table, this complementary dynamic amplifier ranks among the top topologies with respect to power efficiency (F_{circuit} in Table 4).

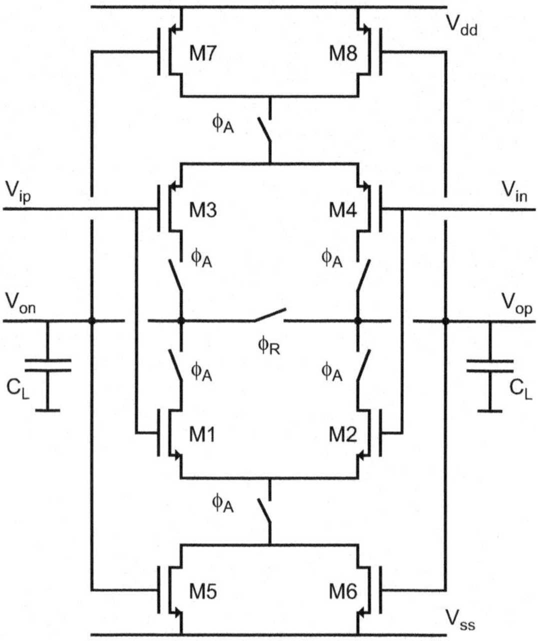

Fig. 22 Complementary dynamic amplifier [24]

8.3 Zero-Crossing-Based Circuits (ZCBC) [20–23]

Zero-crossing-based circuit [20–23] is a very original approach to increase the power efficiency of discrete-time amplifiers. A simplified circuit diagram of the principle of zero-crossing-based circuits (ZCBC) is shown in Fig. 23. Only half of the circuit is shown for simplicity reasons. The "core" circuit driving the load capacitance can be recognized as M1–M4. ZCBC uses feedback, as in the classic solution, but as opposed to a regular amplifier, ZCBC determines whether the output has reached its final value by using a comparator at virtual ground. As soon as signal V_{vir} at virtual ground crosses zero, the comparator trips and switches off the core transistors driving the load capacitance. In the core circuit, either M1–M4 is active, or M2–M3 is active. The core transistors act as current sources that are either ON or OFF, making this a truly integrating system.

The efficiency of this architecture can be very high but is limited by the overhead formed by the comparator. The comparator has to be always ON (as there is no a priori knowledge on when the zero-crossing will happen) and has to be continuous time. This means that there is no clock forcing a decision and hence the comparator is usually an open-loop multistage amplifier. As the noise performance is dominated by the noise performance of the comparator, the comparator power dissipation cannot be negligible. It is exactly this overhead that deteriorates the power efficiency of this system to a certain degree. Table 4 (row 5) lists the estimates of the efficiency parameter.

Fig. 23 The principle of zero-crossing-based circuits

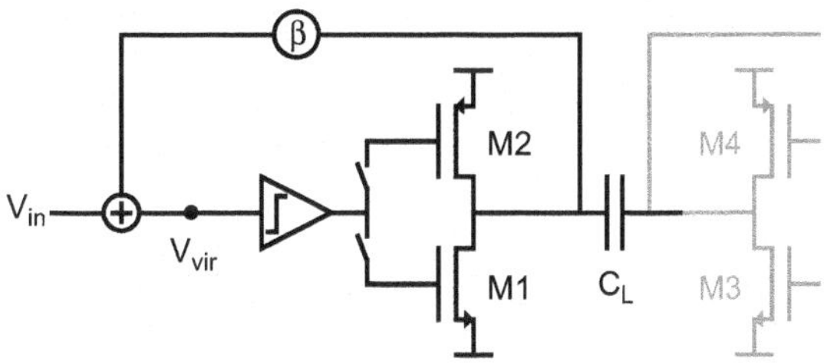

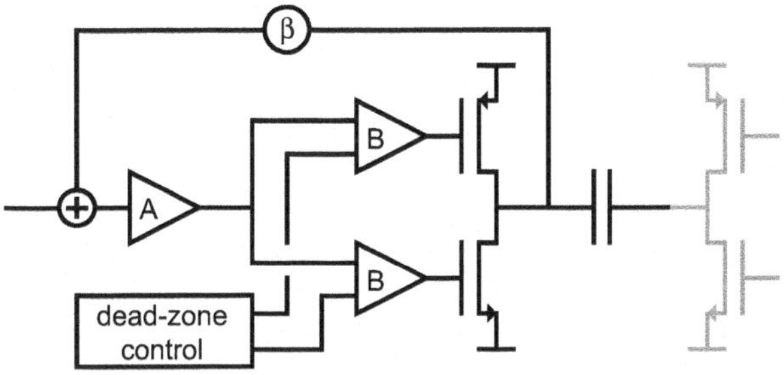

Fig. 24 Ring amplifier schematic

8.4 Ring Amplifier [25, 26]

The ring amplifier [25, 26] is also an interesting approach to achieve high power efficiency in discrete-time amplification. The simplified principle is depicted in Fig. 24. The ring amplifier consists of a minimum of three stages (A, B, and the "core" circuit). It uses feedback, but in contrast to the "classic" high-gain and feedback approach, this amplifier does not use any compensation technique to obtain stability. As a matter of fact, this amplifier is designed to be instable in feedback operation and is allowed to oscillate or "ring," hence the name. This enables this design to have the output swing toward the desired level in a very fast way.

To obtain a useful (and stable) response, the third stage (the "core" circuit) is driven by stage B in such a way that a "dead" zone is introduced in the third stage. Figure 25 shows the transconductance of the third stage, with the "dead zone" indicated. In this dead zone, both output transistors switch OFF, and the transfer function of the third stage enters a region in which the effective gain is (almost) zero. The idea is that at first the amplifier is instable and starts to oscillate. But as soon as the output gets close to the desired value, the third stage enters the dead zone and prevents the ring amplifier to oscillate. The stability of the ring amplifier depends on the width of the dead zone, and hence the dead zone needs to be carefully controlled by dedicated circuitry (Fig. 24).

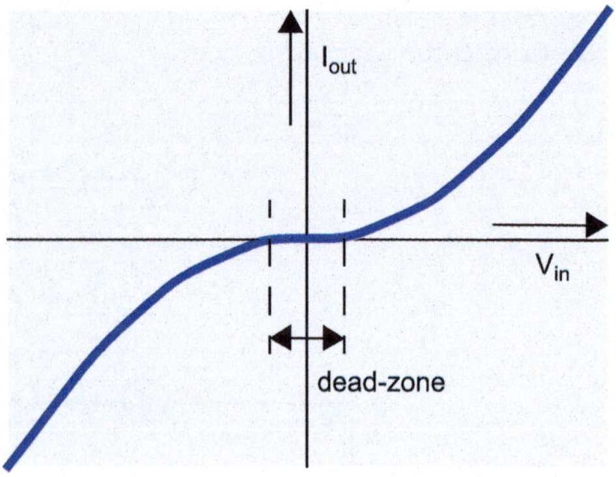

Fig. 25 Third-stage transconductance with the so-called dead zone

There are several advantages to this approach. Since stability is not required, compensation is not necessary and hence all internal nodes can remain fast. This allows for a very fast response and makes the amplifier very power efficient. However, the input-referred noise is primarily determined by the input stage (as in the ZCBC approach), and the power dissipation of this stage (stage A in Fig. 24) cannot be negligible. This is exactly where the overhead of this approach is located. Table 4 (row 4) lists the estimates of the efficiency parameter.

9 High-Efficiency Topologies with Linearization

9.1 The Need for Improved Linearity

As previously discussed (Sect. 8.1), in order to obtain the best possible power efficiency, the core circuit needs to be driven directly from the input and biased in weak inversion (Sect. 4.3). This means that, in most cases, the useful input range (with sufficient linearity) is rather small. At room temperature the peak value of the input swing is about 80 mV. In submicron technologies, with supplies of 1.0–1.2 V, ADC reference voltages are nominally 0.8–1.0 V. Every bit that is resolved by the first quantizer reduces the residue signal by 2×. If the residue amplifier cannot handle more than 80 mV at its input, a minimum of 4 bits have to be resolved by the first quantizer. For higher SNDR ADCs, for which the linearity demands are even more stringent, the minimum number of bits to resolve is even 5 bits. This puts restrictions on the architectural choices of the ADC. In many cases a larger (linear) input range would be desirable and hence the need for linearization techniques for integrating amplifiers. In the following two sections, two different linearization techniques are discussed.

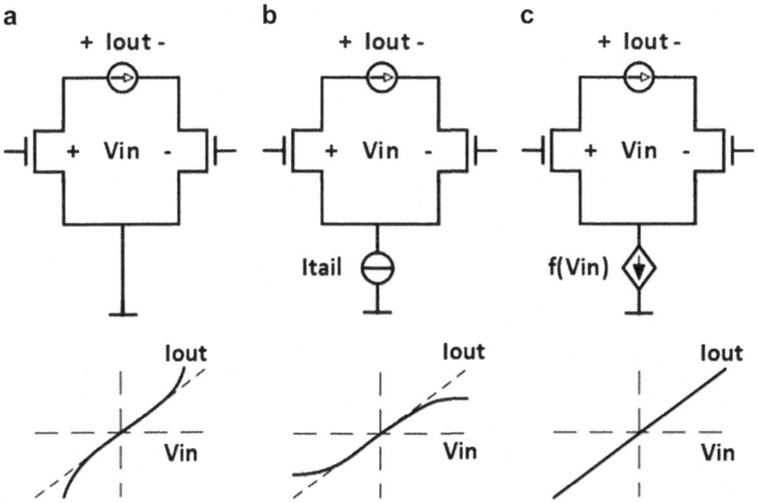

Fig. 26 Linearization principle (**a**) Grounded tail, (**b**) Fixed current-source tail, (**c**) Vin dependent tail-current

9.2 Linearized Integrating Amplifier [27, 29]

A residue amplifier based on the core circuit has been presented in [27, 29], which utilizes an analog linearization technique to achieve a high linearity while adding minimal circuit overhead.

Principle

The basic principle, illustrated in Fig. 26 showing only the NMOS side, is based on two observations:

(a) If the input transistors are biased in weak inversion with the sources tied directly to ground, then the amplifier exhibits an expanding V-I characteristic (Fig. 26a).
(b) If an ideal tail current source is added to restrict the total current flowing through the differential pair, then the amplifier would shift to a compressing V-I characteristic (Fig. 26b).

Between these two opposing distortion paradigms, a midpoint for the tail current characteristic can be found for which the amplifier exhibits a perfectly linear gm (Fig. 26c). If the input transistors are assumed to be biased in weak-inversion saturation region, then this tail current characteristic can be calculated to be:

$$I_{\text{tail}} = G_{\text{m}} \Delta V_{\text{id}} \coth\left(\Delta V_{\text{id}}/2nV_{\text{T}}\right), \tag{69}$$

with $V_{\text{T}} = kT/q$. By using Taylor series expansion, (69) can be expressed as:

$$I_{\text{tail}} = 2nV_{\text{T}} G_{\text{m}} \left(1 + (\Delta V_{\text{id}}/2nV_{\text{T}})^2 + \ldots\right) \tag{70}$$

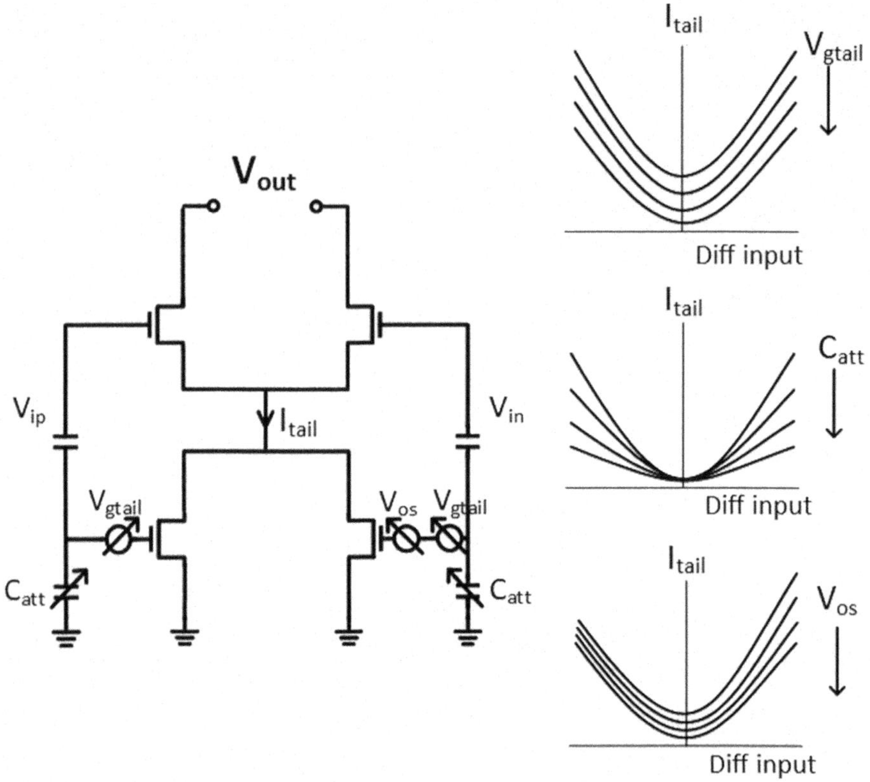

Fig. 27 Linearization parameters

where G_m is the desired amplifier transconductance and ΔV_{id} is the differential input signal. A tail current with the above characteristic would perfectly degenerate the input pair, helping it achieve a perfectly linearized g_m.

The behavior described in Eq. (70) can be approximated by using the sum of the output currents of two weak-inversion transistors driven by ΔV_{id} as shown in Fig. 27. However, as the sum would be extremely sensitive to PVT and Monte Carlo variations, independent circuit parameters are required to tune and match the sum of currents closely to the desired tail characteristic. Hence, this input-driven tail current source is implemented with the following tuning parameters:

(a) Tail bias voltage, which is tuned by using a programmable current DAC into a bias diode.
(b) Input attenuation, which determines the extent with which the tail current modulates with the input signal. This attenuation is implemented by using a programmable capacitor array to ground to tune the drive strength of the input.

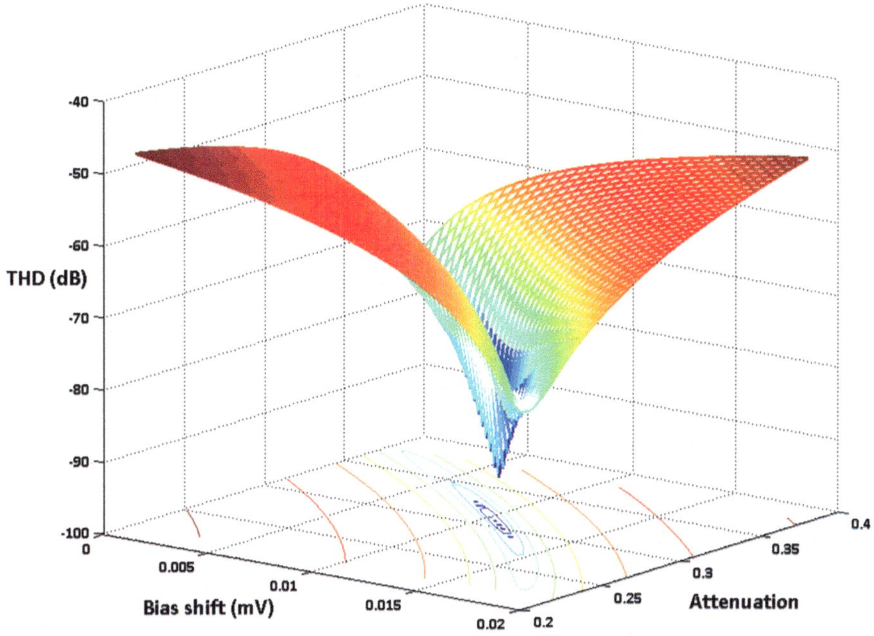

Fig. 28 Simulated amplifier THD versus linearization parameters

These two knobs establish how closely the tail current characteristic approximates the ideal function in (70), hence having a significant effect on the linearity of the amplifier. This is reflected in Fig. 28 which shows the simulated THD of a 250 MHz switched-cap amplifier based on the proposed linearization principle with an input swing of $300\text{mV}_{\text{ppdiff}}$ and a gain of $4\times$, as these two parameters are swept. It can be seen that over the entire search space of combinations of these two parameters, a unique set can be found for which the amplifier distortion exhibits a minimum, where the THD is <-90 dB. Figure 29 shows the deviation of the amplifier output from its ideal value. For an input swing of $300\text{mV}_{\text{ppdiff}}$, the error reduces from nearly -40 dB to -55 dB by sweeping only C_{att} and nearly -90 dB after linearization with both C_{att} and V_{gtail}.

While these two tuning parameters are very effective against odd-order distortion, they only adjust the circuit symmetrically. Due to the high inherent distortion of the WI input pair, any offset or gm mismatch in the amplifier creates a significant even-order distortion. In order to avoid this from limiting the overall amplifier linearity, this mismatch is compensated for by adding an extra tunable offset in the tail current source, as seen in Fig. 17, with the help of an additional programmable current source in the bias diode of one of the tail transistors. This offset provides a knob to correct for the imbalance arising out of any mismatch in the amplifier. It should be noted that this tunable tail current offset does not cancel the inherent offset

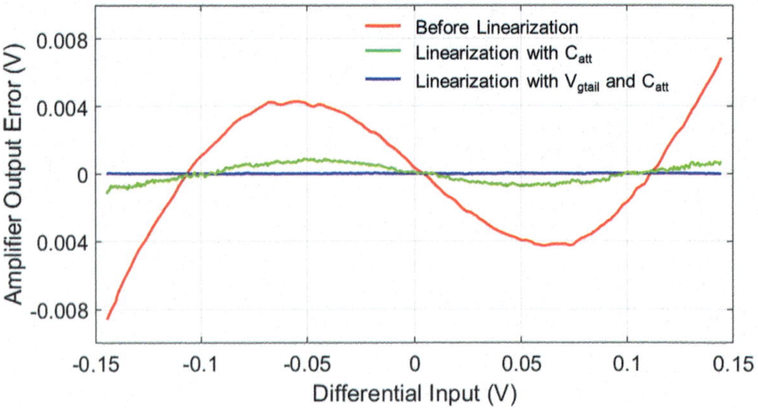

Fig. 29 Amplifier output error voltage before and after linearization

of the amplifier, but merely corrects the even-order distortion by counteracting the imbalance in the amplifier.

Implementation

Based on the proposed linearization technique, an integration-based amplifier was implemented with the input being driven through both NMOS and PMOS sides, as shown in Fig. 30. Since the linearization technique requires the transistors to be biased in deep weak-inversion region, the amplifier power efficiency benefits from an excellent gm/Id ratio. Cascode devices are used to boost the output impedance of the amplifier. The tail current offset is implemented only in the NMOS tail, as it can correct for the entire amplifier imbalance by itself.

Although the integrating amplifier has a tail current source, it does not provide a reliable common-mode (CM) control. Hence, two current sources, tied to the output and driven by a switched-cap CM-feedback loop, are used to regulate the output CM-level. These current sources are one-fourth the size of the input pair to limit their impact on total amplifier noise.

Measured Results

The amplifier was implemented within a 12-bit 3b/stage pipelined ADC which utilized split-ADC calibration architecture. This calibration technique allowed for fast convergence while operating in full background mode. With the help of gain and nonlinearity calibration, the ADC was able to achieve >10.3b ENOB at 280 MS/s, displaying a Schreier FoM of 164 dB with the amplifiers in all five pipeline stages consuming a total of only 400 uW.

Even though this topology exhibits excellent linearity, its power efficiency has not suffered, as can be seen in the comparison in Table 4 (row 8). It shows the same power efficiency as the "complementary dynamic amplifier" discussed in Sect. 8.2, but with improved linearity.

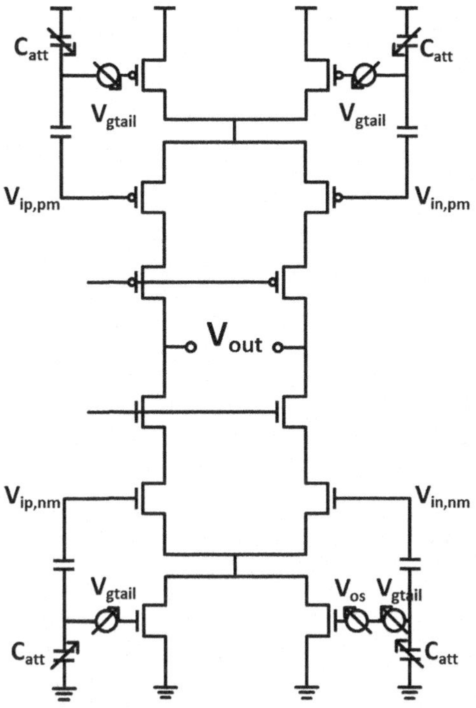

Fig. 30 Proposed integrating amplifier topology

9.3 *Capacitively Degenerated Dynamic Amplifier*

Circuit Description

In [28], a dynamic residue amplifier is presented for pipelined ADCs, as shown in Fig. 31. It employs a linearization technique based on capacitive degeneration to achieve an excellent linearity performance. The amplifier consists of push-pull NMOS and PMOS differential pairs. The C_L and C_{DEG} denote the load and degeneration capacitors, respectively. The amplifier operates in two phases: reset and amplification. During the reset phase, the C_{DEG} capacitors are pre-charged to the supply voltage, while the load capacitors C_L are reset to their common-mode voltage. In addition, the amplifier is switched off to save power by disconnecting the series switches at the NMOS and PMOS sources.

As the amplification phase begins, an input step is applied to the amplifier. The C_{DEG} capacitors are disconnected from the supply and connected to the amplifier in a cross-coupled way (Fig. 31b). During this phase, the C_{DEG} capacitors act as the degeneration capacitor as well as a local supply for the amplifier. The cross-coupled configuration of C_{DEG} capacitors has two benefits. First, it decreases the overall size of C_{DEG} capacitors by $4\times$ for the same amplifier gain. Second, the amplifier exhibits excellent common-mode rejection capability during amplification because only the load capacitors C_L are connected to supply or ground during amplification, ignoring parasitic capacitances. As a result, the output currents I_{LP} and I_{LN} have to

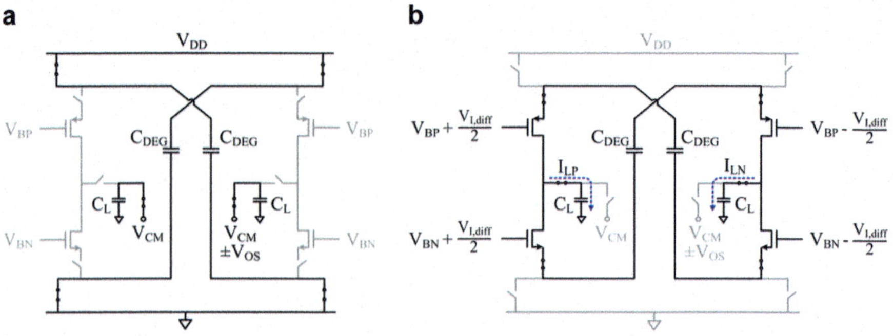

Fig. 31 Principle of operation of the capacitively degenerated dynamic amplifier. (**a**) Reset phase. (**b**) Amplification phase

be equal but opposite in sign ($I_{LP} = -I_{LN}$), allowing only the differential current to flow through the C_L capacitors and the common-mode current ($(I_{LP} + I_{LP})/2$) to be zero.

Linearization Principle
The linearization principle is based on the weak-inversion operation of MOSFETs and can be intuitively explained as follows.

During the amplification period, the amplifier's output voltage changes with time due to the input step but in an input amplitude dependent manner, indicating nonlinearity. Figure 32 illustrates the amplifier's large-signal gain against time for several values of the input step $V_{I,diff}$. At the beginning of amplification, the capacitor acts as a low impedance because of the step input. Since the I-V characteristics of MOSFET is exponential in the weak-inversion saturation region, the amplifier's gain increases with the input signal $V_{I,diff}$, exhibiting an expanding nonlinearity. However, as the amplification progresses, the impedance of the C_{DEG} capacitor gradually becomes higher. The high impedance then degenerates the amplifier more, eventually causing it to exhibit a compressing nonlinearity (i.e., its gain decreases with input). At the transition from this expanding to compressing behavior, the amplifier's gain becomes signal independent [28] given by $A_{opt} = C_{DEG}/2nC_L$, where n = weak-inversion slope factor. Therefore, the amplifier achieves an excellent linearity as can be seen by the THD plot in Fig. 32.

Measured Results
To ensure this linearity performance over PVT, the bias current I_B of the amplifier is made programmable. The proof-of-concept amplifier is fabricated in a 28 nm digital CMOS process and consumes 87 μA from a 1 V supply. Figure 33 shows the measured linearity versus the bias current IB at 43MS/s clock speed with $100mV_{pp,diff}$ input signal and ~4× gain. The amplifier's THD is limited by HD3 as expected with an optimum of −108 dB. Note that even with ±2.5% bias current variation, THD remains better than −80 dB, showing the wide linear range of the proposed amplifier. Figure 34 shows the measured output spectra corresponding

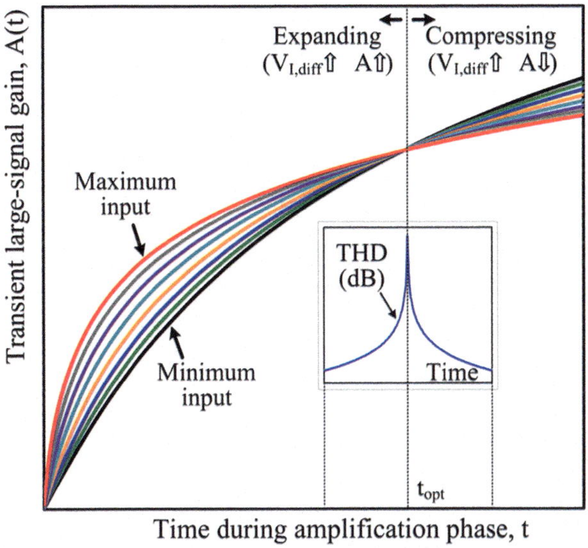

Fig. 32 Illustration of the capacitively degenerated linearization technique

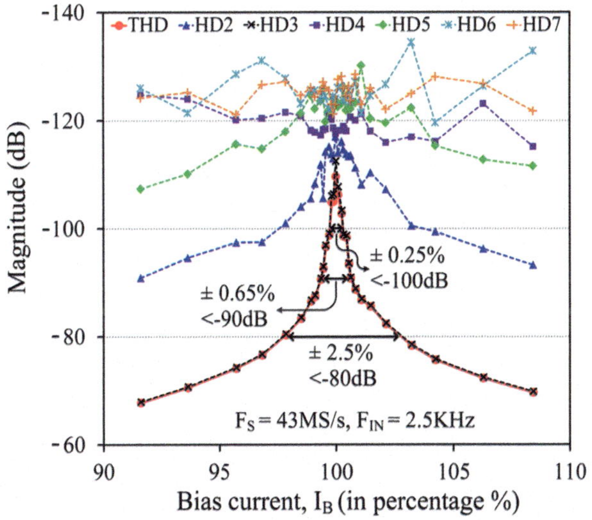

Fig. 33 Measured THD and harmonics as a function of bias current percentage

to the optimum linearity settings. The top graph is showing the results of a low frequency measurement ($F_{\text{in}} = 2.5$ kHz), whereas the bottom graphs shows the equivalent results for a near Nyquist tone ($F_{\text{in}} = 21.5$ MHz). In both cases the measured HD3 is better than -100 dB.

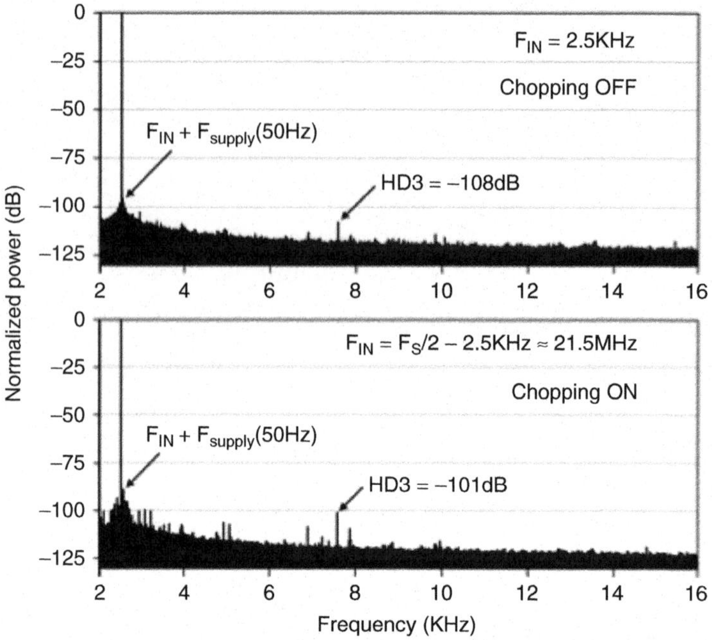

Fig. 34 Measured output spectra at 43MS/s for $100\text{mV}_{\text{pp,diff}}$ input and $400\text{mV}_{\text{pp,diff}}$ output

The results in Figs. 33 and 34 have been obtained in a measurement with an input swing of $100\text{mV}_{\text{pp,diff}}$ and an output swing of $400\text{mV}_{\text{pp,diff}}$. Doubling the signal amplitude to $200\text{mV}_{\text{pp,diff}}$ at the input and $800\text{mV}_{\text{pp,diff}}$ still shows an HD3 better than −86 dB, without any recalibration. Compared to the state-of-the-art dynamic amplifiers [10–15], the dynamic amplifier in [28] with capacitively degenerated linearization technique demonstrates at least 25 dB better linearity while simultaneously allowing over 2× larger output swing.

Although excellent linearity is obtained, the power efficiency is not sacrificed. On the contrary, because of the absence of cascodes or common-mode feedback devices, the voltage efficiency has increased significantly. This is shown in the overview of Table 4, row 9: an excellent number of $F_{\text{circuit}} = 0.25$ is obtained, approximately 50× lower than the reference folded-cascode amplifier.

9.4 Overview

Table 4 shows an overview of estimated values of the efficiency parameters. The expression for power efficiency (Eq. 30) is repeated here for convenience:

$$F_{\text{circuit}} = \text{NEF}\ N_{\tau}\left(V_{\text{gt}}/V_{\text{dd}}\right) / \left(\eta_{\text{cur}}\eta_{\text{vol}}{}^{2}\right)$$

Table 4 Comparison between residue amplifier topologies based on power efficiency

Amplifier technique	N_{eff}	NEF	$\eta_{cur}{}^{-1}$	$\eta_{vol}{}^{-2}$	$F_{circuit}{}^{a}$	[%]	[Ref.]	Comments
Fully settling folded cascode	7	2	4	2.8	75	100	[5]	Reference amplifier
Fully settling push-pull amp.	7	1	2	6.3	44	56	[6, 7]	
Resistive load amplifier	3	1.2	2	6.3	23	29	[30]	Dig. calibration required
Ring amplifier	3	2	3	1.6	14	18	[25, 26]	
Zero-crossing-based circuits	3	1.2	3	1.6	8.7	11	[20–23]	
Dynamic amplifier	2	1	2	3.2	6.2	10	[10–15]	Dig. calibration required
Complementary dyn. amp.	2	1	1	6.3	6.2	8.0	[24]	
Linearized compl. dyn. amp.	2	1	1	6.3	6.2	8.0	[27, 29]	
Capacitively deg. dyn. amp.	2	1	1	1.6	1.6	2.0	[28]	Dig. calibration required
Best possible	2	1	1	1	1	1.3		Ideal design

[a]Operation in weak inversion, at room temperature ($V_{gt} = 0.08$ V) and therefore $\eta_{gm} = 100\%$ was assumed

The numbers for $F_{circuit}$ in this table were calculated assuming a supply voltage of $V_{dd} = 1.0$ V, a weak-inversion operation at room temperature (such that $V_{gt} = 80$ mV), and a differential complementary topology. Under these circumstances, the best possible value for $F_{circuit}$ would be $F_{circuit} = 0.16$ (see bottom row). The various topologies listed in Table 4 are sorted on power efficiency, with descending (improving) $F_{circuit}$. Almost two orders of magnitude improvement have been achieved, reaching values close to the best possible values.

10 Conclusion

A comprehensive method for power estimation of residue amplifiers has been presented. Using this method a definition of power efficiency has been given, which subsequently has been used to analyze recently published, highly efficient residue amplifiers. Design parameters (NEF, N_τ, V_{gt}, η_{cur}, and η_{vol}) have been identified which have a key influence on the power efficiency, and design choices based on power efficiency are discussed. It is shown that the most power-efficient residue amplifier topologies share the same core circuit and differ primarily in how this core circuit is driven from the input. Finally, an overview is given of these topologies, ranked on power efficiency, and it is shown that the latest developments in residue amplifier design are getting close to what is ideally possible. It is also shown that high power efficiency can be combined with high linearity, which supports the notion that there is no direct link between linearity and power dissipation.

References

Classic High-Gain OpAmp with Feedback

1. Lewis SH, Gray PR. A pipelined 5MHz 9b ADC. Digest of technical papers, ISSCC; 1987.
2. Sutarja S, et al. A pipelined 13-bit, 250-ks/s, 5-V analog-to-digital converter. IEEE J Solid-State Circuits. 1988;23(6):1316–23.

Power Estimation Analog Circuits

3. Bult K. The effects of technology scaling on power dissipation of analog circuits. AACD; 2005.

Weak-Inversion Operation

4. Enz C, et al. Charge-based MOS transistor modeling: the EKV model for low-power and RF IC design. Chichester: Wiley; 2006.

Folded-Cascode OpAmp

5. Lee H-S, Gray PR. A self-calibrating 15 bit CMOS A/D converter. IEEE J Solid-State Circuits. 1984;19(6):813–9.

Push-Pull Residue Amplifiers

6. Brunsilius J. et al. A 16b 80MS/s 100mW 77.6dB SNR CMOS pipeline ADC. Digest of technical papers, ISSCC; 2011.
7. Kim J, et al. A 12-b, 30-MS/s, 2.95-mW pipel. ADC using single-stage class-AB amplifiers and deterministic background calibr. IEEE J Solid-State Circuits. 2012;47(9):2141–51.

Incomplete Settling

8. Iroaga E, Murmann B. A 12-bit 75-MS/s pipelined ADC using incomplete settling. IEEE J Solid-State Circuits. 2007;42(4):748–56.
9. Akter MS, et al. A 66 dB SNDR pipelined split-ADC using class-AB residue amplifier with analog gain correction. Conference proceedings ESSCIRC 2015; 2015.

Dynamic Amplifiers

10. Verbruggen B, et al. A 2.6mW 6b 2.2GS/s 4-times interleaved fully dynamic pipelined ADC in 40nm digital CMOS. Digest of technical papers, ISSCC 2010; 2010, p. 296–7.
11. Verbruggen B, et al. A 2.6 mW 6 bit 2.2 GS/s fully dynamic pipeline ADC in 40 nm digital CMOS. IEEE J Solid-State Circuits. 2010;45(10):2080–90.
12. Lin J, et al. A 15.5 dB, wide signal swing, dynamic amplifier using a common-mode voltage detection technique. Circuits and systems (ISCAS), 2011 IEEE international symposium, 15–18 May 2011, p. 21, 24.
13. Verbruggen B et al. A 1.7mW 11b 250MS/s 2∼ interleaved fully dynamic pipelined SAR ADC in 40nm digital CMOS. Digest of technical papers, ISSCC 2012.
14. Verbruggen B, et al. A 2.1 mW 11b 410 MS/s dynamic pipelined SAR ADC with background calibration in 28nm digital CMOS. Digest of technical papers, Symposium on VLSI circuits; 2013.
15. Lin J, et al. An ultra-low-voltage 160 MS/s 7 bit interpolated pipeline ADC using dynamic amplifiers. IEEE J Solid-State Circuits. 2015;50(6):1399–411.
16. Akter MS, Makinwa KAA, Bult K. A capacitively degenerated 100-dB linear 20–150 MS/s dynamic amplifier. IEEE J Solid-State Circuits. 2018 53(4):1115–1126.

Low Power Calibration

17. Sehgal R, et al. A 12b 53 mW 195 MS/s pipeline ADC with 82dB SFDR using split-ADC calibration. IEEE J Solid-State Circuits. 2015;50(7):1592–603.

Cascoded Dynamic Amplifier

18. Goes Fvd, et al. A 1.5mW 68dB SNDR 80MS/s 2x interleaved SAR-assisted pipelined ADC in 28nm CMOS. Digest of technical papers, ISSCC 2014; 2014.

Up/Down Integration

19. Malki B, et al. A complementary dynamic residue amplifier for a 67 dB SNDR 1.36 mW 170 MS/s pipelined SAR ADC. Conference proceedings ESSCIRC; 2014.

Zero-Crossing-Based Circuits

20. Brooks L, Lee H-S. A zero-crossing-based 8b 200MS/s pipelined ADC. Digest of technical papers, ISSCC; 2007.
21. Sepke T, et al. Noise analysis for comparator-based circuits. IEEE J Solid-State Circuits. 2009;56(3):541–53.
22. Chang D-Y, et al. A 21mW 15b 48MS/s zero-crossing pipeline ADC in 0.13μ m CMOS with 74dB SNDR. Digest of technical papers, ISSCC; 2014.
23. Shin S-K, et al. A 12 bit 200 MS/s zero-crossing-based pipelined ADC with early sub-ADC decision and output residue background calibration. IEEE J Solid-State Circuits. 2014;49:1366–82.

Complementary Dynamic Amplifier

24. Verbruggen B, et al. A 70 dB SNDR 200 MS/s 2.3 mW dynamic pipelined SAR ADC in 28nm digital CMOS. Digest of technical papers, Symposium on VLSI Circuits; 2014.

Ring Amplifiers

25. Hershberg B, et al. Ring amplifiers for switched-cap. circuits. Digest of technical papers, ISSCC; 2012.
26. Hershberg B, Moon U-K. A 75.9dB-SNDR 2.96mW 29fJ/conv-step ringamp-only pipelined ADC. Digest of technical papers, Symposium on VLSI circuits; 2013.

Linearization

27. Sehgal R, et al. A 13mW 64dB SNDR 280MS/s pipelined ADC using linearized open-loop class-AB amplifiers. Conference proceedings ESSCIRC 2017; 2017.
28. Akter S, et al. A capacitively-degenerated 100dB linear 20-150MS/s dynamic amplifier. Digest of technical papers, Symposium on VLSI circuits; 2017.
29. Sehgal R, et al. A 13mW 64dB SNDR 280MS/s pipelined ADC using linearized integrating amplifiers. IEEE J Solid-State Circuits. 2018;53(7): 1878–1888.

Open-Loop Amplifiers

30. Murmann B, Boser B. A 12-bit 75-MS/s pipelined ADC using open-loop residue amplification. IEEE J Solid-State Circuits. 2003;38(12):2040–50.

Energy-Efficient Inverter-Based Amplifiers

Youngcheol Chae

1 Introduction

Over the last few decades, important trends in the integrated circuit (IC) industry have followed Moore's "law". Accordingly, ICs have become cheaper, faster, and more efficient. The channel length of a transistor has scaled from 10μm in the 1970s to less than 10 nm today. Feature size scaling in CMOS has enabled digital systems to prevail. However, analog circuits face significant challenges, because operational amplifiers, which are the backbone of many analog circuits, become more difficult to design [1, 2].

Inverters are one of the simplest amplifiers that can be made in CMOS technology [3–5]. Despite their limited performance, CMOS inverters are attracting much attention due to their scalability as digital circuits and the fact that they can be readily designed in scaled CMOS [6–28]. They have also been rediscovered as dynamic amplifiers and are now essential building blocks of energy efficient analog circuits [4, 29, 30]. Many design techniques have been investigated to address performance limitations associated with the use of CMOS inverters [7–28]. As a result, many analog circuits have been implemented only with CMOS inverters, and this trend can be found in lots of cutting-edge results.

This chapter discusses the operation principles and the design of inverter-based amplifiers including various biasing techniques and circuit techniques to improve their performance. It also presents design examples of state-of-the-art inverter-based amplifiers.

This chapter is organized as follows. First, the use of a CMOS inverter as an amplifier is discussed (Sect. 2), and the necessary biasing techniques are presented

Y. Chae (✉)
Electrical and Electronic Engineering, Yonsei University, Seoul, South Korea
e-mail: ychae@yonsei.ac.kr

K. A. A. Makinwa et al. (eds.), *Low-Power Analog Techniques, Sensors for Mobile Devices, and Energy Efficient Amplifiers*,
https://doi.org/10.1007/978-3-319-97870-3_14

(Sect. 3). Advanced inverter-based amplifiers are then presented (Sect. 4). Finally, improved energy-efficient inverters are explained (Sect. 5), followed by a conclusion (Sect. 6).

2 CMOS Inverter as an Amplifier

2.1 Energy Efficiency (g_m/I)

The most commonly used figure-of-merit (FoM) to evaluate the energy efficiency of amplifiers [1] is given by

$$\mathrm{FoM} = \frac{\mathrm{GBW} \cdot C_{\mathrm{L}}}{I_{\mathrm{TOTAL}}} \approx \frac{g_{\mathrm{m}}}{I_{\mathrm{TOTAL}}} \tag{1}$$

where GBW is the gain-bandwidth product in Hz, C_{L} is the load capacitor in pF, and I_{TOTAL} is the supply current in mA. When the GBW is approximated to $g_{\mathrm{m}}/C_{\mathrm{L}}$, the FoM of amplifiers becomes $g_{\mathrm{m}}/I_{\mathrm{TOTAL}}$, which represents the efficiency in realizing a certain transconductance g_{m}. Therefore, energy efficient amplifiers will achieve a higher g_{m} power.

Using this FoM, Sansen [1] showed clear differences in the energy efficiency of different amplifier topologies, as shown in Fig. 1. As described in [1, 2], when a single MOS transistor (MOST) is biased at $V_{\mathrm{gs}} - V_{\mathrm{th}} = 0.2$ V, the FoM becomes about 1500. For the same bias voltage, a simple differential amplifier can achieve about half of this, i.e., FoM = 750. It should be noted that the FoM relationship among amplifiers would be the same as long as the same bias voltage is used for each amplifier, regardless of technology node. The FoM is also the same for a telescopic amplifier, at the expense of output swing. A folded cascode becomes the half of this again, i.e., FoM = 375. Mirrored OTA (mirror ratio $B = 3$) and two-stage Miller OTA ($C_{\mathrm{L}}/C_{\mathrm{C}} = 2.5$) achieve FoMs of 560 and 350, respectively. The reuse of the bias current allows the CMOS inverter to achieve a FoM of 3000, which is doubles the value for a single MOST. Therefore, CMOS inverters provide better energy efficiency than other amplifiers.

When assuming identical parameters between PMOS and NMOS, the $g_{\mathrm{m}}/I_{\mathrm{D}}$ value of CMOS inverter is defined according to the operating region as follows:

$$\text{Weak inversion}: \frac{g_{\mathrm{m}}}{I_{\mathrm{D}}} = \frac{2}{V_{\mathrm{t}}\,(1 + C_{\mathrm{d}}/C_{\mathrm{OX}})} \tag{2}$$

$$\text{Strong inversion}: \frac{g_{\mathrm{m}}}{I_{\mathrm{D}}} = \frac{4}{V_{\mathrm{gs}} - V_{\mathrm{th}}} \tag{3}$$

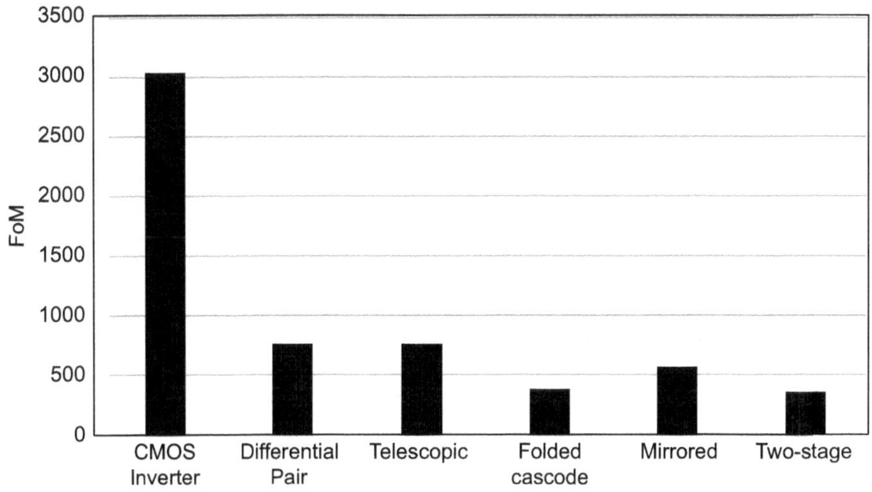

Fig. 1 FoM comparison among amplifiers: from [1]

where V_t is thermal voltage of the MOST and C_d and C_{ox} represent the depletion and the gate-oxide capacitance, respectively. As the inverter is biased deeper in weak inversion, the g_m/I_D is increased by about 3~5 times, depending on the technology. Since the operating speed is limited by the operating region of the MOST, the speed requirement determines the bias region of the inverter and typically low-frequency applications end up in weak-inversion operation.

2.2 *Slew Rate*

The FoM used in the previous section assumes the presence of small signals. The GBW of the amplifier is acceptable for small input signals, but due to the limited slew rate, the settling time may exceed the tolerance limit for large input signals. Therefore, the FoM for a large signal can be given as

$$\mathrm{FoM} = \frac{\mathrm{SR} \cdot C_L}{I_{\mathrm{TOTAL}}} \tag{4}$$

where the SR is the slew rate in V/μs. Therefore, achieving a higher SR with the same power is the goal of an energy-efficient design for large signals.

Krummenacher [5] pointed out the class-AB operation of the CMOS inverter. As shown in Fig. 2, a CMOS inverter features a high slew rate because its input voltage increases the dynamic current to the output load C_L. A CMOS inverter may be considered as a *dynamic amplifier* due to the fact that the bias current is not constant,

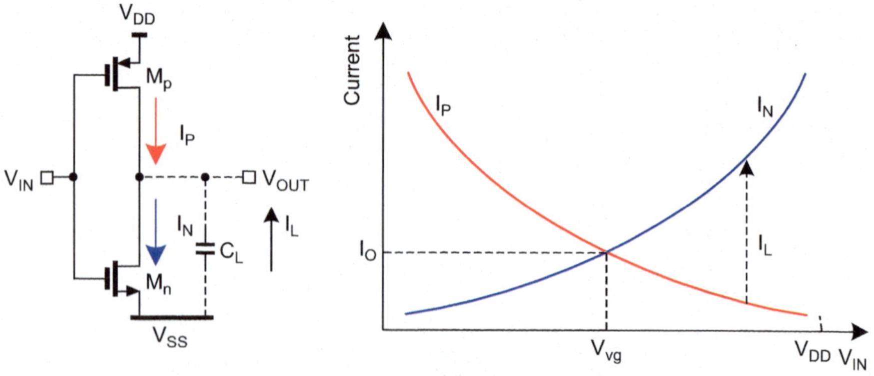

Fig. 2 Class-AB operation of CMOS inverter [5]

but varies during its amplification. This substantially reduces power consumption. In particular, by dynamically changing the operating range of a CMOS inverter between a weak inversion (in steady state) and a strong inversion (during transient), the energy efficiency FoM of a CMOS inverter can be maximized. The detailed operation will be explained in Sect. 3.

3 Biasing Techniques of Inverters

3.1 Signal-Biased Dynamic Inverters

Copeland [3] proposed dynamic amplifiers, whose bias current was not constant but changed during operation, especially when used in switched-capacitor (SC) circuits. Hosticka [4] also recognized CMOS inverters as dynamic amplifiers. Unlike opamps, inverters, however, do not provide virtual ground, simply because they have a single-input terminal. When negative feedback is applied, the input voltage remains close to the offset of the inverter. The amount of charge transferred to a feedback capacitor C_I depends on the inverter offset. Since the inverter has significant offset variation and drift, an offset cancellation technique is essential. Chae [6] proposed an inverter-based integrator, which removes the offset V_{off} and forms virtual ground V_g by using an offset cancellation capacitor C_C as shown in Fig. 3. In the auto-zeroing phase (ϕ_1), the inverter input is connected to the output and stores the V_{off} on the compensation capacitor C_C with the operating point. In the integration phase (ϕ_2), the sampled charge on C_S is transferred to the C_I rapidly due to the signal-dependent output current, allowing class-AB (or C) operation. The energy efficiency of the inverter can be maximized, especially when the supply voltage is selected to be lower than the sum of threshold voltages (i.e., $V_{DD} < V_{THN} + |V_{THP}|$) [5, 7, 8]. During the charge transfer, only one transistor in the

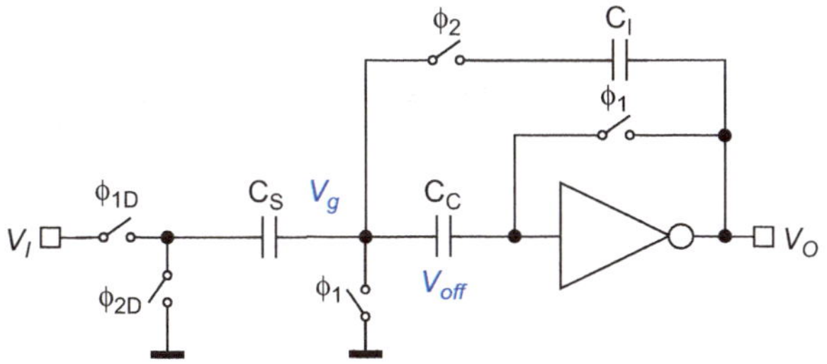

Fig. 3 Inverter-based SC integrator [6]

inverter is in the strong inversion region and thus provides high slew rate, and then both transistors of the inverter return to weak-inversion region in state steady, thus resulting in maximum g_m/I_D. This maximizes the energy efficiency of the inverter for both small and large signals.

The auto-zeroing process removes not only the offset of the inverter but also its low-frequency $1/f$ noise. This improvement is obtained at the cost of increased thermal noise due to noise folding. Therefore, amplifier noise becomes dominant in typical auto-zeroed integrators [31]. To maintain the energy efficiency, this noise penalty should be mitigated by decoupling the amplifier's noise density from its bandwidth. During the auto-zeroing phase, the effective noise bandwidth can be lowered by increasing the value of C_C (e.g., making it much larger than C_s) and thus reducing the amount of noise folding. It should be noted that this does not degrade the settling time during integration phase, because C_C does not load the amplifier due to its series connection. To achieve an improved accuracy, the inverter-based integrator can be refined into a pseudo-differential structure [6]. In this case, a passive CMFB is often adopted to avoid output common-mode drift due to mismatch and charge injection [8, 9, 12, 14–16]. The CMFB detects the signal difference between the detected common-mode voltage and signal ground and provides common charge via its virtual ground node [8, 10].

3.2 *Inverter with Dynamic Biasing*

Without using reference current, the dynamic inverter is relatively vulnerable to process, voltage, and temperature (PVT) variations that cause performance degradation and limit the use of the inverter-based amplifier. Krummenacher [5, 11] reduced the PVT sensitivity with dynamic bias technique as shown in Fig. 4a. During auto-zeroing phase (ϕ_1), the bias current I_O is imposed by a current mirror and it is stored in C_{C1}. It operates like a dynamic inverter for input signals with frequencies much higher than $1/RC_1$. At the same time, C_{C2} samples a bias voltage of a NMOS M_{n1}. The bias current of the inverter is decided by a current source, and the bias voltages of the two input transistors are stored in two separate capacitors. It is not possible to respond to fast changes in the DC component of the input signal, but it can be easily resolved by replacing the resistance R with a switch that is closed

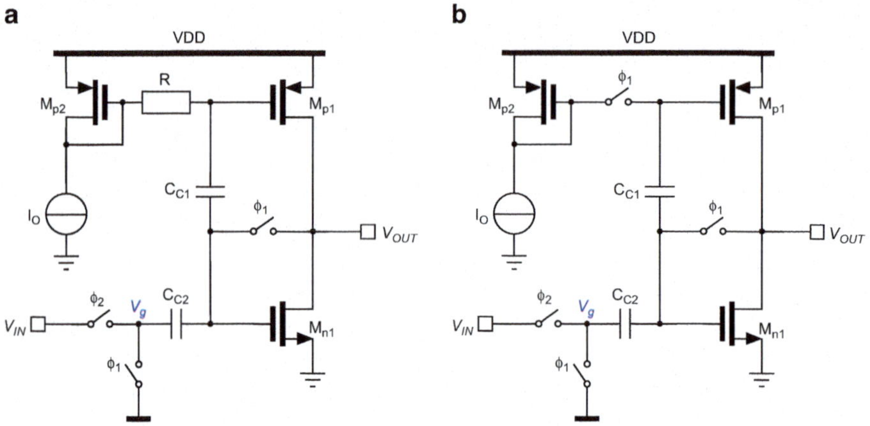

Fig. 4 CMOS inverter with dynamic biasing. (**a**) AC coupled. (**b**) Switched biasing [5]

during the auto-zeroing phase as shown in Fig. 4b. It has the benefit of a well-controlled bias current I_O in standby state compared to the signal-biased inverter, but it should be noted that the additional noise of the bias current is also added to the inverter input, thus suffering excess noise penalty. Michel [12] describes how a dynamic inverter can operate even from a 250 mV supply by using an extra-level shifter and achieves 61 dB DR in a 10 kHz bandwidth.

3.3 *Inverter with Advanced Dynamic Biasing*

The dynamic biasing scheme utilizes two capacitors and a current mirror, which introduces asymmetrical parasitic capacitors at the gates of the two input transistors, resulting in a linearity degradation. This issue can be resolved with advanced biasing techniques employing two balanced capacitors. Hosticka [4] proposed a dynamic biasing technique, which employs two balanced capacitors for the dynamic inverter, and Wang [13] added two current mirrors to reduce PVT sensitivity. Each switching cycle refreshes the bias capacitors C_C, which maintains the required voltage difference between the two nodes of the inverter input. At the input of the inverter, the parasitic capacitor C_P reduces the gain of the inverter by $C_C/(C_C + C_P)$. Therefore, the value of C_C should be increased properly to mitigate the gain degradation.

However, the mismatch between two current sources causes some errors of each operating points resulting in performance degradations. Chae [14] proposed a dynamic CMOS inverter with a floating current source as shown in Fig. 5. The dynamic bias inverter consists of two branches, the main branch and the sub-branch, which are alternatively connected to the input transistors (M_{P1} and M_{N1}). During sampling phase (Fig. 5a), the input transistors are connected to the sub-branch

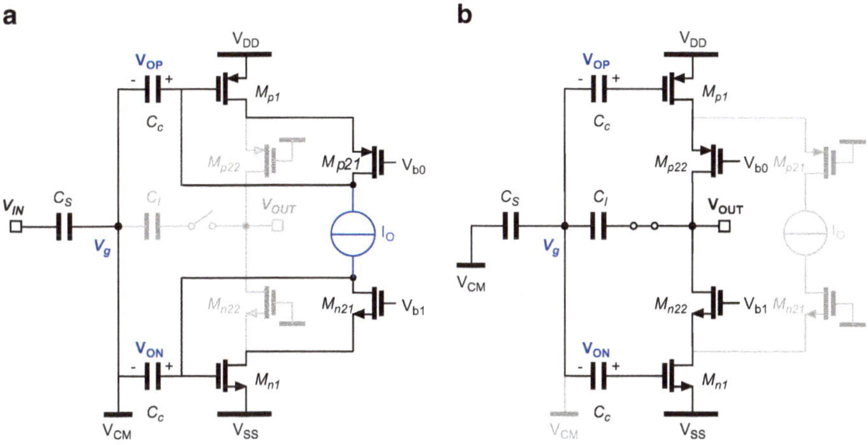

Fig. 5 Dynamic CMOS inverter with floating current source. (**a**) Sampling. (**b**) Integration [14]

with a floating current source I_O and define the operating points with two C_C. This ensures that both M_{P1} and M_{N1} are biased with exactly the same bias currents, and this cancels the inverter's offset and 1/*f* noise. During integration phase (Fig. 5b), the inverter is disconnected from the floating current source and reconfigures as a dynamic inverter with a well-defined bias current. The DC gain of an inverter is typically less than 40 dB and is not sufficient for high-precision circuits. To address this issue, the connection switches are configured as cascode transistors, instead of fully turning on with supply connections. As a result, the dynamic inverter can achieve a high DC gain (>80 dB) over PVT variations and realize a very high-resolution ADC (20 bits). In [14], a 20-bit incremental ADC was realized with the dynamic inverter and achieved 6 ppm INL and 1 μV offset in a conversion time of 40 ms, while consuming only 6.3 μW, thus resulting in the state-of-the-art Schreier FoM of 182.7 dB.

The dynamic switching technique of cascode transistors limits the operating speed. This was not a critical design constraint in [14] because its sampling frequency was 50 kHz, but such a constraint would be hardly applicable to some applications requiring higher sampling frequencies (>tens of MHz). The settling speed of the dynamic inverter is mainly limited by the parasitic gate-source capacitor C_{gs} of the cascode transistor, which should be periodically charged. Lee [15] suggested the use of active compensation capacitors C_{CP} at the cascode transistors to address the speed limitation as shown in Fig. 6a. First, the compensation capacitor C_{CP} is fully discharged. When the phase flips, the C_{CP} is connected to the bias circuit and the charge between C_{CP} and C_{gs} is distributed between each other and the voltage quickly goes near to the target voltage with respect to the ratio of C_{CP} and C_{gs}. Afterward, the rest of voltage to the target voltage (V_{b0} and V_{b1}) can be charged with small bias current. Instead of switching the floating current source, Gönen [16] suggested using the crossbar switch in parallel with the floating current

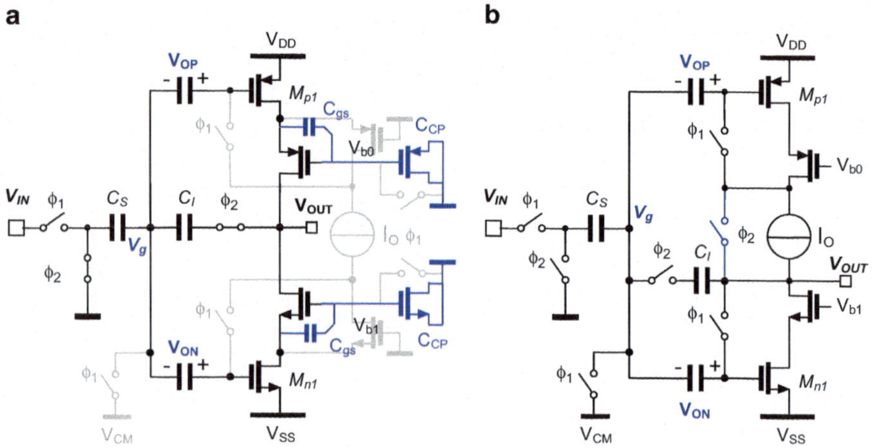

Fig. 6 Dynamic CMOS inverter with floating current source. (a) Active C_{gs} cancellation (integration) [15]. (b) Crossbar switch [16]

source as shown in Fig. 6b. During the integration phase (ϕ_2), the floating current source is simply bypassed by the crossbar switch. Since there is no C_{gs} switching of the cascode transistors in the biasing circuit, its power consumption can be minimized. As a result, this work significantly increases the sampling frequency of the dynamic inverter (>10 MHz). In [16], an 18-bit audio ADC was realized with the dynamic inverter and achieved 109 dB DR and 103 dB SNDR in a 20 kHz bandwidth, resulting in a state-of-the-art Schreier FoM of 181.5 dB.

3.4 Inverter with Adaptive LDO

To address the PVT vulnerability of dynamic inverters, Krummenacher [5] and Nauta [17] proposed the use of an adaptive low dropout regulator (LDO) as shown in Fig. 7. To control its bias current I_O and reduce the effect of supply variations, the V_{DD} of all inverters in the subsystem is maintained at the required bias point with a reference inverter biased with the Io. The V_{DD} is available from the reference voltage by shorting the input and output of the reference inverter. The entire circuit should be supplied at a higher voltage than V_{DD} to allow proper operation of the buffer amplifier.

Christen [18] applied the technique to a SC ΔΣ ADC for MEMS microphone. The LDO improves the power supply rejection ratio (PSRR>78 dB) of the inverter-based ΔΣ ADC, and it can also support a scalable bandwidth by adjusting the internal V_{DD} depending on the sampling frequency. Breems [19] exploited the technique into a CT ΔΣ ADC for wideband FM radio. The loop filter employed inverter-based amplifiers with a dedicated LDO and the ADC achieved the state-

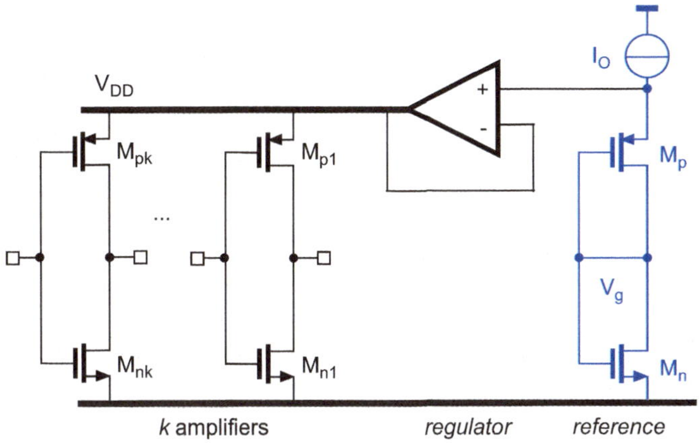

Fig. 7 Inverters with adaptive LDO [5]

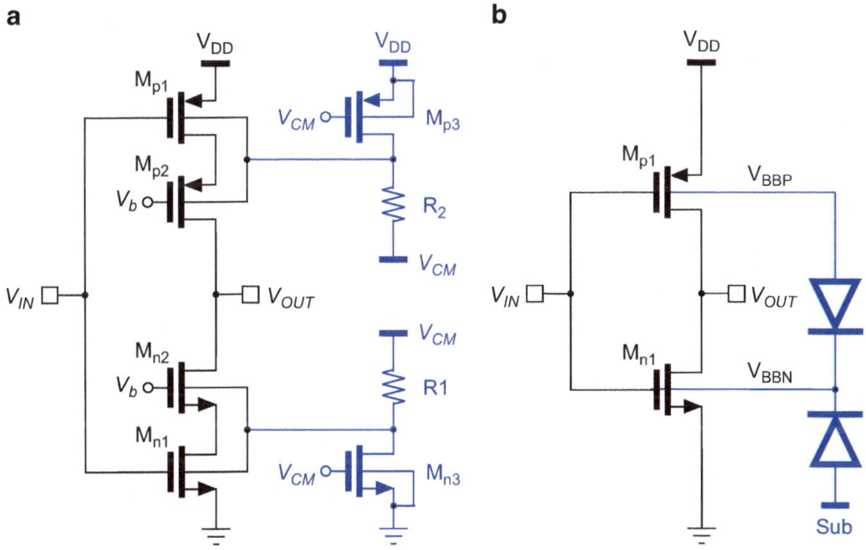

Fig. 8 Body biasing techniques (**a**) using resistors [20] (**b**) using diodes [21]

of-the-art THD of −102dBc and 77 dB SNDR in 25 MHz bandwidth over PVT variations.

3.5 Inverter with Body Biasing

As an alternative, Luo [20] proposed an adjustable body biasing of dynamic inverter as shown in Fig. 8a. Sensing transistors (M_{P3} and M_{N3}) biased in the weak-inversion region are connected to two resistors (R_1 and R_2), which provide a corner-dependent voltage. These voltages can be applied to the body terminals of the main inverter, which regulates the threshold voltage of each transistor and captures the PVT variations. In [20], an audio ADC was realized with class-C inverters and achieved 91 dB SNDR and 98 dB DR at a supply voltage of 0.8 V.

Lechevallier [21] also recognized that it is possible to omit the requirement for tunable supply in inverter-based amplifiers by exploiting body biasing in an ultra-thin buried oxide and body, fully depleted SOI (UTBB FD-SOI) CMOS technology. For 28 nm UTBB FD-SOI transistors, a forward body bias of 3 V is allowed, and this provides approximately threshold variation of 250 mV. As shown in Fig. 8b, the sum of the overdrive voltages of NMOS and PMOS can be kept constant over the supply range of about 250 mV, so that the supply voltage variation can be compensated. As a result, an inverter-based G_m–C filter can maintain the energy efficiency while accommodating supply variations of about 0.3 V without significant performance degradation.

4 Advanced Inverter-Based Amplifiers

With process scaling, the intrinsic gain of inverters continuously drops, making it difficult to design high-performance inverter-based amplifiers. To address this issue, cascoded inverters have been widely used at the expense of reduced output swing [14–16, 20]. This approach is not viable in nanometer CMOS due to the reduced supply voltage. Applying additional techniques allows inverters to maintain their gain and makes them truly scalable in nanometer CMOS.

4.1 Inverter with CDS

Nagaraj [32] describes a correlated double sampling (CDS) integrator (Fig. 9a). The capacitors C_S and C_I perform the integration, whereas C_{CDS} is an auxiliary capacitor used to compensate for the finite gain error and the offset voltage of the amplifier. The CDS technique can be applied to an inverter-based integrator as shown in Fig. 9a [18, 22]. In the sampling phase (ϕ_1), the inverter is switched to form negative feedback through the integration capacitor C_I. The resulting error voltage at the input of the inverter is stored in C_{CDS}. In the integration phase (ϕ_2), the integration is performed with the inverter inserted in series with the C_{CDS}. The gain of the inverter is also boosted recursively. In [22], CDS technique boosts the effective DC gain of the inverter from 35 to 50dB and allows an inverter-based $\Delta\Sigma$ ADC with CDS to achieve 12-bit resolution.

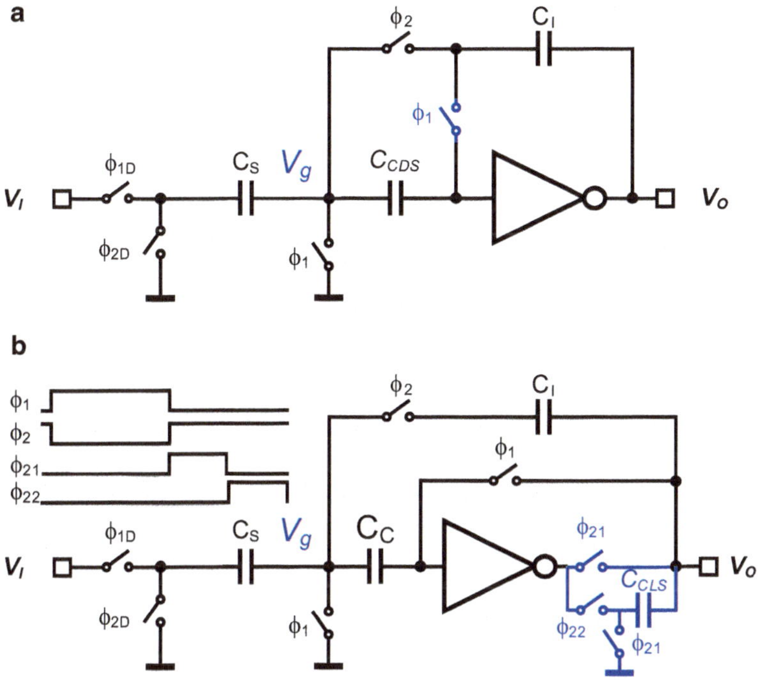

Fig. 9 Inverter-based integrators (**a**) using CDS [22] (**b**) using CLS [24]

4.2 *Inverter with CLS*

Gregoire [23] describes a correlated level shifting (CLS) technique. This estimates the final output voltage during a correlated sample phase and applies it in a second settling phase. The effective gain of the amplifier becomes the square of the open loop gain, because the amplifier is used twice. Zhang [24] applied the CLS technique to an inverter-based SC integrator as shown in Fig. 9b. The integration phase (ϕ_2) is divided into two steps: estimation (ϕ_{21}) and settling (ϕ_{22}). During the estimation phase, the top plate of C_{CLS} is connected to the integrator output while its bottom plate is connected to ground voltage, possibly bias point of the inverter. During the settling phase, the bottom plate of C_{CLS} is connected to the inverter's output, shifting the voltage towards virtual ground and reducing the required output voltage of the inverter. Consequently, the V_{g} shifts towards the target voltage V_{CM}. Therefore, the charge redistribution reduces the amount of residue charge on $\mathrm{C_S}$ and results in a complete charge transfer. In [24], CLS technique boosts the effective DC gain of the inverter by 37 dB (from 42 to 79 dB) and an inverter-based $\Delta\Sigma$ ADC with CLS achieves about 10 dB SQNR improvement compared the conventional one. One interesting point in [24] is that the integrator with CLS cancels offset and $1/f$ noise of the inverter, unlike conventional CLS, because it keeps both C_{C} and C_{CLS}.

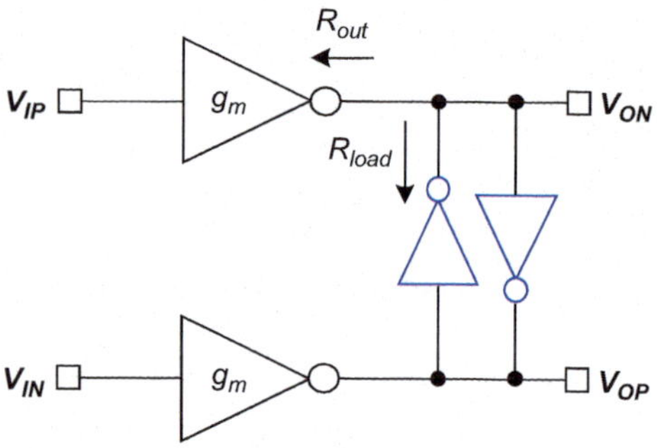

Fig. 10 Inverters with output negative resistor [17]

4.3 Inverter with Negative Resistor

The inclusion of negative resistance has been around for a long time. They can be used at the output of an amplifier to increase its DC gain. Nauta [17] suggested the use of negative resistor at the output of an inverter as shown in Fig. 10. The DC gain of the inverter is $-g_m \times R_{out}$, where R_{out} is the intrinsic output resistance of the inverter. The DC gain can be increased with a negative resistance R_{load} that compensates for R_{out}. The DC gain becomes $-g_m \times (R_{load}//R_{out})$. For $R_{load} = -R_{out}$, the effective gain of the inverter becomes infinite but may become unstable. As long as R_{out} is less than R_{load}, the DC gain of the inverter remains negative. It should be noted that the negative resistance does not sacrifice the GBW of the main inverter. Lee [9] also exploited a negative resistor for the inverter-based SC amplifier to boost the gain of the amplifier by 21 dB from 31 to 52dB.

4.4 Ring Amplifiers

Hershberg [25] offered a new type of amplifier called a *ring amplifier*. The ring amplifier is essentially a cascade of three inverters, but an offset, so-called *dead zone*, is built into the second inverter such that the last inverter is turned off for certain conditions as shown in Fig. 11a. The control of the dead zone allows the cascade of inverters to operate like an amplifier. A ring amplifier can be considered as a dynamic inverter that regulates the settling dynamics through the inherent capacitive feedback of SC circuits. It has the benefit of slew-based charging and also has a near rail-to-rail output swing, because the last stage is a simple inverter that operates in weak-inversion region in steady state. In addition, the required gain can be easily achieved from three inverter stages. There are many ADC papers implemented with ring amplifiers. In particular, Lim [26] proposed a modified ring amplifier as shown in Fig. 11b, where the dead zone is controlled by a resistor instead of a voltage across a capacitor improving the PVT tolerance of the dead zone. To improve the limited CMRR and PSRR in conventional ring amplifiers, a

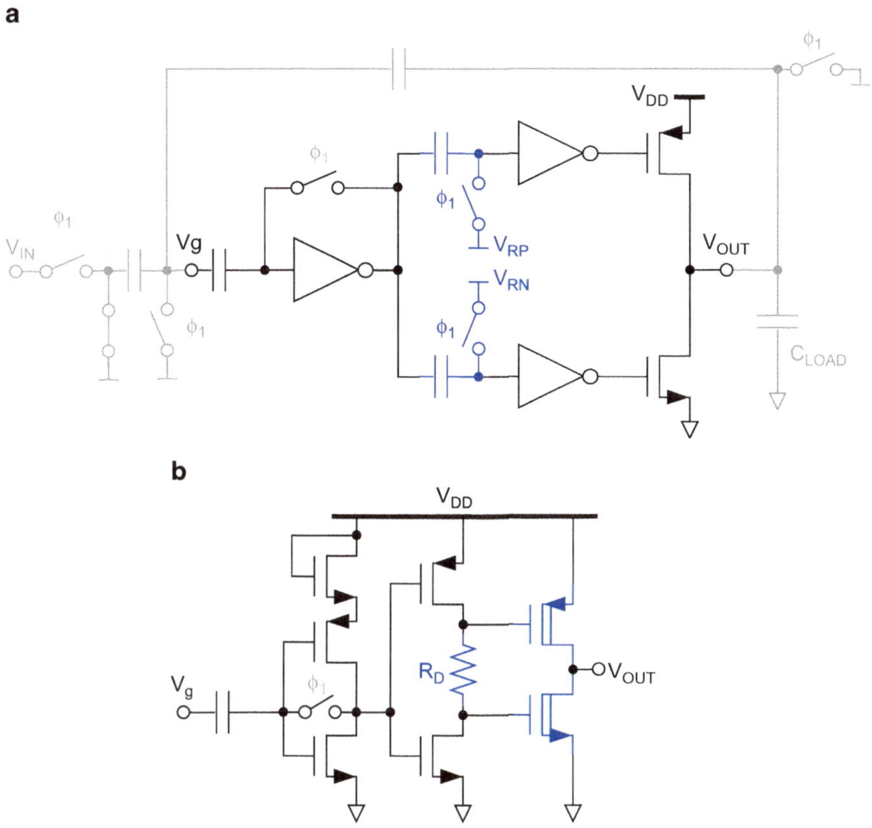

Fig. 11 (**a**) Ring amplifier [25]. (**b**) Modified ring amplifier (amplifier only) [26]

fully differential ring amplifier that replaced the first inverter with a differential pair (class-A) was proposed [27]. One interesting point in [27] is that the addition of a switched load capacitor to the output of the first inverter only at auto-zeroing phase reduces the noise folding of the ring amplifier by about 89%.

4.5 Inverters with Digital Calibration

Digital calibration has been around for a long time to improve linearity, and there are many papers featuring digitally-assisted analog circuits. Nonlinear errors due to finite gain of the inverter also can be mitigated with the help of a deterministic background calibration. Kim [28] proposed the use of three third-order polynomial functions (splines) to calibrate the transfer function of the dynamic inverter. This allows the digital linearization of the inverter that mitigates the distortion up to seventh order with low computational overhead, and the calibration coefficients for dynamic inverters converge within only 0.73 ms. A 12b 30MS/s pipelined ADC using dynamic inverters and digital calibration [28] achieved a SNDR of 64.5 dB, while consuming only 2.95 mW.

5 Improved Energy-Efficient Inverters

CMOS inverters have evolved into other types of amplifier using stacking, g_m-style, and charge-steering to address challenges of nanometer CMOS and maximize energy efficiency. These concepts have been around for a long time, but recently have been revived as attractive techniques that benefit from advances in CMOS technology [29, 30, 33–36].

5.1 *Stacked Inverters*

The stacked topology is used to reduce power consumption [33, 37]. It allows the same DC supply to be shared for multiple AC-coupled signals, via capacitors with isolating DC bias, while reusing the bias current. This idea can be applied to the inverter to further improve energy efficiency. Shen [33] proposed vertically stacking N inverters capable of achieving 2N times current reuse for a single-input signal as shown in Fig. 12. As in [11], a replica biasing ensures that the stacked inverter robust against PVT variations. They recognized that the tail current sources between stacked inverters can be eliminated, but the stacked inverters still preserve CMRR and PSRR. The stacked inverter has some limitations such as supply voltage limit and signal swing reduction, but there are many potential applications. The stacked inverter is suitable for closed-loop capacitive feedback and the AC-coupling to the multiple amplifier input nodes can be realized by splitting the input and feedback capacitors into multiple paths. A bio-amplifier stacking 3-inverters [33] achieved the state-of-the-art NEF of 1.07 under the supply voltage of 1 V.

5.2 *Other Dynamic Inverters*

An effective way to maximize energy efficiency is to remove the unnecessary static current. Lin [34] proposed the use of a dynamic integrator as a gain amplifier. The common-mode voltage detection circuit provides the timing required to terminate the load capacitors from discharging thus allowing the dynamic amplifier to have a steady output voltage. The decision error can be calibrated with background calibration, and dynamic amplifiers have been used to realize the accuracy required in interstage gain for pipeline ADCs [29, 30, 35, 36]. Verbruggen [29] proposed

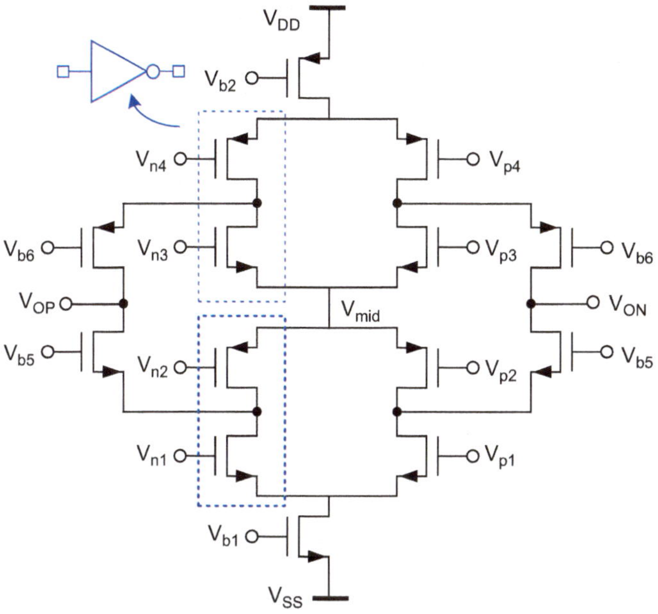

Fig. 12 Stacked inverter: staking 2-inverters [33]

a dynamic inverter-based amplifier as shown in Fig. 13. In the amplification (ϕ_2), an output current difference is integrated in the output capacitor. After a certain time, the switch turns off and it freezes the output voltage. Since the amplifier does not rely on closed-loop settling, it can be designed for a high-speed ADC. A 12b 200MS/s dynamic pipelined SAR ADC [29], implemented in 28 nm CMOS, achieved a 70 dB SNDR, while consuming only 2.3 mW from a 0.9 V supply, resulting in the state-of-the-art FoM of 177.4 dB.

Hosticka [4] proposed a dynamic CMOS amplifier, which consists of a CMOS differential stage whose tail current has been replaced by a capacitor C_O with two switches. The amplifier can be loaded capacitively only at the "integration" phase. This concept evolved into charge steering technique in a close-loop configuration [36]. Akter [30] recognized that the charge steering is still effective in an open-loop configuration and proposed a dynamic inverter as shown in Fig. 14, whose linearity is significantly improved with capacitive degeneration. It employs a cross-coupled capacitor configuration that results in reduced capacitor size and improved CMRR. To compensate for the effects of PVT variations, the dynamic inverter can employ a simple background calibration. A dynamic inverter with capacitive degeneration [30], implemented in 28 nm CMOS, achieved the state-of-the-art THD of −100 dB at a clock speed of 43MS/s, while consuming only 87 μW.

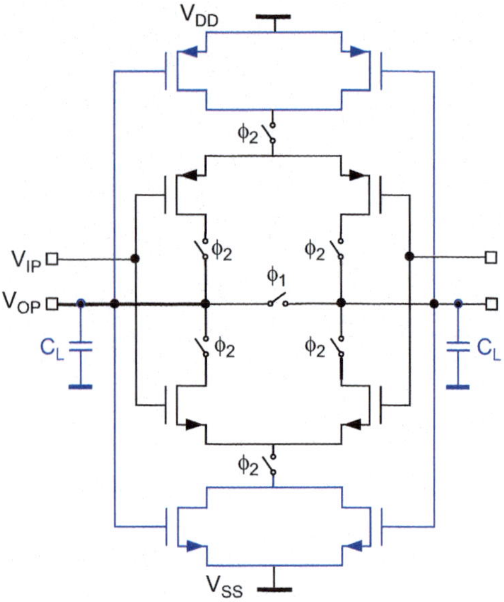

Fig. 13 Dynamic integrator with common-mode control [29]

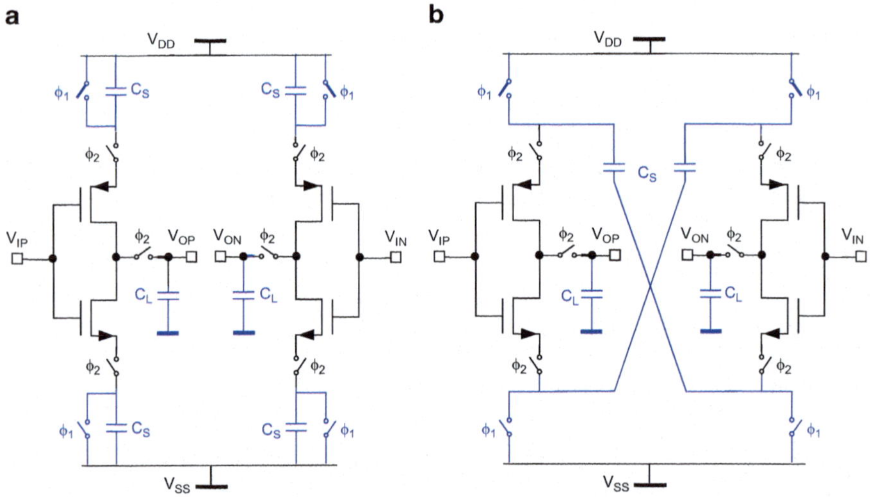

Fig. 14 Dynamic inverter with capacitive degeneration. (**a**) Pseudo differential. (**b**) Fully differential [30]

6 Conclusion

This chapter discusses the operation principle and the design of inverter-based amplifiers including various biasing techniques and circuit techniques to reduce performance degradation due to the use of CMOS inverters. It also describes cutting-edge results in terms of energy efficiency, resolution, speed, and linearity. This result shows that CMOS inverters are gradually replacing traditional opamps and OTAs in many analog circuits and is dominant in nanometer CMOS.

References

1. Sansen WMC. Low-noise energy-efficient amplifier design ISSCC Forum: advanced IC design for ultra low-noise sensing, 2016.
2. Sansen WMC. Opamps, Efficient Sensor Interfaces, Advanced Amplifiers and Low Power RF Systems, Opamps, Gm-Blocks or Inverters?. AACD 2015, Springer; 2016.
3. Copeland MA, Rabaey JM. Dynamic amplifier for MOS technology. Electron Lett. 1979;15(10):301–2.
4. Hosticka BJ. Dynamic CMOS amplifiers. IEEE J Solid State Circuits. 1980;SC-15(5):887–94.
5. Krummenacher F, Vittoz E. Class-AB CMOS amplifier for micropower SC filters. Electron Lett. 1981;17(13):433–4.
6. Chae Y, Han G. Low voltage, low power inverter-based switched-capacitor delta-sigma modulator. IEEE J SolidState Circuits. 2009;24(2):458–72.
7. Chae Y, Han G. A low power sigma-delta modulator using class-C inverter. Symposium on VLSI Circuits, June 2007, p. 240–1.
8. Chae Y, Lee I, Han G. A 0.7-V 36-μW 85 dB-DR audio ΔΣ modulator using class-C inverter. ISSCC, Feb 2008, p. 490–1.
9. Lee I, Han G, Chae Y. A 2mW, 50dB DR, 10MHz BW 5× interleaved bandpass delta-sigma modulator at 50 MHz IF. IEEE Trans Circuits Syst I. 2015;62(1):80–9.
10. van Veldhoven RHM, Rutten R, Breems LJ. An inverter-based hybrid ΣΔ modulator. ISSCC, Feb 2008, p. 492–3.
11. Krummenacher F. Micropower switched capacitor biquadratic cell. IEEE J Solid State Circuits. 1981;SC-17(3):507–12.
12. Michel F, Steyaert MSJ. A 250 mV 7.5 μW 61 dB SNDR SC ΔΣ modulator using near-threshold-voltage-biased inverter amplifiers in 130 nm CMOS. IEEE J Solid State Circuits. 2012;47(3):709–21.
13. Wang J, Matsuoka T, Taniguchi K. A 0.5 V feedforward delta-sigma modulator with inverter-based integrator. In: Proc. ESSCIRC, Sept 2009, p. 328–31.
14. Chae Y, Souri K, Makinwa KAA. A 6.3 μW 20 bit incremental zoom-ADC with 6ppm INL and 1μV offset. IEEE J Solid State Circuits. 2013;48(12):3019–27.
15. Lee S, Jo W, Song S, Chae Y. A 300-μW audio ΔΣ modulator with 100.5-dB DR using dynamic bias inverter. IEEE Trans Circuits Syst I. 2016;63(11):1866–75.
16. Gonen B, Sebastiano F, Quan R, van Veldhoven R, Makinwa KAA. A dynamic zoom ADC with 109-dB DR for audio applications. IEEE J SolidState Circuits. 2017;52(6):1542–50.
17. Nauta B. A CMOS transconductance-C filter technique for very high frequencies. IEEE J Solid State Circuits. 1992;27:142–53.
18. Christen T. A 15-bit 140-μW scalable-bandwidth inverter-based ΔΣ modulator for a MEMS microphone with digital output. IEEE J Solid State Circuits. 2013;48(7):1605–14.
19. Breems, L. et al. A 2.2 GHz continuous-time ΔΣ ADC with −102 dBc THD and 25 MHz bandwidth. IEEE J Solid State Circuits. 2016;51(12):2906–16.
20. Luo H, Han Y, Cheung RC, Liu X, Cao T. A 0.8-V 230-μW 98-dB DR inverter-based ΣΔ modulator for audio applications. IEEE J Solid State Circuits. 2013;48(10):2430–41.
21. Lechevallier J, Struiksma R, Sherry H, Cathelin A, Klumperink E, Nauta B. A forward-body-bias tuned 450MHz Gm-C 3rd-order low-pass filter in 28nm UTBB FD-SOI with >1dBVp IIP3 over a 0.7-to-1V supply. ISSCC, 2015, p. 96–7.

22. Chae Y, Cheon J, Lim S, Kwon M, Yoo K, Jung W, Lee DH, Ham S, Han G. A 2.1 M Pixels, 120 frame/s CMOS image sensor with column-parallel ΔΣ ADC architecture. IEEE J Solid State Circuits. 2011;46(1):236–47.
23. Gregoire B, Moon U-K. An over-60 dB true rail-to-rail performance using correlated level shifting and an opamp with only 30 dB loop gain. IEEE J Solid State Circuits. 2008;43(12):2620–30.
24. Zhang H, Tan Z, Nguyen K. Inverter-based low-power delta–sigma modulator using correlated level shifting technique. Electron Lett. 2017;53(25):1163–4.
25. Hershberg B, Weaver S, Sobue K, Takeuchi S, Hamashita K, Moon Ring U-K. Amplifiers for switched capacitor circuits. IEEE J Solid-State Circuits. 2012;47(12):2928–42.
26. Lim Y, Flynn MP. A 100 MS/s, 10.5 bit, 2.46 mW comparator-less pipeline ADC using self-biased ring amplifiers. IEEE J Solid State Circuits. 2015;50(10):2331–41.
27. Lim Y, Flynn MP. A calibration-free 2.3 mW 73.2 dB SNDR 15b 100 MS/s four-stage fully differential ring amplifier based SAR-assisted pipeline ADC. Symposium on VLSI Circuits, 2017, p. C98–9.
28. Kim JK-R, Murmann B. A 12-b, 30-MS/s, 2.95-mW pipelined ADC using single-stage class-AB amplifiers and deterministic background calibration. IEEE J Solid State Circuits. 2012;47(9):2141–51.
29. Verbruggen B, Deguchi K, Malki B, Craninckx J. A 70 dB SNDR 200 MS/s 2.3 mW dynamic pipelined SAR ADC in 28nm digital CMOS. Symposium on VLSI Circuits, 2014.
30. Akter MS, Makinwa KAA, Bult K. A capacitively degenerated 100-dB linear 20–150 MS/s dynamic amplifier. IEEE J Solid State Circuits. 2018;53:1115–26.
31. Gregorian R. High-resolution switched-capacitor D/A converter. J Microelectron. 1981;12:10–13.
32. Nagaraj K, Vlach J, Viswanathan TR, Singhal K. Switched-capacitor integrator with reduced sensitivity to amplifier gain. Electron Lett. 1986;22(21):1103–5.
33. Shen L, Lu N, Sun N. A 1-V 0.25-μW inverter stacking amplifier with 1.07 noise efficiency factor. IEEE J Solid State Circuits. 2018;53(3):896–905.
34. Lin J, Miyahara M, Matsuzawa A. A 15.5 dB, wide signal swing, dynamic amplifier using a common-mode voltage detection technique. In: Proc. IEEE Int. Symp. Circuits Syst, 2011, p. 21–4.
35. van der Goes F, Ward CM, Astgimath S, Yan H, Riley J, Zeng Z, Mulder J, Wang S, Bult K. A 1.5 mW 68 dB SNDR 80 Ms/s 2 interleaved pipelined SAR ADC in 28 nm CMOS. IEEE J Solid State Circuits. 2014;49(12):2835–45.
36. Chiang S-H, Sun H, Razavi B. A 10-Bit 800-MHz 19-mW CMOS ADC. IEEE J Solid State Circuits. 2014;49(4):935–49.
37. Iguchi S, Sakurai T, Takamiya M. A low-power CMOS crystal oscillator using a stacked-amplifier architecture. IEEE J Solid State Circuits. 2017;52(11):3006–17.

Balancing Efficiency, EMI, and Application Cost in Class-D Audio Amplifiers

Marco Berkhout

1 Introduction

Over the last decade, the maturity of class-D audio amplifiers has increased significantly. The audio performance of modern class-D audio amplifiers has become comparable or better than their class-AB counterparts, and class-D amplifiers have consistently gained ground over class-AB amplifiers in most audio application areas. The driving force behind this development and the key feature of class-D amplifiers is high efficiency. This extends battery life in portable applications and uses up less of the thermal budget in applications where high-output power is required. Unfortunately, the coveted high efficiency comes at the price of electromagnetic interference (EMI) which is an inseparable consequence of the fast voltage and current transients inherent to class-D operation. This results in a complicated trade-off between efficiency, EMI, and the cost of the complete amplifier application including external components. This trade-off revolves around the class-D output stage and modulation scheme. In fact, audio performance of the amplifier, traditionally expressed in terms such as total harmonic distortion (THD) and noise, is of secondary importance. It is basically a matter of organizing sufficient loop gain around the class-D output stage.

M. Berkhout (✉)
NXP Semiconductors, Nijmegen, The Netherlands
e-mail: marco.berkhout@nxp.com

K. A. A. Makinwa et al. (eds.), *Low-Power Analog Techniques, Sensors for Mobile Devices, and Energy Efficient Amplifiers*,
https://doi.org/10.1007/978-3-319-97870-3_15

2 Class-D 101

A class-D amplifier in its simplest form consists of two low-ohmic switches that alternately connect the output node to either a positive or negative supply rail as shown in Fig. 1a. This configuration is called single-ended (SE) since only one side of the loudspeaker is driven. In practice, the SE configuration is not used very often because it requires a dual symmetrical supply or a rather large decoupling capacitor in series with the load. The most commonly used class-D amplifier configuration is the bridge-tied-load (BTL) as shown in Fig. 1b. In the BTL configuration, both sides of the loudspeaker load are driven in opposite (audio) phase. This enables operation from a single supply and the balanced operation cancels out even order distortion.

Efficiency of class-D amplifiers is quantified as the power delivered to the loudspeaker load divided by the power drawn from the supply and depends on the signal level. At zero signal, efficiency is zero by definition and becomes a meaningless metric. Therefore, efficiency of class-D amplifiers is in general specified as maximum efficiency η_{MAX} at full output power, whereas quiescent dissipation (or current) P_Q is specified at zero output. With these two values, it is possible to produce a reliable efficiency versus output power curve.

2.1 Modulation

A class-D output stage can only produce square wave pulses. A signal can be encoded into a pulse train by modulating the transitions. Well-known modulation schemes are pulse-density modulation (PDM), pulse-frequency modulation (PFM), and pulse-width modulation (PWM). Most class-D amplifiers use PWM to drive the output stage.

The textbook method for generating a PWM signal is to compare the input signal with a triangular reference signal as shown in Fig. 2a. PWM has some attractive

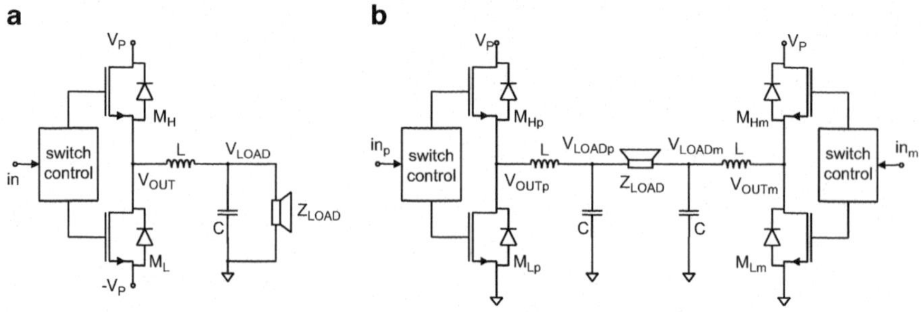

Fig. 1 Class-D amplifiers. (**a**) Single-ended (SE) configuration. (**b**) Bridge-tied-load (BTL) configuration

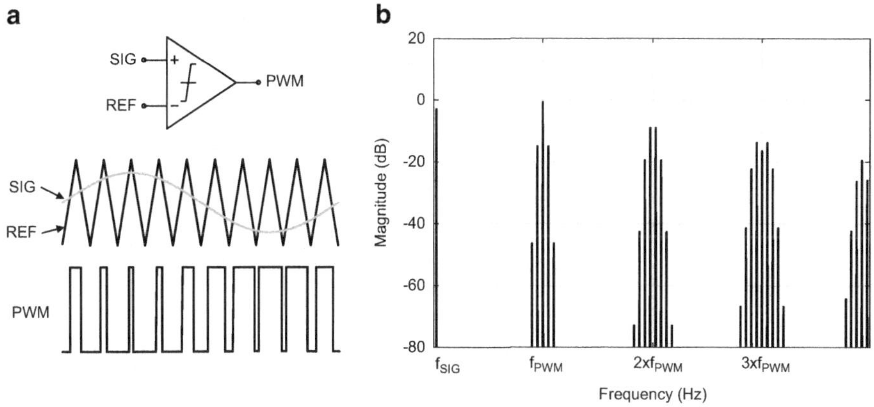

Fig. 2 Pulse width modulation (PWM). (**a**) Natural sampling. (**b**) Frequency spectrum

features that make it the most popular modulation scheme. First, the frequency spectrum of a PWM signal, as shown in Fig. 2b, contains the modulating signal, harmonics of the PWM carrier, intermodulation products of signal, and carrier but no harmonics of the signal [1]. This means that PWM is inherently linear when only signal frequencies are considered. The signal can be recovered undistorted by low pass filtering of the PWM signal. A second advantageous feature of PWM is that it has a constant transition rate that can be chosen relatively low without intermodulation products folding back into the signal bandwidth. Typically, the PWM carrier switches in the 300 kHz–500 kHz range. Having a constant and therefore predictable switching frequency is a useful control knob to mitigate EMI problems. Finally, it is rather straightforward to make a feedback loop around a PWM modulator that is stable over the full output signal range [2]. This is especially true when the implementation of a class-D amplifier is in the analog domain.

PWM can also be performed in the digital domain. In this case, the triangular carrier becomes a staircase, and the pulse-widths become quantized with a resolution that depends on the system clock. The quantization of the pulse-widths results in quantization noise which can be shaped out of band with a $\Sigma\Delta$ modulator as illustrated in Fig. 3.

Here a triangular reference at frequency f_{PWM} is fed through a sample-and-hold circuit that is clocked at a much higher frequency f_{clk} which results in staircase reference that is fed into the $\Sigma\Delta$ loop just before the quantizer.

The resolution of the pulse-widths is determined by the ratio between the clock and PWM frequencies $f_{\mathrm{clk}}/f_{\mathrm{PWM}}$. A higher ratio gives less quantization noise. When the ratio is $f_{\mathrm{clk}}/f_{\mathrm{PWM}}$ equals 2 then the (analog) triangular reference is sampled in the zero crossings only so the staircase signal becomes zero. In that case, the PWM pulse width can only be 0% or 100% of the PWM period time and effectively becomes pulse-density modulation (PDM). As such, PWM can be regarded as an extension of PDM. Figure 4 shows simulated frequency spectra of a third-order PWM $\Sigma\Delta$ Modulator in PWM (black) and PDM (gray) mode. In this example, the

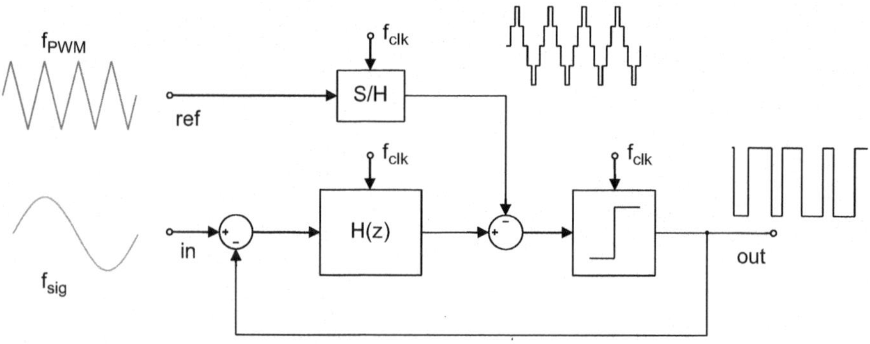

Fig. 3 PWM ΣΔ modulator

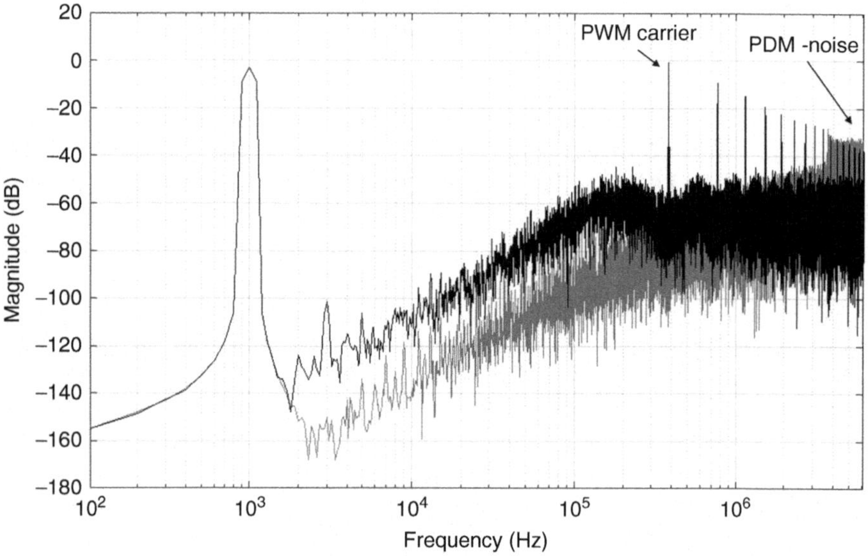

Fig. 4 Frequency spectrum of a third-order ΣΔ modulator in PWM and PDM mode

clock frequency $f_{clk} = 512*48$ kHz $= 24.576$ MHz. For the PWM mode, the PWM frequency is $8*48$ kHz $= 384$ kHz which gives a PWM resolution of 64 (6-bits).

Since both PWM and PDM signals are 2-level signals, the integral energy in both spectra is the same. However, in the PWM spectrum the energy is concentrated in distinct tones at the PWM frequency and harmonics, whereas in the PDM spectrum the energy is distributed at higher frequencies and is free of tones, which is potentially beneficial for EMI. This spectral shape comes at the price of much higher switching activity. Whereas the PWM switches at a moderate and constant 384 kHz, the PDM signal switches around 6.7 MHz. Although there have been some developments of class-D amplifiers based on PDM [3], most class-D amplifiers stick with PWM.

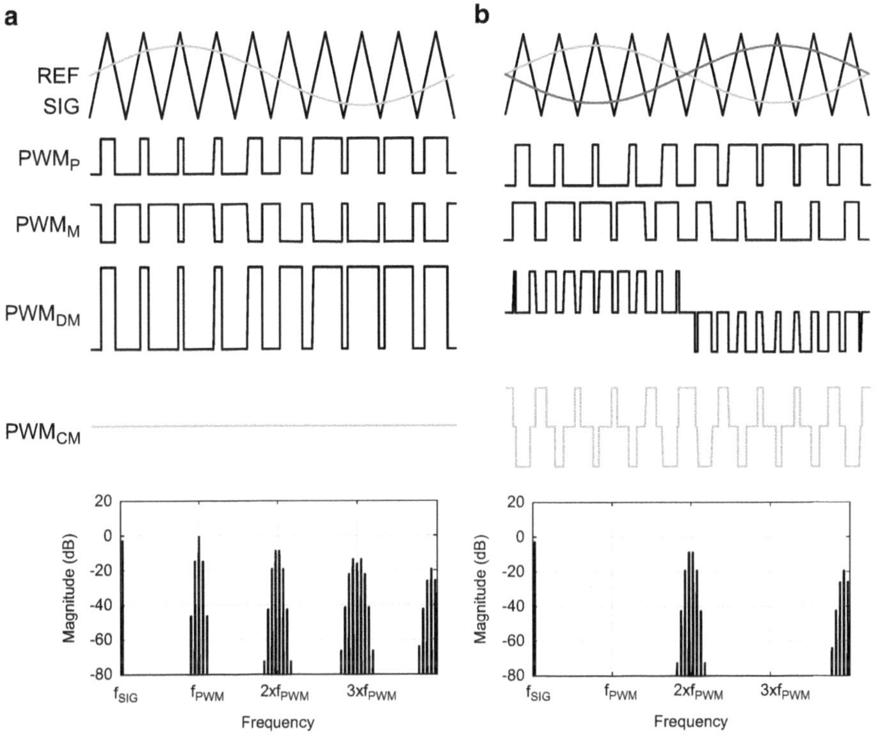

Fig. 5 (**a**) AD-PWM. (**b**) BD-PWM

In the BTL configuration, the phase of the PWM carriers can be chosen independently for each half-bridge. If the carriers are in opposite phase this is called AD-PWM. In AD-PWM, the signals PWM_p and PWM_m that drive the half-bridges are exact opposites as shown in Fig. 5a. The resulting differential-mode PWM_{DM} is identical to single-ended PWM with twice the amplitude. Because in AD-PWM the system is perfectly balanced, the common-mode PWM_{CM} is zero which is often considered a big advantage for EMI.

If, on the other hand, the carriers are in-phase this is called BD-PWM. In BD-PWM, the signals PWM_p and PWM_m are identical when the modulating signal is zero. In this case, the differential-mode PWM_{DM} is also zero. If the modulating signal is nonzero, the differential-mode PWM_{DM} becomes a 3-level signal PWM with twice the transition rate of the reference as shown in Fig. 5b. In the frequency spectrum of a BD-PWM signal, the spectral components around the odd harmonics of the PWM carrier cancel each other, relaxing the requirements on the LC filter. In class-D amplifiers for mobile application such as smartphones, the LC filter is omitted altogether. In such so-called "filterless" applications, the inductance of the loudspeaker load serves as first-order lowpass filter. The drawback of BD-PWM is that it has a large common-mode PWM_{CM} which does have spectral components around f_{PWM} and odd harmonics.

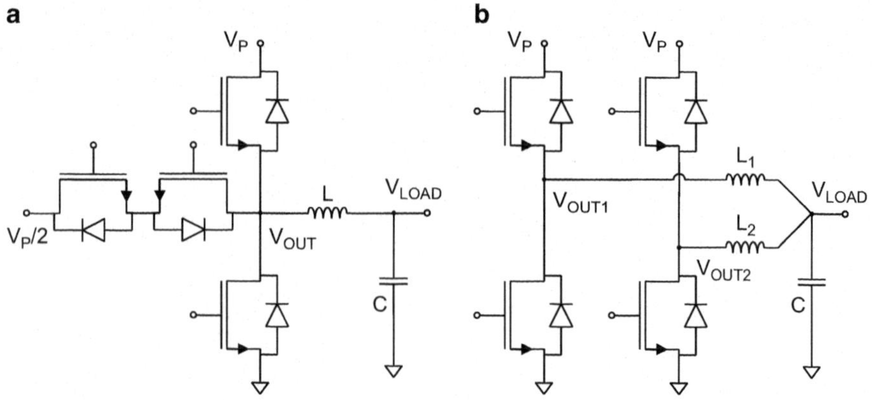

Fig. 6 (**a**) Multilevel class-D. (**b**) Multiphase class-D

A 3-level PWM signal can also be realized with a more complicated SE class-D output stage as shown in Fig. 6a where an anti-series power switch is added between the switching output node V_{OUT} and an additional voltage level at half the supply voltage $V_{\mathrm{P}}/2$. The 3-level PWM signals needed to drive this output stage can be realized with similar modulation schemes as used for BD-PWM. This output stage can produce three voltage levels at the output: 0, $V_{\mathrm{P}}/2$, and V_{P}. Two SE 3-level class-D stages combined in a BTL configuration can produce five (differential) voltage levels: $-V_{\mathrm{P}}$, $-V_{\mathrm{P}}/2$, 0, $+V_{\mathrm{P}}/2$, and $+V_{\mathrm{P}}$. The advantage of a multilevel class-D output stage is that the voltage transients at the output node V_{OUT} are smaller than in a two-level class-D output stage which is beneficial for EMI and relaxes the requirements on the external LC filter, if applicable. This comes at the cost of additional power switches which add to the silicon area.

Figure 6b shows a multiphase or interleaved output stage where a 2-level class-D output stage has been split in two branches that are connected to the load with separate inductors [4]. If the PWM carriers of both branches have opposite phase, then the ripple currents in both inductors also have opposite phase and cancel each other when summed at the capacitor node V_{LOAD}. This reduces the ripple voltage across the load which again is beneficial for EMI but this time at the cost of additional external inductors. The area of the power transistors does not increase since both branches only need to handle half the signal current. Multilevel and multiphase class-D output stages can be extended with more levels and phases, respectively. This will in general improve EMI but increase the number of switches and/or external inductors and increase the application cost.

2.2 Power Loss

An ideal class-D amplifier has a theoretical efficiency of 100% but clearly, in a real implementation, power is lost due to the nonzero on-resistance of the power transistors and finite speed of the transitions. Power loss in class-D amplifiers can be roughly divided in two categories: *conduction loss* and *switching loss*.

Conduction loss results from the on-resistance of the power transistors and the equivalent series resistance (ESR_L) of the external inductor. Assuming all power transistors have the same on-resistance, a class-D power stage can be modeled as a set of ideal switches with a series resistance R_{cond} that is the sum of the on-resistance and ESR_L. In the BTL configuration, the series resistances of both half-bridges add up. The series resistance R_{cond} sets a ceiling on the maximum achievable efficiency, because all current that flows through the load impedance R_{LOAD} also flows through R_{cond}:

$$\eta_{MAX} < \frac{R_{LOAD}}{R_{LOAD} + R_{cond}} \tag{1}$$

The product of on-resistance and area of a power transistor is constant and defines an important, process-specific figure of merit: $R_{on}A$. The $R_{on}A$ increases with the breakdown voltage of the power transistor. Usually, the power transistors occupy between 25% and 50% of the total die area of an integrated class-D amplifier so there is a very clear relation between efficiency and application cost. However, power transistor area does not simply scale linearly with $R_{on}A$. For integrated power transistors, besides on-resistance, also *power density* needs to be considered when determining the power transistor area. The temperature difference ΔT of the power transistor with respect to the heatsink can be estimated as the product of the dissipated power P_{cond} times the *thermal resistance* R_{TH}. Amplifiers for high output power often use power packages where the silicon die is soldered or glued to a (copper) case that allows the attachment of an additional heat sink. In such a situation, the total thermal resistance from junction to ambient R_{TH} is built up with two parts: $R_{TH,C\text{-}AMB}$ and $R_{TH,J\text{-}C}$. The thermal resistance from case to ambient $R_{TH,C\text{-}AMB}$ is determined by the package and the heat sink. The thermal resistance from junction to case $R_{TH,J\text{-}C}$ depends on the power transistor area and the thickness of the silicon die. For large power transistors where length and width are larger than the silicon thickness d, the thermal resistance $R_{TH,J\text{-}C}$ scales approximately with d, the inverse of the area, and the thermal conductivity of silicon κ:

$$\Delta T_{J-C} = P_{cond} \cdot R_{TH,J-C} \approx P_{cond} \cdot \frac{d}{\kappa \text{ Area}} \tag{2}$$

The power dissipation P_{cond} is proportional to the on-resistance R_{on}:

$$P_{cond} = I^2 \cdot R_{on} = I^2 \cdot \frac{R_{on}A}{\text{Area}} \tag{3}$$

Substituting (2) in (3) and rearranging yields:

$$\text{Area} = \sqrt{I^2 \frac{R_{\text{on}}A \cdot d}{\kappa \ \Delta T}} \tag{4}$$

Therefore, for a given maximum chip temperature difference between junction and case, the area of the power transistor scales with the square root of $R_{\text{on}}A$. This corresponds to having a constant power dissipation per area or power density.

The current going through the power transistors is the sum of signal current and ripple current. Conduction loss due to signal current $P_{\text{cond,sig}}$ is proportional to output power P_{OUT}:

$$P_{\text{cond,sig}} = P_{\text{OUT}} \cdot \frac{R_{\text{cond}}}{R_{\text{LOAD}}} \tag{5}$$

and mainly affects maximum efficiency η_{MAX}. Conduction loss due to the triangular ripple current $P_{\text{cond,rip}}$ is typically much smaller than $P_{\text{cond,sig}}$ and is especially relevant for quiescent dissipation P_{Q}.

$$P_{\text{cond,rip}} = \frac{1}{3} R_{\text{cond}} \cdot I_{\text{ripple}}^2 = \frac{1}{3} R_{\text{cond}} \cdot \left(\frac{V_{\text{P}}}{8 L f_{\text{PWM}}} \right)^2 \tag{6}$$

Switching loss occurs during each transition of the output of a class-D amplifier and has two components. The first switching loss component is caused by charging the gates of the power transistors. Each time a power transistor is switched on, a fixed amount of charge Q_{G} flows into the gate to achieve minimal on-resistance. The ratio between charge Q_{G} and transistor area is constant and defines a second important, process-specific figure of merit: Q_{G}/A. Assuming that charge Q_{G} is drawn from the supply V_{P}, the power loss due to gate charging is:

$$P_{\text{gate}} = f_{\text{PWM}} \cdot V_{\text{P}} \cdot Q_{\text{G}} \tag{7}$$

Like $P_{\text{cond,rip}}$, gate charge loss is typically much smaller than $P_{\text{cond,sig}}$ and especially relevant for quiescent dissipation P_{Q}. The second switching loss component is caused by output transitions of the class-D stage. Depending on the direction of the output current, two transition scenarios are possible known as *soft-switching* and *hard-switching*. In the soft-switching scenario, the transition is assisted by the output current. This applies, for example, to a falling transition with negative output current as shown in Fig. 7a. At the start, highside transistor M_{H} is switched on and lowside transistor M_{L} is off. Because of the negative output current I_{OUT}, the output voltage V_{OUT} is just below the supply rail V_{P}. As soon as M_{H} is switched off,

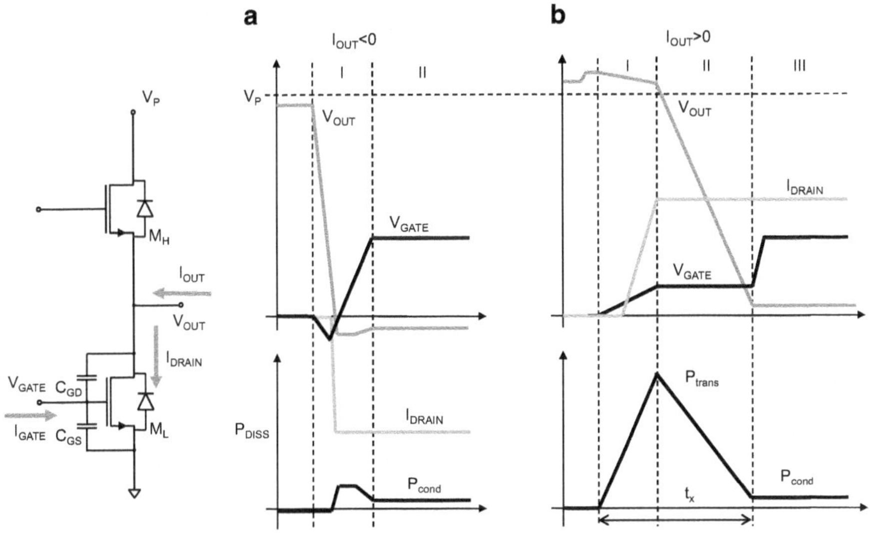

Fig. 7 Power loss in (**a**) soft-switching (**b**) hard-switching

the output current I_{OUT} pulls down the output node V_{OUT}. The output slope in the output transition is determined by the discharge current and gate-drain capacitance of M_{H} and pushes the gate voltage of M_{L} below ground [5]. Once the output voltage drops below ground, current starts flowing through the backgate diode of M_{L} and conduction loss starts as shown in the lower graph of Fig. 7a. Then the gate of M_{L} is charged and conduction loss reduces when M_{L} achieves minimum on-resistance.

In the hard-switching scenario, the output current is opposing the transition. This applies to a falling transition with positive output current as shown in Fig. 7b. At the start, the output voltage V_{OUT} is just above the supply rail V_{P}. After M_{H} is switched off, the output voltage V_{OUT} first increases since the output current now flows through the backgate diode of M_{H}. At the same time, the gate of M_{L} begins charging. When the threshold voltage of M_{L} is reached, the drain current I_{DRAIN} of M_{L} starts to rise. During this time, the voltage across M_{L} is still nearly equal to the supply voltage which means that power dissipation starts to rise as well, as shown in the lower graph in Fig. 7b. Once I_{DRAIN} matches the output current I_{OUT}, the output voltage starts to drop with a slope that is determined by the charge current and gate-drain capacitance of M_{L}. During this time, the current through M_{L} remains nearly equal to I_{OUT} but the voltage across M_{L} drops and so does the dissipation. When the output voltage V_{OUT} drops to one threshold voltage below the gate voltage V_{GATE}, transistor M_{L} enters the linear region and dissipation is counted as conduction loss.

The triangular dissipation peak that occurs during hard-switching is called the transition loss or hard-switching loss and can be expressed as:

$$P_{\mathrm{trans}} = f_{\mathrm{PWM}} \cdot V_{\mathrm{P}} \cdot t_x \cdot I_{\mathrm{OUT}} \tag{8}$$

As can be seen in (8), the loss depends on the duration of the transition t_x which, in turn, depends on the slope of V_{OUT}. Moreover, the transition loss increases with I_{OUT} and therefore with output power P_{OUT}. This can have a significant impact on maximum efficiency η_{MAX}.

The presented breakdown covers the most important power loss contributions in class-D amplifiers [6] and shows that trade-offs are unavoidable. For example, a higher PWM frequency f_{PWM} reduces ripple loss $P_{\mathrm{cond,rip}}$ but increases gate charge loss P_{gate} and transition loss P_{trans}. Increasing die area to reduce on-resistance reduces conduction loss $P_{\mathrm{cond,sig}}$, but at the same time, increases gate charge Q_{G} and gate charge loss P_{gate}.

2.3 External LC Filter

The inductors and capacitors in the external filter are a significant part of the cost of a class-D amplifier. Besides the cost of the components themselves, they also take up valuable PCB real estate [7].

The price of inductors is coupled to the inductance value and the saturation current. The core can only handle a limited magnetic flux which is the product of inductance and current. When a DC current is applied, the inductance value generally decreases due to the magnetic saturation of the ferrite in the core. Figure 8a shows a typical inductance versus current profile for a so-called powder core inductor. Usually the saturation current is specified as the current level where the inductance drops is between 20% and 30%.

The price of capacitors is coupled to capacitance value, voltage rating, and dielectric type. Film capacitors have stable capacitance over temperature and bias voltage but are most expensive. Ceramic capacitors are cheaper but their capacitance value depends more on temperature and bias voltage. A typical capacitance derating curve for an X7R-type ceramic capacitor is shown in Fig. 8b.

The choice of external components impacts both efficiency and EMI. For example: for a given PWM frequency, a larger inductance value results in a lower ripple current which, in turn, reduces conduction loss in the power switches of the class-D output stage and the ESR of the inductors. This is especially beneficial for quiescent dissipation. A larger capacitor reduces ripple voltage at the loudspeaker load or more importantly the cables leading to the loudspeaker. Lower voltage ripple means less EM emission. On the other hand, if the impedance of the capacitor becomes comparable to that of the load for higher signal frequencies, this means that a significant part of the output current of the class-D output stage is sloshing back and forth in the capacitor and does not contribute to the output power but does contribute to the conduction loss and therefore it reduces efficiency.

The voltage dependence of the capacitors also results in differential-mode to common-mode conversion. If AD-PWM is used in a BTL amplifier, the common-mode output voltage is theoretically always zero because the outputs are always perfectly balanced. However, when the amplifier is modulated with an output signal,

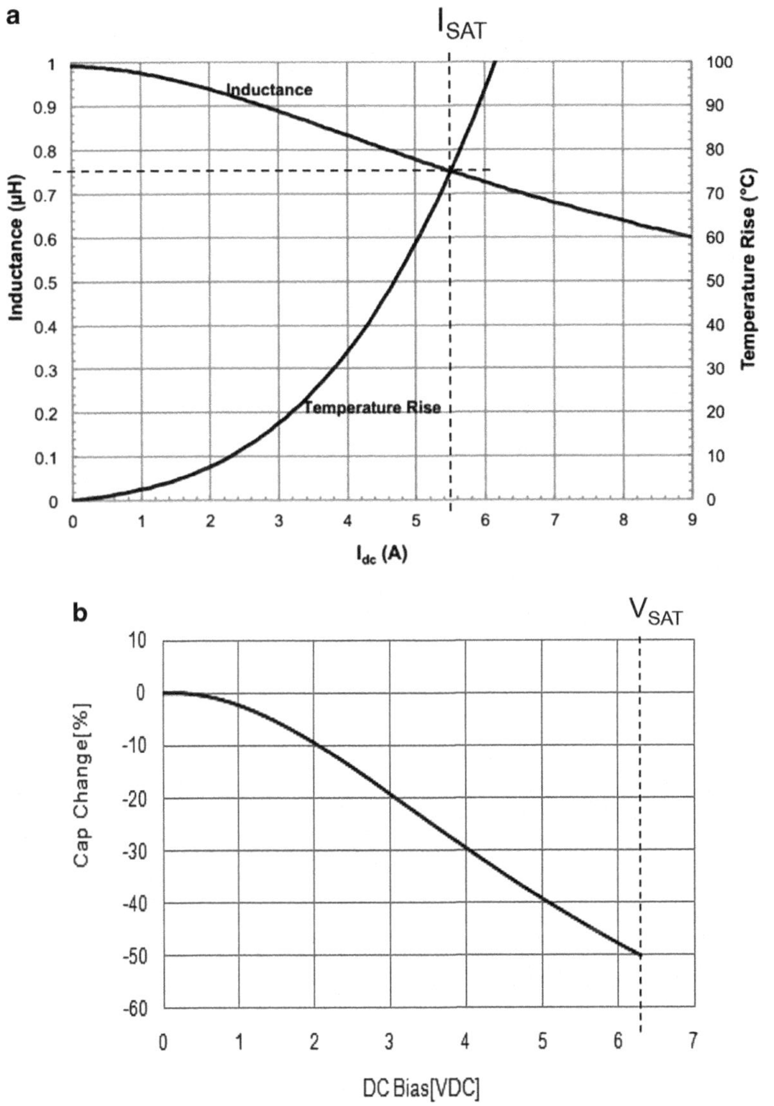

Fig. 8 Typical bias dependence of inductors and capacitors. (**a**) Inductance versus current. (**b**) Capacitance versus voltage

the half-bridge that makes the positive voltage causes that half of the LC filter to have a higher cut-off frequency than the half-bridge that makes the negative voltage. Consequently, the LC filter is not balanced anymore and some of the differential-mode voltage ripple will be converted to common-mode which is bad for EMI.

The bias dependence of the inductors and capacitors result in nonlinear behavior and can significantly impact the distortion of a class-D amplifier. Both the voltage dependence of capacitors and current dependence of inductors can be approximated with a simple expression of the form:

$$f(x) = \frac{1}{1 + x^2} \tag{9}$$

The voltage dependence of a capacitor is then approximately:

$$C(V) = C_{\text{nom}} \frac{1}{1 + \left(\frac{V}{V_{\text{SAT}}}\right)^2} \tag{10}$$

where V_{SAT} is defined as the voltage where the capacitance is reduced by 50% with respect to the nominal value C_{nom}. Note that V_{SAT} is not a standard parameter and is not specified in datasheets. However, in many datasheets, the voltage derating curve is shown from which the value can easily be determined. Likewise, the current dependence of an inductor is approximately:

$$L(I) = L_{\text{nom}} \frac{1}{1 + \frac{1}{3}\left(\frac{I}{I_{\text{SAT}}}\right)^2} \tag{11}$$

where I_{SAT} is defined as the current where the inductance is reduced by 25% with respect to the nominal value L_{nom}. Unlike V_{SAT}, the saturation current I_{SAT} usually is specified in datasheets although the inductance reduction at I_{SAT} varies per manufacturer between 20% and 30%. With eqs. (10 and 11) the total harmonic distortion (THD) caused by nonlinearity in inductance (THD_{L}) or capacitance (THD_{C}) can be approximated as:

$$\text{THD}_{\text{L}} = \omega \frac{L}{R_{\text{LOAD}}} \cdot \frac{P_{\text{OUT}}}{6 R_{\text{LOAD}} \cdot I_{\text{SAT}}^2} = \frac{\omega}{\omega_0 Q} \cdot \frac{P_{\text{OUT}}}{6 R_{\text{LOAD}} \cdot I_{\text{SAT}}^2} \tag{12}$$

$$\text{THD}_{\text{C}} = \omega^2 LC \cdot \frac{P_{\text{OUT}} \cdot R_{\text{LOAD}}}{2 \cdot V_{\text{SAT}}^2} = \frac{\omega^2}{\omega_0^2} \cdot \frac{P_{\text{OUT}} \cdot R_{\text{LOAD}}}{2 \cdot V_{\text{SAT}}^2} \tag{13}$$

where ω is the angular frequency of the signal, ω_{o} is the angular resonance frequency and Q the quality factor of the LC filter. Note that these expressions are valid for a single-ended LC filter; in case of a BTL filter, half the load resistance needs to be used as R_{LOAD}. As can be seen in (12) and (13), THD_{L} is proportional

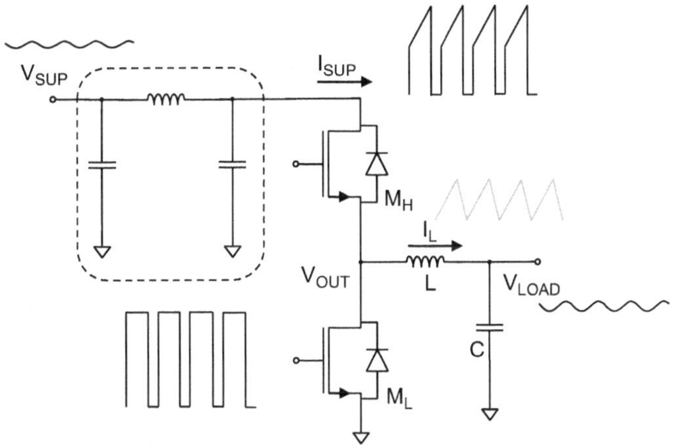

Fig. 9 Transient voltages in class-D amplifiers

to signal frequency, whereas THD_C is proportional to the square of the signal frequency. For practical component values, THD_C is much smaller than THD_L, and the distortion is dominated by the nonlinearity in the inductance. Furthermore, a higher resonance frequency results in lower distortion.

2.4 Electromagnetic Interference

Radiation of electromagnetic energy is a direct consequence of the fast voltage and current transients that are inherent to the high-efficiency class-D operation. EMI problems are mainly caused by common-mode voltages and currents, since differential-mode signals tend to cancel each other away from the radiating antennas. The antennas, in this case, are the cables that connect the loudspeaker to the LC filter but also the cables from the supply need to be considered. When discussing class-D amplifiers, the focus is usually drawn to the switching output node and the LC filter but the supply current too can be a major source of EMI when not properly filtered. In fact, the currents in the supply and ground rails of a class-D amplifier are discontinuous with sharp current slopes as illustrated in Fig. 9. The cost of a supply filter needs to be accounted for when determining the cost of a class-D amplifier application. A supply filter consists of at least a large capacitor but sometimes also a higher order filter is used as shown in Fig. 9.

The EMI performance of Class-D amplifiers is determined using measurement setups that are specified in detail in standards such as the well-known IEC-CISPR25. In these standards, limits are defined for the radiated field strength in frequency bands used for signals from broadcasting stations and mobile communication. Notoriously difficult are the frequency bands used for long-wave (150 kHz–300 kHz) and medium wave (530 kHz–1.8 MHz) AM broadcast, since these frequencies coincide with typical PWM frequencies and their harmonic components. At these relatively low frequencies, the suppression of the LC filter is not so high.

Two strategies to deal with the AM-band are *avoidance* and *spread spectrum.* Avoidance is a strategy where the PWM carrier frequency is tuned such that it does not interfere with the AM channel that is being received. Avoidance can be very effective because the spectrum of PWM signals is concentrated in relatively narrow bands around the PWM carrier and harmonics. A drawback of this approach is that it requires coordination with the AM-tuner in the system which is not always possible. Spread spectrum is a strategy where the PWM carrier is modulated by varying the period of the reference triangle from cycle to cycle. This spreads the energy of the PWM carrier and harmonics over a wider frequency band and makes the frequency spectrum more noise-like.

As a last resort, common-mode chokes can be added in the speaker and supply lines, but although these are very effective in reducing EMI, the added cost is significant.

3 Class-D Architectures

In this section, practical class-D amplifier products and techniques are presented that demonstrate how efficiency, EMI, and application cost are balanced in reality. The market for class-D audio amplifiers has three high-volume domains: mobile, consumer, and automotive. In each of these domains, class-D amplifies must deal with different requirements and constraints resulting in a large variety of architectures.

3.1 Mobile

The mobile domain includes devices such as tablets, wearables, and smartphones. Such devices often have stereo speakers and are active for extended periods of time during video streaming or gaming. Since mobile devices are usually supplied from a battery, the available power is limited. Furthermore, the thermal budget in mobile devices is very much constrained due to the apparent difficulty to remove heat in the absence of a heatsink, whereas the fact that mobile applications mainly use wafer-level chip-scale package (WL-CSP), with relatively high thermal resistance to ambient, does not help either. Consequently, high efficiency is a necessity to extend battery operating time and limit on-chip temperatures. This explains the popularity of class-D amplifiers in the mobile domain. The archetypal mobile Class-D audio amplifier is a mono channel device with single digit output power.

Because of the proximity of the supply and loudspeakers, EMI seems less of a concern and, to keep application cost low, “filterless” is essentially a must-have requirement. This means that mobile class-D amplifiers almost invariably use BTL output stages with BD-PWM. The battery voltage of Li-Ion batteries ranges from 5.5 V when charging down to 2.7 V when almost empty. The supply voltage limits the maximum output power of a class-D amplifier. This limitation has triggered the emergence of so-called smart speaker drivers [8, 9].

A smart speaker driver is a combination of a class-D audio amplifier, a DC-DC boost converter and usually also an arrangement to sense the loudspeaker current. The DC-DC boost converter, usually an inductive converter, supplies the class-D amplifier and guarantees high-output power even at low battery voltage. In fact, the output power of the class-D amplifier can be so high that the loudspeaker is in danger of being damaged. The loudspeakers used in mobile applications need to be cheap and are quite fragile. The speaker membrane can only move so far without being damaged, and heat from dissipation in the voice-coil of the speaker can cause the entire assembly to disintegrate. This is where the current sense comes into play. Both excursion and voice-coil temperature can be predicted using a loudspeaker model. The model needs to be constantly updated with real-time impedance information to make sure that the model matches the environmental conditions of the loudspeaker. To determine the impedance of the loudspeaker, both current through and voltage across the speaker need to be known. The voltage across the speaker can be predicted quite accurately, it is after all the core function of an amplifier to control that voltage. Nevertheless, some products have a dedicated voltage sense arrangement as well [10]. The current through the loudspeaker cannot be predicted and needs to be measured.

With the loudspeaker model that now reliably predicts the behavior of the loudspeaker, output power of the class-D amplifier can be maximized without the risk of damage by using sophisticated speaker protection algorithms. Figure 10 shows a smart speaker driver that includes an embedded DSP that runs the speaker protection algorithm and is also used to improve the quality of the sound by boosting the lower audio frequencies [9]. Alternatively, the impedance information can be fed back to an audio host to run the speaker protection algorithm.

Most class-D output stages use PMOS power transistors as highside switch as shown in Fig. 11a. The $R_{\mathrm{on}}A$ of PMOS devices is typically about three times higher than for NMOS devices, so for the same on-resistance, a PMOS needs three times the area of an NMOS. However, an NMOS power transistor needs a gate voltage higher than the supply voltage which requires external bootstrap or chargepump capacitors that add cost. The gate drive of a PMOS is much simpler. The power transistors are often dedicated high-voltage devices capable of handling the boosted supply voltage that ranges from 5.5 V to 9.5 V [11].

Alternatively, [8] uses an output stage with only standard NMOS devices as shown in Fig. 11b. The boosted supply is distributed over two-stacked NMOS devices to extend the voltage range. Furthermore, an internal bootstrap capacitor is used to generate the gate voltage of the highside NMOS stack. This is made possible without a large area penalty, because the used technology features a high-density MIM capacitor that can be placed on top of the power transistors in the same area.

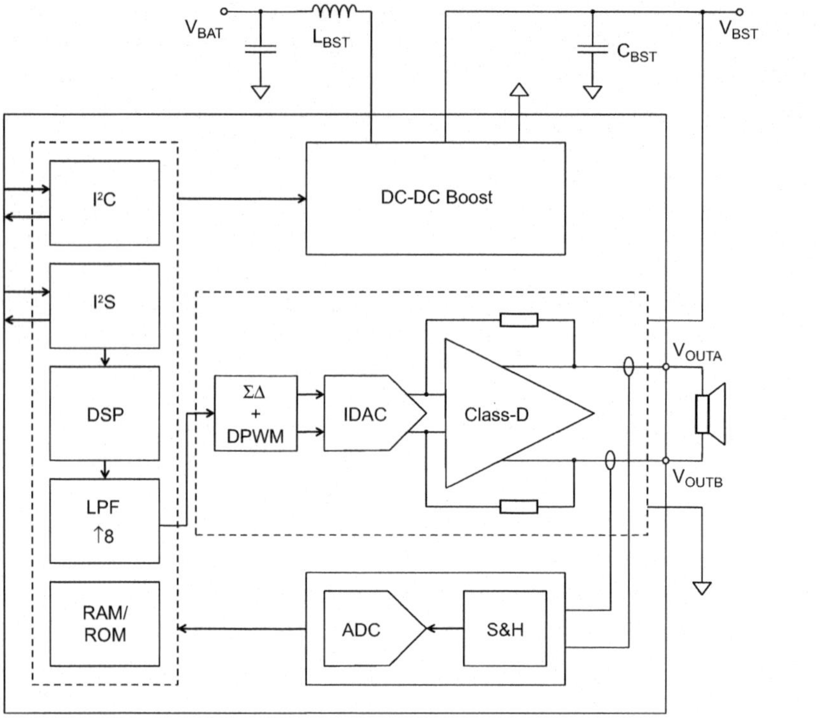

Fig. 10 Smart speaker driver

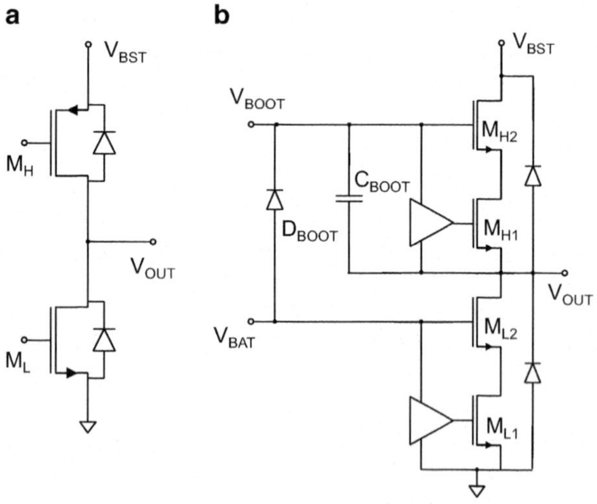

Fig. 11 Class-D power transistors. (**a**) Complementary. (**b**) Cascoded NMOS

Power transistors are quite large to minimize conduction loss at high load currents. At low-load currents, however, the low on-resistance of the power FETs

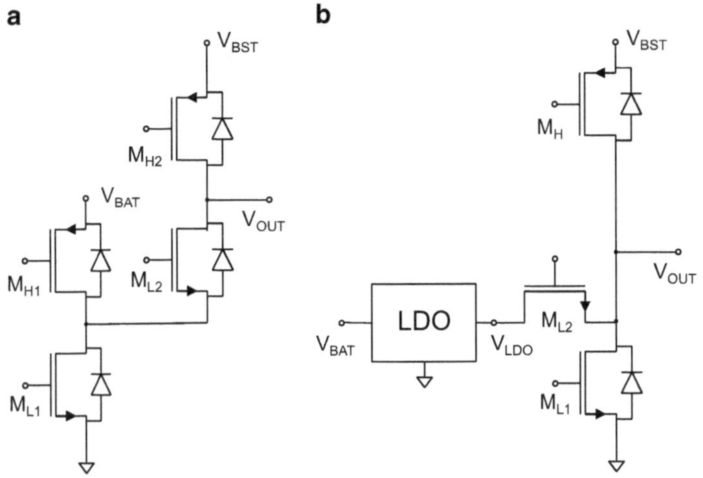

Fig. 12 Multilevel techniques. (**a**) Cascade. (**b**) LDO

does not yield any advantage, and quiescent dissipation is dominated by gate charge loss. This trade-off can be optimized dynamically by splitting up the power transistors in segments that are activated depending on the signal level [5].

In a smart speaker driver, the cascade of the class-D amplifier and the DC-DC booster determines the overall efficiency. Therefore, envelope tracking of the audio signal is used to minimize supply voltage headroom and switching loss of the class-D amplifier. At low signal levels, the DC-DC booster stops switching and just passes the battery voltage to further reduce switching loss in the system. Tight envelope tracking requires that the audio signal is known in advance to allow the DC-DC booster to ramp-up in time. In a system with digital audio input, this is in general not too difficult because the delay of interpolation chains gives sufficient response time. An amplifier with a tracking supply is conventionally called a class-H amplifier. Combined with class-D this becomes class-DH.

A system with a boosted supply rail is a natural fit for a multilevel class-D output stage as shown in Fig. 12a [12, 13]. At low signal levels, the output node V_{OUT} alternates between V_{BAT} and ground only and the boosted voltage V_{BST} is not loaded. For higher signal levels, one half-bridge alternates between V_{BST} and V_{BAT} while the other half-bridge is permanently connected to ground. This approach has many benefits. First, with multilevel the voltage transients at the output are smaller which is good for switching loss and EMI. Second, the booster does not require up-front knowledge of the audio signal since the boosted voltage V_{BST} is always present. This makes the architecture very suitable when analog input signals are used. The price for these benefits is the extra chip area that is needed to make additional power transistors. An amplifier with multiple supply rails is conventionally called a class-G amplifier. Combined with class-D this becomes class-DG.

The class-D output stage in Fig. 12b shows an alternative multilevel configuration where a linear regulator (LDO) is used to generate a very low, e.g., 1 V, supply rail [14]. At low signal levels, the output V_{OUT} switches between ground and the regulator voltage V_{LDO} reducing switching loss and EMI to a minimum. The poor efficiency of the LDO is irrelevant since it delivers negligible power. At high signal levels, the output alternates between V_{BST} and ground. In this configuration, the additional switch M_{L2} does not have to handle high output currents so it can be made relatively small but somehow current flowing back towards the LDO through the body diodes of M_{L2} needs to be prevented.

To improve EMI performance, well-known techniques such as output slope control [5] and spread spectrum modulation [8] are used. In [15], a chopping scheme is presented to suppress the common-mode tone at the PWM frequency that is inherent to BD-modulation. In this chopping scheme, the class-D output alternates between so-called regular frames and chopped frames as shown in Fig. 13. The chopping only affects the common-mode CM while the differential-mode DM remains unchanged.

Instead of using a simple alternating pattern between regular and chopped frames, a noise shaper is used that has NTF notches which suppress common-mode components in the audio band and around the AM reception frequencies.

3.2 Consumer

The consumer domain includes home-audio systems, TV-sets, and the recent wireless speaker systems. The archetypal consumer class-D amplifier is a stereo channel device with several tens to hundreds of Watts per channel output power. Since most consumer applications are supplied from the mains, the available

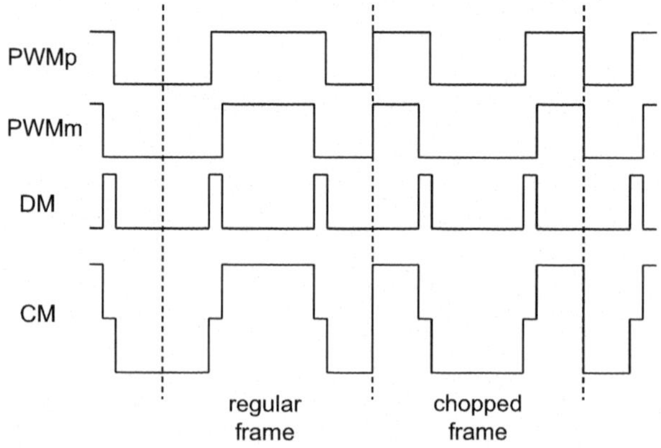

Fig. 13 Common-mode chopping

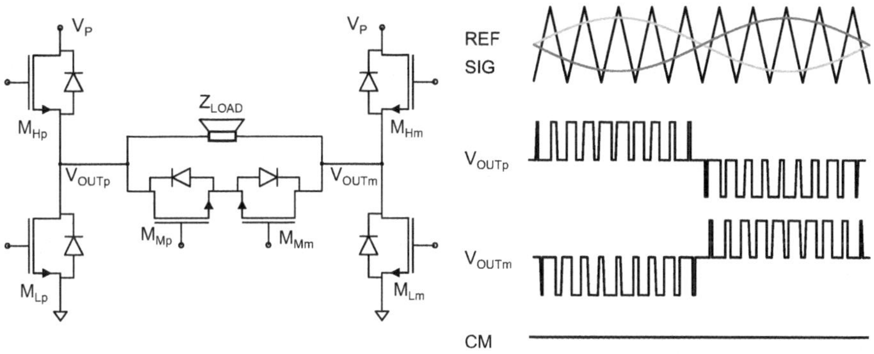

Fig. 14 Zero common-mode filterless class-D output stage

power is essentially unlimited. However, the trend in consumer products is towards more power into a smaller size encapsulation which means the heat caused by power dissipation becomes an issue. Especially, the multichannel home-theater systems that became popular about a decade ago have boosted popularity of class-D amplifiers. Because of the high-output power levels in consumer products, expensive inductors with high-saturation currents are required to limit distortion but also to guarantee robustness. Traditionally, in many consumer applications, the loudspeaker is connected to the class-D amplifier with a relatively long cable where the occurrence of a short-circuit is quite plausible. The speaker cables also make the output filter important for EMI performance. AD-PWM which has the benefit of zero-common-mode has been the preferred modulation scheme in most consumer applications.

Some notable efforts have been made to relax the requirements on or eliminate the LC filter. Figure 14 shows a class-D output stage that combines the advantages of "filterless" BD-modulation with the zero-common-mode of AD-modulation [16]. Two back-to-back power transistors M_{Mp} and M_{Mm} are added to a classical BTL configuration in parallel to the loudspeaker.

When power transistors M_{Hp} and M_{Lm} are switched on, the differential output voltage is $+V_{\mathrm{P}}$. When M_{Lp} and M_{Hm} are switched on, the differential output voltage is $-V_{\mathrm{P}}$. A differential 0 is made by short circuiting the load with M_{Mp} and M_{Mm}. In this latter case, the common-mode voltage is undefined but can be forced to $V_{\mathrm{P}}/2$ by a parallel driver that can be very small since it does not have to deliver any power to the load.

In Fig. 15, a multilevel class-D output stage is shown that elegantly employs a topology that is widely used in power conversion [17, 18]. The output current always flows through a series connection of two power transistors. This would normally result in a 4$\times$ area penalty but this is partly compensated by the fact that, when operating, the voltage across any of the power transistors is never higher than $V_{\mathrm{P}}/2$. This means that in principle power transistor with a lower breakdown voltage and consequently lower $R_{\mathrm{on}}A$ can be used. Each half-bridge can make

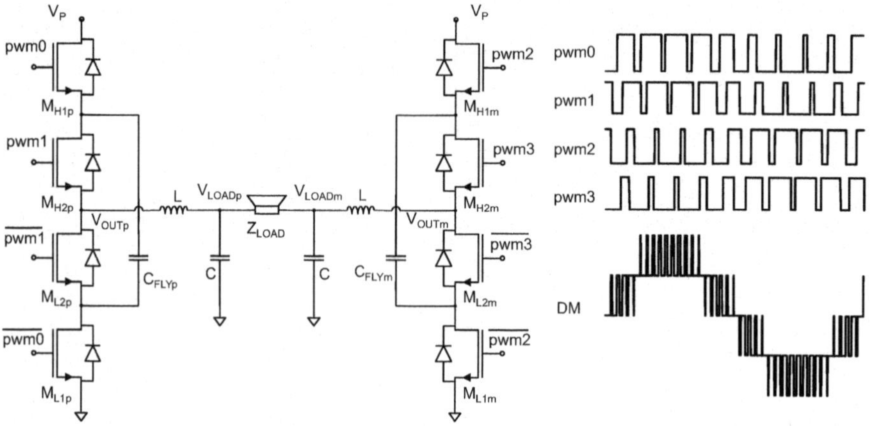

Fig. 15 Multilevel class-D output stage

three output levels: V_P, $V_P/2$, and 0 independently, so in BTL, this results in a five-level differential-mode PWM. This gives a significant reduction in ripple loss and quiescent dissipation. The main drawback is that, per channel, two external capacitors C_{FLY} and consequently also four additional pins are required.

3.3 Automotive

In the automotive domain, class-D audio amplifiers appear in car entertainment systems that can be divided into two subdomains: *head units* and *sound systems*. Head unit audio amplifiers are located in the center of the dashboard and are supplied directly from the 14.4 V car battery. The archetypal automotive class-D audio amplifier is a quad channel device with 25 W (4Ω, 10% THD) per channel output power. For head unit applications, class-D amplifiers need to compete with cheap class-AB amplifiers that need almost no external components and do not cause EMI. The thermal budget in a head unit is limited by the size of standard enclosures while the increase in functionality such as navigation, wireless connectivity has reduced the share available for audio amplification. Automotive EMI requirements are notoriously stringent. On top of the public EMI standards, many car manufacturers demand electronic applications to comply with proprietary emission masks. Apparently, AM-radio is still very much alive in automotive and the proximity of rear window glass antennas and long loudspeaker cables make this a challenging EMI problem.

A solution to both application cost and EMI is to increase the PWM frequency above the AM-band at 2 MHz [19]. This avoids interference with AM reception and high switching frequency allows a significant reduction of inductor and capacitor

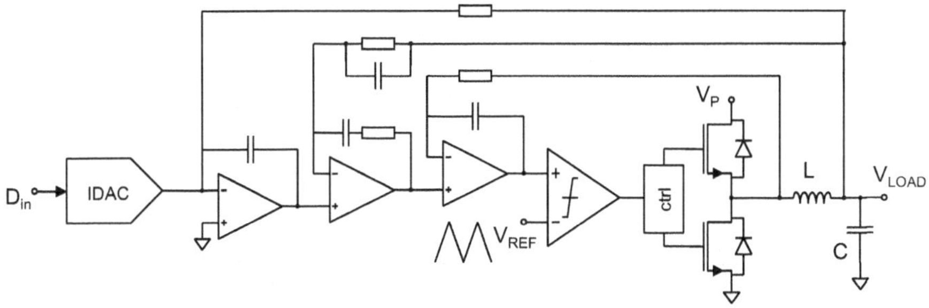

Fig. 16 Multiloop analog feedback after filter

values in the LC filter without an unacceptable increase in ripple loss. The proposed 3.3 μH/1.0 μF filter results in a cut-off frequency of 88 kHz, much higher than traditional values just above the audio band.

Sound systems are typically located in the trunk where the thermal budget is less restricted. Compared to head units, the output power is higher and usually more channels are supported. Regularly, DC-DC boost converters are used to increase the supply voltage and output power. The focus is very much on audio performance but still application cost is important.

The ultimate audio performance can only be achieved when the LC filter is included in the feedback loop. This not only ensures a flat, load independent, frequency response but also compensates for nonlinearities of the output filter which enables the use of smaller, less linear and therefore cheaper components. The stability of such a feedback loop is complicated because the LC filter adds two complex conjugate poles to the loop transfer. The resonance frequency and quality factor of these poles is subject to component spread and loading condition of the filter. In [20], a design is presented that solves the stability with a multiloop architecture as shown in Fig. 16, where feedback is taken before and after the LC filter. This approach has the disadvantage that the loopgain around the inductor is inherently less than the loopgain around the output stage. Furthermore, despite the complexity of the analog loopfilter, the overall loopgain has only a first-order behavior.

Figure 17 shows a single-loop architecture [21] where the full loopgain is available for both output stage and inductor. In this architecture, a custom low-latency ADC in the feedback path enables the use of a fifth-order digital loopfilter H(z) that has at least 50 dB loopgain in the entire audio band.

This results in best-in-class THD performance which is nearly flat over power and frequency. The poles of the LC filter are canceled in the digital domain by a programmable compensation filter LC^{-1}. The design also features a multiphase output stage that improves EMI performance by canceling ripple currents (Fig. 6b).

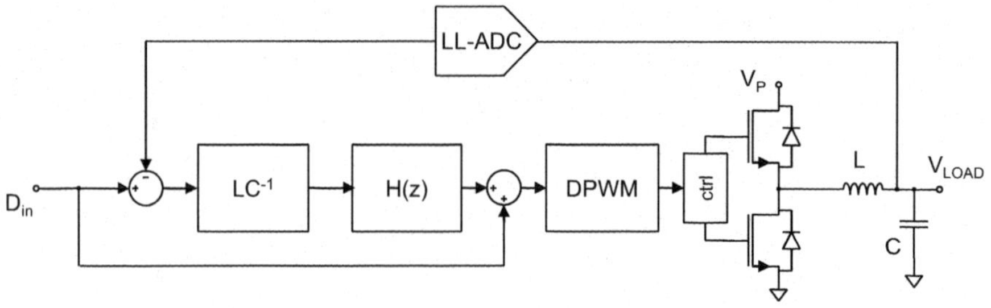

Fig. 17 Single-loop digital feedback after filter

4 Conclusion

Class-D audio amplifiers in different application domains need to deal with application-specific constraints. This has a significant impact on the trade-off between efficiency, EMI, and application cost. In this multidimensional space, no single best architecture exists since an improvement on one aspect generally goes at the expense of another.

References

1. Nielsen K. A review and comparison of pulse width modulation (PWM) methods for analog and digital input switching power amplifiers, presented at the 102nd Conv. Munich: AES; 1997. Preprint 4446.
2. Berkhout M, Dooper L. Class-D audio amplifiers in mobile applications. IEEE Trans Circuits Syst I, Reg Papers. 2010;57(5):992–1002.
3. Gaalaas E, Liu BY, Nishimura N, Adams R, Sweetland K. Integrated stereo ΣΔ class D amplifier. IEEE J Solid State Circuits. 2005;40(12):2388–97.
4. Nielsen K. Parallel Phase Shifted Carrier Pulse Width Modulation (PSCPWM) - A novel approach to switching power amplifier design, presented at the 102nd Conv. Munich: AES; 1997. Preprint 4447.
5. Dooper L, Berkhout M. A 3.4W digital-in class-D audio amplifier in 0.14μm CMOS. IEEE J Solid State Circuits. 2012;47(7):1524–34.
6. Ma H, van der Zee R, Nauta B. A high-voltage class-D power amplifier with switching frequency regulation for improved high-efficiency output power range. IEEE J Solid State Circuits. 2015;50(6):1451–62.
7. Texas Instruments. LC filter design. Application Report SLAA701A. http://www.ti.com/lit/an/slaa701a/slaa701a.pdf.
8. Nagari A, et al. An 8 2.5 W 1%-THD 104 dB(A)-dynamic-range class-D audio amplifier with ultra-low EMI system and current sensing for speaker protection. IEEE J Solid State Circuits. 2012;47(12):3068–80.
9. Berkhout M, Dooper L, Krabbenborg B. A 4Ω 2.65W class-D audio amplifier with embedded DC-DC boost converter, current sensing ADC and DSP for adaptive speaker protection. IEEE J Solid State Circuits. 2013;48(12):2952–61.
10. CS35L32 boosted class D amplifier with speaker-protection monitoring and flash LED drivers, Cirrus Logic, Inc. 2015, online available: http://www.cirrus.com/products/cs35l32.

11. TFA9891 9.5 V boosted audio system with adaptive sound maximizer and speaker protection, NXP Semiconductors. 2016, online available: http://www.nxp.com/docs/en/data-sheet/TFA9891_SDS.pdf.
12. MAX98308 3.3W mono class DG multilevel audio amplifier, Maxim Integrated. 2016, Online Available: http://www.maximintegrated.com/en/products/analog/audio/MAX98308.html.
13. Cerutti C. Multilevel class-D amplifier. U.S. Patent 8,558,617 B2, Oct 15, 2013.
14. CS35L33 boosted class-D amplifier, Cirrus Logic. http://www.cirrus.com/products/cs35l33.
15. Balmelli P, et al. A low-EMI 3-W audio class-D amplifier compatible with AM/FM radio. IEEE J Solid State Circuits. 2013;48(8):1771–82.
16. Siniscalchi PP, Hester RK. A 20 W/channel class-D amplifier with near-zero common-mode radiated emissions. IEEE J Solid State Circuits. 2009;44(12):3264–71.
17. Høyerby M, Jakobsen JK, Midtgaard J, Hansen TH. A 2×70 W monolithic five-level class-D audio power amplifier in 180 nm BCD. IEEE J Solid State Circuits. 2016;51(12):2819–29.
18. MA12070P filterless and high-efficiency +4V to +26V audio amplifier with I2S digital input, Merus Audio. 2017, online available: http://www.merus-audio.com.
19. TAS6424-Q1 75-W, 2-MHz digital input 4-channel automotive Class-D audio amplifier with load-dump protection and I2C diagnostics, Texas Instruments, Inc. 2017, online available: http://www.ti.com/product/TAS6424-Q1.
20. Adduci P, Botti E, Dallago E, Venchi G. Switching power audio amplifiers with high immunity to the demodulation filter effects. J Audio Eng Soc. 2012;60(12):1015–23.
21. Schinkel D, et al. A multiphase Class-D automotive audio amplifier with integrated low-latency ADCs for digitized feedback after the output filter. IEEE J Solid State Circuits. 2017;52(12):3181–93.

A Deep Sub-micron Class D Amplifier

Mark McCloy-Stevens, Toru Ido, Hamed Sadati, Yu Tamura, and Paul Lesso

1 Introduction

Class D is a growth area within the audio space, with manufacturers making use of the increased output power and efficiency for their high-performance amplifiers. In the mobile market, increased battery life and the desire for louder audio is making Class D increasingly important. The technology is now propagating from speaker drivers to earpiece, headset, and haptic use cases.

Class D architectures have progressed from an analog input and analog modulator, Fig. 1a, to the current state of the art where a digital input drives an analog modulator with output monitoring used for speaker protection, Fig. 1b. As mixed-signal design moves to deep sub-micron nodes, selected to enable digital processing features and minimize area, conventional analog circuits do not scale and the analog Class D modulator can have a large impact on the area and power of the device. Digital circuitry scales with process node, making a digital Class D architecture attractive and allowing the design to benefit from the programmability and adaptability of a digital implementation. A digital modulator can be used to drive a DAC, and ADC feedback to the modulator used if the output needs to be sensed to correct errors, Fig. 1c.

This chapter presents a digital Class D architecture [1] that makes use of open-loop and closed-loop configurations to optimize the performance of the amplifier across the full signal range.

M. McCloy-Stevens (✉) · T. Ido · H. Sadati · Y. Tamura · P. Lesso
Cirrus Logic International (UK) Ltd, Edinburgh, UK
e-mail: mark.mccloy-stevens@cirrus.com; toru.ido@cirrus.com; hamed.sadati@cirrus.com; yu.tamura@cirrus.com; paul.lesso@cirrus.com

K. A. A. Makinwa et al. (eds.), *Low-Power Analog Techniques, Sensors for Mobile Devices, and Energy Efficient Amplifiers*,
https://doi.org/10.1007/978-3-319-97870-3_16

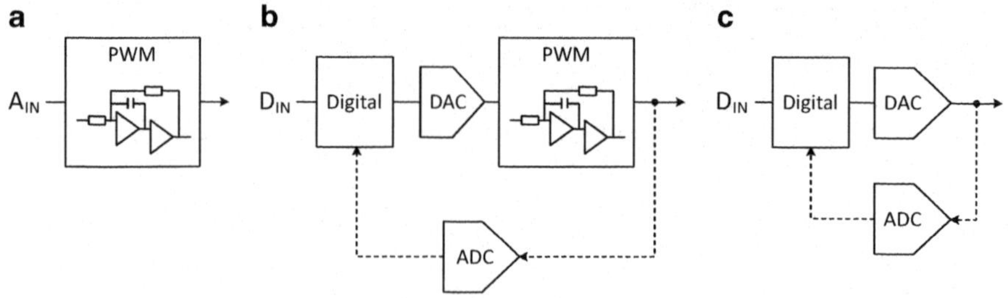

Fig. 1 Analog (**a**), digital input with DAC and analog modulator (**b**), digital (**c**)

Fig. 2 Open-loop Class D

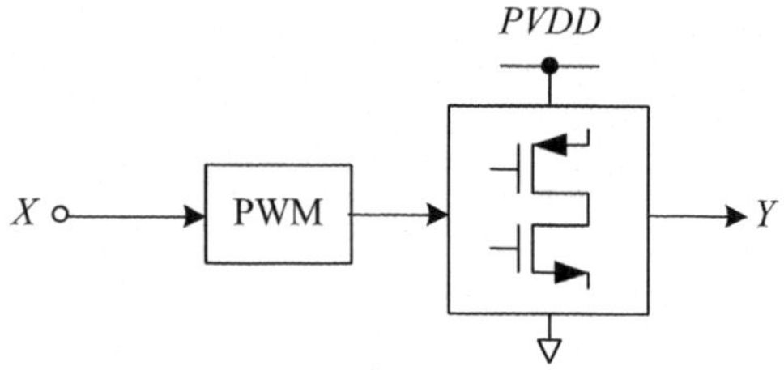

2 Open-Loop Digital Class D

The simplest implementation of digital Class D is an open-loop architecture, but understanding the limitations of this design highlights the need for a closed-loop mode for most use cases.

2.1 Architecture

As shown in Fig. 2, the Class D driver consists of a digital modulator followed by a power stage that is used to drive the output. The power stage is typically connected to a quiet supply that can deliver the necessary power to the output. The simplicity of the architecture means that it is low in area and power consumption.

A huge benefit is that the noise at the output is dominated by the modulator and so high SNR is achievable by selecting the correct modulator design and order. Figure 3 shows a fifth-order sigma-delta modulator design, giving a noise floor of −120dBV. The simulation is of the modulator and does not include the non-linearities of the power stage.

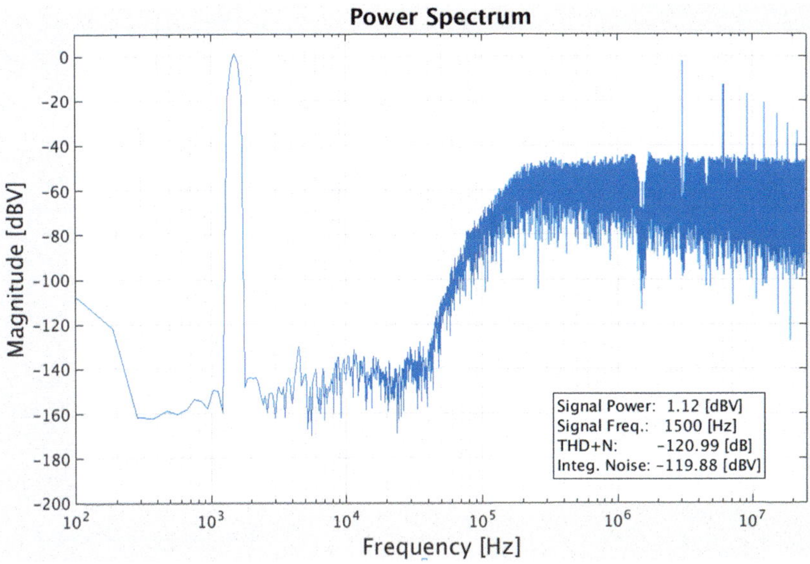

Fig. 3 Open-loop Class D simulation

2.2 *Limitations*

Open-loop Class D performs well at low signals but as the output signal amplitude increases, distortion and power supply noise, introduced by the power stage, can limit the usage of this mode. To understand these effects, a model of the output voltage of a single-ended Class D can be used, Eq. (1). V_{OUT_SE} is a filtered representation of the output and PVDD is the supply to the power stage. S is a dimensionless variable that represents the input signal and has a maximum of $+1$ and minimum of -1, corresponding to 100% and 0% duty cycle respectively, Fig. 4.

$$V_{OUT\_SE} = \frac{PVDD}{2}(1 + S) \tag{1}$$

2.2.1 Power Supply Rejection

To analyze power supply rejection, the power supply (PVDD) can be considered to consist of a supply voltage (V_{DD}) and noise on that supply (V_{noise}), Eq. (2). The error voltage created by noise on the supply, Eq. (3), can be found by substituting

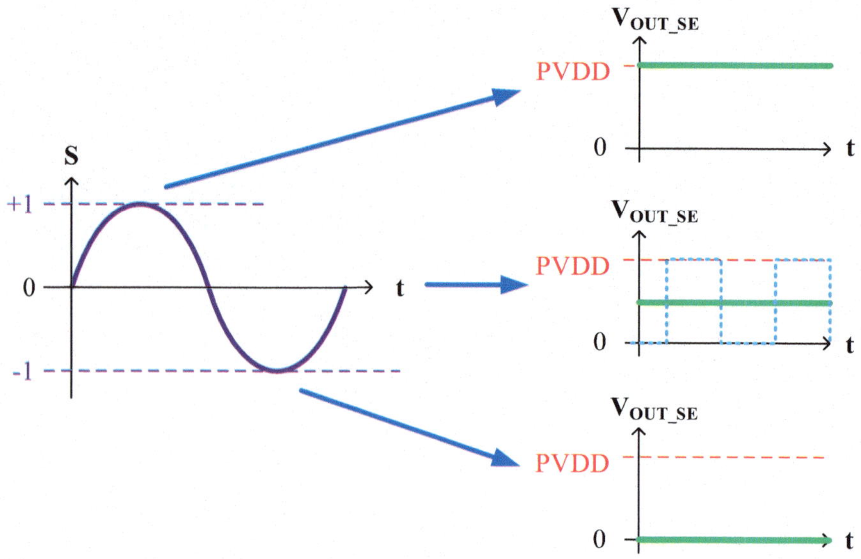

Fig. 4 S used to represent the PWM duty cycle

the expression for *PVDD* into Eq. (1) and isolating the components dependent on V_{noise}. The first error voltage term only varies with supply noise and shows a direct injection of the supply noise to the output. The second error term is dependent on the input signal, and the supply noise will mix with and be scaled by the input signal.

$$\text{PVDD} = V_{\text{DD}} + V_{\text{noise}} \tag{2}$$

$$V_{\text{error\_SE}} = \frac{V_{\text{noise}}}{2} + S\frac{V_{\text{noise}}}{2} \tag{3}$$

In a BTL configuration, assuming perfect matching, the common-mode is cancelled to the differential output, as shown below.

$$V_{\text{P}} = \frac{\text{PVDD}}{2}(1 + S) \tag{4}$$

$$V_{\text{N}} = \frac{\text{PVDD}}{2}(1 - S) \tag{5}$$

$$V_{\text{OUT\_BTL}} = V_{\text{P}} - V_{\text{N}} = \text{PVDD}(S) \tag{6}$$

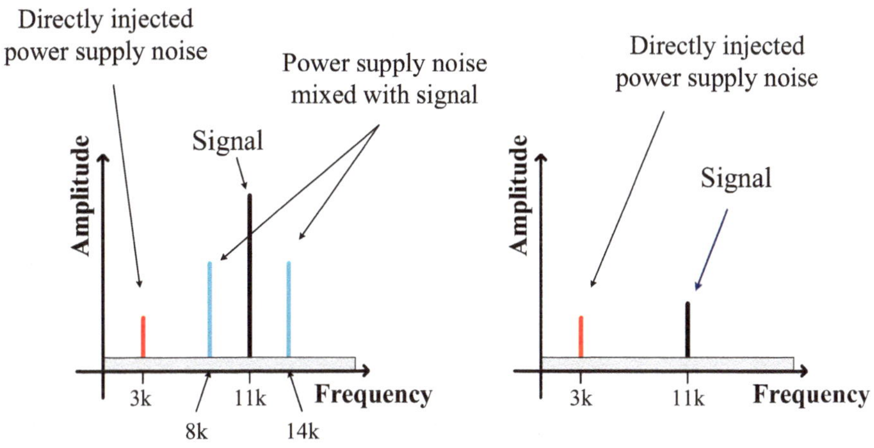

Fig. 5 Supply noise on a BTL output

In a real scenario, there will be some mismatch between the BTL outputs. Using a constant k, that is set to 0 for perfect matching, and multiplying V_N by $(1 - k)$, the output voltage and the error voltage due to supply noise are shown in Eqs. (7) and (8), respectively.

$$V_{\mathrm{OUT\_BTL}} = \frac{\mathrm{PVDD}}{2}\left(k + S\left\{2 - k\right\}\right) \quad (7)$$

$$V_{\mathrm{error\_BTL}} = \frac{k}{2} V_{\mathrm{noise}} + S\left(1 - \frac{k}{2}\right) V_{\mathrm{noise}} \quad (8)$$

As with the perfectly matched BTL analysis, there is a mix term of the power supply noise with the signal but with the gain slightly adjusted by the mismatch of the two outputs. However, a direct injection term now exists from the power supply which will be suppressed by the quality of matching between the two outputs.

The power supply rejection performance of the open-loop Class D can limit its usability. *PVDD* can be regulated to achieve high rejection from the system supply and suppress both power supply error terms to the output. However, at higher signal levels, a PMU between the system supply and *PVDD* will limit the signal output and reduce the efficiency of the amplifier. In addition, the mix term will dominate, and a high level of rejection may be required to meet power supply intermodulation specifications.

Figure 5a shows the effect of a 3 kHz sine wave of supply noise on an open-loop Class D outputting a 11 kHz sine wave. In this case, the magnitude of supply noise is less than the signal, so the mix tones at 8 kHz and 14 kHz are at a lower amplitude than the signal. The direct injection tone at 3 kHz is then further suppressed by the matching between the BTL outputs.

Figure 5b shows the effect with a reduced output signal amplitude, where the mix tones are below the noise floor, but the direct injection tone remains unchanged.

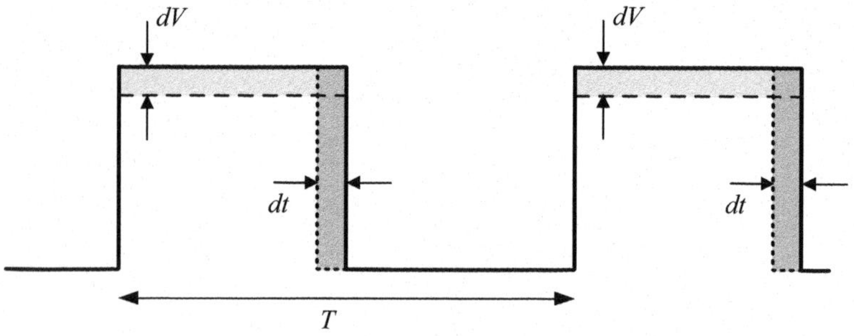

Fig. 6 Single-ended Class D output waveform with errors

2.2.2 Error Analysis

The analysis of power supply rejection in Sect. 2.2.1 focused on a simple gain mismatch between the BTL stages. Taking an alternative approach to analyzing the errors can give further insight into the errors that can exist in the open-loop system.

Considering the output signal of the Class D, shown in Fig. 6, the information is contained in both the duty cycle and the supply voltages at the output. Any change to this from the expectation of the digital modulator will cause an error at the output. For a single-ended output, any fixed error in the timing of a single edge will appear as a fixed offset on the output, and error in the supply will give a signal-dependent and common-mode error, Eq. (9).

To consider mismatch in this scenario, constants m and n can be used as a scale on the errors for time and voltage, respectively, where a value of 1 represents perfect matching between the outputs. Applying the constants to the V_N output and analyzing a BTL configuration gives Eq. (10), which can be used to analyze the impact of errors on the differential output voltage. With fixed timing and voltage errors, there are three common-mode error terms suppressed by matching, one dependent on supply, and a signal-dependent error term.

$$V_{\mathrm{OUT\_SE}} = \frac{\mathrm{PVDD}}{2}(1+S) - \frac{dV}{2}(1+S) - \mathrm{PVDD}\frac{dt}{T} + dV\frac{dt}{T} \tag{9}$$

$$\begin{aligned} V_{\mathrm{OUT\_BTL}} = \mathrm{PVDD}(S) - \frac{dV}{2}(1-n) - \frac{dV}{2}(1+n)\,S \\ - \mathrm{PVDD}\,(1-m)\,\frac{dt}{T} + dV\frac{dt}{T}\,(1-nm) \end{aligned} \tag{10}$$

The errors become more complex considering that the power supply and clock reference can vary and the time and voltage errors, and their matching, could vary with some dependence on power supply and signal. For example, the cause of the voltage error could be due to the drop across the power stage impedance. This will change as the signal, and current drawn, increases leading to a signal dependence on the voltage error that will cause distortion in the output. Error components will also vary depending on the load and the modulation scheme used for the Class D BTL output.

The complex errors discussed above are highly variable and interdependent, so an efficient solution to suppress these errors to the desired level is to add a feedback loop to the design.

3 Closed-Loop Digital Class D

Given a digital input signal, a closed-loop Class D can be created by either using a DAC to drive an analog modulator that has feedback or using a digital modulator with the amplifier output signal fed back to it [2]. In this digital architecture, a digital modulator is used where the biggest analog design challenge is in the feedback.

To make a digital Class D architecture compelling, the power and area of the analog feedback needs to be smaller than that of an analog PWM modulator with similar performance.

3.1 Architecture

The closed-loop Class D uses an ADC to sense the output of the amplifier, as shown in Fig. 7. The digital representation of the feedback signal is compared with the desired output, the error passed through the loop filter and then added to the original digital signal to close the loop.

Considering the PWM and power stage to have a gain G and a gain of $1/G$ in the ADC feedback, the signal transfer function can simplify as shown in Eq. (11).

$$\mathrm{STF_{SE}} = G\frac{1+H}{1+H} = G \tag{11}$$

The noise transfer function of error ε is shown in Eq. (12). This error is suppressed by the loop filter gain in the closed-loop system.

$$\mathrm{NTF_{SE}} = \frac{1}{1+H} \tag{12}$$

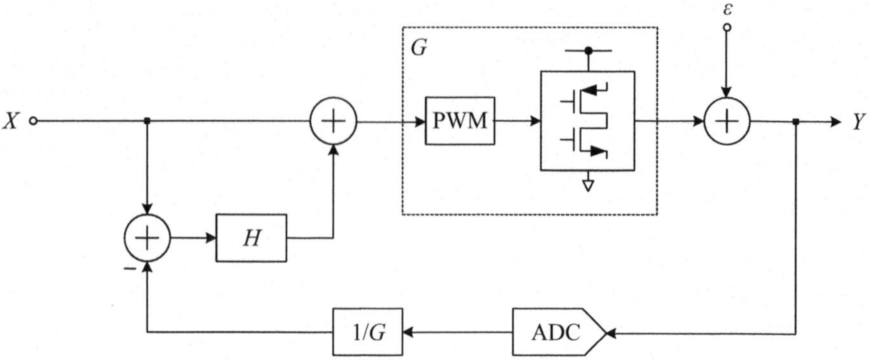

Fig. 7 Closed-loop digital Class D

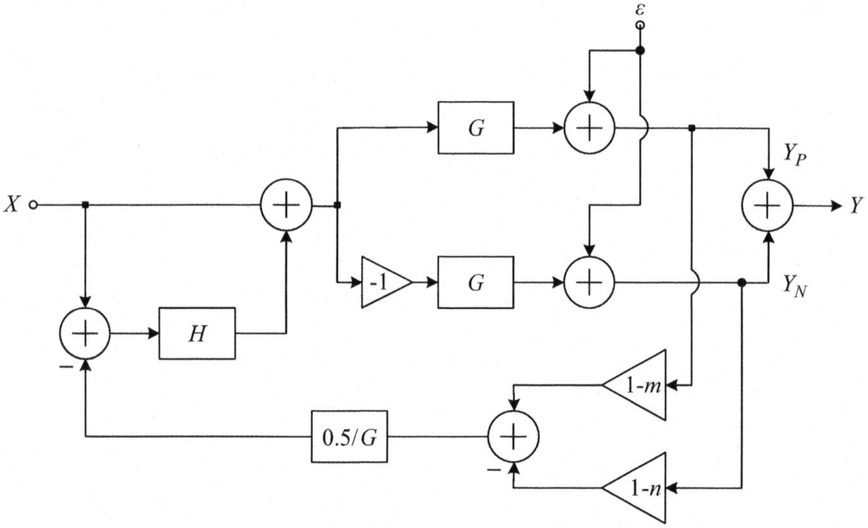

Fig. 8 BTL closed-loop Class D with differential feedback

3.2 BTL with Differential Feedback

A BTL architecture is shown in Fig. 8, where a differential ADC is used to feedback the output and adjust the loop. G is the transfer function of the PWM and power stage, ε is a common mode error, and H is the loop filter transfer function. A limitation of this architecture that needs careful consideration is the matching between the two feedback paths. Gain mismatch is represented in Fig. 8 by using terms m and n, which are set to zero for perfect matching, on the Y_P and Y_N feedback paths, respectively.

The signal transfer function of the differential BTL configuration with a feedback gain of $0.5/G$ is shown in Eq. (13). With no mismatch present, the STF is $2G$. Mismatch between m and n can change the signal gain, which is less of a concern in most applications.

The noise transfer function of common-mode error, ε, is shown in Eq. (14). With no mismatch the common-mode error is perfectly cancelled and the NTF is 0. When there is mismatch present between the two feedback paths, the common-mode error suppression will be limited to the matching between these paths and the closed-loop Class D performance will be limited to that level.

$$\mathrm{STF_{BTL}} = \frac{2G(1+H)}{(1+H(1-\frac{m+n}{2}))} \tag{13}$$

$$\mathrm{NTF_{BTL}} = \frac{H(m-n)}{1+H(1-\frac{m+n}{2})} \tag{14}$$

Matching on deep sub-micron nodes can be sufficient to meet the performance specifications, and additional correction techniques can be utilized to improve the matching, as required. If matching is not sufficient to suppress common-mode errors, a pseudo-differential architecture can be used, where the error suppression is then determined by the loop filter gain. However, the additional power and area overhead of a pseudo-differential architecture should be considered carefully, especially on a deep sub-micron node.

The analysis above is limited to gain matching of the feedback paths in response to a common-mode error. This can be extended to include additional mismatch and error injection mechanisms.

3.2.1 Loop Filter

Given sufficient feedback matching, the loop filter will determine the suppression of the error components due to the power stage. The filter is designed to optimize gain over the frequency range of interest and ensure stability of the system, considering filtering and latency in the feedback path. To ensure stability of the system, the filter bandwidth is reduced to allow for latency in the feedback path. The transfer function of an example implementation of a closed-loop Class D is shown in Fig. 9. The loop filter, H, is a third-order integrator to provide high gain at low frequency and a roll-off to give sufficient suppression for the use case whilst maintaining system stability. The gain of the modulator and power stage can be seen in the STF at low frequency.

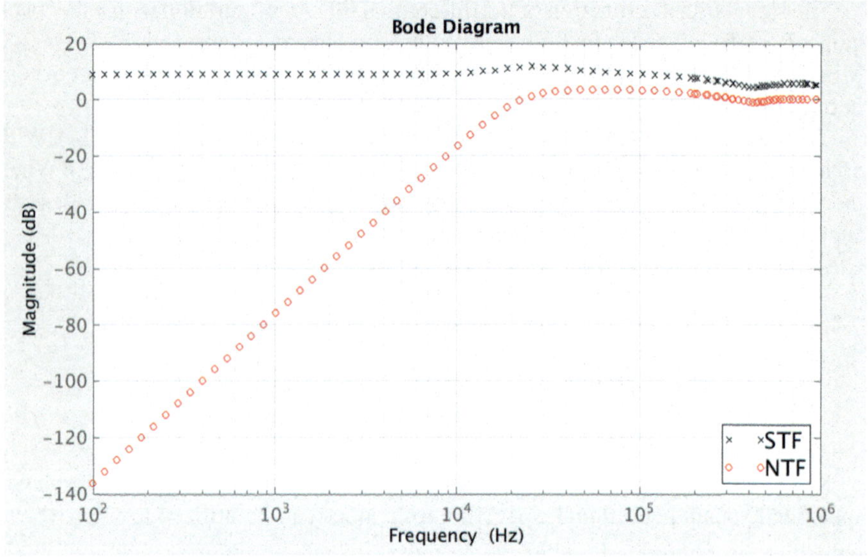

Fig. 9 BTL closed-loop digital Class D transfer function

3.3 ADC

In a closed-loop architecture, the ADC's performance can limit that of the amplifier, and a key challenge is the design of a low power and area ADC. The ADC must be low latency, capable of accepting a PWM signal as its input and have noise and THD performance metrics that satisfy the amplifier requirements.

3.3.1 Design Challenges

To avoid audible hiss in sensitive transducers, high SNR is required in many common use cases in the mobile audio market. To satisfy these cases, a very low noise ADC is required, which can be very power and area hungry. Avoiding the noise limitation of the ADC at low signals in an open-loop/closed-loop architecture can be very beneficial for the ADC design.

The offset of the ADC also requires careful consideration. In the digital closed-loop architecture, the comparison point is happening within the digital and the feedback loop is working to remove errors at that point, not at the Class D output itself. The magnitude of DC offset in the ADC will be seen directly across the load at the Class D output. The offset can cause audible artefacts at startup and shutdown and additional power dissipation at light loads. To address this problem, a servo can be used to remove offset from the ADC.

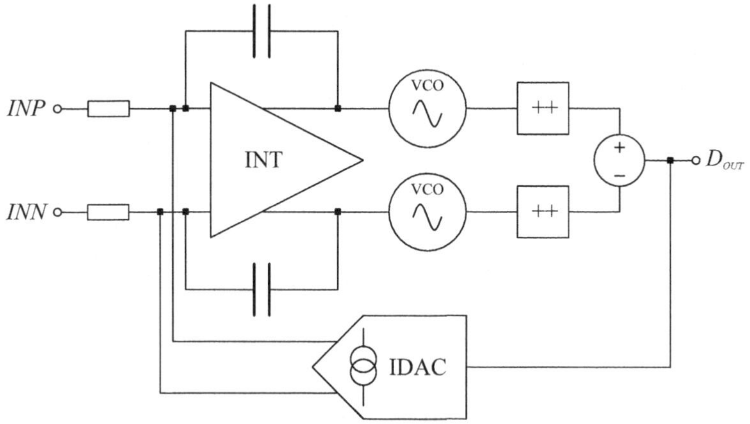

Fig. 10 Second-order continuous-time VCO-based ADC

The Class D output is a switching signal with significant power in the carrier and out-of-band noise from the noise shaping of the modulator. This unwanted frequency content needs to be suppressed by filtering at the front-end of the ADC to avoid the out-of-band energy impacting the performance of the in-band measurement. Latency of the front-end needs to be included in the loop filter design.

3.3.2 Implementation

An ADC implementation well suited to deep sub-micron nodes is a continuous-time VCO-based ADC [3], shown in Fig. 10. This circuit is a second-order sigma-delta-based architecture consisting of a continuous-time integrator, dual VCO quantizer, and a multi-bit feedback DAC. The VCO quantizer has an equivalent transfer function to a first-order sigma-delta modular, noise shaping the quantization noise and leading to a second-order ADC design. VCOs scale well with process node and voltage, making this architecture a good fit for a digital Class D implementation.

3.4 BTL Closed-Loop Digital Class D Simulation

The BTL closed-loop digital Class D architecture is shown in Fig. 11, where G will be defined by the modulation index and power stage gain. Figure 12 shows the output spectrum of an example implementation. It does not include analog noise, and the response of the third-order loop filter can be seen, as discussed in Sect. 3.2.1.

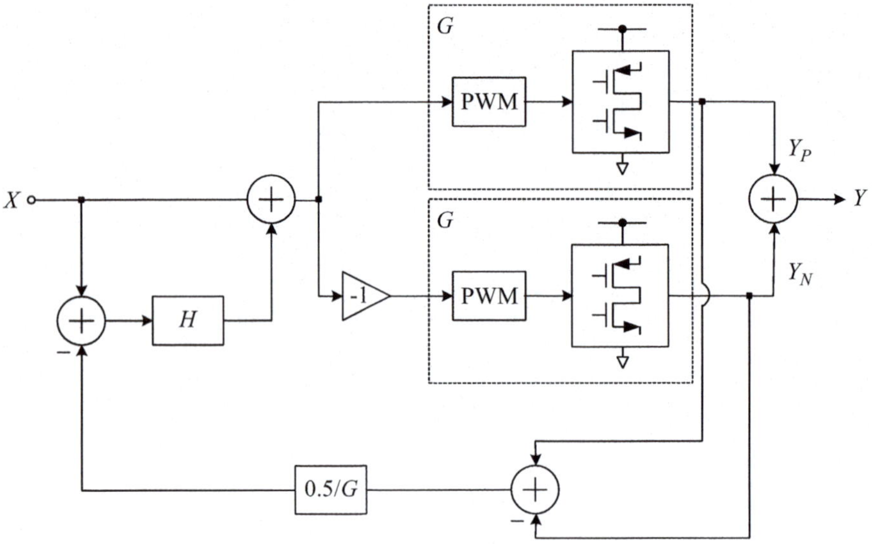

Fig. 11 BTL closed-loop Class D architecture

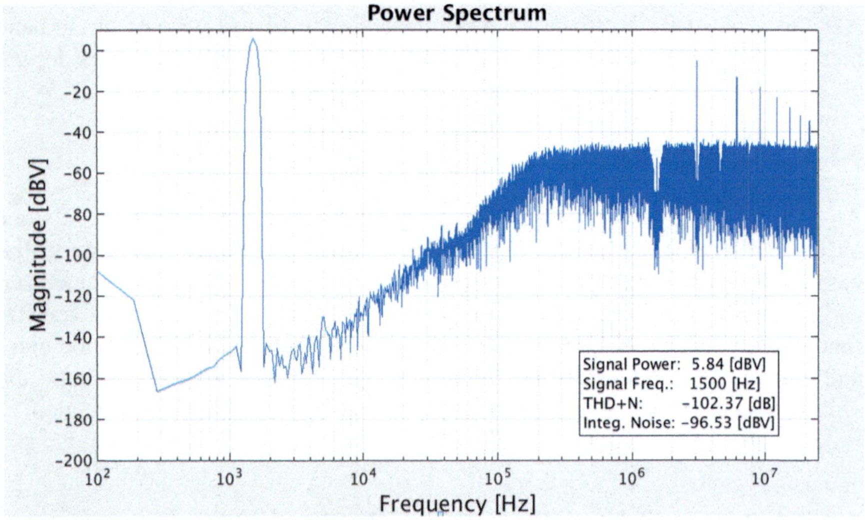

Fig. 12 BTL closed-loop Class D simulation

4 Open-Loop/Closed-Loop

As has been discussed above, open-loop and closed-loop architectures have differing advantages and limitations. By using an architecture that dynamically switches between the two modes depending on output amplitude, Fig. 13, the performance of the digital Class D can be optimized across the full signal range and avoid excessive power and area.

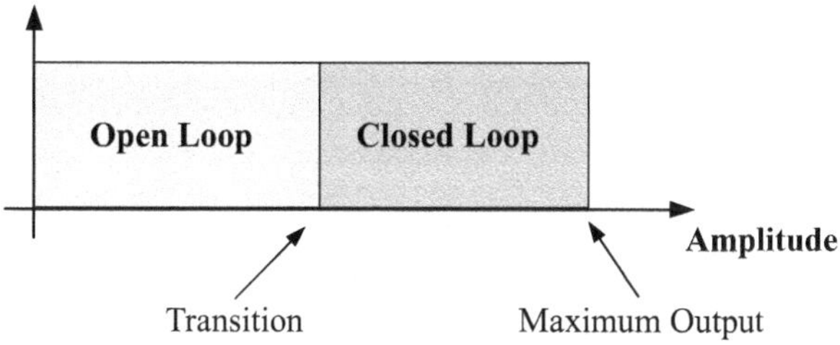

Fig. 13 Dynamic mode switch

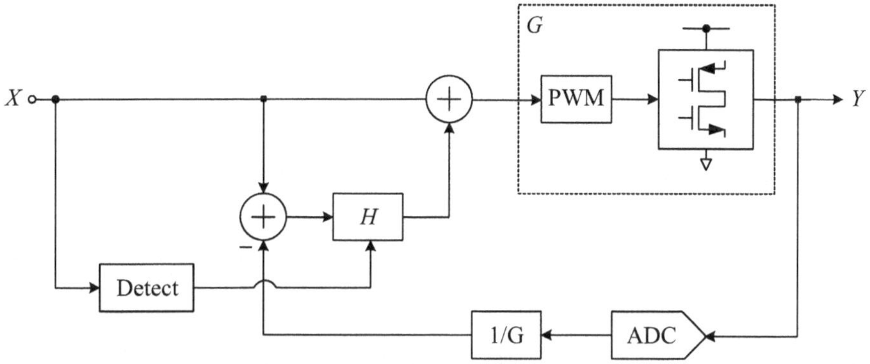

Fig. 14 Open-loop/closed-loop architecture

At low-signal levels, the noise and power consumption are critical requirements to avoid audible noise and increase the standby battery life. Both are key advantages of the open-loop Class D architecture, which should be used at low levels.

When the signal level is higher, error correction to reduce the THD and suppress supply noise is required and a closed-loop Class D configuration can provide this. The noise performance at higher signals can determine the THD+N metric, but this is generally less stringent than the noise performance needed to satisfy SNR requirements. This means that the feedback ADC design does not need to be as aggressive as it would be at lower signals and a low area and power implementation can be used, which is key to making a digital Class D architecture compelling.

As shown in Fig. 14, the detection circuit will monitor the amplitude of the digital signal and adapt the loop filter depending on the level. When the loop filter, *H*, is set to zero, the digital Class D operates in open-loop mode as no error signal is fed back. With the loop filter designed for closed-loop mode errors are suppressed by the gain of the loop filter. The transition between the two states is carefully controlled in the digital, using knowledge of the system, to avoid audible artefacts on the output.

To save power, the ADC can be powered off when in open-loop mode and only enabled when it is needed in closed-loop operation.

The position of the transition point can be tailored depending on the application. Changing at lower signal levels will give high rejection of errors across a larger signal range, whilst transitioning at a higher level will give better efficiency and noise ratio over a wider range.

Table 1 Parametric results

Parameter	Condition	Value	Units
Output power	32 Ω load	120	mW
SNR	1 Vrms output	120	dB
THD+N	70 mW into 32 Ω	−80	dB
PSRR		80	dB
PS-IMD	−4 dBV output at 1 kHz	70	dB

Table 1 summarizes the performance numbers of an example implementation. The SNR and PSRR numbers are dependent on the open-loop mode, whilst the THD+N and PS-IMD are governed by closed-loop performance.

5 Conclusion

A digital Class D architecture has been presented, combining open-loop and closed-loop modes to optimize the performance of the amplifier across the full signal range. The benefits and limitations of the two modes have been discussed to show the significant advantage of using both configurations rather than selecting a single mode. An architecture has been presented that can dynamically change between these two modes of operation avoiding audible artefacts on the output.

To minimize the analog circuitry, and make use of digital functionality, a digital PWM modulator is used to drive the amplifier output through a power stage. In closed-loop operation, the feedback path uses a second-order continuous-time VCO-based ADC, which scales well on deep sub-micron nodes.

The architecture, and implementation presented, is especially suited to deep sub-micron nodes where moving functionality into digital circuits and minimizing the analog design can give area and power advantages.

References

1. Lesso JP, Ido T. Class D amplifier circuit. U.S. Patent 9,628,040 B2, Apr. 18, 2017.
2. Mouton T, Putzeys B. Digital control of a PWM switching amplifier with global feedback. In: Proceedings of the AES 37th international conference. New York; 2009. p. 108–17.
3. Lesso JP, Pennock JL. Analogue-to-digital converter. U.S. Patent 8,742,970 B2, Jun. 3, 2014.

Low Power Microphone Front-Ends

Lorenzo Crespi, Claudio De Berti, Brian Friend, Piero Malcovati, and Andrea Baschirotto

1 Introduction

After the invention of the first microphone in 1876, carbon microphones have been introduced in 1878 as key components of early telephone systems. In 1942, ribbon microphones were developed for radio broadcasting. The invention of the self-biased condenser or electret microphones (ECM) in 1962 represented the first significant breakthrough in this field. Indeed, electret microphones, ensuring high-sensitivity and wide bandwidth at low cost, have dominated the market for high-volume applications until the last decade, when MEMS microphones started to gain popularity [1].

The first microphone based on silicon micro-machining (MEMS microphone) was introduced in 1983. Thanks to the use of advanced fabrication technologies, MEMS microphones offer several advantages with respect to electret devices: better performance, smaller size, compatibility with high-temperature automated printed circuit board (PCB) mounting processes, and lower sensitivity to mechanical shocks. Moreover, MEMS microphones can be integrated together with the CMOS electronics on the same chip or, more commonly, within the same package [2], thus reducing area, complexity, and costs, while increasing efficiency, reliability, and

L. Crespi (✉) · C. De Berti · B. Friend
Synaptics, San Jose, CA, USA
e-mail: Lorenzo.Crespi@synaptics.com

P. Malcovati
University of Pavia, Pavia, Italy

A. Baschirotto
University of Milano-Bicocca, Milan, Italy

K. A. A. Makinwa et al. (eds.), *Low-Power Analog Techniques, Sensors for Mobile Devices, and Energy Efficient Amplifiers*,
https://doi.org/10.1007/978-3-319-97870-3_17

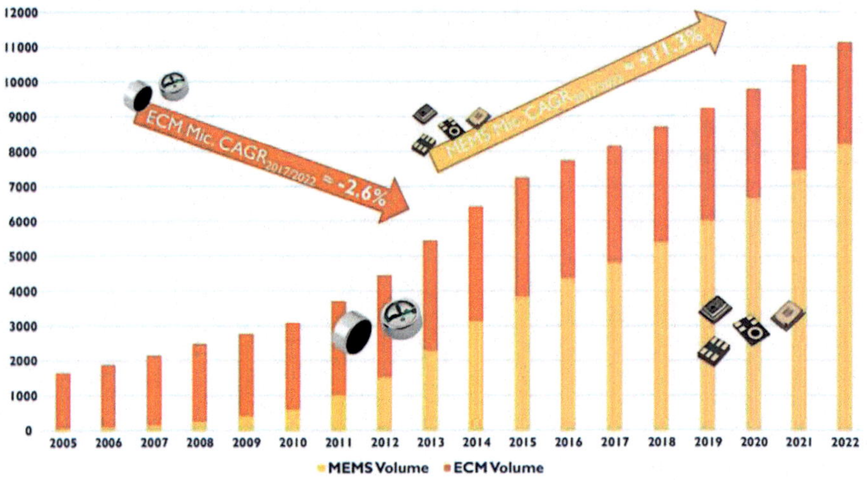

Fig. 1 The microphone market in million units since 2005. (Source: Acoustic MEMS and Audio Solutions 2017 Report, Yole Développement)

performance. As a result, around 2014, MEMS microphones surpassed ECMs in term of sold units, with an annual market size increase of more than 11%, as shown in Fig. 1.

MEMS microphones can be realized by exploiting different transduction principles, such as piezoelectric, piezoresistive, and optical detection. However, more than 80% of the MEMS microphones produced are based on capacitive transduction, since it achieves higher sensitivity, consumes lower power, and is more compatible with batch production.

The front-end circuit is of paramount importance for MEMS microphones, since it represents one of the most significant competitive advantages with respect to ECMs. Therefore, the development of high-performance front-end circuits has always progressed in parallel with the evolution of MEMS microphones [3–11]. This has led to a steady reduction of their power consumption, while maintaining or even improving their audio performance, such as signal-to-noise ratio (SNR), dynamic range (DR), and total harmonic distortion (THD). This trend is mainly driven by portable applications, whose audio-related functionality has expanded significantly. For example, voice interfaces are becoming pervasive. A growing number of people now talk to their mobile devices, asking them to send e-mails and text messages, to search for directions, or to find information on the internet. These functions require continuous listening, thus introducing severe constraints on the power consumption of the microphone modules. Low power consumption is, therefore, the key design goal of modern front-end circuits for MEMS microphones.

2 Capacitive Microphones

A microphone is a sensor that translates a perturbation of air pressure, i.e., sound, into an electrical quantity. In a capacitive microphone, pressure variations cause the vibration of a mechanical mass, which is transformed into a capacitance variation. Sound pressure is typically expressed in dB_{SPL} (sound-pressure-level).

A sound pressure of 20 μPa, corresponding to 0 dB_{SPL}, is generally accepted as the auditory threshold (the lowest amplitude of a 1-kHz signal that a human ear can detect). The sound pressure levels of a face-to-face conversation range between 60 dB_{SPL} and 70 dB_{SPL}. This rises to 94 dB_{SPL} if the speaker is at a distance of 1 inch from the listener (or the microphone), which is the case, for example, in mobile phones. Therefore, a sound pressure level of 94 dB_{SPL}, which corresponds to 1 Pa, is used as a reference for acoustic applications. The performance parameters for acoustic systems, such as the SNR, are typically specified at 1-Pa and 1-kHz. Some additional examples of typical SPL levels are shown in Fig. 2.

2.1 *MEMS Microphones*

A MEMS capacitive microphone, whose simplified structure is shown in Fig. 3, basically consists of two conductive plates at a distance x. The top plate, in this

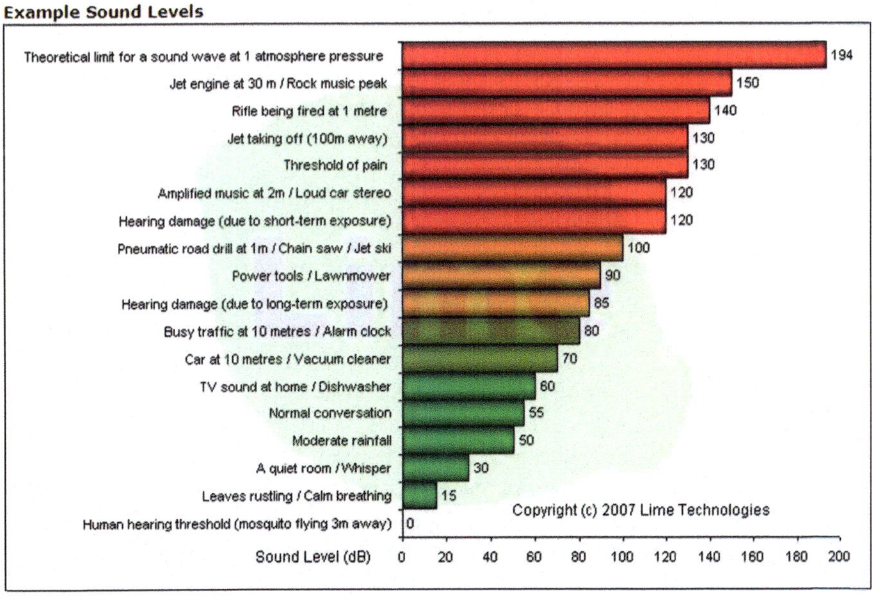

Fig. 2 Example sound levels in dB_{SPL}

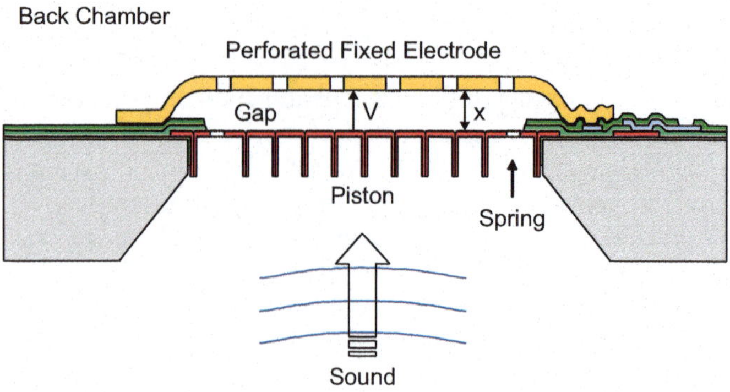

Fig. 3 Basic structure and working principle of a MEMS capacitive microphone

case, is fixed and cannot move, while the bottom plate is able to move in response to sound pressure, producing a variation of x (Δx) with respect to its steady-state value (x_0), proportional to the instantaneous pressure level (P_S). Different arrangements of the electrodes and fabrication solutions are possible, but the basic principle does not change [12–18].

The capacitance of a MEMS microphone is then given by

$$C\left(P_S\right) = \frac{\varepsilon_0 A}{x\left(P_S\right)} = \frac{\varepsilon_0 A}{x_0 + \Delta x\left(P_S\right)} \tag{1}$$

where A is the area of the smallest capacitor plate and ε_0 is the vacuum dielectric permittivity.

The MEMS microphone capacitor is initially charged to a fixed voltage V_B, with a charge $Q = C_0 V_B$, where C_0 is the capacitance value in the absence of sound ($x = x_0$). Therefore, assuming a linear relation between the sound pressure variation ΔP_S and the displacement Δx ($\Delta x = -k\Delta P_S$), the capacitance variation leads to a voltage signal (ΔV) across the microphone, given by

$$\Delta V = \frac{Q}{C(P_S)} - \frac{Q}{C_0} = \frac{Q\Delta x}{\varepsilon_0 A} = -\frac{kC_0 V_B}{\varepsilon_0 A}\Delta P_S = -\kappa \Delta P_S \tag{2}$$

where κ denotes the sensitivity of the microphone. In order to avoid degradation of the voltage signal ΔV, the input impedance of the front-end circuit must be extremely large, thus ensuring that Q remains constant.

In practical implementations, a MEMS microphone is not just a capacitor—some additional parasitic components also have to be taken into account. The equivalent circuit of an actual MEMS microphone is shown in Fig. 4.

Besides the variable capacitance $C(P_S)$, the equivalent circuit includes two parasitic capacitances, C_{P1} and C_{P2}, connected between each plate of the MEMS microphone and the substrate, as well as a parasitic resistance R_P, connected in parallel to $C(P_S)$. The value of these parasitic components depends on the specific

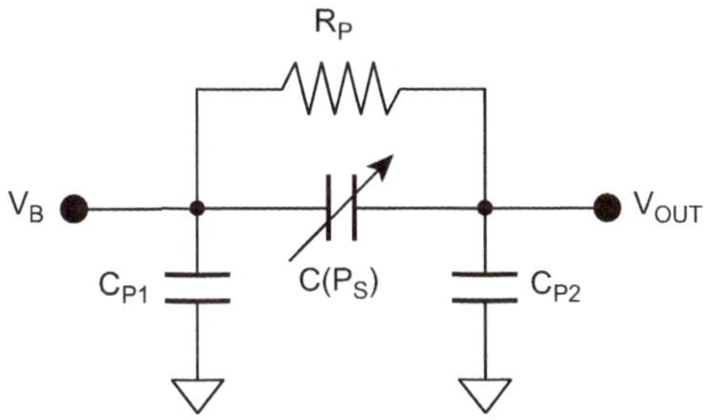

Fig. 4 Equivalent circuit of an actual MEMS capacitive microphone

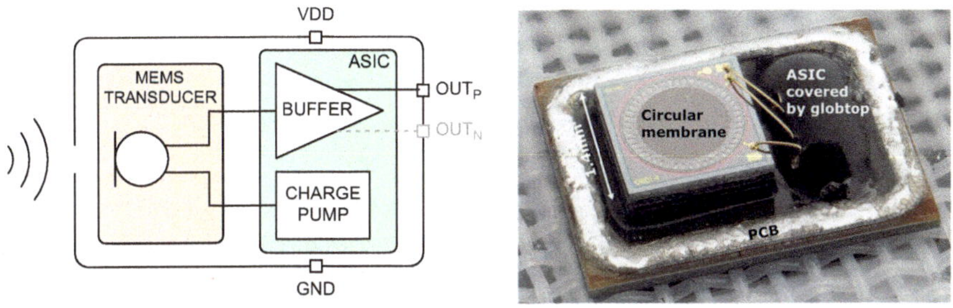

Fig. 5 Typical commercial MEMS microphone module

implementation of the microphone, but typically C_{P1} and C_{P2} are in the order of a few pF, while R_P is in the GΩ range.

2.2 MEMS Microphone Modules

The extremely large source impedance of a capacitive MEMS sensor makes its output signal very susceptible to EM interference and attenuation by routing parasitics. In most systems, it would thus be impractical to route the unbuffered MEMS sensor output, via wires or PCB traces, to the System-on-Chip (SoC) responsible for digitizing and processing it.

A MEMS microphone sensor is typically co-packaged with a small ASIC including biasing and buffering circuits, as shown in Fig. 5. A charge-pump-up converts the supply voltage V_{DD} to generate the MEMS bias voltage V_B. Since the sensor sensitivity is proportional to its bias voltage as shown in (2), V_B is set to a relatively high voltage, typically in the 8–12 V range. V_B is limited on the high side to a critical voltage called the pull-in voltage, at which the MEMS membrane collapses and the device ceases to operate properly.

A simple low-noise amplifier with a very high input impedance then generates a buffered version of the microphone signal, which can be routed via wires or PCB traces to the processing SoC. In its simplest form, this amplifier could be implemented by using a single FET transistor. The output of the microphone module is typically single-ended, but balanced differential outputs are becoming more commonly available, in consideration of the higher performance with negligible additional power consumption.

2.3 Performance of MEMS Microphone Modules

Performance of commercial microphone modules is generally specified by the following key parameters:

Sensitivity The rms voltage produced at the microphone output in response to a 94-dB_{SPL}, 1-kHz sinusoidal input, expressed in dB_V. For modern MEMS sensors, microphone sensitivity typically ranges from -32 dB_V to -42 dB_V.

Sensitivity Tolerance This is a particularly critical parameter for microphone arrays, where mismatched gains can degrade performance of beam-forming and other voice processing algorithms. State-of-the-art MEMS microphones typically achieve $\pm 1\%$ sensitivity matching. This is a significant improvement over ECM microphones that are usually rated at $\pm 3\%$.

Signal-to-Noise Ratio (SNR) The ratio between the output produced by a reference 1-kHz signal at 94 dB_{SPL} and the residual output noise floor with no input, integrated over 20 Hz–20 kHz band with A-weighting. Many recent MEMS microphones achieve SNRs in the 60–70 dB range, with best-in-class modules now approaching SNRs of 75 dB. The best ECM microphones still hold a slight advantage over MEMS devices in this category, reaching up to 80 dB SNR at the expense of much larger physical dimensions.

Acoustic Overload Point (AOP) The sound pressure level at which microphone THD equals 10%. It indicates the maximum acoustic level that the microphone can process without drastically distorting the signal. Typical AOP levels for current MEMS microphones are 120–130 dB_{SPL}, with some microphones now achieving 135–140 dB_{SPL}. The trend in recent years has been toward rapidly increasing AOPs. While the benefit of reaching AOPs larger than the human threshold of pain (see Fig. 2) may seem questionable, at least in the context of consumer electronic products, a high AOP is actually very useful to prevent microphone saturation from wind noise, proximity to a powerful loudspeaker, or from low-frequency thump-like signals, which can occur in a car interior during door closing, or while a train is going through a tunnel, and so on. A temporary microphone saturation can be disruptive to adaptive voice-processing algorithms, such as the ones used in acoustic noise cancelling (ANC) headphones, and should be avoided.

Distortion (THD or THDN) Typically measured at 1 kHz, and at different sound pressure levels, depending on the manufacturer, THD typically ranges from 1% to 0.04%.

Output Impedance Typically, in the 50–1000 Ω range.

Power Supply Rejection (PSRR or PSR) Both indicate the capability of the ASIC to reject spurious noise on the supply voltage; the main difference is that the PSRR is expressed as a dB ratio, while the PSR is expressed in dB_V or dB_V A-weighted ($dB_{V\text{-}Aw}$). Test conditions vary among manufacturers, but generally a 217 Hz or 1 kHz, 100-mV_{pp} square wave or sine wave is injected as supply noise. The typical range for PSRR is 45–75 dB.

3 Microphone Front-End Architecture and Specifications

The interface circuit for a MEMS module reads out an analog signal and converts it to the digital domain. The system diagram for a typical front-end circuit for a MEMS capacitive microphone module is shown in Fig. 6, for both single-ended and differential microphones. The circuit consists of a programmable-gain preamplifier (PGA) followed by an analog-to-digital converter (ADC). The input of the preamplifier is typically AC-coupled to remove the DC voltage at the microphone output. The RC network created by the AC coupling can also be useful as a high-pass filter (HPF) to filter out low-frequency noise, such as those generated by wind and other undesirable acoustic sources.

In the case of a single-ended microphone output, it is best to AC couple the ground terminal of the microphone to the negative input of the PGA, in order to reject common-mode interference that may couple into the wiring or the PCB traces. A series resistor on the ground line is often used to equalize the impedance level on the negative line, which improves RFI rejection [19]. Series ferrite beads

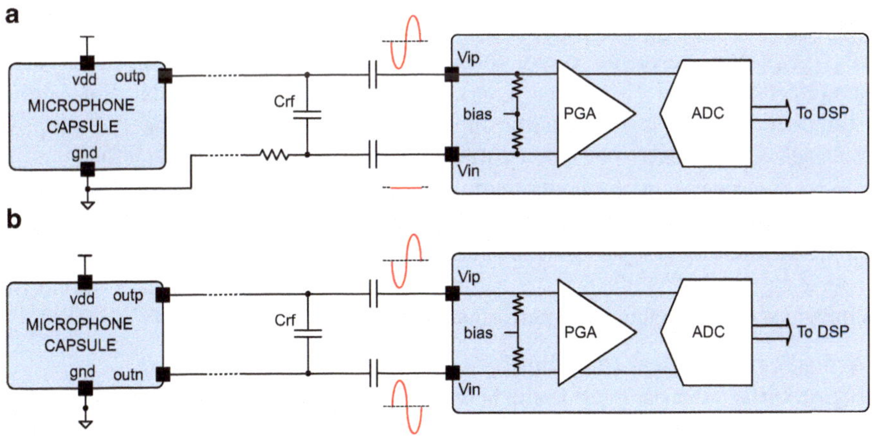

Fig. 6 Typical block diagram of the front-end circuit for a MEMS microphone module: (**a**) single-ended microphone; (**b**) differential microphone

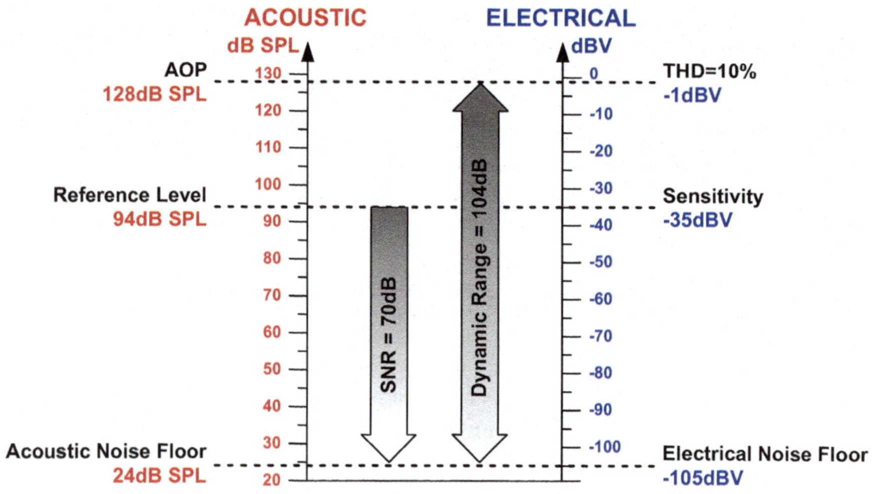

Fig. 7 Microphone parameters in acoustic vs. electrical domain

and/or a small RF shunt capacitor are also commonly placed to reduce RF noise in traces [20].

3.1 *Interface Requirements*

In general terms, the fundamental requirement of a microphone interface is to digitize the analog signal from the microphone *without significantly degrading its quality*. Since the microphone module is usually selected by the system manufacturer based on various criteria (cost, performance, physical dimensions, manufacturability, business relationships, etc.), it is imperative for a general-purpose microphone interface to be able to efficiently couple with a wide range of state-of-the-art commercial microphones modules. The following section describes how the key microphone parameters can be translated into electrical specifications for its interface circuits. The performance quality of different microphone front-ends is of course in trade-off with the power consumption. Generally, the higher the power consumption, the better is the performance.

Acoustic to Electrical Domain

Figure 7 illustrates the relationship between microphone sensitivity, SNR, AOP, and DR, in both acoustic and electrical domains, for a hypothetical microphone with −35-dB sensitivity, 70-dB SNR, and 128-dB_{SPL} AOP.

In Fig. 8, the SNR/sensitivity/AOP specifications for available MEMS modules from various manufacturers are collected and translated into noise floor and voltage

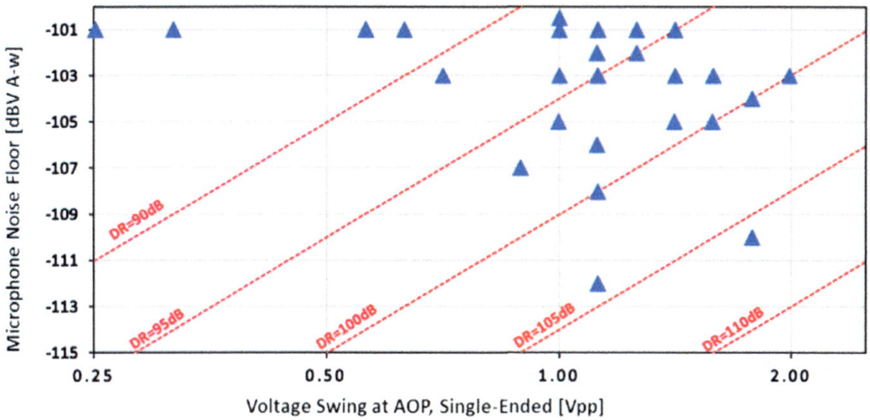

Fig. 8 Voltage swing and electrical noise floor for commercial MEMS microphones

swing at the microphone output. The voltage swing is shown as peak-to-peak single-ended, as this is the most useful information to determine headroom requirements for the preamplifier. From this chart, a few key parameters for the interface circuit can be extracted:

Max Input Voltage Swing While conventional ECM (and earlier MEMS) microphones typically produce a signal in the order of 100 mV_{pp} or less, recent MEMS microphones with high AOP and sensitivity can generate a significantly larger signal, in the order of 1–2 V_{pp} single-ended or 2–4 V_{pp} differential. A general-purpose microphone interface should be able to handle such signal without distorting; depending on the circuit architecture, this can entail using a higher supply voltage for the input stage of the preamplifier relative to the rest of the interface circuitry.

Input-Referred Noise and Dynamic Range Many high-end MEMS microphones have an output noise floor close to $-105\ dB_{V\text{-}Aw}$, with the best in class reaching up to $-112\ dB_{V\text{-}Aw}$. Therefore, a high-performance microphone interface should have an input-referred noise lower than $-118\ dB_{V\text{-}Aw}$, in order to avoid degradation of the overall SNR and DR (this, of course, requires higher power).

Preamplifier Gain The preamplifier buffers the signal from the microphone and scales its amplitude to match the full-scale of the ADC. In principle, a fixed preamplifier gain is sufficient; however, meeting all worst-case requirements for voltage-swing and input-noise simultaneously is a very challenging proposition. Handling 2 V_{pp} full-scale with a –118 $dB_{V\text{-}Aw}$ noise floor requires an ADC dynamic-range of 115 dB, which can be expensive in terms of die area and power. To alleviate the ADC requirements, a preamplifier with variable gain is generally employed to compensate for different microphone sensitivities. The low-end of the preamp gain range is determined by the largest microphone signals, as discussed in the previous paragraph. Assuming an ADC full-scale of 1 V_{rms} differential, and a max input swing of 2 V_{pp} single-ended, a minimum preamplifier gain of 3 dB is adequate. At the high-end, preamplifiers have traditionally implemented gains in the 20–40 dB range; however, given the recent increase in microphone AOP levels, this is no longer possible. As shown in Fig. 8, most modern MEMS microphones can generate at least 0.5–1 V_{pp} near AOP, which limits the max usable gain to 12–15 dB. The preamplifier gain steps should be 3 dB or less to allow tailoring the interface characteristics to the specific microphone used in the system.

AC Versus DC Coupling AC coupling is prevalent because it blocks the unknown DC voltage across the microphone with no power consumption or performance impact. This is typically implemented with an external and expensive capacitor in the order of a few μF to keep the high-pass pole in the order of 1 Hz. DC coupling is recently being introduced for applications that have stringent constraints for PCB area or BOM cost. A few solutions have been proposed to implement DC coupling [21–24]. However, a trade-off between power consumption, SNR performance, and/or die area is generally unavoidable when designing DC-coupled preamplifiers. This chapter focuses on AC-coupled interfaces.

Input Impedance The source impedance of MEMS microphones typically ranges from 200 Ω to 1 kΩ (or 2.2 kΩ if ECM mics are included). Even MEMS microphones with low-output impedance are often current-limited and unable to drive their peak signal into heavy resistive loads. To avoid significant attenuation and distortion of the microphone signal, a general-purpose preamplifier must present an input impedance in the order of 10 kΩ or larger. The presence of an AC-coupling capacitor on the microphone inputs adds further restrictions to the preamplifier input impedance, due to the HPF formed with the input resistance of the stage.[1]

Linearity Given that most microphones are limited to ≥0.04% THD (−68 dB), the linearity requirement for the interface circuit is fairly relaxed compared to other parameters. A THD < −75 dB is typically sufficient for most applications.

[1]To achieve 20 Hz cutoff frequency with a 10-kΩ input resistance, the AC-coupling caps must be of the order of 1 μF. While 1-μF ceramic capacitors are widely available even in very small form factor, their large voltage coefficient can create a significant nonlinearity at low frequencies. For this reason, it is strongly preferable to utilize capacitors in the order of 10 nF, which requires a preamplifier input impedance in the order of 1 MΩ.

In the following, the circuit and system solutions for each block (PGA and ADC) will be introduced, emphasizing the trade-off between power consumption and performance.

4 Preamplifier Design

A conventional preamplifier consists of a resistive feedback operational amplifier with a large input resistor, as shown in Fig. 9.

This architecture is used in many commercial products, since it is quite simple and ensures good linearity even with large input signals, but it has several limitations to realize wide gain range. Indeed, to avoid attenuating the microphone signal, the input resistor should be large, thus requiring an even larger feedback resistor. As a result, both the preamplifier area and input referred noise become excessive. To overcome these limitations, a convenient solution is to use a preamplifier based on a transconductance input stage, as shown in Fig. 10, thus achieving both high-input impedance and high gain range without requiring large resistors that contribute noise.

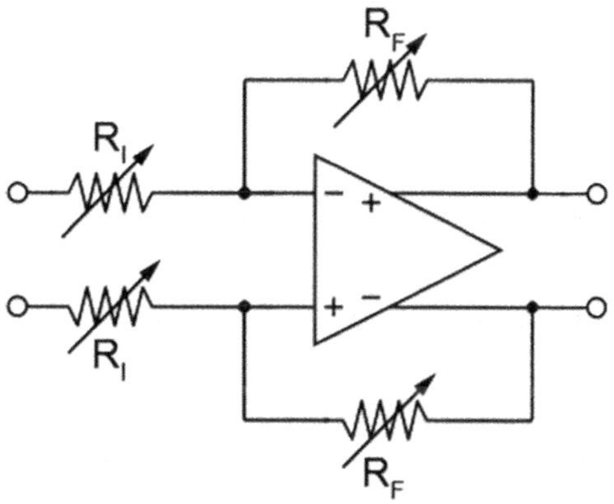

Fig. 9 MEMS microphone preamplifier with resistive feedback

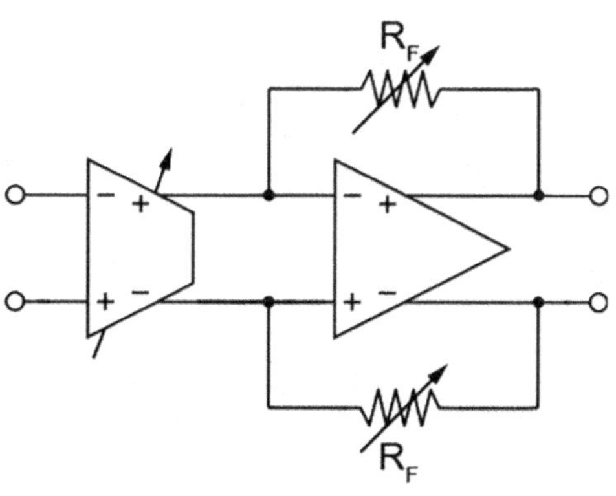

Fig. 10 MEMS microphone preamplifier with transconductance input stage

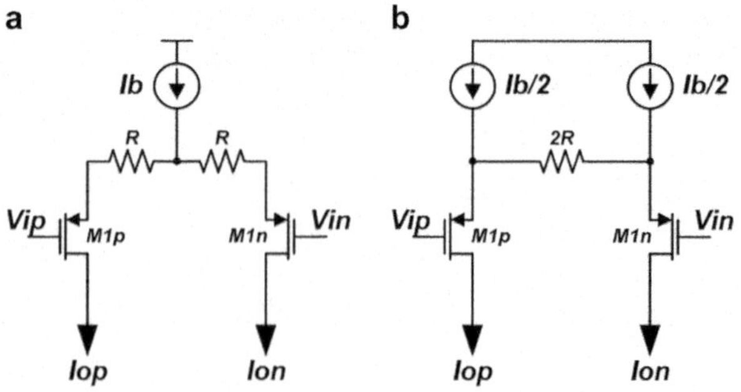

Fig. 11 Simple implementation of transconductor with lumped tail current (**a**) and with split tail current (**b**)

4.1 State of the Art: Transconductance Amplifier

The efficiency of the circuit of Fig. 10 depends on the implementation of the input transconductance stage, which must combine low power consumption with wide gain programmability.

In its simplest form, a linearized transconductor can be implemented as a source-degenerated differential pair biased at a constant current I_b. Its total transconductance is $G_m = g_m/(1 + g_m R)$, which can be approximated to R when $R \gg 1/g_m$.

Figure 11 shows two implementations of a source-degenerated differential pair. The two solutions provide the same input/output transfer function, but version (b) is often preferred because of its improved voltage headroom, given the fact that current I_b does not flow through the degeneration resistors. However, version (a) presents a fundamental advantage noise-wise: the noise current associated with the bias current I_b splits equally between I_{op} and I_{on} when $V_{ip} \approx V_{in}$ (small signal conditions) and becomes a common-mode noise component that is rejected by the following trans-resistance stage. On the other hand, in version (b), the two tail currents produce uncorrelated noise currents which are added to the differential signal current. Moreover, their mismatch would produce offset. This makes structure (a) the better choice for audio preamplifiers.

At full-scale signal conditions, the two circuits are almost equivalent as the noise from I_b is steered completely into I_{op} and I_{on} and is added to the differential signal current. With a full-scale sinusoidal input, version (a) retains a 3-dB advantage over version (b).

To further enhance the transconductor linearity, transistors M_{1p} and M_{1n} can be supplemented with feedback structures that decrease their output resistance and generate a more accurate copy of $V_{ip} - V_{in}$ voltage across resistor R_1. A well-known example based on the super-source-follower (SSF) is shown in Fig. 12b. This simple circuit is very effective in this application, and relative to (a), it biases

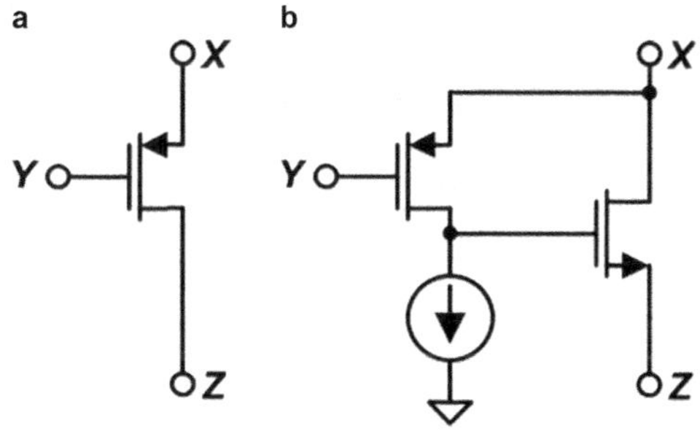

Fig. 12 From single transistor to Type-II current conveyor

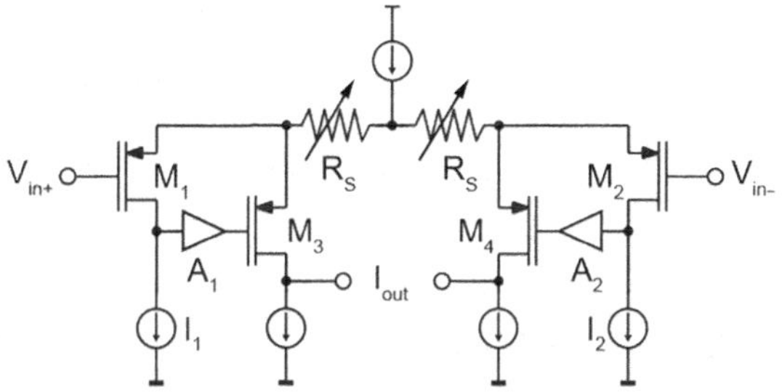

Fig. 13 Schematic of the transconductance stage for a MEMS microphone preamplifier

the input transistor M_1 at constant current, thus maintaining a signal-independent V_{gs} and dividing the impedance on node X by a factor $g_m r_0$.

A good example of a MEMS microphone preamplifier based on this technique has been proposed in [25] and is shown in Fig. 13. Transistors M_1 and M_2, current sources I_1 and I_2, and inverting amplifiers A_1 and A_2 form an active feedback loop for improving linearity. The effective transconductance of the stage is determined by the source degeneration resistances R_S ($g_m = 1/R_S$). Compared to a conventional degenerated differential pair, the linearity and gain accuracy of this transconductor are enhanced by an additional factor $g_{m1,2}A_{1,2}R_{X,Y}$, where $g_{m1,2}$ is the transconductance of M_1 and M_2, $A_{1,2}$ is the gain of the inverting amplifiers, and $R_{X,Y}$ is the impedance at node X or Y. With these additional design parameters, the input-referred noise, the linearity, and the gain accuracy can be optimized independently. The noise effect of M_3, M_4, and R_S is the same as in a conventional degenerated differential pair, but the high-loop gain of the active feedback loop helps to reduce the input-referred noise of all the components except transistor M_1, M_2, I_1, and I_2. Compared to a conventional transconductor, this circuit achieves better linearity and gain accuracy with equal or lower power consumption.

The THD + N of a preamplifier based on the scheme shown in Fig. 13, featuring a gain range from 22 dB to 42 dB is illustrated in Fig. 14. This preamplifier consumes 350 μW.

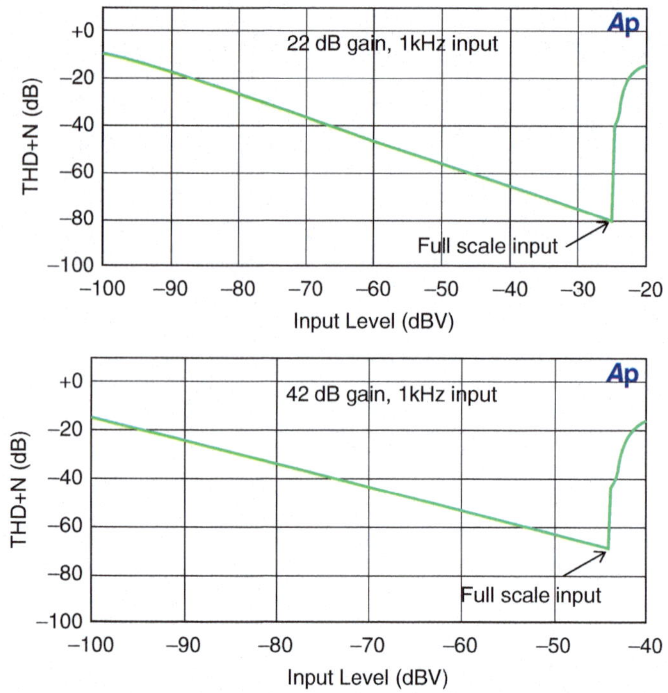

Fig. 14 Measured THD + N of a MEMS microphone preamplifier with transconductance input stage

4.2 Improving the Transconductance Amplifier

"Class-H" Adaptive Biasing

Further improvement in terms of efficiency can be achieved with adaptive biasing techniques, which allow the average power consumption of audio circuits to be reduced and takes advantage of the bursty nature of voice/audio signals. Some authors have proposed a bandwidth-adaptive preamplifier [26]. Examples of amplitude-adaptive amplifiers have been proposed in [27].

A conventional source-degenerated transconductor is biased in Class-A, with a constant current equal to or larger than the peak output current. However, when the incoming signal has small amplitude, the biasing current can be temporarily reduced without incurring any performance penalties. The amount of instantaneous bias current is controlled by an envelope detector circuit which tracks the amplitude of the input signal. This principle can be seen as the current-domain analog of traditional class-H voltage amplifiers. An envelope detector that can be used to adjust the tail current of the main transconductor is shown in Fig. 15.

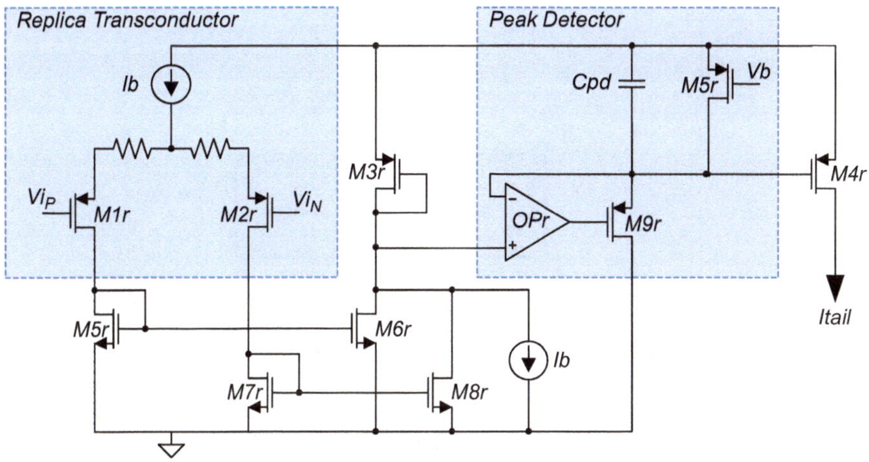

Fig. 15 Envelope detector used to generate the transconductor bias current

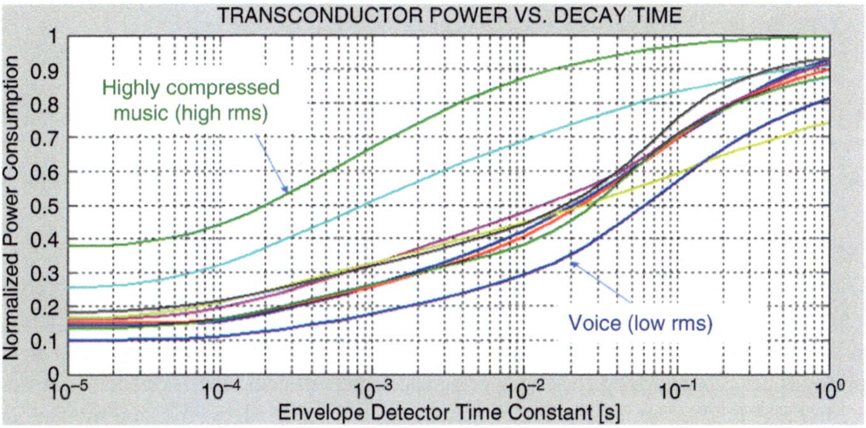

Fig. 16 Transconductor power vs. envelope detector time constant

A scaled version of the main transconductor generates a differential current proportional to the input signal, which is then rectified, converted to voltage-mode by transistor M_{3r}, and processed by a peak detector with a long decay time-constant in the millisecond range. The leaky element of the peak detector is implemented by a long-channel p-channel transistor M_5 biased in deep sub-threshold region. The output of the peak detector is then converted back to current I_{tail} by transistor M_{4r}.

A long time constant in the peak detector is useful to filter audio-band components from current I_{tail}, which could degrade overall THD due to the finite CMRR of the main transconductor. However, a trade-off exists between THD and power efficiency: a longer time constant keeps the PGA operating at high bias currents for a larger percentage of time. Figure 16 shows the theoretical transconductor power

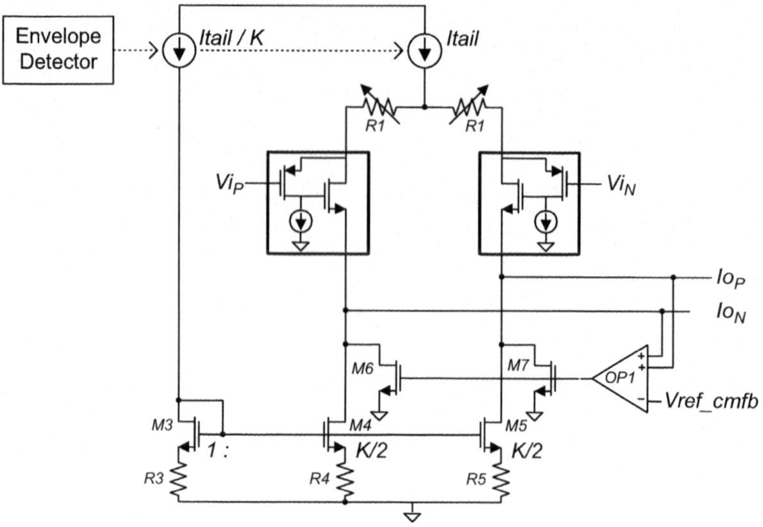

Fig. 17 Top level of transconductance amplifier

consumption vs. time constant, for various speech and music signals, normalized to the power consumption of an ideal Class-A transconductor. For a 10-ms time constant, the power savings from Class-H operation range from 12% (green curve, highly compressed music) to 71% (blue curve, speech).

Main Transconductor Circuit

The overall circuit for the transconductor is shown in Fig. 17. The variable tail current from the envelope detector is mirrored by source-degenerated n-channel transistors M_3, M_4, and M_5 to remove the common-mode component of I_{tail} from the output currents. Since the mirroring operation unavoidably introduces errors, a residual common-mode current exists and is cancelled by the common-mode feedback loop formed by OP_1, M_6, M_7.

Transconductor Gain-Selection Switches

The PGA gain is selected by switching the amount of degeneration resistance R_1. This optimizes noise vs. signal amplitude and, hence, maximizes efficiency. The switched resistor array is shown in Fig. 18. Since the switches are in series with the poly resistors and carry signal-dependent current, the linearity of the switch resistance directly impacts the THD performance of the PGA.

The voltage on the switch source V_{tail} is a rectified and level-shifted version of the input signal, which makes it impractical to implement the ON switches with p-channel transistors biased at $V_g = 0$, unless an extremely large W/L is chosen. Instead, the gate of the ON switches is biased at voltage $V_{bON} = V_{tail} - R_{LS}I_{bLS}$, therefore achieving a constant-V_{gs} biasing that makes the switch resistance nearly constant across signal swing. Current I_{bLS} is chosen to be $\ll I_b$.

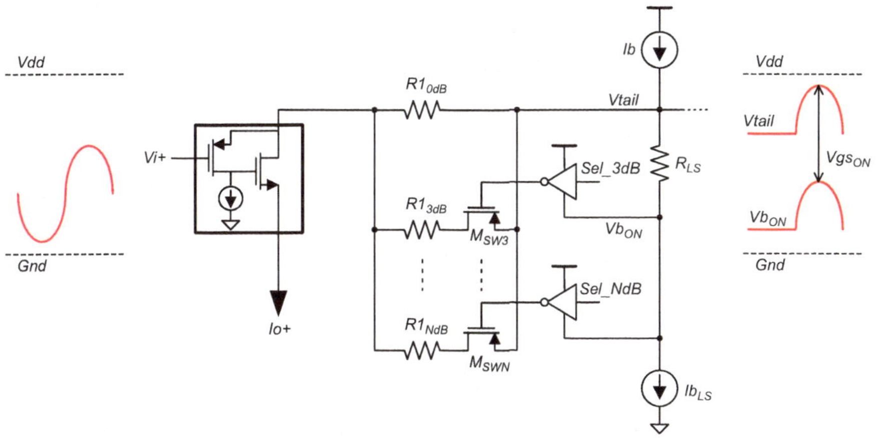

Fig. 18 Biasing of gain-selection switches

Supply Voltage Selection

Power consumption in the PGA can be minimized by selecting the most appropriate supply voltage for a given PGA gain setting. In most battery-powered systems, at least two power supplies are available: the battery itself (with a typical value of 3.7 V for Li-ion batteries) and one or more regulated supplies whose voltage depends on technology selection.

Low PGA gain is used for highly sensitive microphones that can output as much as 2 V_{pp} single-ended. In this case, the battery voltage should be used to maximize headroom. One problem with this approach is that the battery voltage is variable and generally quite noisy, due to its connections to DC/DC converters, RF power amplifiers, etc. Unless the tail current of the transconductor is designed to achieve very high PSRR, it is advisable to insert an LDO between the battery and the PGA supply.

Only the transconductor stage needs the higher supply voltage; the trans-resistance stage that follows can always be operated at the lower supply voltage.

For gains of 12 dB (signal $\leq$ 0.25 V_{rms}) or more, the signal swing is low enough to allow operation of the transconductor at 1.8 V.

The DC bias voltage at the transconductor input must be adjusted with the supply voltage, in order to keep the signal swing centered in the linear region of the transconductor.

Current Sources with Variable Source-Degeneration Resistors

A trade-off between noise and headroom exists when sizing the source degeneration resistors used for the noise-sensitive current sources: for a given current level, higher degeneration resistance means lower $1/f$ noise and higher voltage headroom.

When the PGA operates in its lowest gain setting (high-sensitivity microphone), the large signal swing requires using a minimal amount of resistive degeneration. This is acceptable since the input-referred noise can also be increased in large signal conditions. As the gain increases, the headroom requirements become more relaxed, while the noise requirements become more stringent, and it is appropriate to progressively increase the amount of source degeneration resistance.

5 A/D Converter

The ADC in MEMS microphone front-end circuits is typically implemented with a $\Sigma\Delta$ Modulator ($\Sigma\Delta$M), which exploits oversampling to achieve the required DR. In particular, continuous-time (CT) $\Sigma\Delta$Ms represent the most promising solution for minimizing power consumption, since they require operational amplifiers (op-amps) with lower bandwidth with respect to switched-capacitor (SC) $\Sigma\Delta$Ms, which have been traditionally used. The Schreier figure of merit, defined as $\mathrm{FoM_S} = \mathrm{DR} + 10\log(B/P)$, B being the bandwidth and P the power consumption, is a useful indicator to compare different ADC solutions. Figure 19 shows the values of $\mathrm{FoM_S}$ of recently published ADCs as a function of the Nyquist frequency, $F_N = 2B$.

5.1 *State of the Art: Continuous-Time $\Sigma\Delta$ Modulator*

In the audio field (B = 20 kHz), best-in-class performance ($\mathrm{FoM_S}$ = 180 dB) has been achieved with the third-order CT $\Sigma\Delta$M with 15-level quantizer, whose block diagram is illustrated in Fig. 20. It achieves excellent efficiency thanks to several circuit and system choices as follows [28].

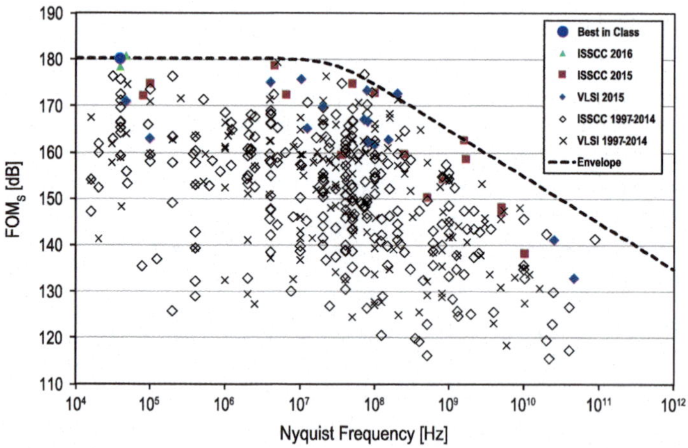

Fig. 19 ADC state of the art based on $\mathrm{FoM_S}$ from [29]

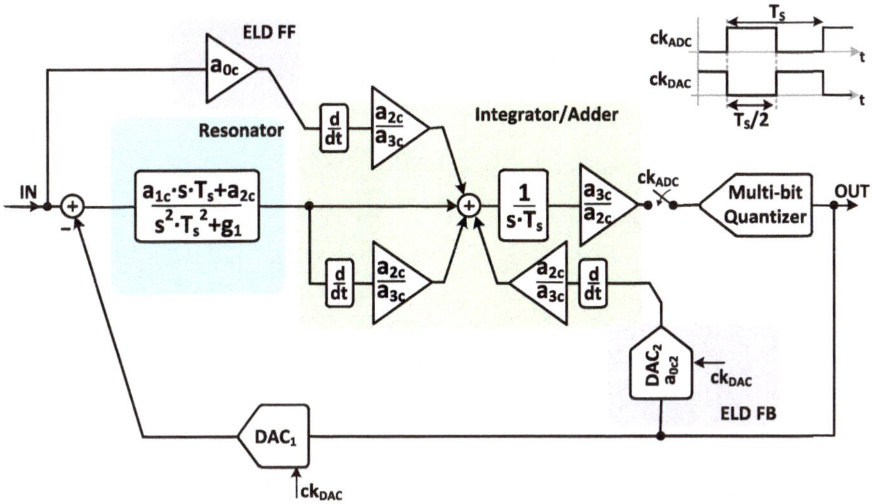

Fig. 20 Block diagram of the CT $\Sigma\Delta$M

The loop filter of the CT $\Sigma\Delta$M consists of a resonator (second-order transfer function) followed by an integrator. A local feedback DAC around the quantizer (DAC_2) and a dedicated feedforward path are used for compensating the excess loop delay (ELD). The feedforward paths of the loop filter and the local ELD feedback are differentiated and added at the input of the integrator, in order to avoid an active adder at the input of the quantizer. The multi-bit quantizer drives a 15-level DAC (DAC_1) with dynamic element matching (DEM) to close the main feedback loop of the CT $\Sigma\Delta$M.

The schematic of the active-RC implementation of the CT $\Sigma\Delta$M is shown in Fig. 21. The resonator is implemented using a single op-amp, and no active adder is used at the input of the quantizer, thus, requiring only two op-amps for implementing the third-order loop-filter transfer function. The local feedback DAC for ELD compensation is implemented with a SC structure, whereas the main feedback DAC is realized with a three-level (−1, 0, 1) current-steering topology, which guarantees minimum noise for small input signals. Indeed, with the three-level topology, the unused DAC current sources are not connected to the resonator input and, hence, they do not contribute to the CT $\Sigma\Delta$M noise. The multi-bit quantizer is realized with 14 identical differential comparators and a resistive divider from the analog power supply for generating the threshold voltages. The values of the passive components used for implementing the CT $\Sigma\Delta$M are summarized in Table 1. The value of R_i has been chosen as low as 47 kΩ to fulfill the thermal noise requirements, while R_1, R_3, R_4, C_1, C_2, C_f, and C_4 are obtained consequently to achieve the desired CT $\Sigma\Delta$M coefficients. Eventually, resistor R_i can be removed if the preamplifier is realized with a transconductor which provides directly an output current. Both op-amps are realized with a two-stage, Miller compensated topology in which transistor size and

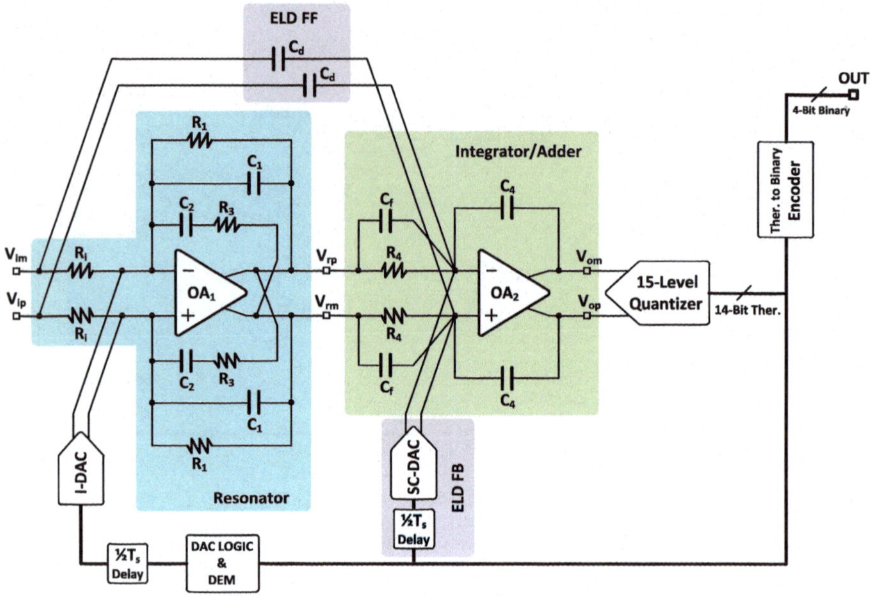

Fig. 21 Schematic of the active-RC implementation of the CT ΣΔM

Table 1 Values of the passive components used for implementing the CT ΣΔM

Resistor	Value	Capacitor	Value
R_i	47 kΩ	C_1	18.5 pF
R_1	5.7 MΩ	C_2	18.7 pF
R_3	57 kΩ	C_f	2.1 pF
R_4	1 MΩ	C_4	1 pF

bias current are sized to fulfill the noise requirements (the values in the second op-amp are scaled with respect to the first one, since its noise contribution is negligible).

The CT ΣΔM has been fabricated using a 0.16-μm CMOS technology. The micrograph of the 0.21-mm^2 chip is illustrated in Fig. 22. Figure 23 shows the measured SNDR as a function of the input sinusoidal signal amplitude at 1~kHz. The full-scale input signal (0 dB_{FS}) corresponds to 1 V_{rms} differential. The achieved DR is 106 dB (A-weighted), corresponding to an ENOB of about 17 bits, whereas the peak SNDR is 91.3 dB. The change of slope in the SNDR curve for input signal amplitudes larger than −17 dB_{FS} is due to the increased current-steering DAC noise when more than 1 three-level DAC element is used (acceptable for the microphone application, where the performance for large input signals is limited by the microphone itself).

The CT ΣΔM output spectra obtained with −60 dB_{FS} and − 1 dB_{FS}, 1-kHz input signals are shown in Fig. 24. As expected, at −1 dB_{FS}, the noise floor increases of about 10 dB with respect to −60 dB_{FS}, due to the increased DAC noise. Figure 25 shows the measured inherent anti-aliasing properties of the CT ΣΔM. The spectral

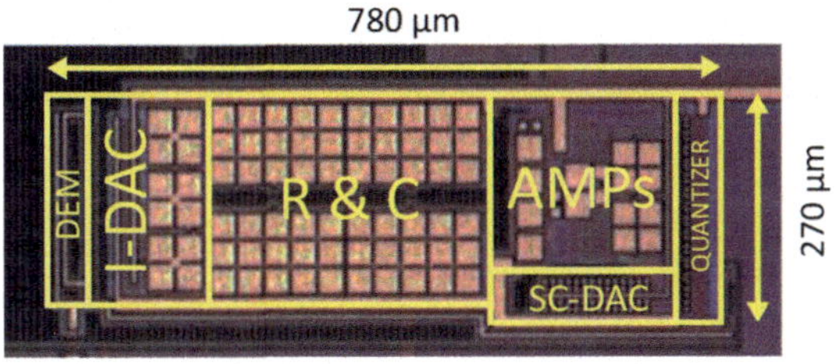

Fig. 22 Chip micrograph of the CT $\Sigma\Delta$M

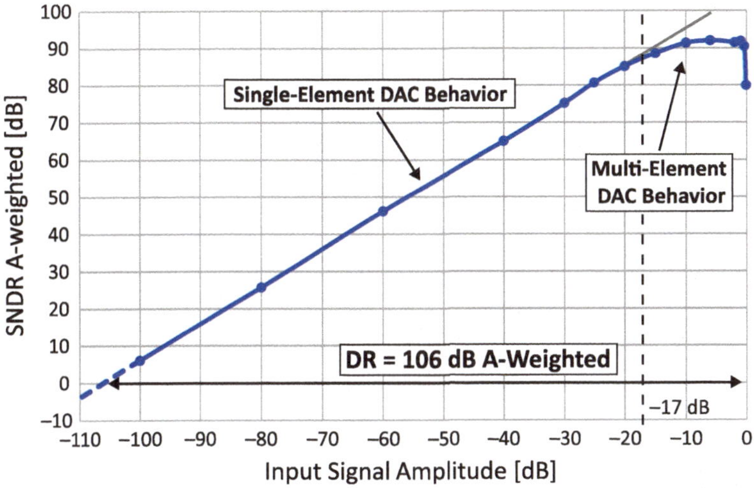

Fig. 23 Measured SNDR of the CT $\Sigma\Delta$M vs. input signal amplitude

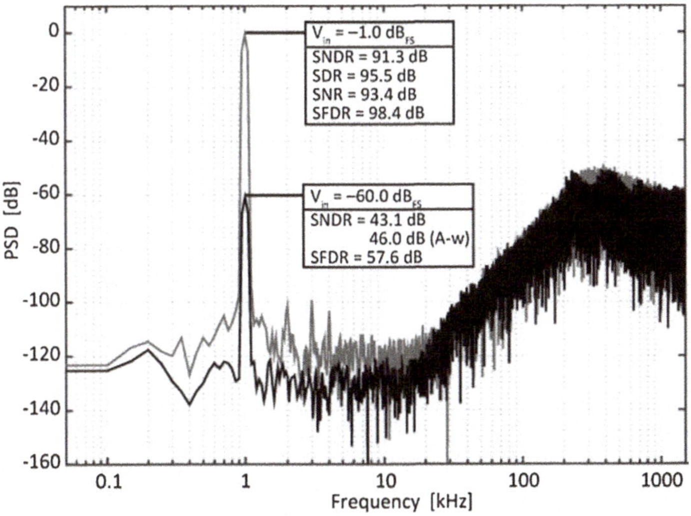

Fig. 24 Measured output spectra of the CT $\Sigma\Delta$M with -60 dB_{FS} and -1 dB_{FS}, 1-kHz input signals

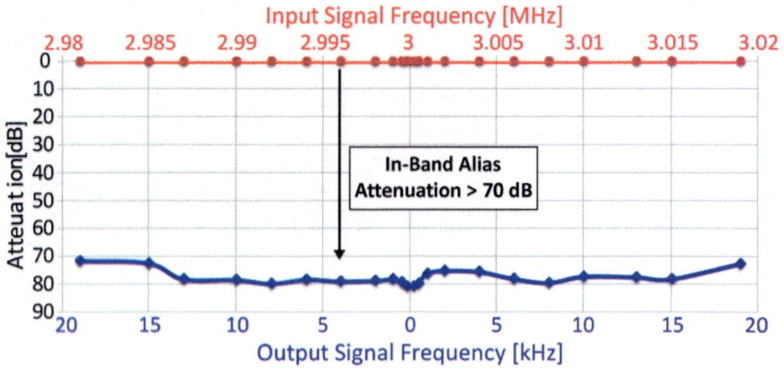

Fig. 25 Measured anti-aliasing properties of the CT $\Sigma\Delta$M

Table 2 Performance summary of the CT $\Sigma\Delta$M

Parameter	Value
Technology [nm]	160
Architecture	CT
Supply voltage [V]	1.6
Power consumption (*P*) [mW]	0.39
Bandwidth (*B*) [kHz]	20
Oversampling ratio (OSR)	75
Area [mm^2]	0.21
Peak SNDR [dB]	91.3
Dynamic range (DR) [dB]	103.1
Dynamic range A-weighted (DR$_A$) [dB]	106
Schreier figure of merit (FoM$_S$) [dB]	180

components around f_s are aliased back to the audio band, but with an attenuation of more than 70 dB, in excess of the application requirements. This value is typical of a CT $\Sigma\Delta$M based on the CIFF topology.

The analog section of the third-order CT $\Sigma\Delta$M consumes 350 μW, while the digital blocks (i.e., DEM and thermometer-to-binary converter) consume 40 μW, both from a 1.6-V power supply and during conversion. The FoM$_S$ is 180 dB. Table 2 shows a summary of the performance achieved by the CT $\Sigma\Delta$M.

5.2 *Future Trends*

Further efficiency improvements in microphone front-ends are under development, and some of them are reported here.

Higher Quantizer Resolution to Decrease Sensitivity to Clock Jitter
One major drawback of CT-$\Sigma\Delta$Ms with respect to SC architectures is the increased DR degradation in the presence of clock jitter. In fact, in CT-$\Sigma\Delta$Ms the jitter on the

clock used by the feedback DAC produces an equivalent noise component, which is directly added to the input signal, while this is not the case in SC structures, in which the clock jitter only affects the input signal sampling.

In first approximation [30], for a multibit CT-$\Sigma\Delta$M, the expected value of the signal-to-jitter-noise ratio (SJNR) is given by:

$$\mathrm{SJNR} = 10 \cdot \log_{10}\left[\frac{\left(2^{N}-1\right)^{2}}{16 \cdot \mathrm{OSR} \cdot J_{\mathrm{RMS}}^{2} \cdot B^{2}}\right] [\mathrm{dB}], \tag{3}$$

where J_{RMS} is the standard deviation of the clock jitter and N the number of bits of the quantizer. According to (3), a straightforward solution for reducing the performance degradation due to jitter is to increase the number of bits in the quantizer. However, if the quantizer is implemented with a conventional flash ADC, this would result in a more complex structure, larger power consumption, and larger silicon area.

Given the large OSR used for audio converters, tracking ADCs are a convenient solution to achieve high resolution while reducing power and area compared to classic flash ADCs, however, they can perform a proper conversion only if the input signal remains in the tracking range [31]. Wrong or missed conversions in a tracking ADC employed as quantizer in a $\Sigma\Delta$M ADC can lead to instabilities and oscillations.

In SC-$\Sigma\Delta$M, an anti-aliasing filter is required in the input path, and usually such filters are designed with a cut-off frequency just above the audio bandwidth. Therefore, if the tracking ADC can operate with a full-scale input signal at the cut-off frequency of the anti-aliasing filter, input signals at higher frequency will always stay in the tracking range since they are attenuated by the filter itself. In CT-$\Sigma\Delta$Ms, the input signal is attenuated only by the loop-filter, which has a cut-off frequency one order of magnitude higher. A conventional tracking ADC, therefore, should be designed with a larger tracking range, thus increasing power consumption and area.

A solution to this problem can be a tracking ADC that is able to convert audio-band signals with full resolution, while performing only a coarse conversion when an input signal that exceeds the tracking range is applied, thus ensuring stability for the CT-$\Sigma\Delta$M.

The analysis of this solution can start referring to Fig. 26. It is worth noting that the sample-and-hold circuit (S&H) operates at the rising edge of clock Ck, while the feedback DAC is clocked at the rising edge of Ckn. Therefore, there is a delay time of half sampling period ($T_{\mathrm{S}}/2$) in the feedback loop. Having such delay is a common solution in CT-$\Sigma\Delta$M, because it can relax the speed requirement of the quantizer.

A tracking ADC for a CT-$\Sigma\Delta$M is shown in Fig. 27. The number of comparators N_{tk} is a function of the final desired resolution of the tracking ADC (N_{ADC} levels), the audio bandwidth (B), and the sampling period (T_{S}). To a first approximation, N_{tk} is given by:

$$N_{\mathrm{tk}} = 2 \cdot \mathrm{round}\left[N_{\mathrm{ADC}} \cdot \pi \cdot B \cdot T_{\mathrm{S}}\right] \tag{4}$$

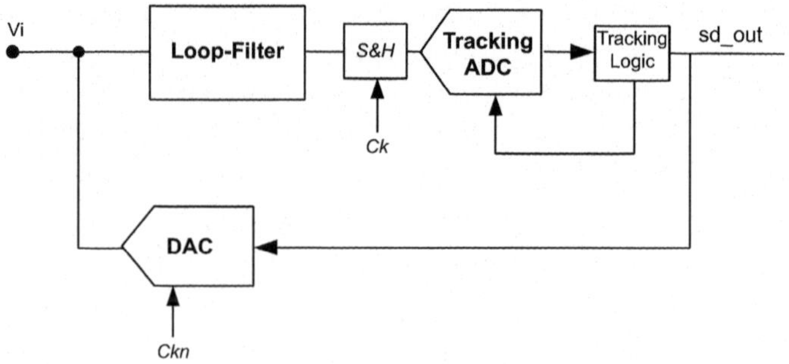

Fig. 26 CT-$\Sigma\Delta$M with tracking ADC

Fig. 27 Block diagram of the tracking ADC

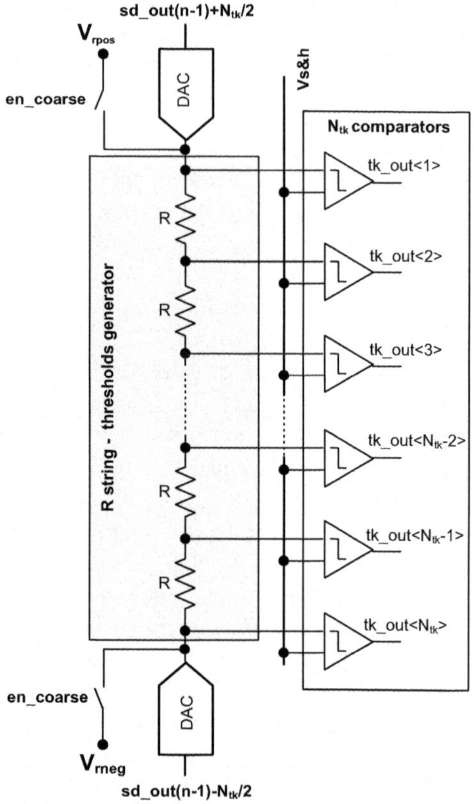

The comparator thresholds can be generated with a resistor string. The voltage drop for each resistor R is equal to V_{FS}/N_{ADC}, where V_{FS} is the full-scale value of the signal to be converted.

The upper and lower ends of the resistor string are connected to two complementary DACs. Each DAC generates a voltage that is a function of the CT-$\Sigma\Delta$M's

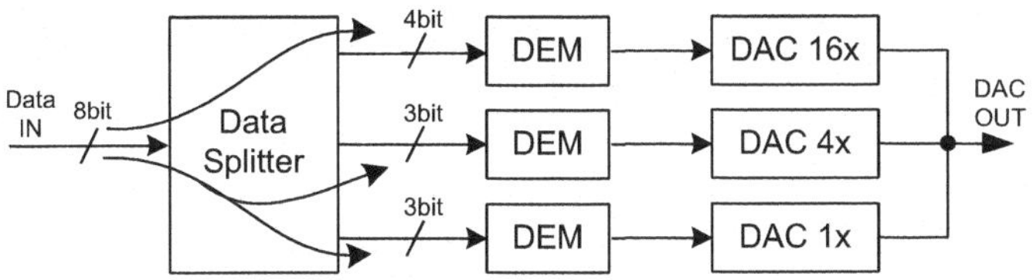

Fig. 28 Data segmentation for 8-bit DAC

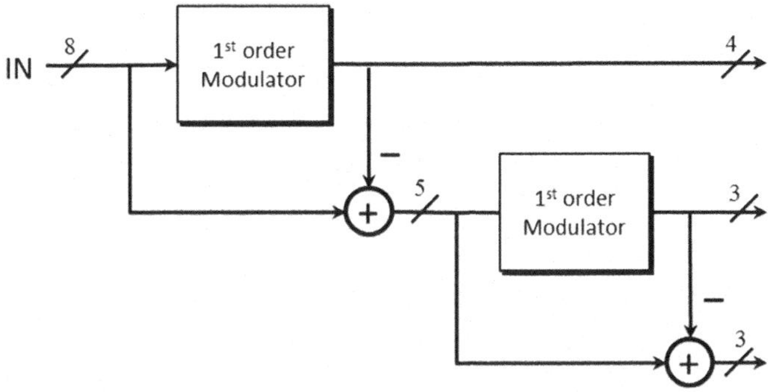

Fig. 29 Block diagram of an 8-bit 3-way data splitter

output previous, keeping the voltage drop across the resistive string constant and centered on the signal under conversion. The output of the CT-$\Sigma\Delta$M can thus be reconstructed from the previous conversion and the current output of the tracking ADC. If the tracking ADC output is at the limit of the tracking range (i.e., it is $+N_{tk}/2$ or $-N_{tk}/2$), a second coarse conversion is performed in the same conversion time window of $T_S/2$. The coarse conversion is performed by shorting the ends of the resistive string to V_{rneg} and V_{rpos}, where $V_{rpos} - V_{rneg} = V_{FS}$. If the result of this conversion is out of the tracking range, the Tracking Logic forces the use of coarse conversions in successive conversions, until the input signal returns in the tracking range.

Adaptive DEM in Feedback DAC

Increasing the number of quantizer bits has the drawback of increasing the complexity of the feedback DAC, particularly of the DEM logic. In order to reduce the DEM complexity, a technique known as segmentation (or noise-shaped splitting) can be used, in which the N bit digital signal at the output of the quantizer can be segmented in to multiple digital signals, each having less than N bits, so that each smaller segment can be processed and recombined with the other segments [32]. An example of this technique applied to an 8-bit digital signal is shown in Fig. 28. The data splitter can be realized as a cascade of first-order digital $\Sigma\Delta$Ms, as shown in Fig. 29. There are two main drawbacks of this technique that limits the achievable DR. The first one is the effect of thermal noise, considering that the signal is processed by the DAC with the highest weight, while the DACs with

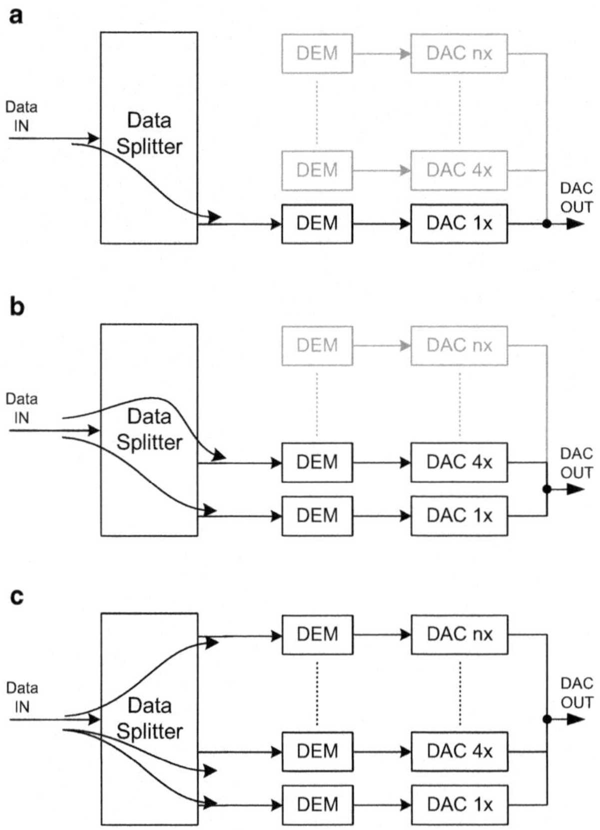

Fig. 30 Adaptive DEM

smaller weights are processing the quantization noise. Since the thermal noise is proportional to the weight of the DAC, the noise floor is dominated by the thermal noise generated by the DAC with the highest weight, even for small amplitude output signals, thus limiting the DR. The second drawback is the gain error between the DACs: the DAC-to-DAC error is shaped only by a first-order high-pass transfer function, thus again limiting the DR. Therefore, advanced layout techniques are required to minimize the mismatch between the DACs, increasing the complexity and the design area.

Another drawback is related to the power consumption and is correlated to the already mentioned fact that the signal is processed by the DAC with the highest weight: Even if the output signal is small, i.e., it is contained within few DAC levels, it is actually the result of the subtraction of a large signal generated by the DAC with the highest weight and the smaller quantization signals generated by the DACs with lower weights.

This means that all the DACs must be always active, i.e., the power consumption for small signal is comparable to the power consumption at full-scale. A solution to these problems is the use of an adaptive DEM scheme, in which [33], the segmented DAC can be dynamically reconfigured. An envelope detector tracks the amplitude of the digital signal at the input of the DAC. When the signal can be expressed with only the lowest-weight DAC ($1\times$), the other segments are bypassed, and their DACs are turned off, as shown in Fig. 30a. Likewise, when the signal can be expressed only with the first- and second-lowest weight DACs ($1\times$ and $4\times$), the other segments are bypassed and their DACs are turned off, as shown in Fig. 30b. Finally, when the signal amplitude requires the DAC with the highest weight to be used, all the segments are turned on, as shown in Fig. 30c. The number of possible operational states is equal to the number of segments.

This solution overcomes several drawbacks of the previous technique. In small-signal operation (i.e., when only the DAC $1\times$ is used), the thermal noise is lowered compared to large signal operation, increasing the DR. Moreover, the noise and distortion from DAC-to-DAC gain error is avoided, since only one DAC is used. Similar considerations can be made for the mid-level signal operation (i.e., when the segmentation is applied only to DACs $1\times$ and $4\times$). Finally, a dynamic "Class-H"-like power consumption is achieved: for each operational state, the power consumption is given only by the DAC elements that are actually in use, while the other DAC elements can be turned off. This means that the power consumption is greatly reduced in the presence of small signals.

References

1. Hsu YC, et al. Issues in path toward integrated acoustic sensor system on chip. In: Proceedings of IEEE sensors; Lecce, Italy; 2008. p. 585–8.
2. Malcovati P, Maloberti F. Interface circuitry and microsystems. In: Korvink J, Paul O, editors. MEMS: a practical guide to design, analysis and applications. Dordrecht: Springer; 2005. p. 901–42.
3. Bajdechi O, Huijsing JH. A 1.8-V $\Delta\Sigma$ modulator interface for an electret microphone with on-chip reference. IEEE J Solid-State Circuits. 2002;37:279–85.
4. Chiang CT, Huang YC. A 14-bit oversampled delta-sigma modulator for silicon condenser microphones. In: Proceedings of IEEE IMTC; Singapore; 2009. p. 1055–8.
5. Pernici S, et al. Fully integrated voiceband codec in a standard digital CMOS technology. IEEE J Solid-State Circuits. 2004;39:1331–4.
6. van der Zwan EJ, Dijkmans EC. A 0.2-mW CMOS $\Sigma\Delta$ modulator for speech coding with 80 dB dynamic range. IEEE J Solid-State Circuits. 1996;31:1873–80.
7. Zare-Hoseini H, et al. A low-power continuous-time $\Delta\Sigma$ modulator for electret microphone applications. In: Proceedings of IEEE ASSCC; Beijing, China; 2010. p. 1–4.
8. Jawed SA, et al. A 828-mW 1.8-V 80-dB dynamic-range readout interface for a MEMS capacitive microphone. In: Proceedings of ESSCIRC; Edinburgh, UK; 2008. p. 442–5.
9. Picolli L, et al. A 1.0-mW, 71-dB SNDR, fourth-order $\Sigma\Delta$ interface circuit for MEMS microphones. Analog Integr Circuits Sig Process. 2011;66:223–33.
10. Le HB, et al. A regulator-free 84-dB DR audio-band ADC for compact digital microphones. In: Proceedings of IEEE ASSCC; Beijing, China; 2010. p. 1–4.

11. Citakovic J, et al. A compact CMOS MEMS microphone with 66-dB SNR. In: IEEE ISSCC digest of technical papers; San Francisco, USA; 2009. p. 350–1.
12. Weigold JW, et al. A MEMS condenser microphone for consumer applications. In: Proceedings of IEEE MEMS; Istanbul, Turkey; 2006. p. 86–9.
13. Scheeper PR, et al. A review of silicon microphones. Sensors Actuators A. 1994;44(1):1–11.
14. Bergqvist J, Gobet J. Capacitive microphone with a surface micromachined backplate using electroplating technology. J Microelectromech Syst. 1994;3(2):69–75.
15. Kasai T, et al. Novel concept for a MEMS microphone with dual channels for an ultrawide dynamic range. In: Proceedings of IEEE MEMS; Cancun, Mexico; 2011. p. 605–8.
16. Leinenbach C, et al. A new capacitive type MEMS microphone. In: Proceedings of IEEE MEMS; Wanchai, Hong Kong, China; 2010. p. 659–62.
17. Martin DT, et al. A micromachined dual-backplate capacitive microphone for aeroacoustic measurements. J Microelectromech Syst. 2007;16(6):1289–302.
18. Zou QB, et al. Design and fabrication of silicon condenser microphone using corrugated diaphragm technique. J Microelectromech Syst. 1996;5(3):197–204.
19. InvenSense Application Note AN-1003. Recommendations for mounting and connecting InvenSense MEMS microphones, Online.
20. Knowles Application Note AN-16. SiSonic design guide, Online.
21. Nicollini G, et al. A high-performance analog front-end 14-bit CODEC for 2.7-V digital cellular phones. IEEE J Solid-State Circuits. 1998;33:1158–67.
22. Barbieri A, Nicollini G. 100+ dB A-weighted SNR microphone preamplifier with on-chip decoupling capacitors. IEEE J Solid-State Circuits. 2012;47:2737–50.
23. Croce M, et al. Cap-less audio preamplifiers for silicon microphones. In: Proceedings of IEEE sensors, Orlando, FL, USA; 2016. p. 943–5.
24. Croce M, et al. MEMS microphone fully-integrated CMOS cap-less preamplifiers. In: Proceedings of IEEE PRIME, Giardini Naxos, Taormina, Italy; 2017. p. 37–40.
25. Jiang X, et al. A low-power, high-fidelity stereo audio CODEC in 0.13-μm CMOS. IEEE J Solid-State Circuits. 2012;47:1221–31.
26. Du D, Odame KM. A bandwidth-adaptive preamplifier. IEEE J Solid-State Circuits. 2013;48:2142–53.
27. Tsividis Y, et al. Internally varying analog circuits minimize power dissipation. IEEE Circuits Device Mag. 2003;19:63–72.
28. De Berti C, et al. A 106-dB A-weighted DR low-power continuous-time $\Sigma\Delta$ modulator for MEMS microphones. IEEE J. Solid-State Circuits. 2016;51:1607–18.
29. Murman B. ADC performance survey 1997–2017. Online. http://web.stanford.edu/~murmann/adcsurvey.html.
30. De Berti C, et al. Colored clock jitter model in audio continuous-time $\Sigma\Delta$ modulators. In: Proceedings of IEEE NEWCAS, Grenoble, France; 2015. p. 14B5/1–4.
31. Dörrer L, et al. A 3-mW 74-dB SNR 2-MHz continuous-time delta-sigma ADC with a tracking ADC quantizer in 0.13-μm CMOS. IEEE J Solid-State Circuits. 2005;40:2416–27.
32. Nguyen K, et al. A 108-dB SNR, 1.1-mW oversampling audio DAC with a three-level DEM technique. IEEE J Solid-State Circuits. 2008;43:2592–600.
33. Crespi L., et al. Audio digital-to-analog converter with enhanced dynamic range. US Patent Application No. 62/425,510, 2016.

Challenges of Digitally Modulated Transmitter Implementation at Millimeter Waves

Khaled Khalaf, Steven Brebels, and Piet Wambacq

1 Introduction

The insatiable need of consumers worldwide for higher data rates in wireless communication brings the frequencies of operation toward the millimeter wave spectrum (between 30 GHz and 300 GHz). Today's cell phones, tablets, laptops employs different communication standards that operate below 6 GHz. Operation at much higher frequencies in silicon technologies is facilitated by the increased speed of advanced technology nodes. By going to the mm-wave frequency range, more bandwidth is available for the same fractional bandwidth. This facilitates communication at multi-gigabits per second. For example, the 802.11ad [1] WiFi standard defines four channels of 2.16 GHz between 57 GHz and 66 GHz. After the amendment of 802.11-2016 [2], coded datarates of up to 8.1 Gbps are possible using a single channel with 64-QAM modulation. More extensions are being defined (e.g., 802.11ay) with even higher datarates catching up with the high bandwidth demands of the consumer market. Several other applications can also benefit from the high datarates, such as wireless streaming of high-definition videos, AR/VR headsets, small-cell backhaul communication, and 5G, which is adopting mm-waves for its future networks.

K. Khalaf · S. Brebels
imec, Leuven, Belgium

P. Wambacq (✉)
imec, Leuven, Belgium

Vrije Universiteit Brussel (VUB), Department of Electronics and Informatics (ETRO), Brussels, Belgium
e-mail: Piet.Wambacq@imec.be

K. A. A. Makinwa et al. (eds.), *Low-Power Analog Techniques, Sensors for Mobile Devices, and Energy Efficient Amplifiers*,
https://doi.org/10.1007/978-3-319-97870-3_18

2 Phased Arrays for mm-Wave Communication

Operating in the mm-wave range also comes at the cost of higher attenuation, where the Friis transmission equation estimates a decay of the signal power in proportion to the square of its wavelength. To overcome signal losses at mm-waves, higher power levels need to be radiated to reach the same communication distance. This can be achieved either in a passive way by increasing the antenna gain or in an active way by increasing the transmitted power.

Figure 1 shows an example of increasing the antenna gain by connecting the output of each front-end to multiple antenna patches forming a multi-patch antenna element. Antenna arrays are more area efficient at mm-waves since the area is proportional to the square of the wavelength. Antenna gain in this case is limited by antenna feeding losses from the chip, which is proportional to the number of antenna patches.

Another way of increasing the radiated output power is by increasing the output power. With the limited supply in advanced technology nodes, the output power per amplifier is quite limited. Therefore, parallel amplifiers are used to transmit simultaneously in air and increase the total radiated power. Figure 2 shows a direct conversion example architecture with a transmit-receive front-end array, featuring T/R switch to share the same antennas between transmit and receive modes. The array gain is increased with the number of paths, the transmitted signal also becomes more directive, where.

$$\mathrm{EIRP} = P_{\mathrm{PA}} + G_{\mathrm{element}} + 20 \times \log\left(N_{\mathrm{elements}}\right). \tag{1}$$

This leads to a thinner beam that can start to miss its target due to misalignment. Instead of mechanically steering the equipment for alignment, the transmitted signal from each front-end can be progressively delayed in order to electronically steer the beam in other directions. As shown in Fig. 2, shifting the signal phase is usually used as an approximation of true time delay for beamsteering, which works well for most application scenarios and is limited by the signal bandwidth, array size, and scanning angle.

Phased arrays are also used with multiple patch antennas to maximize the radiated output power. Each antenna element is composed of several antenna patches. Therefore, Eq. (1) becomes:

$$\mathrm{EIRP} = P_{\mathrm{PA}} + G_{\mathrm{patch}} + 10 \times \log\left(N_{\mathrm{patches}}\right) + 20 \times \log\left(N_{\mathrm{elements}}\right). \tag{2}$$

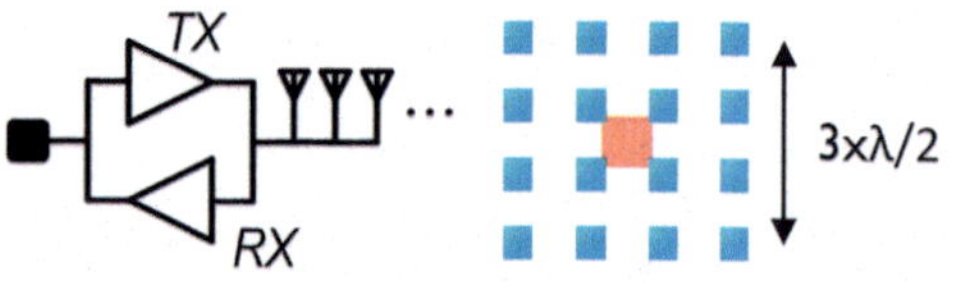

Fig. 1 Increasing antenna gain by increasing the number of patches

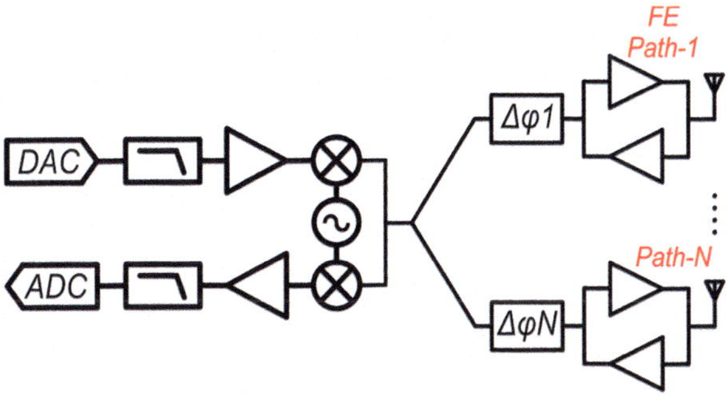

Fig. 2 An RF beamforming transceiver architecture showing antenna gain increase by increasing the number of elements

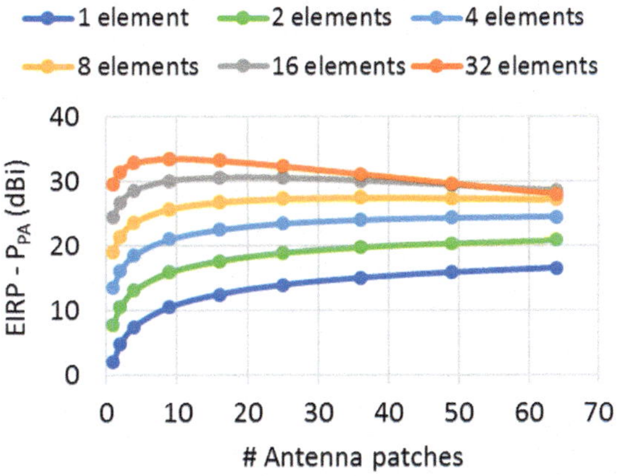

Fig. 3 Improvement of EIRP from PA output power by increasing the number of antenna patches for different number of active antenna elements

Equation (2) is used in Fig. 3 showing EIRP for different number of antenna elements and different number of patches per element. As mentioned earlier, antenna feeding losses limit the antenna gain improvement, leading to an optimum number of patches beyond which the antenna feed losses are higher than the antenna gain. This is more obvious with more active elements as it results in a large number of patches. For example, the maximum useful number of antenna patches per element for a 32-element chip is about 9 (i.e., 3×3 patches per element resulting in a total of 288 patches). In this calculation, a square distribution of the antenna patches is assumed with a single patch gain of 4 dB, and 0.2 dB/mm loss is assumed for the antenna PCB material. The Manhattan distance is used to calculate the distance between the chip and the outer patch, where.

$$\text{Distance to outer patch} = \lambda/2 \times \left(\sqrt{N_{\text{patches}}} - 1\right). \tag{3}$$

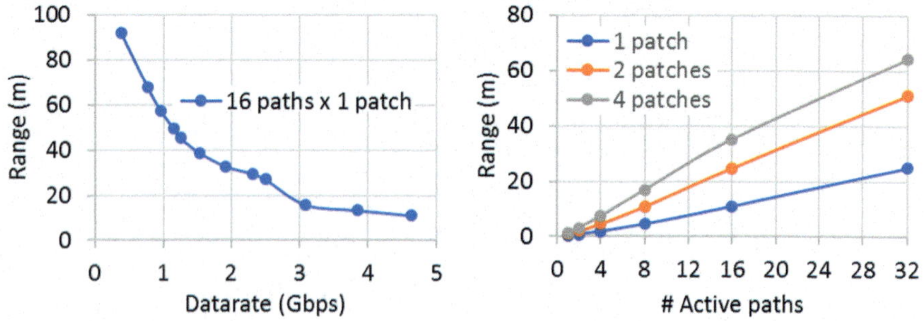

Fig. 4 Link budget analysis showing the communication range of a TRX with (left) 16 active paths and a single antenna patch and (right) different number of active paths and antenna patches for MSC-12 with 4.62 Gbps

The patches are placed at a pitch distance of $\lambda/2$ for optimal scanning performance. An extra 2 dB margin for the vertical connection between the chip and the antenna is also considered in this analysis for the feeding loss.

Figure 4 shows a link budget example for an indoor scenario using the phased array system as in Fig. 2 with an assumed PA output compression power of 9 dBm. The number of paths and the number of patches in the receiver are the same as the transmitter. For a 16-path TRX with 1 patch per path, the maximum communication range for 4.6 Gbps (MCS-12 in 802.11ad) is around 11 meters. For applications requiring more datarates or more distance, more patches per path can be used. As shown in Fig. 4, 4 patches per path with 32 active paths can reach a distance of about 64 meters.

One drawback of having more gain in the antenna element (e.g., by using more patches per path) is the reduced beam scanning angle. For example, four patches (2×2 array) per element has a scanning range of $\pm 25°$ compared to $\pm 45°$ with a single patch. A square array of 16 patches has a scanning range of only about $\pm 10°$. Therefore, outdoor applications that require both wide scanning angle and more than 100 meters multi-Gbps communication distance can use more than 32 active paths.

3 Efficient Transmitter Architectures

Phased arrays, required for most high datarate mm-wave applications, use parallel front-ends to overcome the signal losses. This causes an increase in the power consumption, where the transmitter usually contributes the most. The power amplifier (PA) is usually the most power-consuming block in the transmit chain. Therefore, increasing the PA efficiency helps controlling the thermal behavior of the system and saving the battery lifetime for mobile applications.

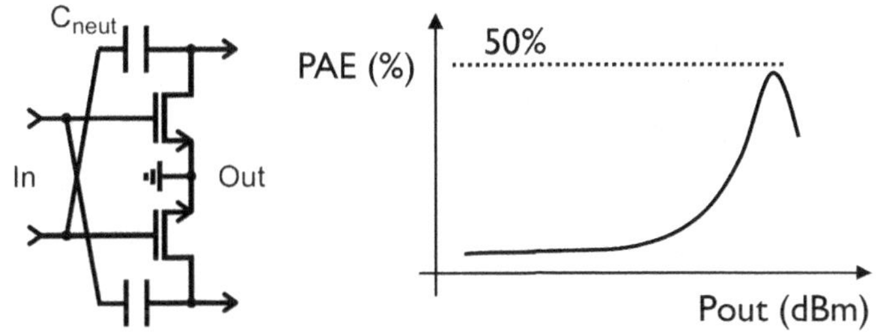

Fig. 5 An example of a class-A power amplifier with its efficiency behavior versus output power

High-order modulations, for example up to 64-QAM, are used to increase the communication throughput while utilizing the same bandwidth. Therefore, the modulated signal consists of different amplitude levels with a certain peak-to-average-power level (PAPR) depending on the modulation scheme. This requires the average transmitted power to be reduced allowing its peak to be amplified linearly. For example, a power back-off of up to 6 dB is used for 64-QAM modulation signals. Nonlinear amplification causes signal distortion, which can be quantified by the error vector magnitude (EVM) and contributes to an increase in the bit error rate (BER) (i.e., poor signal quality).

Power efficiency of an amplifier is usually a function of its power level (see Fig. 5). The figure also shows an example design of a mm-wave power amplifier operating in the linear class-A. The differential operation is used to enhance the output power by 3 dB without affecting the amplifier efficiency. Moreover, neutralization is used to cancel the effect of gate-drain capacitance, making the device unilateral ensuring a stable operation over the whole frequency band and increasing the amplifier's power gain. Practical implementations of the class-A power amplifier measure efficiency values of less than 5% although up to 30% efficiency is achievable at maximum output power. An implementation example at 60 GHz in 40 nm-LP process is shown in [3], where the amplifier is integrated in a full TX chain and uses an on-chip output balun to match its output impedance to 50 Ω. The amplifier achieves 4.9% power-added efficiency (PAE) at 5 dB back-off from the output compression point out of a maximum value of 32.7% in saturation.

Several architectures exist in literature and tested at mm-waves that help increasing the power amplifier's back-off efficiency. This includes lower-class operation, envelope tracking, adaptive bias, Doherty and Outphasing solutions, as well as digitally modulated architectures, such as I-Q or polar transmitters. The following sections discuss some of these solutions, followed by a table summarizing different mm-wave realizations. The discussion focuses on the 60 GHz frequency band for consistency but the main concepts can also be mapped to other mm-wave frequencies of operation.

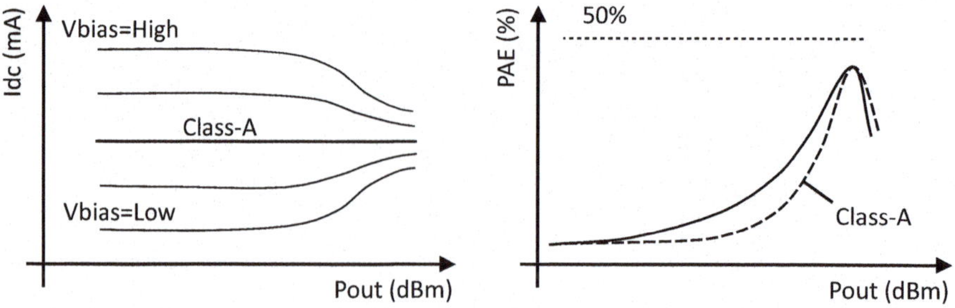

Fig. 6 A class-AB current and efficiency behavior versus output power

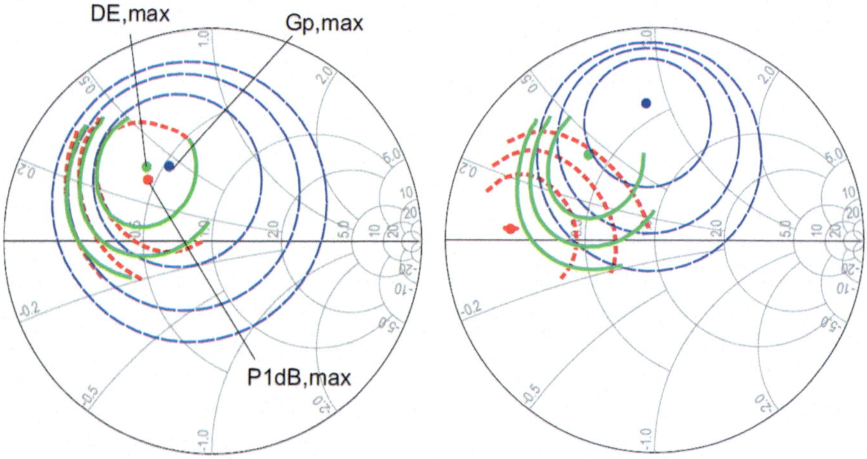

Fig. 7 Load pull simulations on a neutralized push-pull amplifier stage in (left) class-A and (right) deep class-AB

3.1 Analog Solutions

One straightforward solution to increase the power back-off efficiency is to operate the amplifier in class-AB using a lower bias voltage. In this case, the power consumption scales with the output power, improving the back-off efficiency (see Fig. 6).

A trade-off should be made in the selection of the output impedance to ensure correct operation in both classes. For example, Fig. 7 shows power, gain, and efficiency circles of a neutralized push-pull amplifier at two different bias voltages representing class-A and deep class-AB operations. The optimum impedance values change according to the class of operation, which should be considered during the design.

At a lower bias voltage, the amplifier gain is also reduced. This leads to an optimal bias value for maximum PAE. An implementation example is also shown

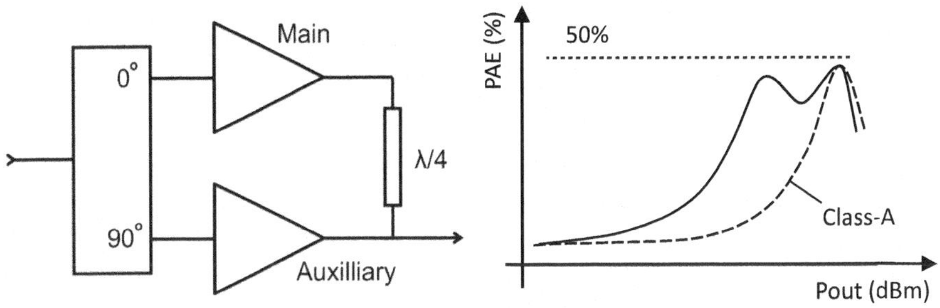

Fig. 8 An example of a Doherty power amplifier with its efficiency behavior versus output power

in [3], where the 5 dB back-off efficiency moves from 4.9% in class-A to 7.4% in class-AB, representing around 50% efficiency improvement.

Another way of increasing the efficiency at power back-off is by disabling part of the amplifier while adapting the output impedance accordingly. The Doherty architecture (see Fig. 8) uses a parallel PA that turns off at back-off and actively changes the output impedance of the main PA, maintaining a high efficiency value spanning over a wider power range below the maximum. The classical Doherty approach uses λ/4 transmission lines at the PA output for its operation. At 60 GHz, this is close to 600 μm. Other implementations try to be more compact and replace transmission lines with transformers.

An example implementation of transmission-line based Doherty implementation is shown in [4], where the chip is fabricated in a 130 nm SiGe BiCMOS process. The amplifier achieves a 6 dB back-off efficiency of 13%, reduced from a maximum value of 23.7% in saturation. This is limited by the finite output impedance of the auxiliary (peaking) amplifier, resulting in a deviation of the back-off efficiency from its ideal peak response of a Doherty architecture.

A transformer-based Doherty implementation at 72 GHz is shown in [5] using a 40 nm bulk CMOS technology. An asymmetrical transformer-based combiner and an additional LC-based matching/tuning network are used to optimize the efficiency and power performances. Again the back-off efficiency is affected by the finite output impedance of the auxiliary amplifier as well as the limited quality factor of the passive output matching network. The PA achieves 7% PAE at 6 dB power back-off and a maximum value of 13.6% in saturation.

3.2 *Digitally Modulated Solutions*

Since high efficiency is present at maximum power (i.e., in saturation), digitally modulated architectures provide a solution to use a variable envelope signal while still driving the amplifier in saturation. In this case, a constant envelope signal is used to drive the amplifier, while the different amplitude levels are generated

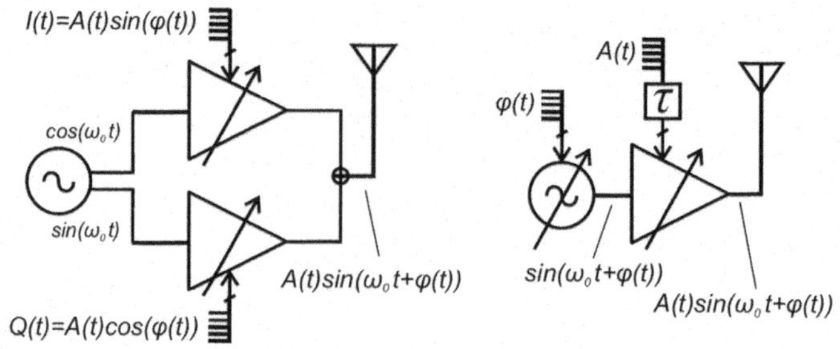

Fig. 9 Digital PA solutions with (left) I-Q and (right) polar architectures

by switching the parallel cells of the amplifiers. Two high-level architectures can implement this functionality, either based on I-Q or polar coordinates. As shown in Fig. 9, a digital power amplifier (DPA) represents the core of both architectures, where the signal modulation also occurs. In the I-Q architecture, the carrier signal drives two DPAs while their strength (represented by the total width) is controlled by the baseband digital I-Q symbols. In the polar architecture, a polar representation is first extracted from the baseband I-Q symbols (e.g., using a CORDIC algorithm). Only the phase information modulates the carrier signal before driving the DPA. This allows all the DPA cells to be driven in saturation at their maximum output power levels. The amplitude information then directly controls the amplifier to reconstruct the envelope variations.

Switching the PA cells to generate amplitude variations also reduces the average efficiency since the load is ideally fixed. For example, the PA drain efficiency can be represented as follows:

$$P_{\text{out}}/P_{\text{DC}} = (i_{\text{out}})^2 \times RL/\left(I_{\text{DC}} \times V_{\text{DC}}\right). \tag{4}$$

Therefore, if half of the cells are switched off for 6 dB power back-off, i_{out} and I_{DC} are halved, leading to a 50% ideal reduction in efficiency at a fixed load. This is still higher than other implementations, where a 30% maximum efficiency amplifier leads to an ideal value of 15% average efficiency at 6 dB back-off. This is at least $3\times$ higher than a class-A amplifier.

4 Challenges in Digitally Modulated Implementations

In both architectures, the DPA represents an RF-DAC, where signal aliases are present at its output at frequency offsets that are multiples of the baseband signal sampling rate. Since bandpass filters would be too lossy to implement at the DPA output, aliases can be overcome by oversampling the baseband signal. Oversampling factors of more than $4\times$ are required at 60 GHz to push the first alias below the

802.11ad spectral mask. This leads to sampling rates of more than 7 GS/s. Extra challenges are also discussed in the following sections.

4.1 Digital I-Q Transmitters

As shown in Fig. 9, the I-Q architecture includes two amplifiers running at 90° phase-shifted carrier signals that need to be combined. The power combiner can either be of an isolating type (i.e., isolating the two amplifiers from each other) or nonisolating. Nonisolating combiners are usually preferred for their higher efficiency. However, dynamic (i.e., code-dependent) output impedance variations of one amplifier are seen by the other, resulting in a highly complex 2D calibration procedure. AM-AM and AM-PM calibrations are usually required in the DPA to get a linear, constant phase performance over the code.

An example of an isolating combiner is the Wilkinson combiner, where an ideal 3 dB extra output power is expected compared to its single port input power, assuming that both input ports are excited in phase. Assuming a perfect lossless Wilkinson at the center of the band, where the input and output characteristic impedances are real and all equal to Z_c, then the combined output power can be represented as $|V_1 + V_2|^2/(4 \times Z_c)$. V_1 and V_2 are the voltage phasors at the input of the combiner. The total available power at the input of the combiner is $(|V_1|^2 + |V_2|^2)/(2 \times Z_c)$. Therefore, the power combining efficiency can be represented as follows:

$$P_{out}/P_{in,total} = |V_1 + V_2|^2/\left[2 \times \left(|V_1|^2 + |V_2|^2\right)\right]. \quad (5)$$

For $V_1 = 1$ and $V_2 = j$ (90° phase-shifted input signals), the maximum ideal power combining efficiency (i.e., without losses or mismatch) for a Wilkinson combiner becomes 50%. This is equivalent to an ideal 3 dB loss at the TX output.

For a nonisolating power combiner, the ideal addition of two equal I and Q signals lead to a $\sqrt{2}$ higher magnitude compared to an increase of $2\times$ if both signals have the same phase. This corresponds to an ideal increase of 1.5 dB in the output power compared to a 3 dB power increase if two PAs are added in-phase, leading to around 30% efficiency drop. This represents a worst-case efficiency instance, where the signal phase is 45° with equal I and Q components. The best-case instance is when the signal phase is at 0° or 90° with only an I or Q signal component. In order to further evaluate the nonisolating approach, a simulation is performed on two DPAs combined at the output using a direct connection in the current domain (see Fig. 10). A 28 nm bulk-CMOS technology is used for the simulation. The DPA consists of parallel unit cells of common-source (CS) amplifiers in a differential configuration and switched at the source. The total size of the single-ended CS amplifier is 128 μm. The switching device total width is twice as large as the CS amplifier.

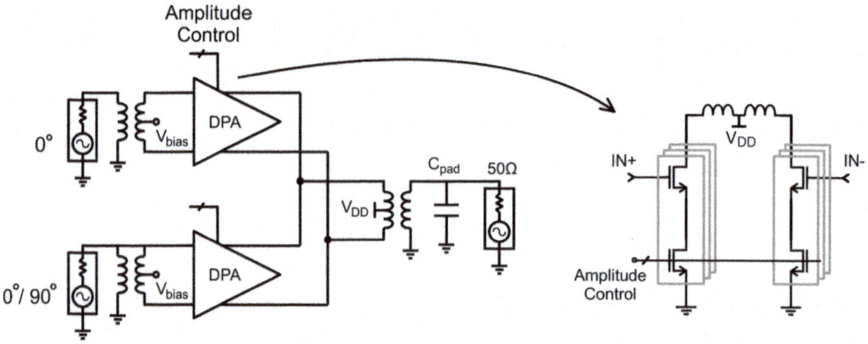

Fig. 10 (left) DPA combining testbench and (right) its DPA schematic

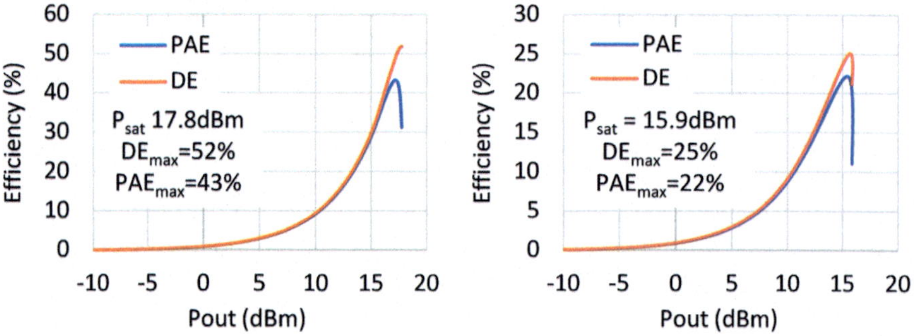

Fig. 11 DPA efficiency simulation using the testbench of Fig. 10 excited with signals of phase difference of (left) 0° and (right) 90°

The simulation results are shown in Fig. 11. When both amplifiers are fed with the same signal, a maximum PAE of 43% is achieved. When one amplifier is fed with a 90° phase-shifted signal, representing an output signal with 45° phase, the maximum PAE drops to 22%. Therefore, the average PAE over all the signal phases is between 22% and 43% with an average PAE efficiency loss of less than 50%.

Another way of combining the I and Q amplifiers is through air. An example implementation of spatial I and Q combination at 60 GHz is presented in [6] in 65 nm bulk-CMOS. As shown in Fig. 12, the chip implements a four-way beamforming digital I-Q transmitter with an effective amplifier resolution of 7 bits excluding the sign bit. The chip runs with 4× oversampling ratio (i.e., 7 GS/s) and uses twofold interpolation in the DPA in order to further reduce the amplitude of the first alias at 7GHz offset from the transmitted signal. Therefore, the output signal after interpolation is transmitted at an effective rate of 14 GS/s.

The design shows high PA efficiency values, where the maximum reported drain efficiency is 28.5% and the average is 16.5%. At 6 dB back-off, the drain efficiency goes to 14%. However, these values are measured with a probed setup, where the loading effect between the I and Q amplifiers is not visible. This is expected to be

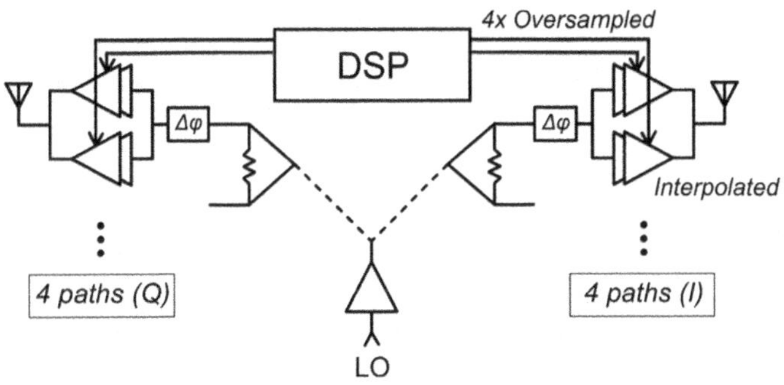

Fig. 12 Digitally modulated I-Q based phased array solution of [6]

seen in a wireless measurement, where the isolation between the I and Q paths is broken through the antennas. Therefore, spatial combination can still be considered as a nonisolating power combination, where any extra losses of on-chip combination are replaced with field combination effects in air.

4.2 Digital Polar Transmitters

Figure 9 shows the polar transmitter concept, where the constant-envelope phase-modulated signal drives the PA in saturation to make use of the maximum efficiency, while the amplitude information modulates the PA by controlling its total width digitally, leading to an efficiency performance that scales linearly with amplitude. In this case, phase modulation is represented by a VCO/DCO direct modulation, which adds to the design challenges by putting extra limitations on the VCO linearity and tuning range.

An easier way for phase modulation is by mixer up conversion. An implementation example at 60 GHz is shown in [7] in a 40 nm bulk CMOS technology. As shown in Fig. 13, The I-Q baseband digital symbols are first upsampled, filtered, and converted to polar representation. The phase signal gets upconverted in I-Q mixers after being converted to the analog domain and filtered. Filtering reduces aliases of the phase signal, which improves the combined signal alias performance at the PA output. This allows a reduction in the required sampling rate. In this chip, a sampling rate of 10 GS/s is used compared to an effective rate of 14 GS/s in [6], which is required with the absence of an alias filtering opportunity in the I-Q architecture. This is implemented by a signal oversampling factor of 6, but could also be implemented with an oversampling factor of 3 together with using twofold interpolation.

The amplitude signal, extracted after the digital baseband signal is converted to polar representation, the amplitude signal directly feeds the DPA with a speed of 10 GS/s and a resolution of 4 bits after applying a delay to get synchronized with the

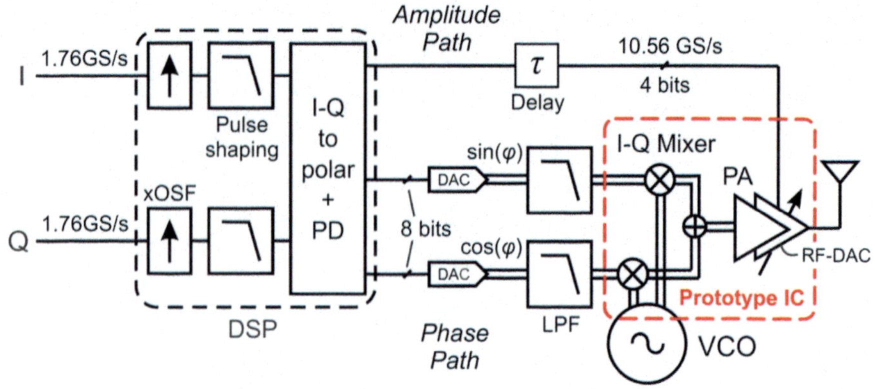

Fig. 13 Mixer-based digitally modulated polar transmitter block diagram [7]

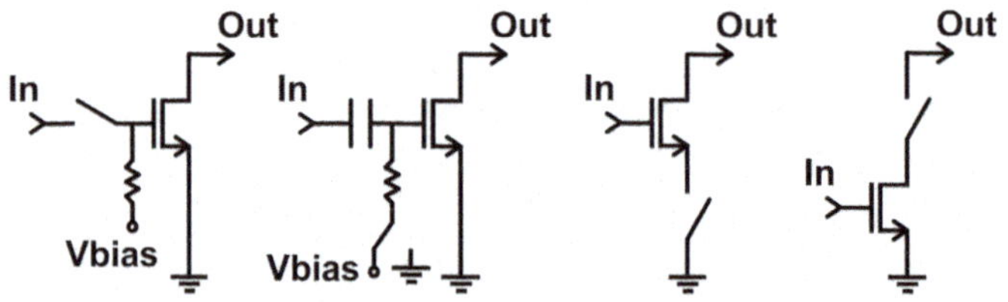

Fig. 14 Switching options for the DPA unit cell

phase signal. The full signal has an effective resolution of 5 bits after adding a sign bit, represented in the phase signal. This is enough for 16-QAM modulation schemes with target single-carrier (SC) TX EVM values between −19 dB and −23 dB but is not sufficient for 64-QAM with target SC TX EVM values lower than −26 dB.

Several implementation options exist for the DPA. Figure 14 shows the single-ended representations, where the PA unit cell can either be switched at the gate, source, or drain. Introducing a physical switch at the gate side reduces gain of the last stage, which severely affects PAE. Another approach to switch the PA from its gate is to do it only in DC. As shown in Fig. 14, the AC signal is always fed to all the cells, whereas the gate bias of each cell is switched between the bias voltage and 0 V. This can be implemented with an inverter supplied by the required bias voltage (i.e., Vbias). One drawback of this approach is that the switching speed is limited by the biasing resistor. Another drawback is that the biasing cell delivering Vbias needs to provide enough DC current to accommodate the switching dynamics of the switching inverter, which affects the PA efficiency.

Another approach is to switch the PA cell at the source. This has a smaller effect on efficiency, since the switch works as a degenerating device rather than being introduced in the main signal path. The switch is also not applied to a biasing point, allowing the control lines to be directly connected to the switch, increasing its speed capability. This is confirmed by the implementation of [7], where the PA achieves

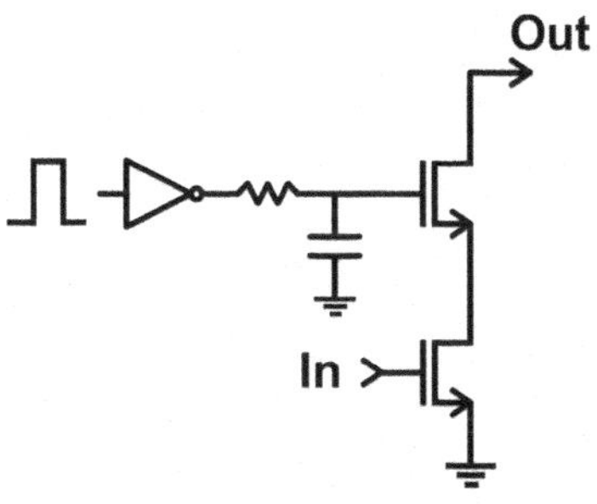

Fig. 15 DPA cell of [8] with drain switching

a maximum drain efficiency of 29.8% and a QPSK average efficiency of 15.3%. At 6 dB back-off, the drain efficiency goes to 12.3%, which can be improved to 50% of the maximum value (i.e., 15%) with a redesign of the output matching network to be optimized for the back-off code rather than the maximum code. The DPA switches at a sampling speed of 10 GS/s limited by the measurement equipment rather than the design. One drawback in this implementation is the signal leakage from input to output in *off* state. This affects the DPA linearity and cannot be easily calibrated. For example, the QPSK TX EVM is improved from −20.7 dB to −23.6 dB and the 16-QAM TX EVM from −16.5 dB to −18.1 dB after calibration, still limited by the signal leakage.

The third switching option for the PA unit cell is drain switching. In this case, the switch works as a cascode or stacked device. This improves signal leakage, where the switch disconnects the input and output paths when turned off. However, the DPA efficiency is also degraded. An example implementation is shown in [8], where the chip is fabricated in 28 nm bulk-CMOS technology. As shown in Fig. 15, the switch represents a stacked device in the *on* state. In order to maintain a signal swing at the gate of the stacked device for the correct operation of the stacking configuration, a bias resistor is used after the switching inverter. This limits the switching speed, in this case to 5 GS/s. Moreover, if the stacked device requires a bias voltage different from the supply, the switching inverters will also require more current to be supplied from a biasing block, affecting the DPA efficiency. The DPA is implemented with 6 bits and achieves EVM values of about −27 dB thanks to the isolating stacked switch, which reduces signal leakage and allows signal modulations up to 64-QAM. The maximum DPA drain efficiency is 17.7% (PAE 15.6%), whereas the drain efficiency at 6 dB back-off is 6.9%.

5 Comparison of Transmitter Architectures

In this section, different approaches are compared using a combination of published results and qualitative analysis.

Table 1 Literature comparison of several TX PA architectures around 60 GHz

Refs.	Topology	Freq. (GHz)	Tech.	Supply (V)	P_{sat} (dBm)	PAE_{PA} max	PAE_{PA}6dB[a]
[9]	Class-A/AB	60	28 nm	0.9	13, 11.3	29, 28[b]	2.5, 9[b]
[10]	Envelope tracking	44	GaAs	6	33	10	7–8.4
[11]	Adaptive bias	60	55 nm BiCMOS	1.2	9.7	16	6
[5]	Doherty (transfo.)	77,72	40 nm	0.9	16.2,21	12,13.6	5.7,6
[4]	Doherty (TL)	60	130 nm SiGe	2.5	17.5	23.7	13
[12]	Outphasing	60	40	1	15.6	25	9.2
[6]	Digital Cartesian	60	65 nm	1	9,6	28.5[a,b]	14[a,b]
[7]	Digital polar	60	40 nm	0.9	10.8	29.8[a,b]	12.3[a,b]

[a]Estimations used
[b]Drain efficiency

5.1 PA Comparison from Literature

Table 1 shows a PA comparison of few selected mm-wave TX implementations around the 60 GHz frequency range. The architectures in [6, 7, 12] require additional signal processing that is not considered in the PA efficiency. The implementations in [5, 10, 12] show potential bandwidth limitations, while the ones in [6, 7, 11] show potential linearity limitations. Therefore, the challenge of bringing a high back-off efficiency together with a large bandwidth and low distortion still remains. The implementation in [4, 6, 7] are of particular interest as they show a back-off efficiency of more than 10%. The implementation of [4] is a transmission-line-based Doherty PA, where the presence of transmission lines leads to an increase in chip area. The implementations in [6, 7] represent the digital implementations discussed before, where the digital I-Q has extra efficiency degradation after combining the I and Q paths, whereas the digital polar efficiency has extra system challenges, such as the signal bandwidth extension and synchronization between the amplitude and phase paths.

5.2 Qualitative Comparison for Phased Arrays Transmitters

Since the digital implementations require additional signal processing, it is beneficial for the evaluation to have an idea about the changes in the whole system. Figure 16 shows an example of phased array transmitter architectures based on the digital I-Q and polar implementations. RF beamforming is used in this example,

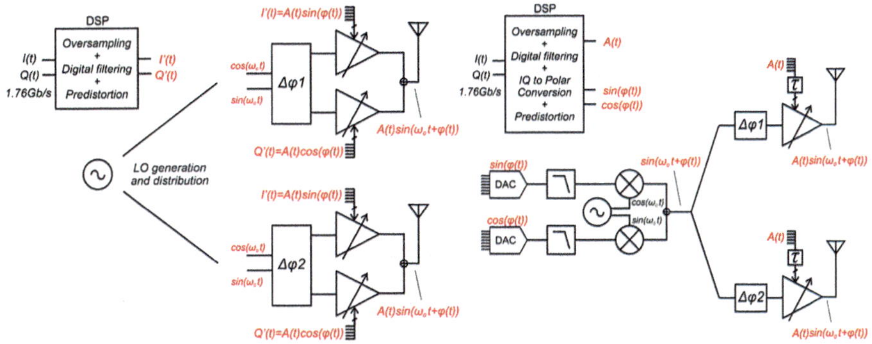

Fig. 16 Phased array transmitter architecture example for digital (left) I-Q and (right) polar systems

which is suitable for a large number of front-ends to limit the calibration complexity of mixer nonidealities in both the transmit and receive modes. The analog blocks used in the mixer-based digital polar TX architecture are similar to an analog TX approach except for the digital control required in the PA. The digital processing has more functionality, such as the IQ-to-polar conversion and the digital predistortion of AM-AM and AM-PM effects of the DPA, and is running at a higher speed. The analog baseband also processes a higher bandwidth signal, representing only the phase information compared to the full variable envelope signal in order to allow the use of PAs in saturation with higher efficiency. In the case of a digital I-Q transmitter, the architecture is simpler since an upconversion path is not required and the LO carrier is directly fed to the front-ends. However, the digital processing still includes extra AM-AM and AM-PM predistortion and runs at a higher speed to overcome signal aliases.

Table 2 shows the different components each system requires compared to an analog solution (e.g., with a class-A, AB, or Doherty PA). The conventional signal path includes digital oversampling and filtering, followed by a DAC and then the analog baseband including an anti-aliasing filter and the upconversion mixer. Upsampling is still required in the DSP of an analog solution to keep a frequency margin for the anti-aliasing filter after the DAC. In a digital solution, an oversampling factor of $8\times$ is required to bring the first alias down to -30 dBc. Another way of implementing such a high oversampling factor is using interpolation, where an oversampling factor of $4\times$ and a twofold interpolation can also be used. In the digital polar approach, a filter in the phase path can reduce the first alias after being combined with the amplitude information, leading to a reduction in the required speed. The number of digital lines in the polar system is higher as the phase information is already based on an I-Q system. The high-speed digital calculations are usually made at a lower frequency then multiplexed to the desired speed. Therefore, a serializer is required in the digital implementations. The digital signal also needs to feed the DPAs, and therefore buffers are required to bridge the long distances. Also due to the imperfections of the DPA, AM-AM

Table 2 Hardware comparison between analog and digital transmit architectures

	TX architecture	Analog	Digital I-Q	Mixer-based digital polar
Digital	Oversampling ratio	2×	8×	6×
	Approximate resolution	$N_I + N_Q = 2N$	$N_I + N_Q = 2N$	$NPH_I + NPH_Q + N_{AMP} = 3N$
	Extra functions	NA	Serializer, buffers, AM-AM(/AM-PM) predistortion	Serializer, buffers, AM-AM/AM-PM predistortion, CORDIC
	Predistortion complexity	Up to 2^{2N+1}	Up to 2^{N+1}	
	Extra delay elements	Fine	Coarse + fine	
Analog	Upconversion path bandwidth (MHz)	880	NA	2500
	PA power combining	No (A/AB) Yes (Doh.)	Yes	No
	PA efficiency	Low (A) Med. (AB) Med. (Doh.)	Medium	High

and AM-PM predistortion is usually required in the digital domain. The digital I-Q system requires a 2-D calibration, largely adding to the digital complexity. If the DPA phase also needs to be calibrated, the system complexity is even increased since phase information needs to be extracted and applied synchronously with the I-Q information to the system, lending several additional drawbacks from the polar system. Synchronization is required in the digital architectures. This is at least between the I and Q paths and between the late and early paths in the I-Q system. Only fine resolution delay optimization is required here since the four paths have similar lengths. On the other hand, the amplitude and phase signals in the polar system take different paths, leading to a large difference in their length that requires a coarse and fine-tuning delay element for synchronization. The I-Q architecture does not require a baseband path since the digital signal directly modulates the DPA. The polar architecture, on the other hand, the signal bandwidth expands as an effect of the conversion from an I-Q to polar system. This requires around 3× higher baseband circuit bandwidth and DAC speed in the phase path assuming an SNR floor of 40 dB and a DPA resolution of 6 bits [7]. For the PA, power combining of at least two amplifiers is required in both the digital I-Q and Doherty architectures. This adds to the area occupied by the PA and usually also has an effect on its efficiency. Examples of PA efficiencies are shown in Table 1. The values reported for transformer-based Doherty are not yet attractive compared to a class-AB solution. The TL-based Doherty approach has a relatively high-efficiency value without additional digital complexity but requires an increase of the chip layout area. The digital I-Q solution has less analog but more digital functionality. The reduced I-Q combination efficiency, which may result in a PA efficiency value close to a class-AB, and the high-order calibration complexity causes the approach to have less chances of reducing the transmitter power consumption. The polar solution has the highest PA efficiency but also the most additional analog overhead and digital functionality. Although it has more digital functionality (e.g., CORDIC and coarse delay elements) than the I-Q architecture, the calibration complexity is much lower, and it can run at a lower speed. In the RF beamforming architecture, and apart from the digital buffers to each DPA, the additional overhead in the polar solution compared to an analog one is in common blocks to all the phased array front-ends. Therefore, their contribution to the overall power consumption may lead to an overall advantage of the polar architecture.

An example implementation of a DSP for 60 GHz polar TX in 28 nm bulk-CMOS technology is shown in [13]. The chip implements 4× upsampling, CORDIC polar conversion, and predistortion functionalities and consumes less than 40 mW at 7 GS/s.

6 Conclusions

Millimeter wave frequencies attract future high datarate communication for its wide available bandwidth. High-gain antennas are inevitable to overcome signal losses at mm-waves. Passive gain by increasing the number of antenna patches is limited by the feeding loss toward the chip. Phased arrays use multiple front-ends to increase the module gain actively. Therefore, several analog and digital transmitter architectures are explored for their efficiency advantage, especially at power back-off. Transmitters operating around 60 GHz are considered for reference. The TL-based Doherty power amplifier reaches 13% efficiency at 6 dB power back-off. However, it includes transmission lines that are not area efficient. Replacing transmission lines with transformers lead to a reduction in efficiency, where 6% is reported at 72 GHz. Digital architectures are also considered as they operate the power amplifier in saturation utilizing its maximum efficiency at maximum code and an ideally linear efficiency behavior over the code. In a digital I-Q architecture, an isolating power combiner has an ideal 50% efficiency reduction, while a nonisolating one has a phase-dependent efficiency with up to 50% reduction at 45° together with a two-dimensional calibration complexity. The digital polar architecture has an additional CORDIC digital functionality and wider baseband bandwidth requirement. However, with a high-efficiency front-end, the architecture has a potential to provide an overall transmitter power consumption advantage in phased arrays, where the front-end efficiency dominates.

References

1. IEEE Std. 802.11ad-2012. IEEE Standard for Information technology—Telecommunications and information exchange between systems Local and metropolitan area networks—Specific requirements – Part 11: Wireless LAN Medium Access Control (MAC) and Physical Layer (PHY) specifications—Amendment 3: Enhancements for very high throughput in the 60 GHz band. 2012.
2. IEEE Std. 802.11-2016. IEEE Standard for Information technology—Telecommunications and information exchange between systems Local and metropolitan area networks—Specific requirements – Part 11: Wireless LAN Medium Access Control (MAC) and Physical Layer (PHY) Specifications. 2016.
3. Vidojkovic V, Szortyka V, Khalaf K, et al. A low-power radio chipset in 40 nm LP CMOS with beamforming for 60 GHz high-data-rate wireless communication. In: IEEE ISSCC digest of technical papers; 2013. p. 236–7.
4. Greene K, Sarkar A, Floyd B. A 60-GHz dual-vector doherty beamformer. IEEE J Solid State Circuits. 2017;52(5):1373–87.
5. Kaymaksut E, Zhao D, Reynaert P. Transformer-based Doherty power amplifiers for mm-wave applications in 40-nm CMOS. IEEE Trans Microwave Theory Tech. 2015;63(4):1186–92.
6. Chen J, Ye L, Titz D, et al. A digitally modulated mm-wave Cartesian beamforming transmitter with quadrature spatial combining. In: IEEE ISSCC digest of technical papers; 2013. p. 232–3.
7. Khalaf K, Vidojkovic V, Vaesen K, et al. Digitally modulated CMOS polar transmitters for highly-efficient mm-wave wireless communication. IEEE J Solid State Circuits. 2016;51(7):1579–92.
8. Dasgupta K, Daneshgar S, Thakkar C, et al. A 25 Gb/s 60 GHz digital power amplifier in 28nm CMOS. In: IEEE ESSCIRC; 2017. p. 207–10.
9. Mangraviti G, Khalaf K, Shi Q, et al. A 4-antenna-path beamforming transceiver for 60GHz multi-Gb/s communication in 28nm CMOS. In: IEEE ISSCC digest of technical papers; 2013. p. 246–7.

10. Yan JJ, Presti CD, Kimball DF, et al. Efficiency enhancement of mm-wave power amplifiers using envelope tracking. IEEE Microwave Wireless Compon Lett. 2011;21(3):157–9.
11. Serhan A, Lauga-Larroze E, Fournier J-M. Efficiency enhancement using adaptive bias control for 60GHz power amplifier. In: IEEE NEWCAS; 2015. p. 1–4.
12. Zhao D, Kulkarni S, Reynaert P. A 60-GHz outphasing transmitter in 40-nm CMOS. IEEE J Solid State Circuits. 2012;47(12):3172–83.
13. Huang Y, Li C, Khalaf K, et al. A 28 nm CMOS 7.04 Gsps polar digital front-end processor for 60 GHz transmitter. In: IEEE ASSCC; 2016. p. 333–6.

Printed by Printforce, the Netherlands